state of stress *in the earth's crust*

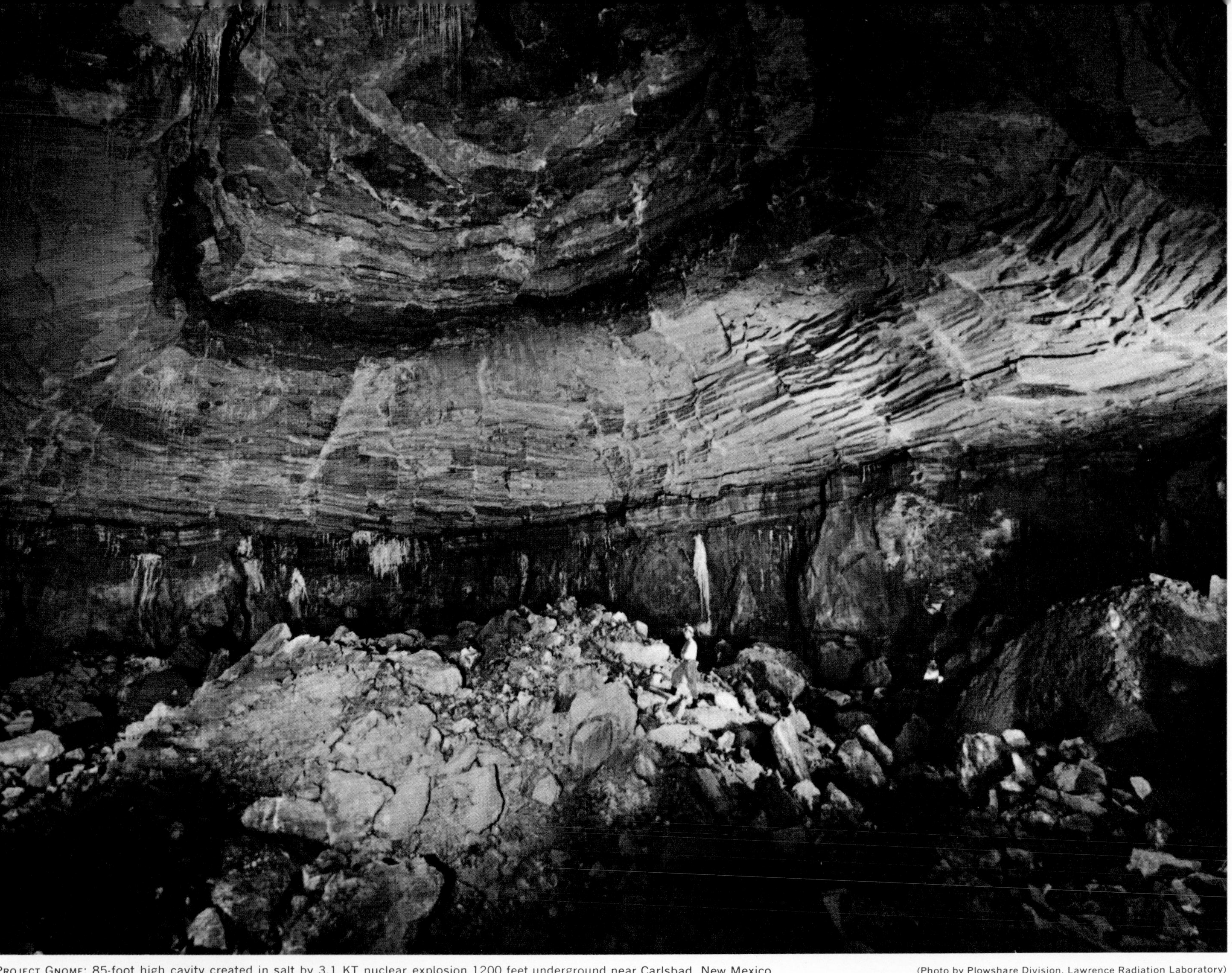

PROJECT GNOME: 85-foot high cavity created in salt by 3.1 KT nuclear explosion 1200 feet underground near Carlsbad, New Mexico. (Photo by Plowshare Division, Lawrence Radiation Laboratory)

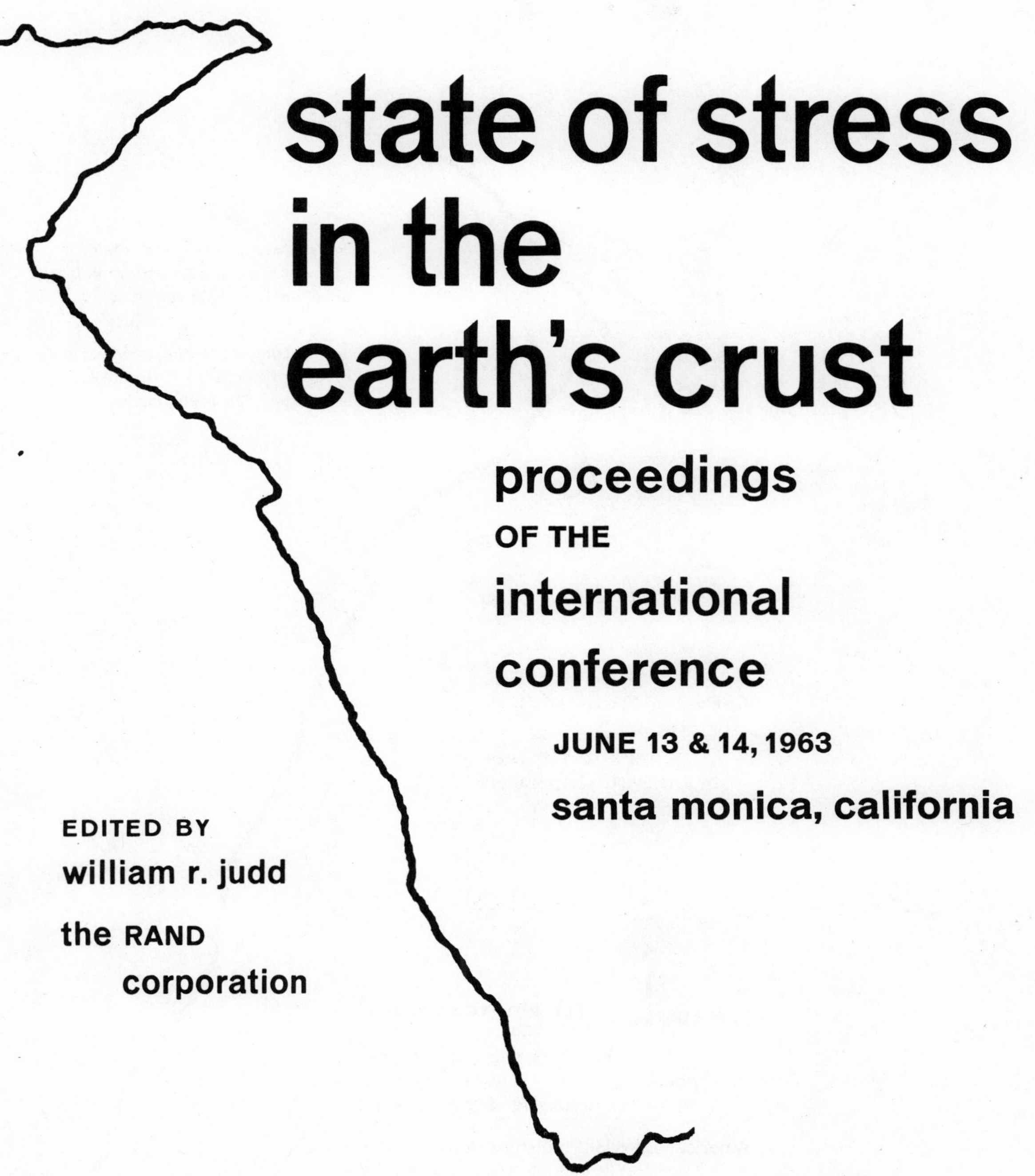

state of stress in the earth's crust

proceedings
OF THE
international conference

JUNE 13 & 14, 1963
santa monica, california

EDITED BY
william r. judd
the RAND corporation

american elsevier publishing company, inc.
NEW YORK, 1964

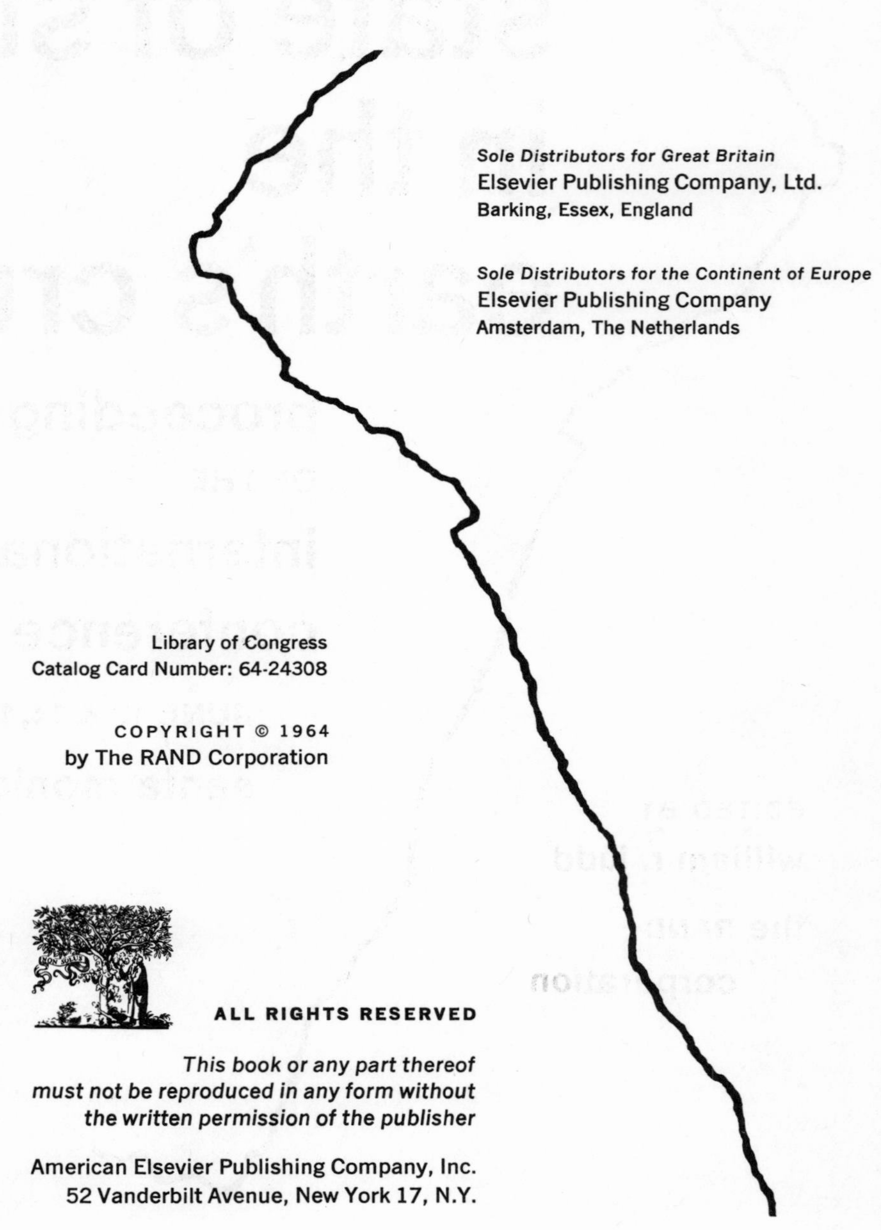

Sole Distributors for Great Britain
Elsevier Publishing Company, Ltd.
Barking, Essex, England

Sole Distributors for the Continent of Europe
Elsevier Publishing Company
Amsterdam, The Netherlands

Library of Congress
Catalog Card Number: 64-24308

American Elsevier Publishing Company, Inc.
52 Vanderbilt Avenue, New York 17, N.Y.

Printed in the United States of America

organizing committee

WILLIAM R. JUDD, *Committee Chairman*
Head, Basing Technology Group
Aero-Astronautics Department
The RAND Corporation
Santa Monica, California

DON U. DEERE
Professor
Departments of Geology and Civil Engineering
University of Illinois
Urbana, Illinois

VICTOR DOLMAGE
Consulting Engineering Geologist
Vancouver, British Columbia, Canada

WILBUR I. DUVALL
Applied Physics Research Laboratory
U.S. Bureau of Mines
College Park, Maryland

CHARLES L. EMERY
President, CIM Consultants, Ltd.
Assistant Professor
Departments of Mining Engineering and Mathematics
Queen's University
Kingston, Ontario, Canada

NILES GROSVENOR
Assistant Professor
Mining Department
Colorado School of Mines
Golden, Colorado

JOHN W. HANDIN
Senior Research Geologist
Exploration and Production Research Division
Shell Development Company
Houston, Texas

ROBERT H. MERRILL
Senior Physicist
Denver Mining Research Center
U.S. Bureau of Mines
Denver, Colorado

EUGENE C. ROBERTSON
Theoretical Geophysics Branch
U.S. Geological Survey
Silver Spring, Maryland

DART WANTLAND
Head, Geophysics Branch
Division of Engineering Geology
Office of the Chief Engineer
U.S. Bureau of Reclamation
Denver, Colorado

FOREWORD

IN JUNE 1963 The RAND Corporation was host to the International Conference on the State of Stress in the Earth's Crust, organized by the Committee on Rock Mechanics of the Engineering Geology Division of the Geological Society of America.

The topic of the conference was indeed an important one. Glamour and mystery have lured man and his scientific instruments into the earth's surrounding atmosphere and beyond, into the distant reaches of space, and our society has dedicated many billions of dollars to this exploration. But man and his lowly ancestors have spent over a million years earth-bound, and for centuries his chief goal has been to explore and to map the surfaces of our planet. The participants in this conference, more than most other scientists, know how little we understand of the history, changes, and ultimate sources of earth's present structure. It is a significant fact that this conference set as its goal an assessment of the state of our knowledge of the earth's crust, to call attention to the many gaps in our understanding.

There have been many theories concerning the forces and processes that were at work in the formation of earth. Few things can be more important to man in the coming centuries than a more detailed understanding of the earth's crust and its hidden mysteries, for—at least for as long a time as is of interest to us—this earth is all we have, and what we do with it, how we use its resources and its potentially useful constituents, will certainly determine the future of mankind and his way of living.

The conference attracted an impressive number of distinguished and knowledgeable scientists, and this volume records their papers and discussion.

The RAND Corporation
Santa Monica, California

F. R. COLLBOHM
President

PREFACE

THE PREFACE is generally where the author (or editor) explains the why and wherefore of the book. However, in this volume a considerable portion of such an explanation is included in the introductory paper. Thus, this preface is confined to the miscellany that was of interest but did not comfortably fit elsewhere in the book.

The acknowledgments generally conclude a preface; however, we consider our acknowledgment of sufficient importance to earn a position at the beginning of this preface. Our reason is that the conference, and thus this book, would not have been possible without considerable financial as well as moral support. Thus, on behalf of the organizing committee, I wish to acknowledge and thank the several sponsors of this conference: the Advanced Research Projects Agency of the Department of Defense (more familiarly known by the acronym "ARPA"); the Defense Atomic Support Agency ("DASA"), also of the Department of Defense; the National Science Foundation; and the U.S. Air Force Project RAND—the major research contract of The RAND Corporation, a nonprofit corporation engaged principally in basic research on problems affecting the national security of the United States.

The basic philosophy of the organizing committee was that this should be an actual conference on fundamental problems and not just a symposium of miscellaneous papers; thus, approximately 50 per cent of the available time was allotted to open discussion. To avoid extraneous discussion, the registrants were given advance questionnaires concerning the preprint papers (which had been supplied to each registrant about thirty days prior to the conference). Additional discussion from the floor was accepted insofar as time permitted. A second committee decision was to extend speaking and panel invitations to a very limited number of individuals known to be well versed in one or more of the conference topics. To maintain continuity in the conference program, these invitees were asked and agreed to prepare papers on specific subjects selected by the committee. The only constraint imposed on the authors was that they avoid case history details except where absolutely necessary to explain a fundamental point.

It was with some hesitation that I acquiesced to the use of the term "Editor" on the title page. Actually, in this task I received considerable assistance from the Panel Chairmen, Drs. D. U. Deere, C. L. Emery, and J. W. Handin, and we purposely minimized the editing of the original papers to avoid publication delays and interference with each author's writing style. The discussions, however, were extensively edited to delete extraneous thoughts and to rephrase hastily spoken comments into a more readable form. As a result of this latter editing, a discussant may find that one of his more enlightening remarks was deleted; however, I wish to assure him that any such deletion was entirely unintentional.

As indicated in the introductory paper, the conference dealt primarily with the subject material many of us would term "rock mechanics." Our reasons for avoiding the specific use of this term in the conference title are discussed in the opening paper. The book is divided into four parts to facilitate its use as a reference work and perhaps aid in its application as a textbook. (The latter possibility has frequently been mentioned to the committee because of the current lack of an English-language textbook in rock mechanics.) Part I attempts to depict the unusually broad scope of the conference subject material—broad in the sense that it is of interest, for example, not only to the tectonic geologist engaged in fundamental research on the structure of the earth, but also to the civil engineer concerned with placement of a concrete dam on a rock foundation. Part II discusses the present fundamental theories used by the many scientific and engineering disciplines concerned with stress in rock, whether imposed in a laboratory or in a mountain massif. Part III describes the current theories and instruments used to obtain quantitative measurements of stress conditions in the crust. Part IV presents the application of the information presented in the preceding papers; this part, as would be true of each of the other parts, could justify a complete volume. Unfortunately, owing to conference time limitations, descriptions are given of only a few of the many major applications that are possible.

April 30, 1964
Santa Monica, California

WILLIAM R. JUDD

CONTENTS

PART I--THE SCOPE

CONTENTS

ABSTRACT

Rock mechanics research and application involves many scientific disciplines and engineering professions. Therefore, this paper discusses the links and interactions between the seemingly diverse topics later discussed in this book. Examples are presented of the major problems facing rock mechanics, including the apparent fallacies in currently used theories. The major theses of the paper are (1) the advisability of using the theory of elasticity as merely the starting point for rock mechanics analyses and (2) the desirability of extended research on the effects of loading-rate variations in testing of rock.

RÉSUMÉ

Les recherches et les applications portant sur la mécanique des roches impliquent de nombreuses disciplines scientifiques et de nombreuses professions de génie civil. Cette conférence va par conséquent discuter des liens et des intéractions qui existent entre les sujets, apparemment divers, discutés plus loin dans ce livre. Des examples de problèmes majeurs qui confrontent la mécanique des roches sont ici présentés, ainsi que les pièges apparents des théories actuellement en usage. Les thèses principales de la conférence sont les suivantes: (1) il est recommandé d'utiliser la théorie de l'élasticité uniquement comme point de départ dans les analyses de mécanique des roches et (2) il est désirable de pousser les recherches sur les effets des variations du taux de chargement dans les essais sur la roche.

AUSZUG

Felsmechanische Forschung und Anwendungen umfassen viele wissenschaftliche Gebiete und Ingenieurberufe. Diese Arbeit behandelt daher die Verbindungen und Wechselbeziehungen zwischen den anscheinend verschiedenen, in diesem Band behandelten Themen. Beispiele der wichtigeren Probleme, die der Felsmechanik entgegenstehen, werden gebracht, einschliesslich der augenscheinlichen Trugschlüsse in derzeit verwendeten Theorien. Die hauptsächlichen Leitgedanken dieser Arbeit sind (1) die Empfehlung, die Elastizitätstheorie lediglich als einen Anfang für felsmechanische Berechnungen anzusehen und (2) der Wunsch nach ausgedehnterer Forschung über die Auswirkungen von Änderungen der Belastungsgeschwindigkeiten bei Felsversuchen.

ROCK STRESS, ROCK MECHANICS, AND RESEARCH

William R. Judd*

Introduction

THE BASIC OBJECTIVE of the International Conference on State of Stress in the Earth's Crust is most briefly outlined by placing a rock specimen in front of each reader and asking: "What is this?" The diversity of the answers can be predicted by a quick glance at Table 1, which summarizes the background of the individuals who registered for the conference.

A careful study of these widely divergent interests should make it obvious why it is difficult to answer this opening question by a simple definition. And yet, progress in any science requires that communication be established between individuals and organizations doing work that can be of mutual benefit to the science. The Committee on Rock Mechanics of the Geological Society of America** realized that the first step in good communication requires definition of the parameters of rock mechanics.

Definition

The first obstacle the committee faced was agreement on a correct name for the science. For example, should it be called "rock engineering" or "rock physics"? These and similar terms appeared to be too restrictive for both the theoretical and practical applications of the subject. However, the term "rock mechanics" seems to be sufficiently comprehensive. The parameters of this science are broadly

---

*The RAND Corporation.

**This committee became the National Academy of Science Committee on Rock Mechanics on July 1, 1963.

Table 1

NATURE OF REGISTRATIONS FOR CONFERENCE

Profession or Specialty	Per Cent	Representation by Organization, Institution, etc.	Per Cent	Country (By Order of Number of Registrants)
Engineering geology and geological engineering	22.7	Architect-engineers, research corporations, consulting engineers and geologists, geophysical and geological companies, testing laboratories, etc.	32.9	1. United States
Civil and structural engineering	12.4	Universities and colleges	28.1	2. Canada
Structural geology	10.0	Federal agencies	18.6	3. France
Rock mechanics	10.0	Petroleum companies	6.0	3. Mexico
Geology	9.7	Mining companies	3.6	4. Italy
Soil mechanics and foundation engineering	7.7	Aircraft companies	3.0	5. Japan
Mining engineering and mining research	6.8	City and county agencies	3.0	6. Australia
Geophysics	4.4	State agencies	3.0	6. Austria
Physics	3.5	Cement companies	1.8	6. South Africa
Mechanical engineering and mechanics	2.9			6. West Germany
Seismology and engineering seismology	2.4			7. England
Materials engineering	1.8			7. New Zealand
Petroleum engineering and research	1.5			7. Panama
Economic geology	0.9			7. Portugal
Petrology and petrofabrics	0.6			7. Sweden
Chemistry	0.3			7. Switzerland
Computer science	0.3			7. Venezuela
Geochemistry	0.3			
Hydrology	0.3			
Lunar geology	0.3			
Mathematics	0.3			
Nuclear geology	0.3			
Rock physics	0.3			
Surveying	0.3			

expressed in the following definition adopted by the committee:*

> Rock mechanics is the theoretical and applied science of the mechanical behavior of rock; it is that branch of mechanics concerned with the response of rock to the force fields of its physical environment.

Every word in this definition was carefully evaluated to ensure a minimum of misunderstanding. No attempt was made to qualify the term "rock." As the final goal of a study in rock mechanics is its practical application, the individual so engaged must decide whether the structural material he is studying is best analyzed by theories being developed under the name of rock mechanics, or whether the material better conforms to commonly accepted principles of soil mechanics.

The Scope of Rock Mechanics

The communication problem is also emphasized if you ask: "Why did the committee adopt the name 'State of Stress in the Earth's Crust' for the conference?" Use of the term "rock mechanics" would not have been satisfactory because of the present lack of common understanding on what the science of rock mechanics actually encompasses. On the other hand, the obtaining or the use of rock apparently has one common factor; that is, whether rock is excavated, extracted, re-used, or loaded by a structure, the stress in the rock has to be considered. Although it is important to know whether this stress is caused by nature or by man, there always is the common and basic problem of how this stress condition will influence the ultimate use of the rock material. Furthermore, since man's use of rock ranges from the construction of a surface building on a rock foundation to mines in excess of 10,000 ft in depth and to the boring of holes some 5 mi deep in a search for oil

---

*The committee agreed to this definition on May 27, 1964. It replaces the following tentative definition presented at the conference: "Rock mechanics is the study of the rheology of geological materials with emphasis on those materials that, in practice, are regarded as rock rather than as soil." Although we used "rheology" as defined by Eirich,[1] there were still too many objections and misunderstandings in this application of the word.

(and possibly deeper in Project Mohole[*]), it would seem that whether man-made or natural, the stress conditions in the earth's crust would be a basic concern of what many of us term "rock mechanics." The following outline was developed by our committee as a concise method of defining the scope of rock mechanics:

A. Fundamentals

B. Measurements

1. Laboratory methods ⟷ 1. Static loading
2. Field methods ⟷ 2. Dynamic loading

C. Applications

1. Surface foundations, man-made excavations, and natural slopes
2. Underground openings (including boreholes)
3. Rock as a construction material
4. Comminution
    a. Drilling
    b. Blasting
    c. Crushing
5. Subsidence studies
6. In structural geology

Empirical versus Theoretical Considerations

The Empirical Case

Another factor common to practically all aspects of rock mechanics can be described by the question: "How will rock react when put to man's use?" The absence of adequate theory too often forces the practitioner in rock mechanics to base his design on what he has actually seen or what has been reported to him--for example, the slow outward movement of the wall of a surface excavation in rock that has to be constrained by rock bolts (Fig. 1), or the slow crushing of tunnel supports (Fig. 2), or the explosive release or slow spalling of rock in

---

[*]Present plans indicate that about 30,000 ft of drilling will be required; however, about 10,000 ft of this will be in ocean water.

Fig. 1—Rock bolts supporting slab in surface excavation (U.S. Bureau of Reclamation photo).

Fig. 2—Crushing of supports and lagging in tunnel (U.S. Bureau of Reclamation photo).

a tunnel, or unpredicted laboratory results such as the triaxial test result shown in Fig. 3 (an increase in volume of the specimen occurred without apparent fracturing), or unpredicted tensile stress in an arch dam because of unexpected deformation in its rock abutments, or the costly slowdown of oil-well drilling because the rock refuses to be cut or crushed by the bit, or the refusal of rock to break into required sizes in an ore crusher or when blasted for dam riprap.

Faced with empirical evidence of failures that apparently are not explainable by theory, the designer may attempt to prevent failure by substantially increasing the amount of man-made structural materials; that is, he will use more and heavier steel for tunnel supports, such as the closely spaced steel ribs shown in Fig. 4; or, because of his uncertainties as to the bearing capacity of the underlying rock, he will use larger areas of concrete to spread the footing loads for a building; or he will use heavier balls or teeth and possibly higher speeds in the rock crusher. The drill bit will be forced to cut by increasing its weight and speed. These are all examples of the attempt to prevent failure by using man-made brute force to oppose nature rather than attempting to work with her.

It is true that such methods, based almost entirely on empirical data, often succeed. This type of solution, however, may not be good engineering, because lack of knowledge as to what caused the failure can result in inefficient and expensive overdesign. Thus, a major area for research in rock mechanics today is the determination of the mechanism of failure. How often has the rock been reduced to its microscopic components to learn what really happened at failure? How often has the design that failed been analyzed to learn its weaknesses? How often has there been an attempt to uncover <u>all</u> of the factors that might have influenced failure? In other words, have those of us attempting to understand and use rock mechanics done what the good engineer faced with a structural failure does--learn from that failure? It has been said that many of the great steps forward in engineering have been the result of intensive research on an engineering failure. Can this be said about rock mechanics?

DEFORMED LIMESTONE CORE TESTED IN TRIAXIAL COMPRESSION

Loads

Lateral: 22,100 psi
Axial: 125,100 psi (computed from original cross sectional area)
Axial: 79,000 psi (computed from final cross sectional area)

Duration of Test: 2 hours and 5 minutes

Source of Core: Marble Canyon Dam Site, Arizona (Colorado River above the Grand Canyon)

Physical Characteristic	Before Test	After Test	Remarks
Diameter (max)	5.9 in.	7.4 in.	25% increase
Height	12 in.	8.8 in.	27% decrease
Volume	328 in.^{3}	354 in.^{3}	8% increase
Density	173 lb/ft^3	159 lb/ft^3	8% decrease
Dynamic modulus of elasticity E	10.88×10^6 psi	1.07×10^6 psi	1/10 original modulus
Average velocity of longitudinal vibration	17,070 ft/sec	5,580 ft/sec	1/3 original velocity

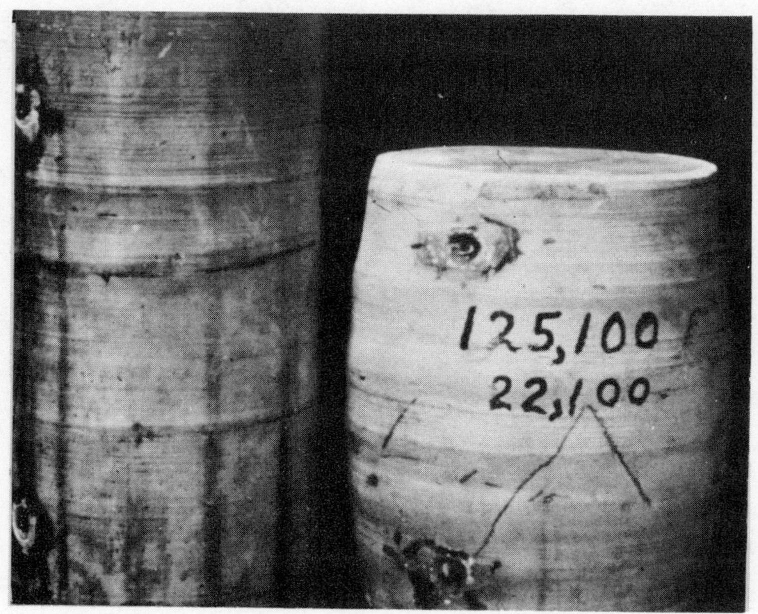

Fig. 3—Volume Increase in Triaxial Test on Core (U.S. Bureau of Reclamation photo).

Current Failure Research*

Fortunately, basic research and failure research in rock mechanics are not entirely lacking. For example, the Civil Engineering Department of the University of Glasgow, Scotland, is attempting to determine the failure criteria of rocks tested in triaxial compression. The University of Witwatersrand Mining Engineering Department, Johannesburg, is investigating ground movements around and in a deep stope during mining, as well as surface subsidence resulting from the excavation of tabular

---

*Information provided on this topic was obtained from personal correspondence with the institutions and organizations mentioned.

Fig. 4—Closely spaced tunnel supports.

ore bodies in hard rock. The Iron Ore Company of Canada, Newfoundland, is performing crater blasting experiments to determine the effect of repeated large-scale blasts on underground installations to obtain the limiting factors in size and proximity of proposed production blasts. To develop more efficient opening designs, the University of California at Berkeley is studying a new method of finite element analysis to determine the stresses in rock around different-shaped openings under various loads. Among other rock mechanics research, the Department of Mines and Technical Surveys at Ottawa, Canada, is studying the structural and physical properties of rock during comminution. The Jersey Production Research Company, Oklahoma, is attempting to determine tectonic stresses by analyzing well-fracturing data. To assist deep drilling for oil and gas, and for other purposes, high-pressure and temperature experiments are being performed on rock at several research establishments including the Fuel Department of the Geological Survey of Japan; the Shell Development Company, Texas; Harvard University, Massachusetts; and the Geophysical Department of Australian National University, Canberra. The Snowy Mountain Hydroelectric Authority, Australia, has become noted for its basic research on the design of rock bolt systems that will supplement or eliminate tunnel supports.(2)

The preceding are merely a few random samples of literally hundreds of rock mechanics research projects throughout the world. However, these examples illustrate one of the major objectives of the conference: Can we initiate the steps required to overcome present difficulties in communication between scientists engaged in widely diversified applications of rock mechanics?

Some Possible Solutions

Originally I asked if we were taking full advantage of our past failures? The balance of the papers in this volume indicates we are well on our way to doing so. For example, Dr. Brace points out that the appearance of a laboratory specimen after tests may not be indicative of how failure actually occurred; that is, he found it more informative to empirically trace the history of a failure by stopping the tests at preselected intervals before the final failure. Drs. Müller and Talobre and their colleagues did not find it good engineering to design stable rock slopes by requiring stronger and stronger retaining walls or flatter and flatter slopes--instead they formulated theories that included quantitative compensations for geological weaknesses and proceeded to successfully place their theories into practice. Dr. Emery discusses a successful method for visually demonstrating that stress causes permanent strain in certain directions or patterns in the rock; thus, possible errors caused by extrapolation from indirect measurements can be minimized. To ascertain if our measurement techniques could be improved, Mr. Merrill and his co-workers discuss the type and need for careful field evaluation tests.

Furthermore, many of these papers demonstrate that in a relatively new science, correct solutions to problems may require the researcher to deviate from established theory borrowed from other sources. Admittedly, such deviation often is difficult because many of us have a natural tendency to conform to pre-existing standards; however, those who have had the foresight to engage in the relatively new science of rock mechanics also must have the courage to examine objectively the basic criteria that have come into accidental use over a period of many years--particularly the criterion that rock always will respond to elastic theory.

In Underground Opening Design

The designs of many underground excavations are still based on the hydrostatic theory proposed by A. Heim in 1878.[3] A summary of his approach is given in Fig. 5.[4] This mode of analysis has not been open to serious challenge until very recent years. Now, however, owing to the progress in techniques of rock strain measurements, there has been increasing evidence of stress conditions that are anomalous if compared to hydrostatic theory.

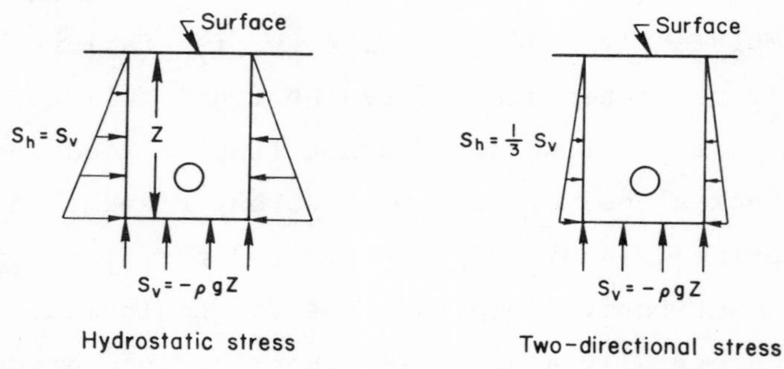

Fig. 5—Theoretical stress distribution on underground openings (Ref. 4).

The development of such devices as the measuring bolt illustrated in Fig. 6 has permitted us to examine the actual stress history of an opening from the time an underground chamber excavation is started until many months after it is completed. Thus, if there is a partial or complete failure of the opening, the stress history obtained from such measurements provides information that permits us to minimize the possibility of future similar failures. Unfortunately, as yet we

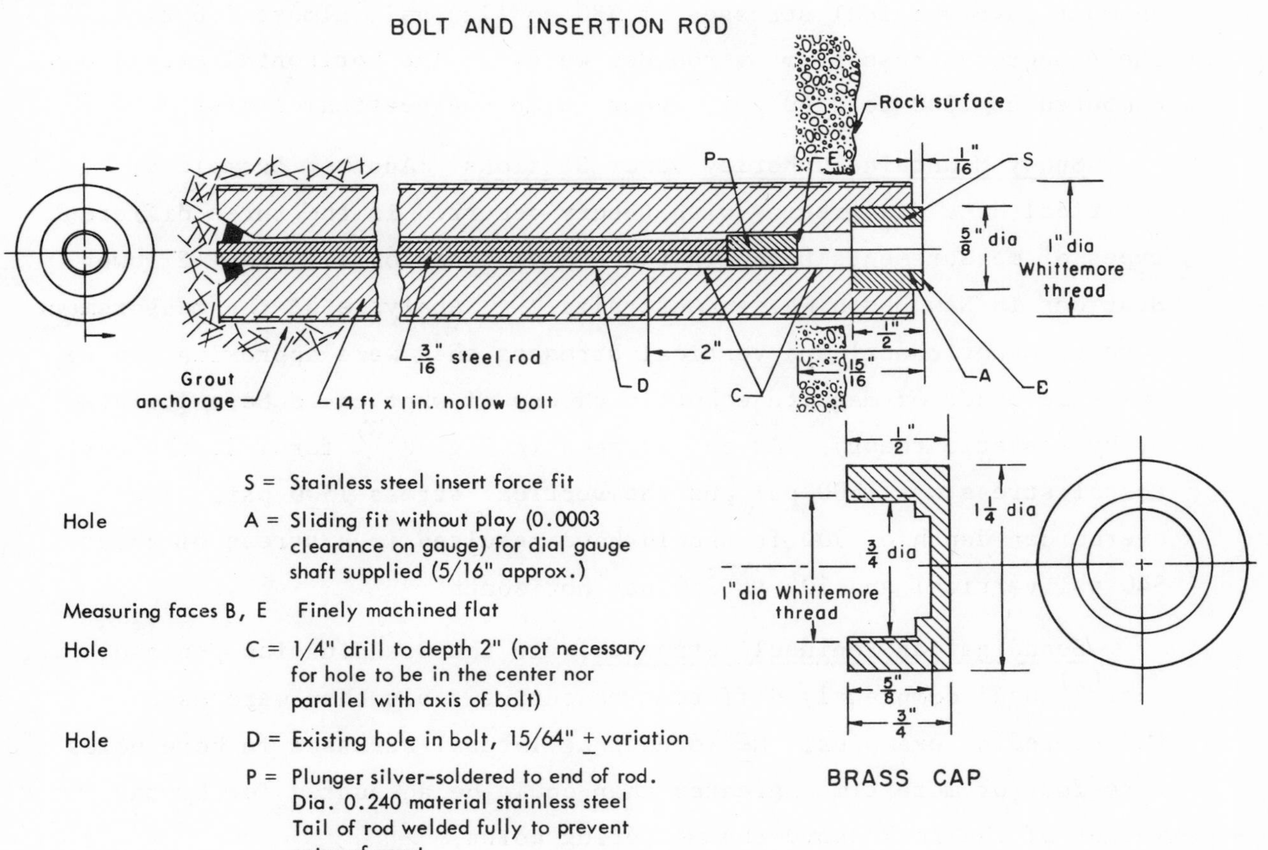

Fig. 6—Bolt for measuring rock deformation--Snowy Mountain Authority, Australia.

have only a few well-documented stress histories on large openings. And, some of these show stress anomalies that cannot be explained by the conventional static theories of stress distribution. Furthermore, as briefly discussed in subsequent paragraphs, these data provide increasing evidence that the condition of hydrostatic stress distribution around an opening may be an exception rather than a common occurrence.

Poatina Power Station. Figure 7 illustrates a stress anomaly that occurred during excavation of the Poatina Power Station in Tasmania.[5] The station is about 500 ft below the surface; therefore, use of conventional theory and a Poisson's ratio of 0.25 for rock (see statement on page 48) would result in computed vertical stresses of about 560 psi and horizontal stresses of 140 psi. However, even after the authors[5] allowed for as much as 25 per cent error because of anisotropy in the layered rock, computations from flatjack measurements in the actual chamber gave vertical stresses of 930 to 1120 psi, almost 1.6 times the expected stress from overburden weight. The horizontal stress was computed as 1600 to 2160 psi, about twice the vertical stress.

Snowy Mountain Authority Power Stations. Another example of empirical data disagreeing with theory occurred in the three different types of measurements taken in the excavations for T-1 and T-2 Power Stations in New South Wales, Australia.[6] Analysis of the measurement data gave horizontal and vertical stresses that were approximately of the same order of magnitude but in excess of what could be calculated by hydrostatic theory. As can be seen in Fig. 8[6] for T-2, the horizontal stress was 1800 psi and the vertical stress 1500 psi. The overburden depth of 700 ft should have resulted in a stress of about 840 psi vertical and 100 to 250 psi horizontal.

Scandinavian Peninsula Stresses. In the Scandinavian peninsula Hast[7] used completely different measuring tools than were used in the preceding examples. He found that lateral stresses in bore holes were four or more times greater than could be accounted for by the weight of the rock above the measuring point.

Picote Power Station. Stress distribution in the underground chamber of the Picote Power Station in Portugal is shown in Fig. 9.[8]

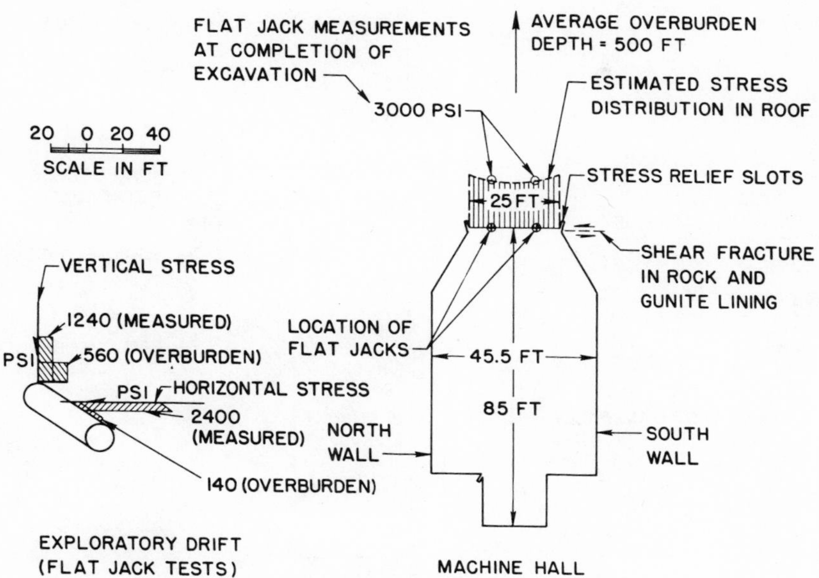

Fig. 7—Stress distribution--Poatina Underground Power Station, Australia (Ref. 5).

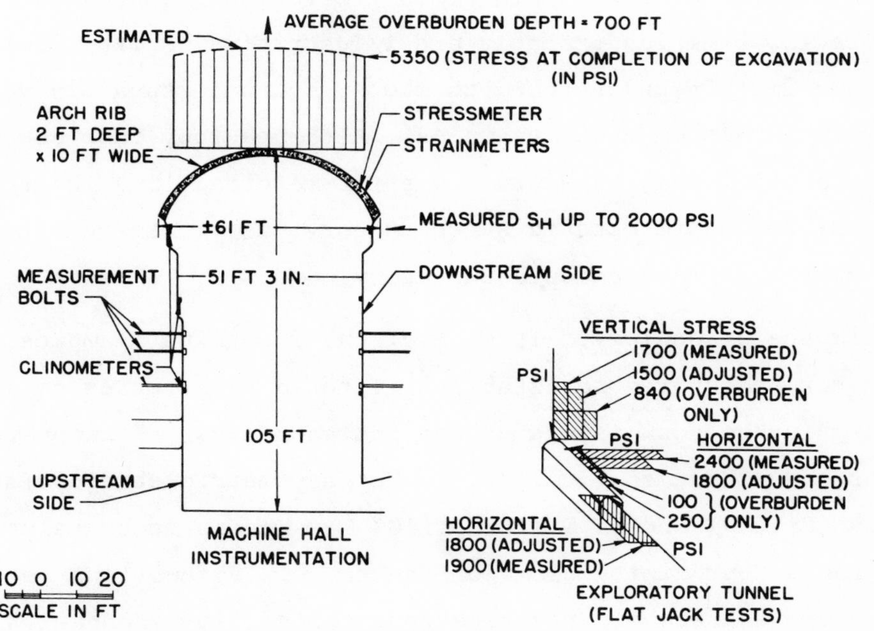

Fig. 8—Stress distribution--T-2 Underground Power Station, Australia (Ref. 6).

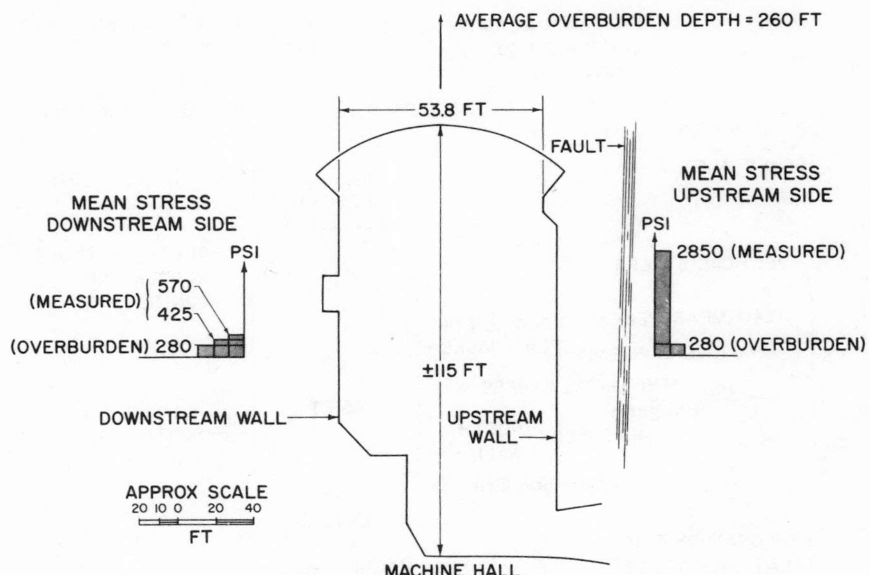

Fig. 9—Stress distribution--Picote Underground Power Station, Portugal (Ref. 8).

Serafim's calculations based on actual measurements showed that the horizontal stresses were in excess of what could be calculated from the overburden weight. He further discovered that one side of the chamber had stresses almost six times greater than the opposite side.

These few examples probably are illustrative of many similar stress anomalies in underground chambers, but unfortunately very few data are available on other examples. Although they cast doubt on the acceptance of rock as a truly elastic material, and the results do not conform to hydrostatic stress theory, as yet there are not enough data to support a more acceptable theory.

Influence of Geology. In each of the foregoing examples, the investigator endeavored to establish if residual stresses or possible anisotropy in the rock layers caused the anomalies. It appears that both causes may have some validity. The asymmetric stress distribution in Picote (Fig. 9) and in T-1 (Fig. 10[9]) appears to be related to the orientation and proximity of major faults. As a fault can be considered as a zone of weakness or area of stress relief, it is conceivable that a fault could induce asymmetry in chamber stresses.

The importance of considering such geological weaknesses in the design is illustrated in Fig. 11. The crack in the concrete is at the

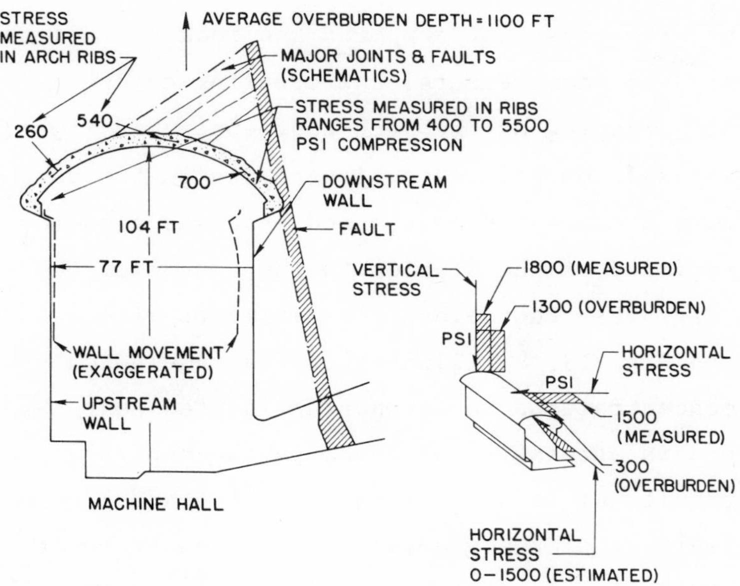

Fig. 10—Stress distribution--T-1 Underground Power Station, Australia (Ref. 9).

Fig. 11—Cracked concrete at base of roof support arch--T-2 Underground Power Station, Australia.

toe of one of the arch ribs that supports the rock roof of the power station. According to measurements, the cracking evidently was caused by the slow inward crushing of the arch abutment; this inward movement apparently was a result of high lateral stresses in the surrounding rock. At T-1 Power Station the unbalanced stresses also may have been caused by the proximity of the station to the bottom of the V-notched valley shown in Fig. 12. Photoelastic experiments with this type of shape, as shown in Fig. 13,[9] illustrated that the notch could induce unusual stress concentrations in an opening located near the notch.

The imbalance in stress in the Scandinavian peninsula is attributed by Hast to the result, at least in part, of the gradual rebound of the peninsula. The melting of the glacial ice that once covered Scandinavia resulted in the gradual unloading of the terrain, and thus rebound could occur. Although detailed geological studies are not complete, the Poatina stress pattern may have been influenced by major structural discontinuities in the stratigraphy as shown in Fig. 14.[5] In addition, a vertical cross section of the chamber would show it is excavated through at least three different types of sedimentary rock. As these rock layers each have different elastic constants, some variability or asymmetry could be expected in the stress pattern around the chamber opening. Some support for this latter assumption is indicated in work by Clutterbuck in 1958.[10] By theory and by three-dimensional photoelasticity, she found that stress distribution could be significantly dependent on a difference in elastic constants in a multilayer medium. It also is of interest that she notes that differences in Poisson's ratios had less influence than errors introduced by other factors.

As a final note on this particular phase of rock mechanics, it is interesting to speculate whether the usual elastic theory would prove a mine opening of the size shown in Fig. 15 to be stable. The rock is a steeply dipping and fractured siliceous dolomite. Despite the fracturing, the some 1400 ft of rock weight over the opening, the span in excess of 200 ft, frequent blasting for ore, and no supports, there have been only nominal rock falls during the two or more years the opening has been in existence.[11,12] And, to further confound theory, there is another opening of similar size in this same mine!

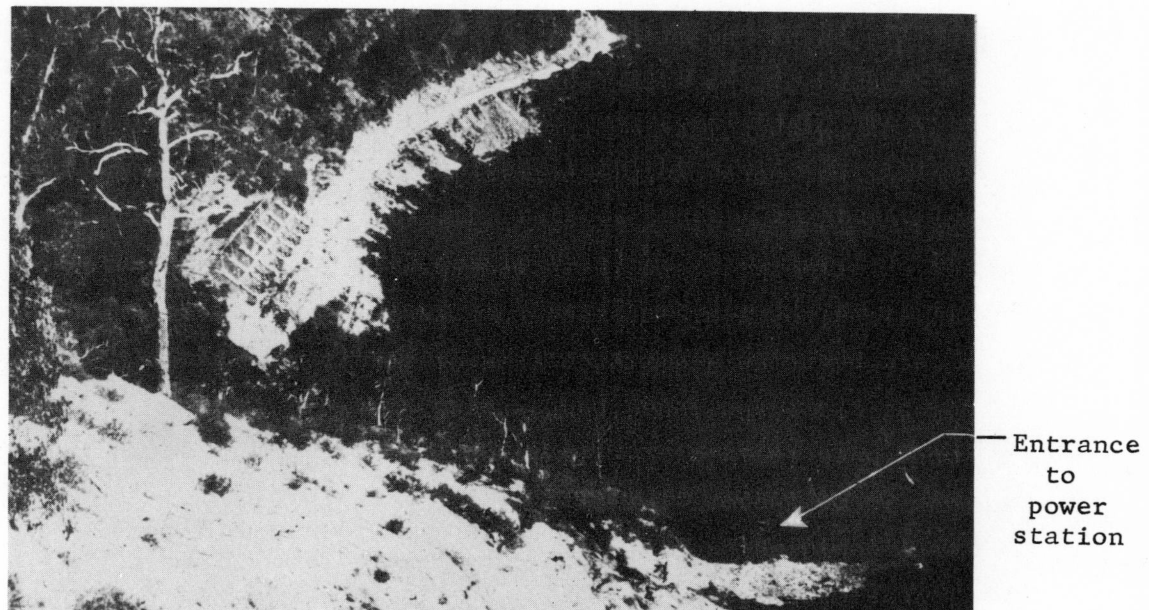

Fig. 12—V-notched valley at T-1 Power Station, Australia. (Station excavation is about 1100 ft beneath the photographer; structure in left center is the switchyard for the station.)

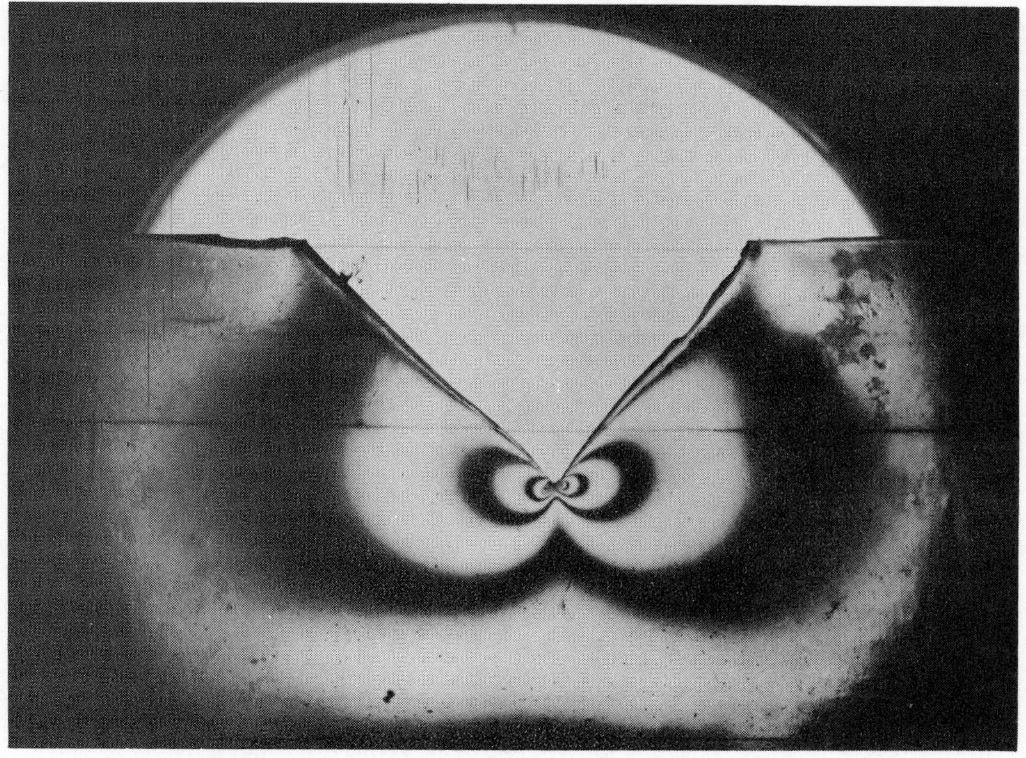

Fig. 13—Stress fringes around V-shaped notch simulating valley at T-1 Power Station (adapted from Ref. 9, Plate 5).

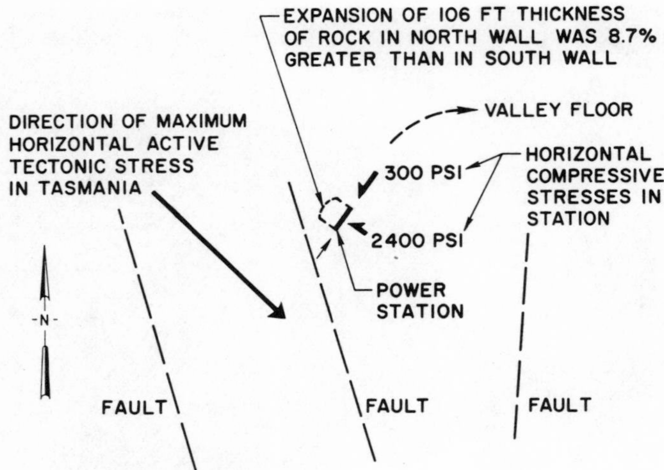

Fig. 14—Aerial map showing relation of geology to stresses in Poatina Power Station, Australia (Ref. 5).

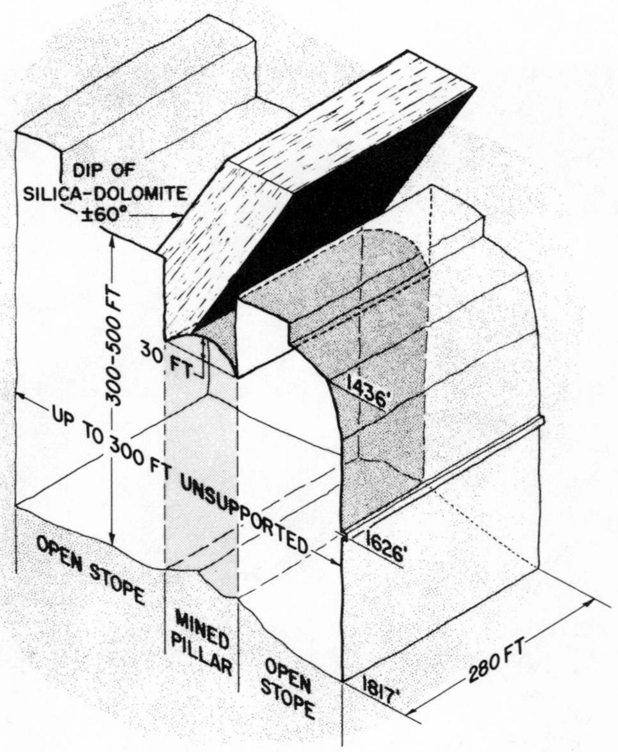

Fig. 15—Open stope at Mount Isa Mine, Australia (rough sketch) (Refs. 11, 12).

Correlation of Static and Dynamic Load Tests

The preceding examples are illustrative of only a few of the many contradictions between hypothetical assumptions and empirical data that occur in rock mechanics. This paper will emphasize a few of what I regard as the more outstanding uncertainties in rock mechanics. Many others are discussed in more detail in the remainder of the book. The examples previously presented were concerned only with stresses resulting from static forces; however, similar uncertainties also occur in the "moving" or dynamic world of rock mechanics.

Rock in the Laboratory and In Situ

A serious and unsolved problem in rock mechanics today is how to correlate the results of static and dynamic tests of rock.* As many researchers know from personal experience, all too frequently there is considerable discrepancy between moduli obtained by static load tests and those obtained in dynamic load tests. For example, if we try to use that anathema of some scientists--logic--to predict results, we could conclude that a specimen should react differently in a laboratory than if the same rock is tested in its natural habitat, that is, *in situ*. Unlike a rock mass *in situ*, the lab test specimen (Fig. 16) is regarded as an integral, intact piece of rock free from obvious structural discontinuities. Thus, it would appear logical that the lab test would result in a higher compressive strength and Young's modulus than would be obtained in field tests on the *in situ* rock (which generally has natural fractures and other gross weaknesses). To date, my experience has been that sometimes logic is correct, but more often it is not. And, as I will discuss later, attempts to make statistical comparisons between field and lab tests result in considerable

---

*As used in this paper, a test performed under static load conditions is one wherein the load is applied relatively slowly, such as in an unconfined compression test machine. Conversely, the dynamic load condition is regarded as a test where a very rapid motion is imparted to the rock or rock molecules, such as by a hammer, dynamite, or sonic waves.

Fig. 16—Intact drill core for laboratory test.

scatter in the data. To solve this problem, it appears the first question to be answered is whether present methods of lab and field testing are truly comparable.

The small size of the laboratory specimen makes it relatively simple to impose a uniaxial or triaxial load on it, or to propagate a sonic or shock wave through it. However, to extend our analogue, we compare the lab test with static-type field tests that impose a load over a considerably larger area of rock. Unlike the lab specimen, the field load may be imposed on macrogeological weaknesses not present in the lab specimen, such as joints, fractures, and faults. However, as shown schematically in Fig. 17, for a practical evaluation of the field test, we must recognize that the stresses induced by the completed engineering structure may involve a considerably greater number, and possibly different types, of geological weaknesses than were influenced by the field test load. In other words, the field load test usually can be regarded as measuring the gross effect of only a few geological weaknesses.

The Seismic Test Method. Test methods using dynamic forces, such as the seismic method illustrated in Fig. 18, have been developed as a

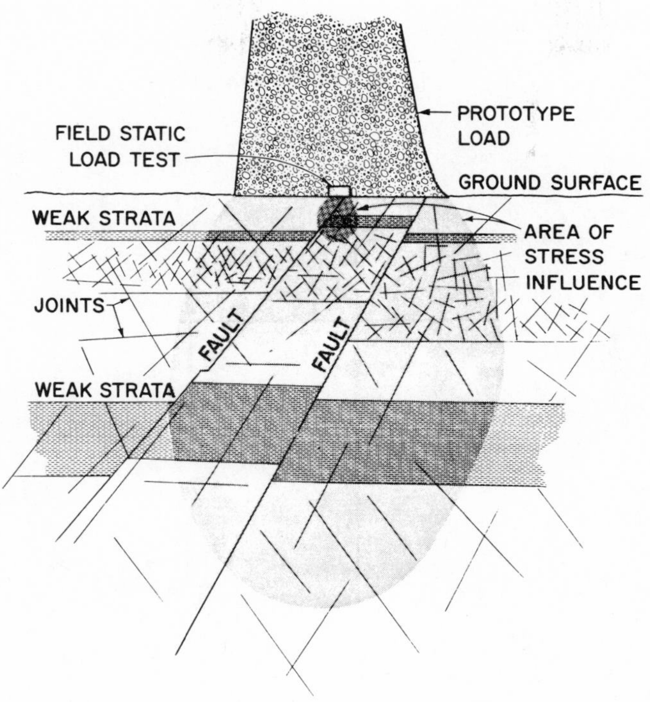

Fig. 17—Influence of field load test and prototype load on geological anomalies.

possible means of solving the preceding difficulty in extrapolating from a field test to prototype effects. The seismic method would appear to evaluate a rock mass equivalent in size to that which will be stressed by the completed structure. As discussed by Mr. Wantland in his paper, one dynamic field method frequently used measures the propagation velocities of the longitudinal wave (P-wave) and the transverse wave (Q- or S-wave) through the rock. (The S-wave is also referred to as the shear wave.) But this method poses the question as to whether the dynamic forces (from an extremely rapid impact) will induce a stress comparable to that imposed by the slowly applied static load of the completed structure.

Statistical Comparison. A statistical evaluation of results of dynamic and static tests, with both types being performed in the laboratory (Fig. 19), results in a reasonably good correlation. However, although the correlation is within 92.9 per cent of being linear, there still are significant deviations. For design purposes, a predicted

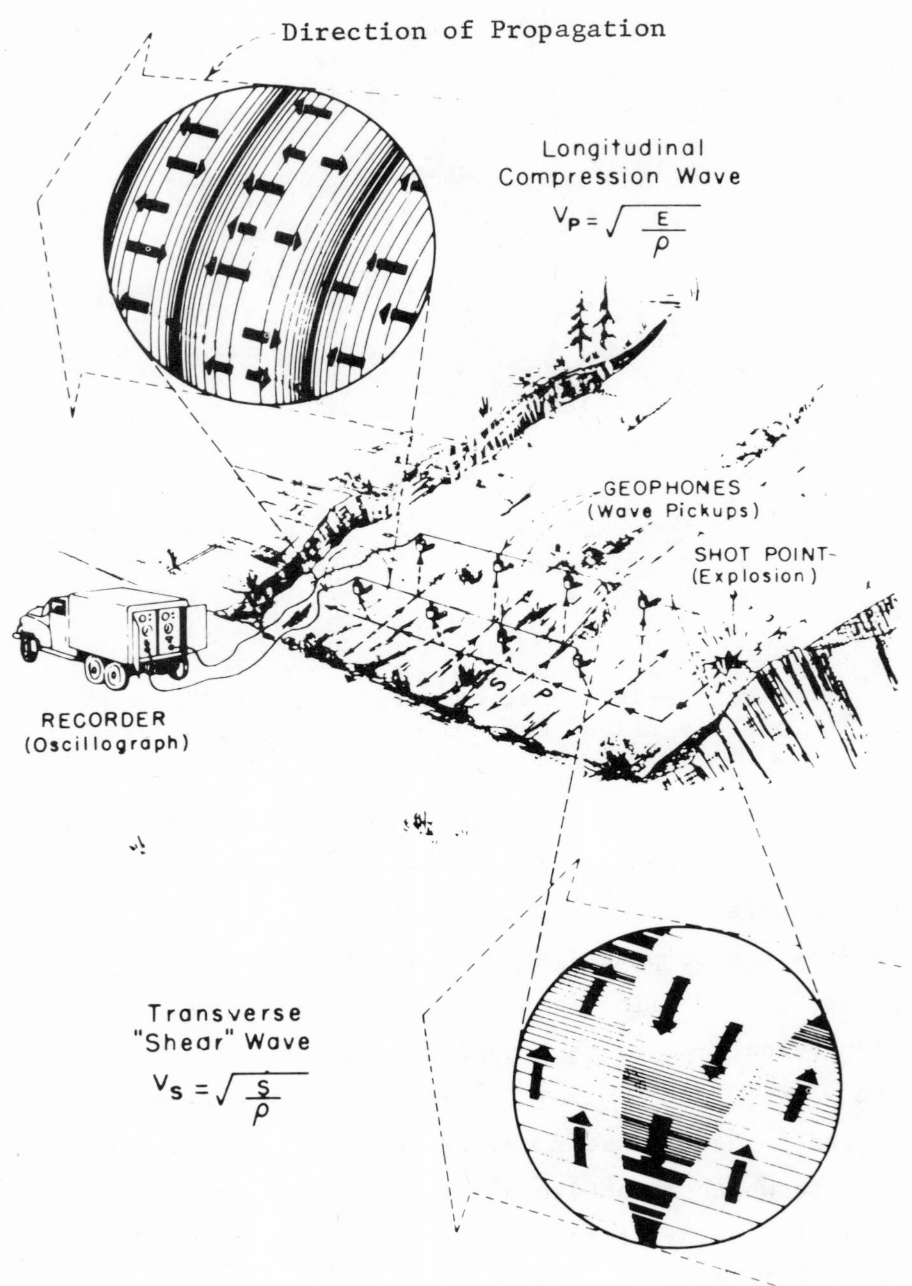

Fig. 18—Seismic method of determining Young's modulus of elasticity (Ref. 13).

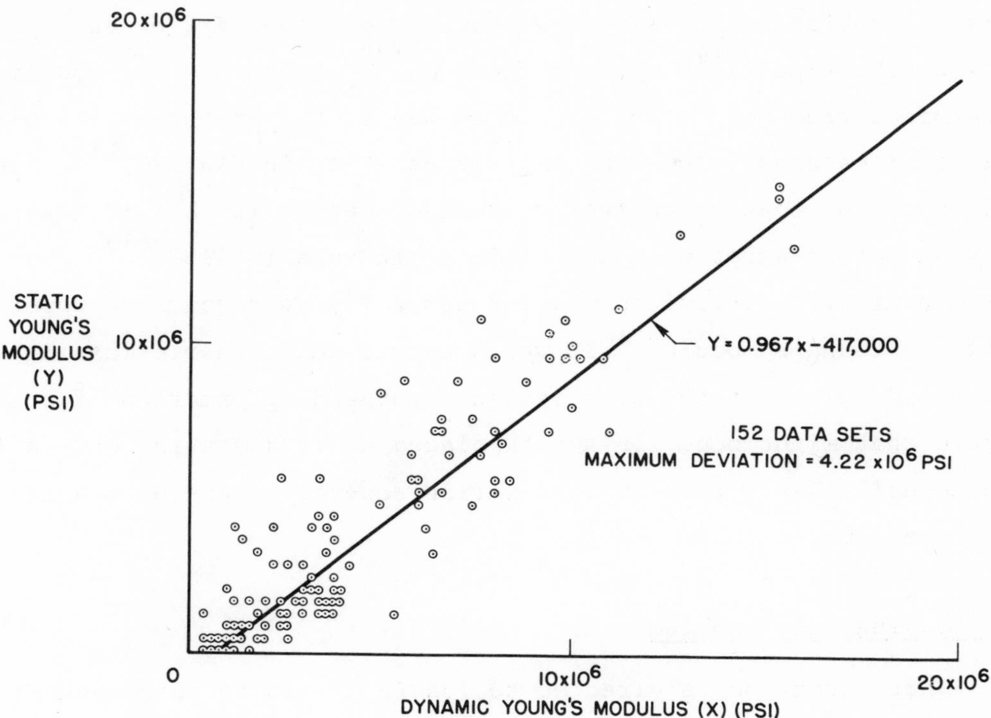

Fig. 19—Statistical comparison of dynamic and static Young's modulus (rocks of all types).

deviation in higher moduli may be acceptable, whereas equivalent deviations in lower moduli values may impose unacceptable errors in design. In this regard, Dr. Serafim discusses the influence of rock moduli on the design of concrete dams. Significant deviation errors in moduli of the rock that is to serve as an abutment for a concrete arch dam seldom can be tolerated. The reason is that tensile stresses in arch dams must be held to a minimum; unpredicted and excessive tensile stresses may cause severe cracking of the concrete. Thus, if the rock abutments were to yield (or deform) considerably more than had been compensated for in the design, the stresses in the dam might change from compressional to tensional and cracks would occur in the arch.

Empirical Comparisons. But is this comparison between different laboratory methods of load application also valid in field testing? As previously mentioned, it seems logical to assume that the dynamic method of field testing evaluates the properties of the entire rock mass that is to be loaded by the prototype structure. However, for this assumption to be valid, numerous comparisons are necessary between the results of dynamic field or lab tests and the moduli determined from the actual strain caused by the prototype. I was able to locate only the few comparisons shown in Fig. 20.[14] The only apparent conclusion from this wide scatter is that in general they appear to agree with Kujundzic's findings in 1957;[15] that is, the dynamically determined modulus generally represents an upper limit to Young's modulus. It would appear that dynamic loading is too rapid to permit the slow readjustment and deformation of rock fabric that might occur in a rock placed under a static load--a loading that usually has a rate of application several orders of magnitude less than the dynamic rate.

The Loading Rate Problem

Thus attention is directed to loading rates in rock mechanics tests. It is a problem I consider of utmost importance, because we have yet to establish if any available loading test reliably simulates prototype conditions. This question has been studied by several investigators including D. W. Phillips,[16] the U.S. Bureau of Mines,[17] and the Canadian Department of Mines.[18]

Effect of Varying Static Load Rates. Some 15 years ago, Phillips found (Fig. 21) that if beams of sandstone or shale were loaded at the rate of 1075[*] psi/min, at a mean stress of 1500 psi, the rock had a Young's modulus of about 1,250,000 psi. However, if the loading rate was increased to 1568 psi/min, the elastic modulus at 1500 psi mean stress had increased some 60 per cent to slightly over 2,000,000 psi. Further research showed that specimens placed under a constant load for several days finally would fail; the interesting aspect of this finding was that the amount of the constant load was less than that required to cause failure in conventional rapid loading tests.

---

[*]The printing of the original article is such that a question occurs on whether this value is "1075" or "1.075." The text accompanying the figure in the original article states a value of "about 1 lb/sq in./min," whereas the value given on the curve in the original printing appears to be 1075 lb/sq in./min.

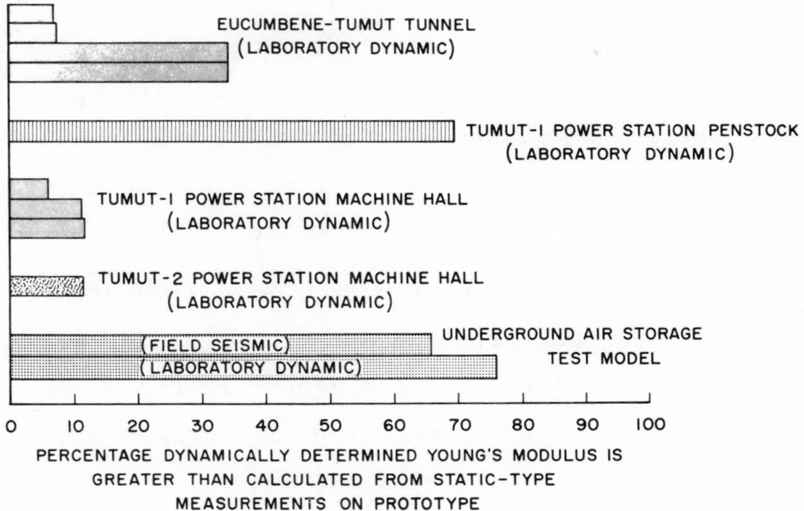

Fig. 20—Comparison of laboratory- and prototype-derived Young's modulus (Ref. 14).

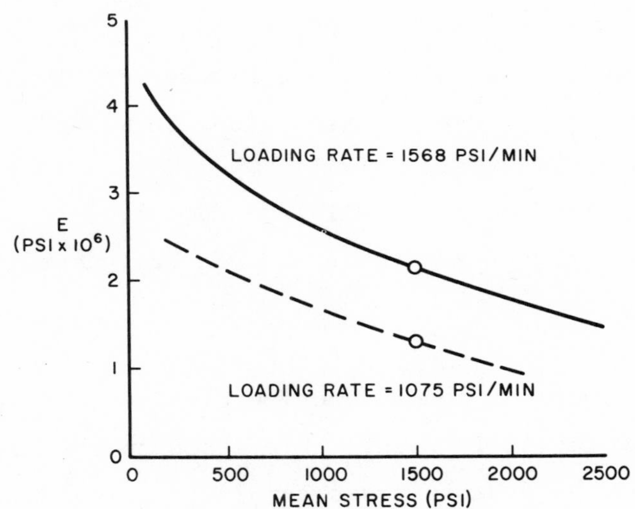

Fig. 21—Effect of loading rate on Young's modulus (Ref. 16).

The criticality of the amount of load per unit of time does not appear to be established except possibly within relatively narrow limits. For example, for the relatively rapid loading rates of 100, 200, and 400 psi/sec used by the U.S. and Canadian Bureaus of Mines, there were no significant differences in the ultimate strength values. However, this finding must be viewed in the light of Phillips' discovery that significant differences in the ultimate strength did occur when significantly higher load increments but much slower rates of loading were used. Thus, to return to our basic problem it appears that not only are wide deviations possible in comparing static and dynamic test results, but they also are possible merely between types of static load tests. The question then must be asked: "Why?"

Effect of Porosity and Rock Fabric. Some investigators such as Belikov(19) concluded that differences between dynamically and statically determined moduli for rocks may depend on the porosity of the material. This infers a direct relationship between porosity and density or specific gravity. Therefore, the value of Young's moduli obtained from the approximate formula generally used for dynamic field tests,

$$E = V^2\rho \ ,$$

where V = velocity of propagation of the P-wave, and ρ = rock density, should be distinctly influenced by variations in ρ. (The more exact formula is $E = V^2\rho\ [(1 - 2\nu)(1 + \nu)/(1 - \nu)]$; however, the effect of Poisson's ratio, ν, is considerably less than the expected accuracy in this type of measurement and therefore can be ignored for most field evaluations.)

In general, it would appear that the loading rate in a dynamic test is too rapid to permit nondestructive closure of pore spaces and readjustment of the fabric in rock. This leads to the assumption that a modulus determined by dynamic tests should closely approximate that which can be derived from tests on the individual mineral constituents of a rock. Although Belikov believes that the influence of the constituent minerals is considerably lessened by increases in porosity, the lack of correlation between specific gravity and porosity shown in the plot in Fig. 22 for 346 data sets indicates that there must be an influence overriding that of porosity.

A slow application of load, such as by a flatjack or laboratory compression machine, should permit nondestructive fabric readjustments

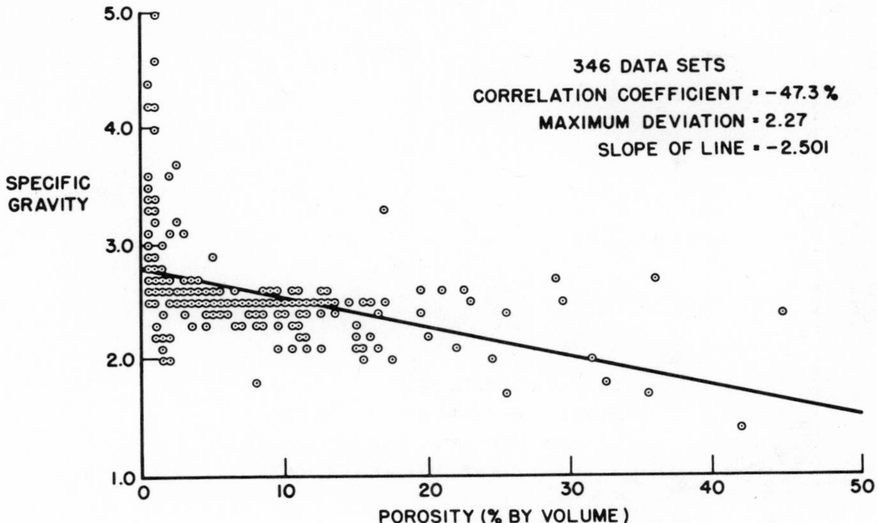

Fig. 22—Statistical comparison of porosity and specific gravity (rocks of all types).

in the specimen. This subsequent slow deformation of the specimen should result in a lower Young's modulus than obtained by dynamic methods.

It now is interesting to examine Fig. 23 to see if a statistical analysis will support our logic regarding the effects of dynamic loading. We find there is not a linear relation between porosity and dynamically determined Young's modulus--at least insofar as is shown by the 261 data sets on Fig. 23. The correlation factor is less than 62 per cent.

In an attempt to cast light on this question, the propagation velocity of the P-wave was compared to the specific gravity of the rock. As shown in Fig. 24, for 345 sets of laboratory and field data there was only a 63 per cent correlation coefficient with a curious grouping of data between specific gravities of 2.4 and 2.8.

We thus arrive at one of the many possible applications of statistical analysis of rock properties. It would appear that within a relatively broad range of specific gravities, P-waves can be expected to propagate at anywhere from about 5000 to 18,000 ft/sec, regardless of rock type. Comparing this finding with the value of 2.7 that often has been used as an average specific gravity for the earth's

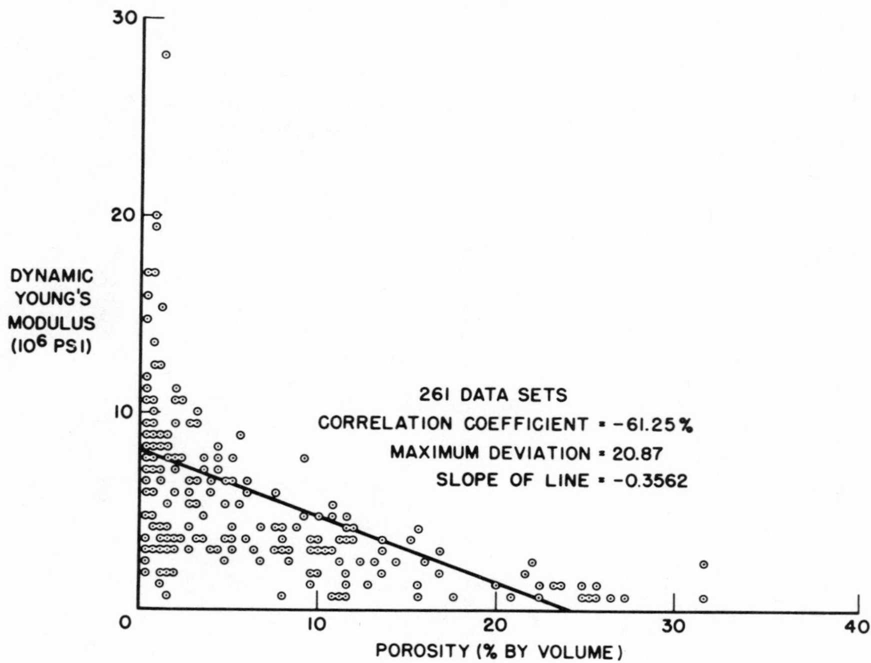

Fig. 23—Statistical comparison of porosity and dynamic Young's modulus (rocks of all types).

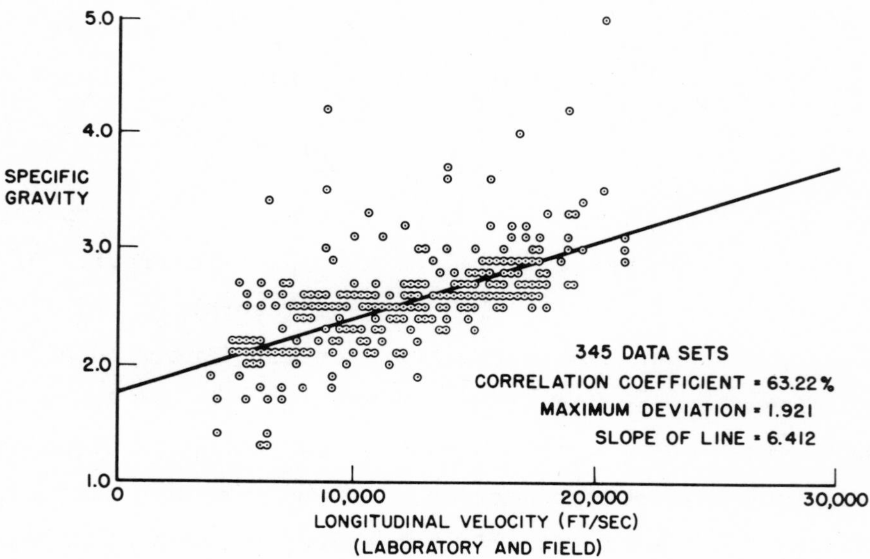

Fig. 24—Statistical comparison of longitudinal velocity of wave propagation and specific gravity (rocks of all types).

crust[20] leads to an interesting question: Unless materials are being tested that have specific gravities below about 2.0 or greater than about 3.0, is the order of accuracy in dynamic measurements sufficient to justify a refined value for ρ in $E = V^2\rho$?

The Prototype Loading Rate. At this point we can ask if significant progress has been made toward finding a means of correlating field static and field dynamic tests. The plot in Fig. 25 indicates that at present only one assumption appears acceptable: The dynamically determined E generally is higher than the static E. Furthermore, as I have previously indicated, the rate of loading in static tests should be carefully considered (as in Phillips' experiments), because there is a significant reduction in Young's modulus if the increases in loading times are large enough to result in an over-all very low loading rate.

As rock mechanics is primarily an applied science, the question must be posed whether this loading rate problem is worthy of further study. Such a question can best be answered by examining the usual laboratory test methods to determine the design modulus for a dam abutment. These tests generally apply a static load to the specimen at rates of 200 to 400 psi/sec. But this loading rate seldom is comparable to the actual prototype loading rate as can be shown by analyzing construction records for large dams. For example, the concrete in Kortes Dam (Fig. 26) in Wyoming was placed at a rate that resulted in the foundation rock being loaded at about 5.4 psi in 7 days.[21] When the dam was complete, it had taken some two years to place about 260 psi load on the foundation rock!

Rock Creep

This discussion of loading rates now brings us to the important subject of rock creep and strain rate. These factors are of increasing importance to rock mechanics: By developing correct theories of creep it may become possible, under many circumstances, to satisfactorily explain why rock very often does not react in accordance with Hooke's law. Also, there is an unanswered question as to whether a load imposed

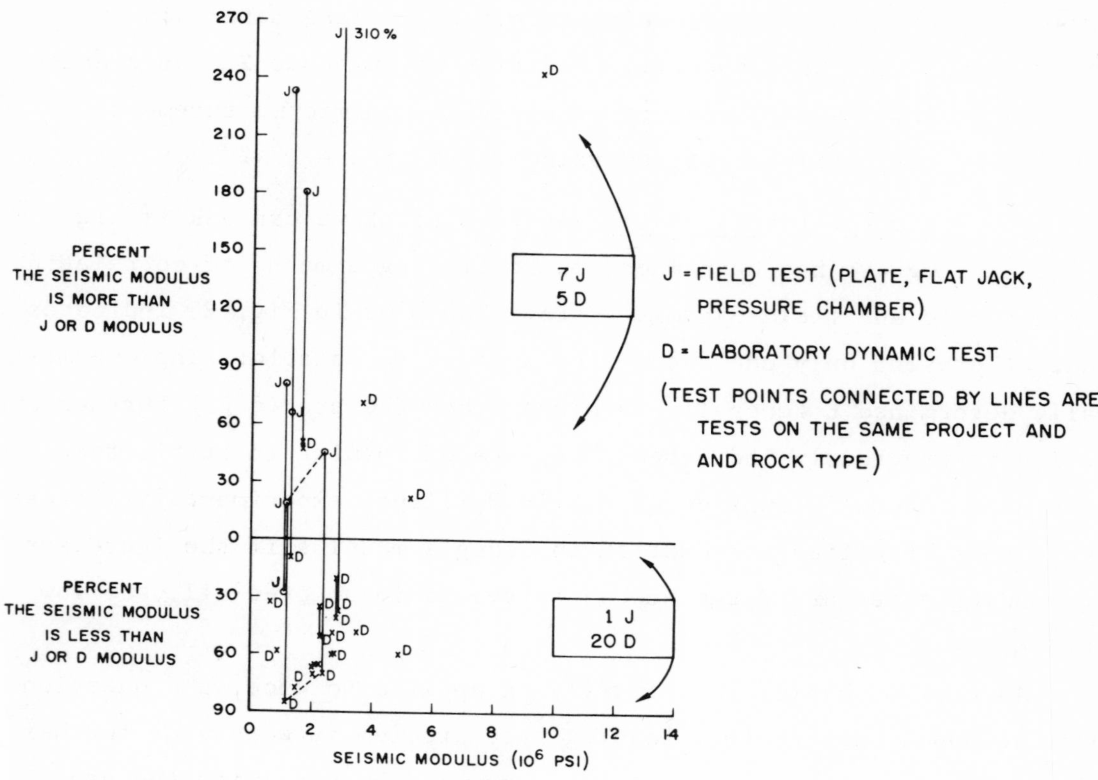

Fig. 25—Comparison of dynamic laboratory modulus with seismic field modulus.

for a long time period results in fatigue or creep failure of rock as it does in metals. Acceptable answers to these problems probably will require that rock be regarded, not as an elastic, but as a viscoelastic material. For example, is rock a truly elastic material if empirical data can and do show that (1) even under moderate loads rock undergoes some permanent deformation (this author has yet to see a stress-strain curve for rock that has a hysteresis loop that returns to 0 on the x-axis when the rock is unloaded), (2) temperature influences this deformation rate, and (3) the rate of load application is critical? Dr. Robertson, in his discussion, explores the criticality of these factors and defines their effect on rock. To do so, he must and does regard rock as a viscoelastic material, and presents a theory for rock behavior that introduces the very important factor of strain rate. Thus, it appears that significant progress is being made in providing a theoretical explanation for the strength reduction under long time loads that was demonstrated by Phillips.[16]

Fig. 26—Kortes Dam--first pour of concrete (U.S. Bureau of Reclamation photo).

Set and Petrofabrics. Another facet of the loading rate problem is that strain rate criteria for loading tests can be correctly evolved only if differentiation is made between (1) deformation resulting from alteration in the petrofabric of an integral rock specimen (Fig. 27A) and (2) deformation caused primarily by macrogeological weaknesses (Fig. 27B). Both of these effects can result in permanent deformation that often is measured as the percentage of "set" that occurs (Fig. 27C).

The analytical difficulty is that the percentage of set in a lab specimen may result from completely different causes than the percentage measured in a field load test. As shown in the "before-and-after" photomicrographs in Fig. 28, permanent deformation in a lab specimen may be the result of fabric changes within the integral piece of rock. However, in field tests, the initial set may be the result of movement along a fault, or the closure of macroscopic openings such as fractures and joints in the rock. Some of the permanent deformation in the lab specimen may be regarded as somewhat similar if the loading causes closure of pore spaces within the lab specimen.

It is important to distinguish between these two phenomena if the

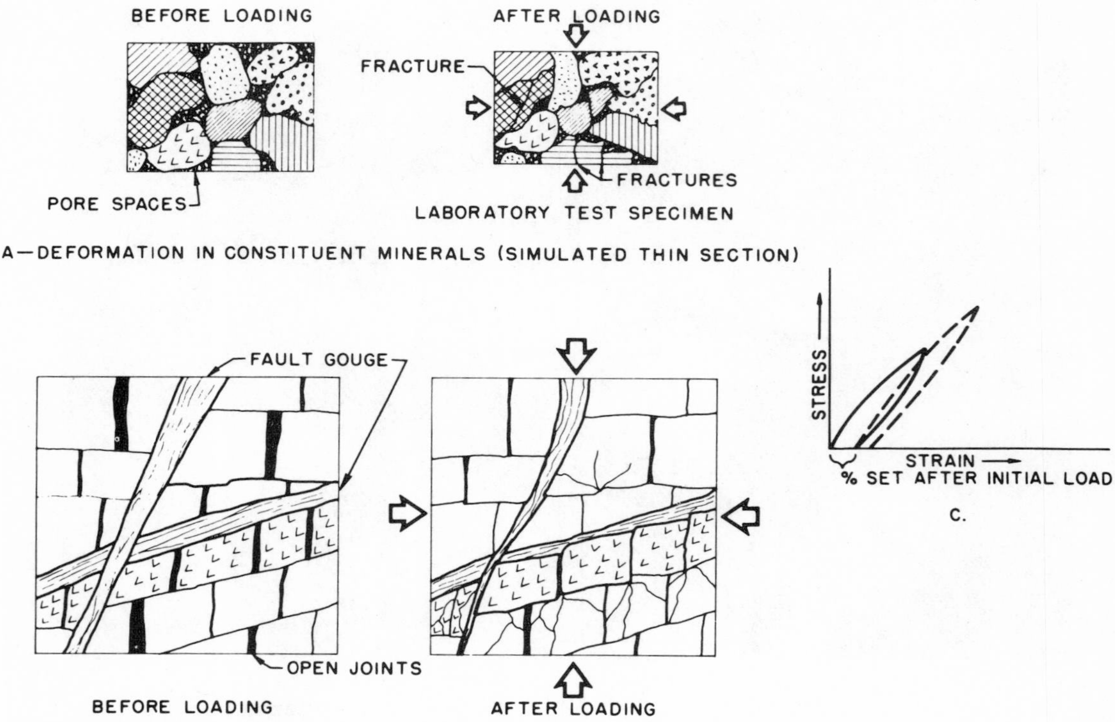

Fig. 27—Effect of geology on rock deformation (schematic).

purpose of the tests is to provide the correct design moduli for the rock. For example, in the case of field set, the controlling factor may be the coefficient of friction between surfaces of rock blocks. Increased surface roughness may cause increases in bulk modulus because of decreased deformation of the rock mass. However, there may be a decrease in modulus if clayey material or moisture is present between the block surfaces. Some factors that can influence laboratory set are illustrated in the microphotograph in Fig. 29. These factors are pore spaces, the nature of grain boundaries, and the integral strength of an individual rock grain or crystal. The major question thus to be answered by rock mechanics research is whether the prototype load will cause only the phenomenon termed field set, or whether this load will be of sufficient magnitude and duration to cause, in addition, actual fabric readjustment within the integral rock in the mass under stress. These two possibilities can also be described in a different manner;

(A) Before

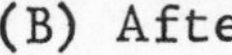

(B) After

Fig 28—Effect of microgeology on rock deformation; photomicrographs of thin sections before and after loading (crossed Nicols at 150 magnifications in original photo) (U.S. Bureau of Reclamation photo of Marble Canyon limestone).

Fig. 29—Effect of microgeology on rock deformation; photomicrograph of thin section before loading (crossed Nicols at 150 magnifications in original photo) (U.S. Bureau of Reclamation photo of Marble Canyon limestone).

that is, the laboratory test on an integral rock specimen may provide a Young's modulus representative of that intact piece of rock. However, the field test may result in a Young's modulus that is representative of the rock mass as a whole with all of its inherent weaknesses; this latter modulus often is referred to as the "deformation" modulus.

This discussion points to a very important and yet almost untouched area in rock mechanics research--and that is the determination of rock strength by petrofabric analysis. Dr. Friedman, in his paper, not only points out the implications of petrofabrics from a microscope point of view, but directs attention to the possibilities implied in petrofabric analysis of gross geologic features.

A Recent Example

The importance of the loading rate question as well as the need for research into correlation problems was demonstrated in a recent field test of the model (Fig. 30) of an underground, high-pressure, gas storage tank.[22] Laboratory tests on intact rock specimens resulted in an average modulus of approximately 3,500,000 psi with a very low percentage of set in the initial load cycle. However, the known presence of open fractures in the rock at the field test site made it necessary to consider adequate safety factors in the design of the field model. The final design of the model was based on the assumption that the rock fractures would be filled by a thorough cement grouting of the rock surrounding the test chamber. The model chamber was built and placed under test. As shown in Fig. 31, the T-1 steel liner of the model failed when the internal air pressure (being used as a loading medium) reached approximately 60 per cent of the proposed ultimate test load.

After the failure, cores were obtained of the concrete lining and the rock around the chamber (Fig. 32). Strain readings during the test had been obtained from instrumented probes extending into the rock around the chamber (Fig. 30), from gauges in the concrete lining and on the steel lining, and from extensometers on diametrical rods within the chamber (Fig. 33). The presence of ungrouted openings in the rock cores indicated an inadequate grouting program. A comparison of the strain data with the appearance of the concrete and rock cores indicated that the failure probably was caused by the gross movement of one or more large masses of rock into open fractures. Some crushing of the rock did occur immediately adjacent to the chamber lining but this could have been caused by differential stress resulting from the aforesaid mass movement of rock.

This research was supplemented by an attempt to simulate the field test with a three-dimensional photoelastic analysis. An epoxy resin (Hysol) was used to encase an aluminum model of the test chamber. After the resin had hardened, the aluminum was dissolved by acid; this resulted is a stress-free block of epoxy for testing. The model then was submerged in oil, and hydraulic pressure, dimensionally equivalent to the field test pressure, was imposed on the <u>interior</u> of the epoxy model. The resulting stress pattern then was frozen into the epoxy and the block was sliced into 1/4-in. plates for photoelastic analysis. The results still are being analyzed, but one example is shown in Fig. 34.

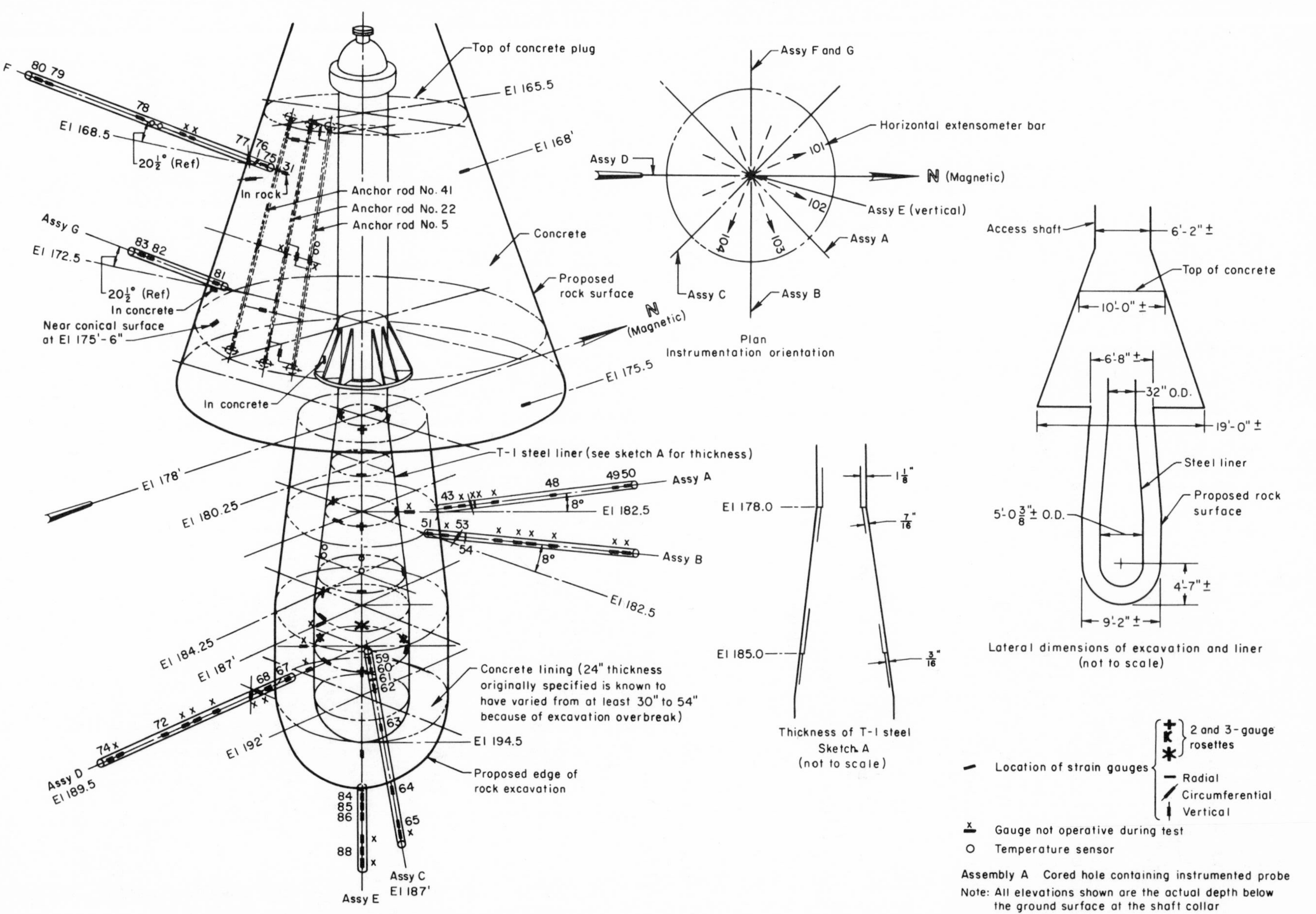

Fig. 30—Design and instrumentation of proposed underground high-pressure test to determine in situ rock modulus (data from unpublished drawings of The Marquardt Corporation).

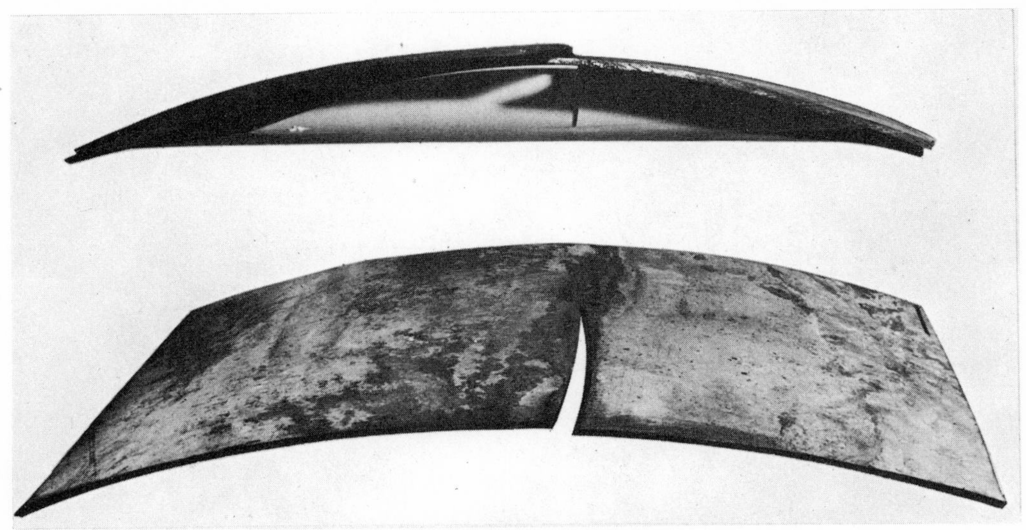

Fig. 31—Rupture in steel liner of high-pressure test chamber.

Fig. 32—Crushing of rock caused by high pressure in test chamber (The Marquardt Corporation photo).

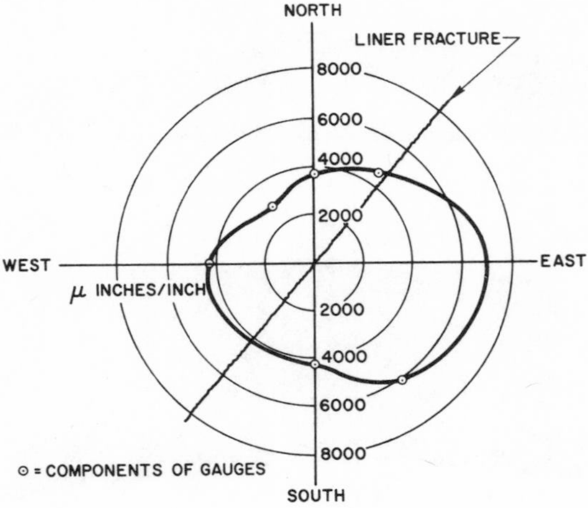

Fig. 33—Strain on steel liner at 2500-psi pressure (polar plot) (Ref. 22).

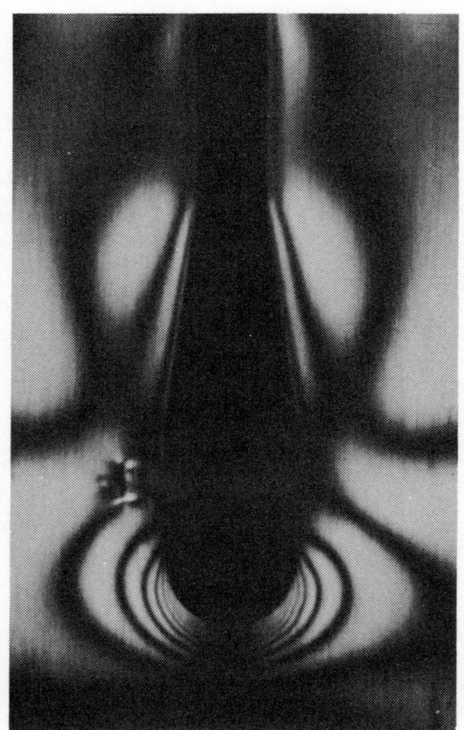

Fig. 34—Stress fringes in 3-dimensional photoelastic model (0° orientations of polarizer and analyzer); irregularity on left side of cylinder apparently from defect in the plastic.

It appears that the maximum stress occurred at the hemispherical bottom of the chamber. The lining rupture in the field model had the greatest width across the hemispherical bottom; however, the strain gauge readings indicate that the maximum stresses may have occurred opposite the cylindrical portions of the chamber.

As a final note on this particular example, I wish to emphasize that the field test was regarded as successful (even though ultimate test load was not reached), because (1) it provided sufficient information to design possible prototypes; (2) as shown in Fig. 35, it is a

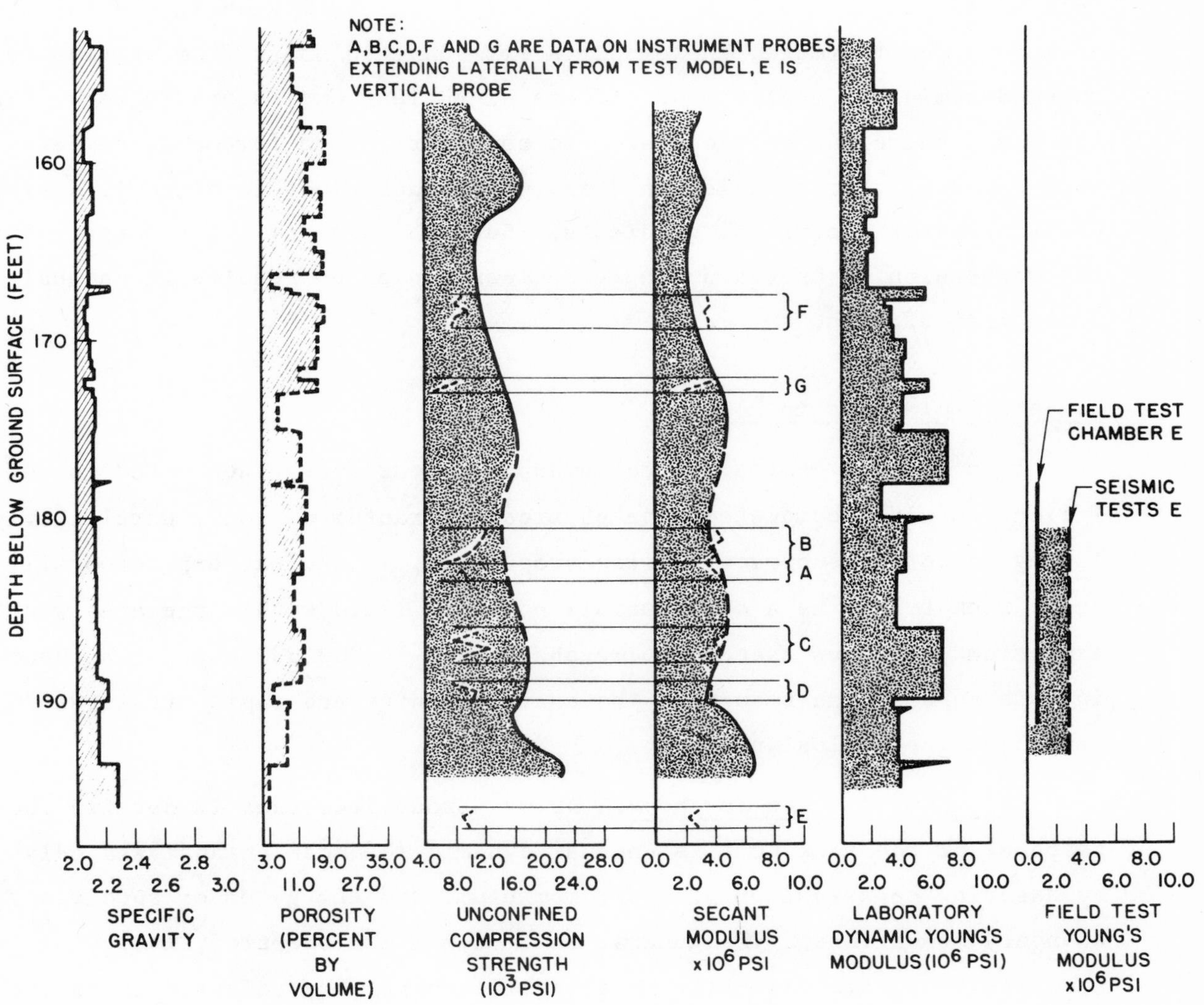

Fig. 35—Comparison of field and laboratory data on rock properties (high-pressure test) (Refs. 22, 23).

unique opportunity to compare rock mechanics data obtained by several different methods; and (3) of most importance to rock mechanics, it is providing an unequaled opportunity of thoroughly researching a rock failure.

A New Regime in Rock Mechanics

This discussion so far has been confined to what is one end of the spectrum of rock loading. It now is desirable to look at the other end of the spectrum. This end will introduce relatively new terms into rock mechanics such as shock tubes, hydrodynamics, megabars, and equations of state. Furthermore, the emphasis will have to shift from statics to thermodynamics to analyze the effects of nuclear explosions in rock. In the past, there has been occasion to consider rather low-order temperature regimes. For example, in deep tunnel operations it often is desirable to predict geothermal gradients. But, in this new high-energy regime, the conversion of energy during a nuclear explosion results in unusual geothermal effects.

Nuclear Explosions in Rock

In a nuclear explosion, the amount of heat energy generated is sufficient to alter completely the physical character of rock, particularly in regard to its supposed elastic properties. A nuclear explosion will cause rock to act as a compressible nonideal fluid within the area where explosion pressures exceed one megabar (over 14,000,000 psi). The heat in this zone is the result of the extremely high and rapid pressures caused by the explosion.

Equation of State for Rock. As Dr. Brode discusses in detail, the response of the rock to this unusual dynamic load can be analyzed only by assuring conservation of mass, momentum, and energy under specific boundary conditions. The general thermodynamic properties, including such items as compressibility and specific heat, are referred to as the equation of state for the rock. In reality, this equation, as shown for basalt in Fig. 36, is a sophisticated stress-strain diagram. The lower

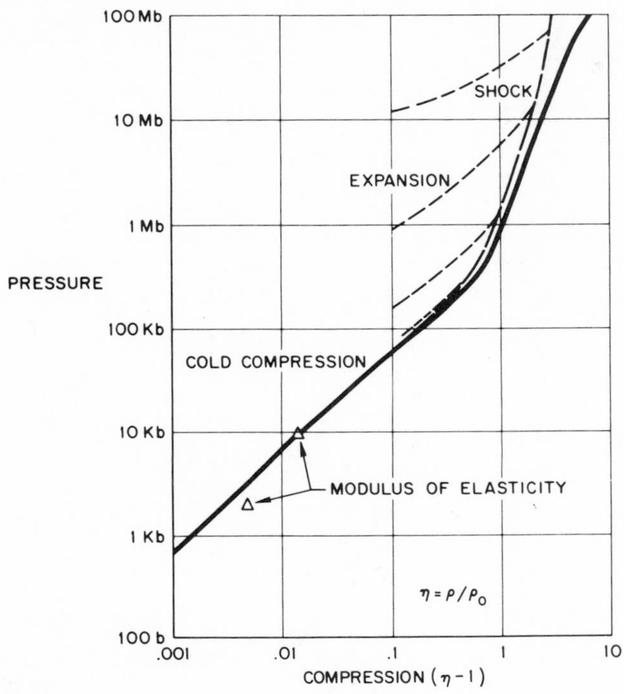

Fig. 36—Equation of state (schematic) for basalt (prepared by H. Brode).

part of the curve roughly resembles the stress-strain condition for rock when the loading regime is still in the cold stage; that is, there is no unusual heat generation. However, as the shock increases in magnitude, the compressive forces generate increasing amounts of heat, and the curve develops a shape considerably different from the familiar stress-strain diagram.

The Loading Rate Problem Reappears. Thus, again we have dramatic evidence that the question of loading rate is of prime importance in rock mechanics--particularly at this end of the load spectrum as megabar loads are being applied in micro- to milliseconds of time. The criticality of this rate is important to the simulation of kilo- or megaton nuclear blasts by the use of large quantities of conventional chemical explosives. As shown schematically in Fig. 37, there is a distinct difference in the manner in which peak stress attenuates with distance from the explosion. Close-in to the nuclear explosion the reaction of rock as a compressible fluid requires the application of hydrodynamic theory to determine the stress distribution, whereas the

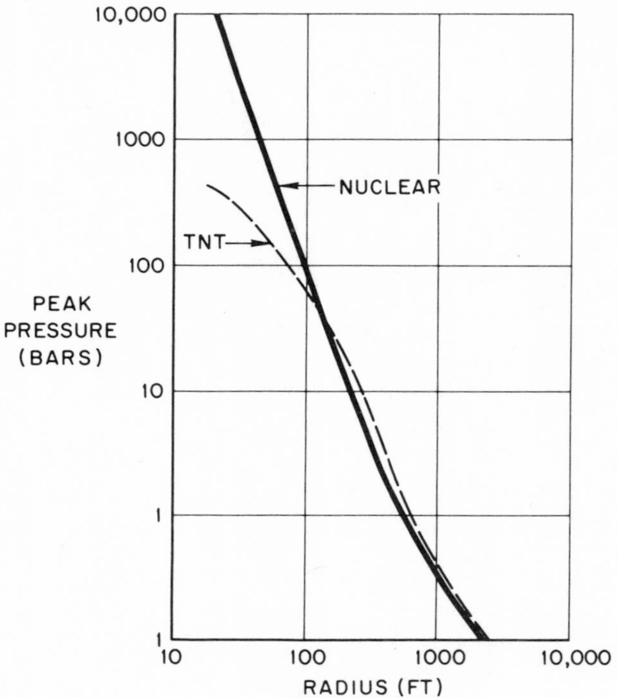

Fig. 37—Schematic comparison of pressure effects of chemical and nuclear explosions (prepared by H. Brode).

rock close to a high-yield chemical explosion is still physically a solid, although it may be severely fragmented.

A major problem in this regime of rock mechanics is illustrated by Fig. 38--as yet we do not have the theory that will accurately define the boundaries of that zone where the rock ceases to act as a compressible fluid and responds as a viscoelastic material, and where there is insufficient shock attenuation for the rock to more or less react as an elastic material. Even if the rock in this transition zone is crushed, the crushed particles may react as a viscoelastic aggregate. Proof for theories developed in regard to this problem will require that empirical data be available from either laboratory or field tests. This will require the laboratory technician in rock mechanics to develop apparatus capable of generating dynamic pressures in the multimegabar range. Possible types of such test equipment have been described by Kormer,[24] Krupnikov,[25] and others who have generated up to 10 megabars pressure in experiments on metals. Further-

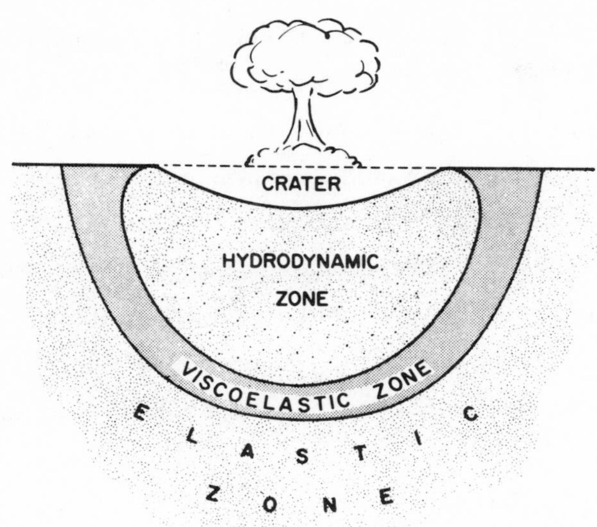

Fig. 38—Pressure zones from a nuclear explosion on the surface.

more, it is necessary for the design of measuring instruments to keep pace with these developments. And such design will have to consider not only the unusually high pressures, but also temperatures on the order of 30,000°C and higher.

A Statistical Analysis Approach

This discussion now has taken a brief look at random but specific problems in rock mechanics. It now is desirable to take a broader look. The predominant difficulty in rock mechanics is that we must evaluate a material that has unusually diverse physical properties or as said concisely, but perhaps somewhat understated, by Heiskanen and Meinesz in 1958: "...the behavior of the materials of the Earth under the effect of stress fields is often more complicated than the elastic deformation according to Hooke's Law."*

Occasionally rock appears to react as an elastic solid, but more often its reaction will be that of a material that is anelastic and anisotropic. In nature we more often find it as a discontinuum than

*Ref. 20, p. 18.

as a continuous media. And, not infrequently, its response to tests will be somewhere between these descriptions. These wide variations have led me, in desperation almost, to see if statistical analysis would provide a few clues on rock response to stress. The philosophy of my approach is based on a statement attributed to Disraeli that there are three kinds of lies: Lies, Damned Lies, and Statistics! Therefore, by permitting the computer to evaluate objectively all pairs of data, I have tried to avoid preconceived ideas as to how certain physical property parameters of rock should interrelate.

An Average Poisson's Ratio

An excellent example of a current preconceived idea is the frequent acceptance of 0.25 for Poisson's ratio in approximate calculations of stress distribution in rock. The validity of this value is somewhat dubious when a simple arithmetic mean of 1254 Poisson's ratios is derived from laboratory and field static test results. The mean was 0.13 and the range was 0.0 to 0.610. The arithmetic mean of 218 Poisson's ratios from dynamic test results was 0.172 with a range of 0.0 to 0.94. To lend further emphasis to this particular point, these means were computed only after all negative values for Poisson's ratio had been discarded (and there were a large number of such values). (It also is interesting to contemplate how any value over 0.5 could still be defined as Poisson's ratio!)

The Findings and Present Problems

As in any statistical analysis, the ultimate reliability of our study will depend on having an adequate sample of factual data; that is, there must be enough data sets from enough different rock types to cover the possible ranges of physical properties, and equally important, to provide a correct weighting for these ranges. A major difficulty in using statistical analysis in rock mechanics is the lack of accepted standards for rock property tests. For this reason, our initial computer runs used the data from reasonably standard tests by the U.S. Bureaus of Mines and Reclamation and the U.S. Geological Survey.

The primary interest in our study is to determine the degree of correlation possible between laboratory and field or <u>in situ</u> test results. A secondary interest is to determine which physical properties will provide the most reliable parameters for evaluation of rock reaction under stress. The first computer runs, which still are in progress, are to determine if pairs of physical property parameters can be correlated by linear regression methods; that is, will a straight line from a least squares plot provide the best correlation between the data sets. Figure 39 summarizes our results to date. The most interesting one is the previously mentioned high degree of correlation between the laboratory-determined dynamic Young's modulus and the laboratory-determined static Young's modulus (Fig. 19). (The close relationship shown on Fig. 39 between laboratory-dynamic Young's modulus and laboratory-dynamic shear modulus is to be expected; each data set is derived from measurements of the velocity of P- and S-waves propagated at the same time from the same source through the same rock media.)

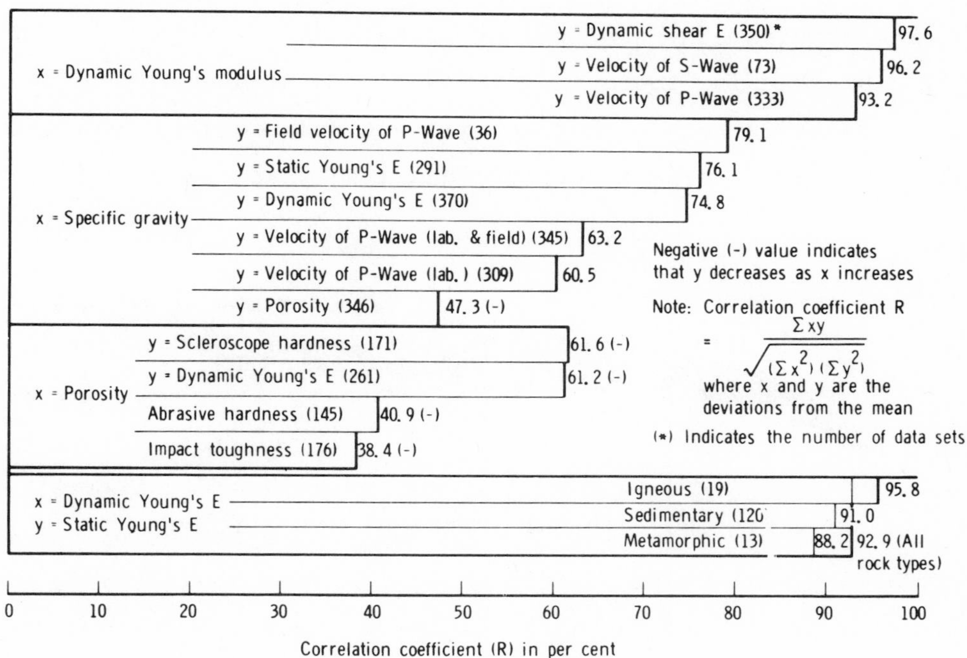

Fig. 39—Linear-regression comparisons of various rock properties.

Yet to be determined are the extent of the deviations and the percent confidence that can be used in predicting one parameter from the other. Furthermore, these data are abstracted from only about 900 data sets, whereas the final study contemplates two to three times this number. The additional data, however, have to be obtained from widely scattered sources and may exhibit considerable scatter because of the lack of test standardization.

After it has been determined which data pairs cannot be reasonably correlated by linear regression, attempts will be made to compare three, four, five, and more parameters by linear regression. Those combinations of physical parameters that cannot be successfully correlated in this manner then will be subjected to a curvilinear analysis. Hopefully, the final study will have determined which physical properties of rocks can and cannot be correlated and with what degree of reliability. Furthermore, if we can establish which physical properties give the best evaluation of reactions of rock under design loading, we would be in a better position to select the best laboratory and/or field test methods for any particular project. In any event, hopefully our findings will clearly delineate how useful or useless it is to apply elastic theory to rock mechanics. My own feeling in this latter regard is illustrated in Fig. 40; that is, there is increasing empirical evidence that our major advances in rock mechanics will not be made until--like the snake--we shed the skin of elastic theory and find what really will work.

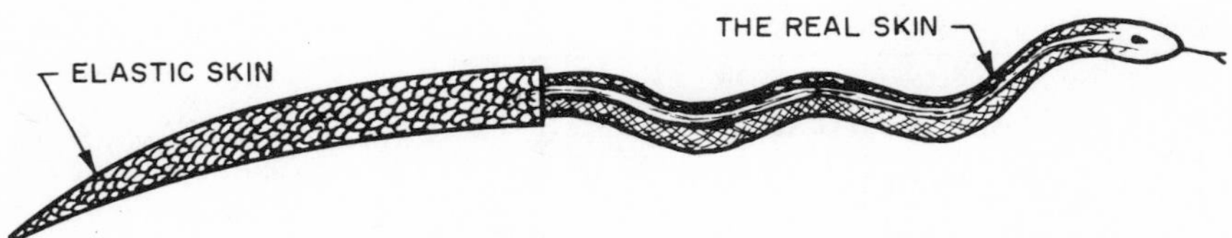

Fig. 40—"Shedding" the elastic theory.

Conclusion

All of the major problems in rock mechanics studies can be traced to the undeniable fact that rock is a heterogeneous material that obeys only the laws of nature, and we have yet to define these laws. However, recognition and acceptance of this approach will greatly stimulate progress in this new science. We then will be able to take major strides in assessing the possibilities of correlating laboratory and field tests. Each piece of rock or rock mass will have to be examined both microscopically and macroscopically. The geological defects will have to be quantitatively assessed. The geologist and the engineer will have to blend their talents to produce quantitative and thus useful results. And, this blending will require full appreciation of the mathematics and physics that must be introduced to find the solutions.

It has been only within the past few years that we have started to recognize the problems in rock mechanics. And, as in every science, the problems must be known before the solutions can be found. Thus, this rapid introduction to laboratory, field, and statistical problems in rock mechanics is intended to better establish not only why the earth's crust is in a state of stress, but also why those of us who engage in rock mechanics frequently are in the same state!

REFERENCES

1. Eirich, F. R., Rheology, Theory and Applications, Vol. 1, Academic Press, Inc., New York, 1956, pp. 1-8.
2. Lang, T., "Theory and Practice of Rock Bolting," Trans. Am. Inst. Mining Met. Engr., Mining Div., Vol. 223, 1962.
3. Heim, A., Untersuchungen über den Mechanismus der Gebirgsbildung, im Anschluss an die geolische Monographie der Tödi-Windgällen-Gruppe, B. Schwabe, Basel, 1878.
4. Obert, L., W. I. Duvall, and R. H. Merrill, "Design of Underground Openings in Competent Rock," U.S. Bur. Mines Bull. 587, 1960.

5. Endersbee, L. A., and E. O. Hofto, "Civil Engineering Design and Studies in Rock Mechanics for Poatina Underground Power Station, Tasmania," Preprint 1711, Australian Inst. Engr. Conf., May, 1963.
6. Personal communication with Snowy Mountain Authority, Australia, 1962.
7. Hast, N., "The Measurement of Rock Pressure in Mines," Sveriges Geol. Undersokn. Arsbok 52, 3, Stockholm, 1958, pp. 170-171.
8. Serafim, J. L., "Internal Stresses in Galleries," Papers 7th Congr. Large Dams, Sec. R-1, Question No. 25, Rome, 1961.
9. Moye, D. G., "Rock Mechanics in the Investigation and Construction of T-1 Underground Power Station, Snowy Mountains, Australia," Geol. Soc. Am., Eng. Geol. Case Histories, No. 3, May, 1959, pp. 13-44.
10. Clutterbuck, M., "The Dependence of Stress Distribution on Elastic Constants," Brit. J. Appl. Phys., Vol. 9, August, 1958, pp. 323-329.
11. "400,000-Ton Underground Pillar Blast," Mimag (Mount Isa Mines Ltd., Australia), Vol. 15, No. 7, August, 1962, pp. 14-18.
12. Summary of Operations, Mount Isa Mines, Ltd., Mount Isa, Queensland, Australia (no date).
13. Judd, W. R., and D. Wantland, "Influence of Geotechnical Factors on Arch Dam Design," Proc. XXth Intern. Geol. Congr., Sec. XII, Mexico, 1959, pp. 191-214.
14. Unpublished data from The Marquardt Corporation and the Snowy Mountain Authority.
15. Kujundzic, B., "Testing Mechanical Features of Rocks," Zbornik Radov (Beograd), No. 5, 1957 (OTS 60-21655), pp. 5-24.
16. Phillips, D. W., "Tectonics of Mining," Colliery Eng., August, 1948, pp. 278-282.
17. Duvall, W. I., L. Obert, and S. L. Windes, "Standardized Tests for Determining the Physical Properties of Mine Rock," U.S. Bur. Mines Rept. Invest. 3891, 1946.
18. Hardy, H. R., Jr., "Standardized Prodedures for the Determination of the Physical Properties of Mine Rock under Short-period Uniaxial Compression," Dept. Mines Tech. Surv., Ottawa, Mines Branch Tech. Bull. TB 8, December, 1959.

19. Belikov, B. P., "Elastic Properties of Rocks," Union Géodésique et Géophysique Internationale, Assoc. de Séismologie et de Physique de l'Intérieur de la Terre, Comptes Rendus des Séances de la Douzième Conférence, réunie à Helsinki, July 25--August 6, 1960, pp. 222-223.

20. Heiskanen, W. A., and F. A. Vening Meinesz, The Earth and Its Gravity Field, McGraw-Hill Book Co., Inc., New York, 1958, p. 7.

21. Kortes Dam and Powerplant--Technical Record of Design and Construction, U.S. Bureau of Reclamation, Denver, Colorado, December, 1959.

22. Unpublished reports by The Marquardt Corporation, Van Nuys, California, 1962-1963.

23. Ege, J. R., and R. B. Johnson, Consolidated Tables of Physical Properties of Rock Samples from Area 401, Nevada Test Site, U.S. Geol. Surv. Technical Letter Pluto-21, August 21, 1962 (and other similar Pluto Technical Letters in 1961 and 1962).

24. Kormer, S. B., A. I. Funtikov, V. D. Urlin, and A. N. Kolesnikova, "Dynamic Compression of Porous Metals and the Equation of State with Variable Specific Heat at High Temperatures," Soviet Phys. JETP, Vol. 15, No. 3, 1962, pp. 477-488.

25. Krupnikov, K. K., M. I. Brazhnik, and V. P. Krupnikova, "Shock Compression of Porous Tungsten," Soviet Phys. JETP, Vol. 15, No. 3, 1962, pp. 470-476.

ABSTRACT

Editor's Note: The author was not requested to prepare an abstract of this paper as it was intended to be and is a résumé of current theories regarding the composition and structure of the earth. Brief comments are presented on the more significant controversies in these subjects. The discussion is directed toward providing the reader with an explanation of the origin and distribution of stresses in the earth's crust. Thus, it emphasizes the need for considering the geological situation whenever man finds it necessary to use any portion of the crust.

RÉSUMÉ

Note de la rédaction: On n'a pas demandé à l'auteur de préparer un résumé de cette conférence, car celle-ci était destinée à être--et est--un résumé des théories actuelles concernant la composition et la structure de la terre. De brefs commentaires sont présentés sur les controverses les plus importantes portant sur ces sujets. La discussion est dirigée de manière à fournir au lecteur une explication de l'origine et de la distribution des efforts dans la croûte terrestre. Elle insiste ainsi sur la nécessité de prendre la situation géologique en considération chaque fois que l'homme à besoin d'utiliser une partie quelconque de la croûte terrestre.

AUSZUG

Bemerkung des Schriftleiters: Es war vom Verfasser nicht verlangt worden, einen Auszug dieser Arbeit zu geben, da diese von vorneherein als eine Zusammenfassung der gegenwärtigen Theorien über Beschaffenheit und Gefüge der Erde vorgesehen worden war. Über die bedeutenderen Streitfragen dieser Gebiete werden kurze Bemerkungen gegeben. Diese Aussprache soll dazu dienen, dem Leser eine Erläuterung über den Ursprung und die Verteilung der Spannungen in der Erdkruste zu geben. Die Notwendigkeit, die geologischen Gegebenheiten zu berücksichtigen, wenn immer es notwendig wird, Teile der Erdkruste in den Dienst der Menschheit zu stellen, wird daher in dieser Arbeit betont.

MEGAGEOLOGICAL CONSIDERATIONS IN ROCK MECHANICS

Francis Birch*

Introduction

WHEN THE CONFERENCE COMMITTEE gave me this subject, I had some misgivings. The prefix "mega," when attached to units of measurement as in "megabar," "megameter," "megaton," is convenient in eliminating six zeros and fashionable in showing suitable respect for the space age. Used with other terms, it may signify simply "large," as in "megalithic" or "megapolis." But geology is the science of the earth; what is "megageology" the science of? At this stage of my ruminations, I received the preprints of the papers to be presented at the conference and discovered to my relief that the prevailing philosophy was that of Humpty-Dumpty: "When I use the word, it means just what I choose it to mean, neither more nor less." So I shall give my own meaning to this subject. I shall try to look at the problems of stress in the earth through the wrong end of the telescope, seeing only large features and leaving the troublesome details to you.

The Concept of Stress

It may not be out of place to refer briefly to the artificial nature of the concept of stress in the interior of a body. Stresses in solids are not directly measurable; they appear as a part of the process of determining deflections or displacements, including, finally, yield or rupture, when bodies are subjected to specified loads. The

*Department of Geological Sciences, Harvard University, Cambridge, Massachusetts.

mental image of interaction between the parts of a body is indispensable, but it may be well to remind ourselves occasionally of the large element of assumption usually introduced. In place of real crystalline aggregates, we substitute a uniform mathematical continuum without voids, flaws, or local variations, and having only certain average properties of the real medium. Even worse, we have to choose among many possible kinds of relations between stress and strain, or load and deformation, usually settling for the most tractable rather than the most realistic. The earth may be adequately treated as elastic for small stresses of short periods; for loads lasting as long as geological periods, it is necessary to consider some type of anelastic behavior, of which many different kinds are conceivable.

Since the earth as a whole is virtually free from external forces, except for the small tidal effects of sun and moon, the only forces are body forces, arising mainly from the gravitational field. We can seldom make a clear distinction between the load and the structure that supports it. Furthermore, when we say that a load is supported, we must specify the interval of time involved; do we mean support for a thousand years, a million, a thousand million? A degree of permanence more than satisfactory for the engineer may still be consistent with movements that seem feverish on the geological time scale. Over such times, nonmechanical processes may become dominant; besides fracture or plastic flow, we must consider the possibility of melting, recrystallization, and chemical alteration. There is no well-defined initial condition from which to start; as far back as we can go we find the marks of an earlier history.

The Structure of the Earth

The earth's crust is hardly more than a surface film. Its mechanical state as well as composition must be controlled by the interior, and a hasty summary of the large-scale structure seems in order. Most of the details about the interior have been derived from the travel times of the seismic body waves; reduced to velocities (Fig. 1), these show

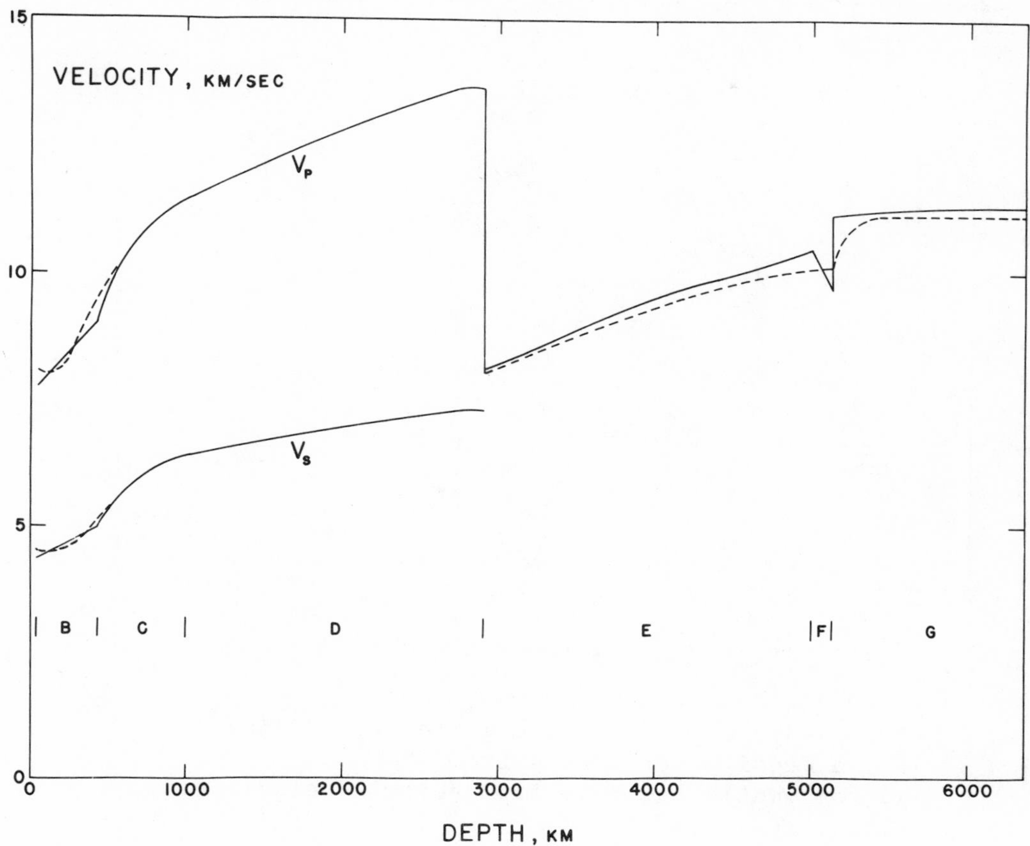

Fig. 1—Velocities of seismic body waves and internal layers: B, upper mantle; C, transition layer; D, lower mantle; E, outer core; G, inner core. Solid line after Jeffreys (Ref. 1), broken lines after Gutenberg (Ref. 2).

the main feature to be the division between mantle and core. The mantle contains about two-thirds and the core one-third of the whole mass. Velocity and density change drastically at the mantle-core boundary, and with the aid of experimental data now extending to core pressures, the classical identification of the material of the core as mainly iron has been confirmed. Equations of state for most of the metals have been determined to several megabars with the method of shock waves; perhaps the most satisfactory way of demonstrating the chemical difference between mantle and core is by a "sound velocity versus density" plot, on which we may also put the corresponding quantities for mantle and core (Fig. 2). The core is evidently composed of elements of the

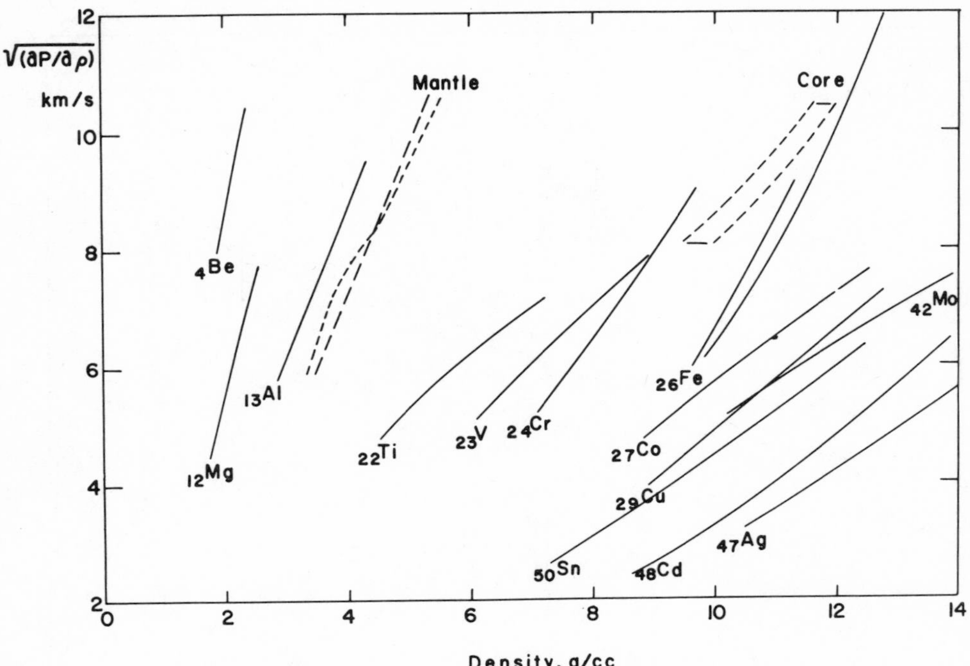

Fig. 2—Hydrodynamic sound velocity, $(\partial P/\partial \rho)^{\frac{1}{2}}$, versus density for metals from shock compressions, and for mantle and core from seismic velocities (Birch, Ref. 3).

transition group, of which iron is by far the most abundant, while the mantle values are in the region of the light elements. The results for metals have been supplemented by shock-wave studies on rocks (Fig. 3) that show that at mantle pressures the densities are close to those required for the mantle (and far below what is required for the core). I think we may firmly shut the door on speculations that core and mantle might have similar chemical compositions. The chemical differentiation between mantle and core is the major fact about the earth's constitution. As such an arrangement can hardly be the original one, the segregation of iron from an initial state of dispersion becomes the principal event of earth history.

The details of this process are obscure, though interesting suggestions have been offered. The descent of the heavier material and rise of the lighter released a large amount of gravitational energy, most of which was converted to heat. The core separated as a liquid phase, and I believe that the upper mantle formed at the same period,

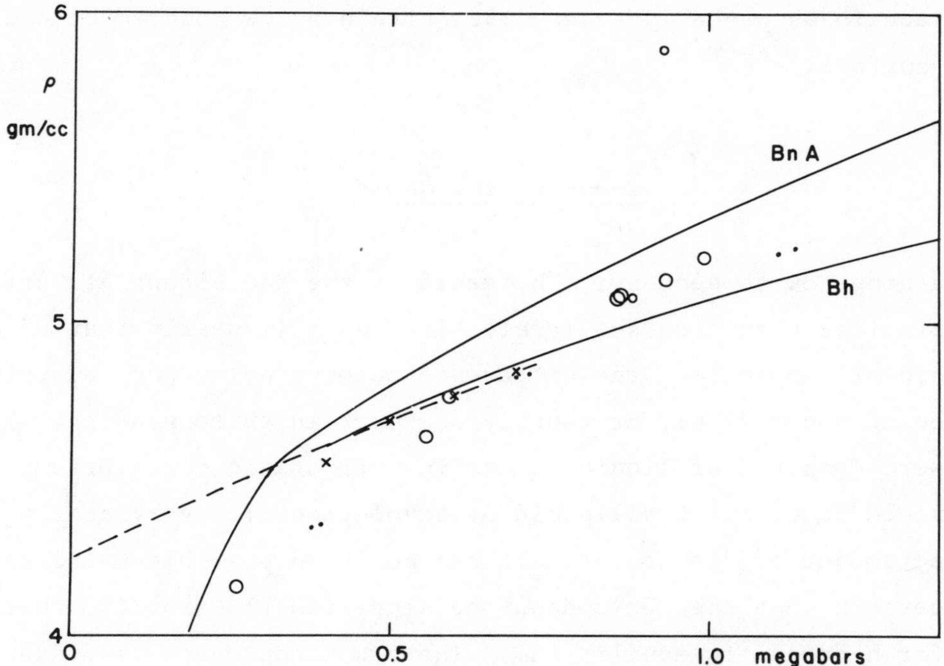

Fig. 3—Density versus pressure for a variety of rocks (after McQueen, Ref. 4, and Wackerle, Ref. 5) and several possible density distributions for the lower mantle. The small circle near the upper margin is for an iron-rich dunite (hortonolite); the crosses are quartz; and the other symbols represent samples of ordinary dunite, pyroxenite, eclogite, and diabase.

with partial melting leading to an upward concentration of the low-melting silicate fraction. Subsequent differentiation of the upper mantle has produced the crust, now amounting to about 1 per cent of the whole mass. There are reasons for believing that the lower mantle consists of high-pressure forms of the oxides of magnesium, silicon, and iron, and that the transformation of the silicates of the upper mantle to these denser forms gives a satisfactory explanation of the rapid rise of seismic velocities between the depths of 300 and 900 km. The upper mantle (Bullen's Layer B) and the transition layer (Layer C) probably contain the dynamic regions of the earth. The lower mantle, so far as we know, is inert. The deepest earthquakes occur at about 700 km, near the base of the transition layer; basaltic magmas probably originate within the upper mantle; and the adjustment to redistribution of surface loads probably takes place within a few hundred kilometers of the surface.

Shape of the Geoid

Turning now to geodesy: The geoid is the equipotential surface that coincides with mean sea level. Its shape is nearly that of an ellipsoid of revolution, the difference between polar and equatorial radii being about 22 km, or roughly one part in three hundred. If the earth were composed of liquid layers in hydrostatic equilibrium, the geoid would be an exact ellipsoid of revolution with very nearly the observed amount of flattening. It has not been possible until recently to be certain that the geoid departed significantly from its theoretical shape for hydrostatic equilibrium. The new techniques associated with artificial satellites now indicate a small discrepancy: The second-degree harmonic of the external potential, which expresses the main equatorial bulge, is not exactly what it should be for hydrostatic equilibrium, but, curiously, it seems that the equipotential surfaces are *more* flattened than for a fluid earth. This observation has been interpreted as showing a delay in adjustment from an earlier period when the day was shorter, perhaps some tens of millions of years ago. Another possible interpretation and what is directly indicated is anomalous density (with respect to the hydrostatic layered earth) somewhat lower under the poles and greater under the equator. The nonhydrostatic stresses associated with this second-degree departure from fluid equilibrium may be of the order of 100 bars and extend throughout the mantle.

There are also harmonics of higher degree in the external potential that could not exist for fluid equilibrium, notably the third-degree harmonic (Fig. 4). Although its maximum amplitude as geoidal disturbance is only 16 m (hardly producing a "pear-shaped" figure), the fact that this harmonic exists at all is additional evidence for widespread, and probably deep, density disturbances.

Coefficients of zonal and tesseral harmonics of the external potential have been derived from the satellite observations, from measurements of gravity on the earth's surface, and from various combinations of these two sources. Probably only those of low degree are adequately determined as yet. So far as they go, it is noteworthy

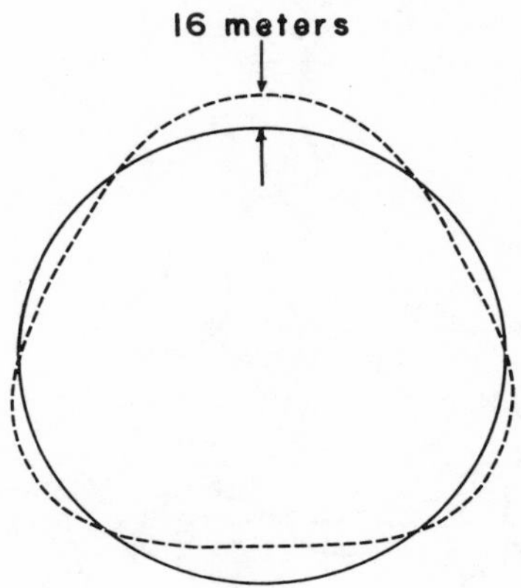

Fig. 4—Displacement of the geoid for the third-degree zonal harmonic referred to an ellipsoid with flattening 1/298.2; exaggeration about 600,000.

that they show little relationship to the corresponding coefficients in the spherical harmonic expansion of the topography. With terms to the fifth degree in the topography, continents are roughly indicated (Fig. 5); with terms of the fourth degree in the external potential (Fig. 6), there seems to be little correspondence between geoidal height and continental outline. This is perhaps not surprising in view of the general near-surface compensation of surface relief, and because the principal term in the expansion of the topography--the first-degree term--is excluded by theory from the expansion of the external potential. On the other hand, if large-scale internal density variations were controlled by the surface configuration, or vice versa, some similarity in the coefficients of low degree might have been expected. Whether the indicated density variations are related to past or to future positions of continents is an open question. The stresses and density differences can be estimated with the aid of hypotheses about distribution; if the density perturbation is distributed

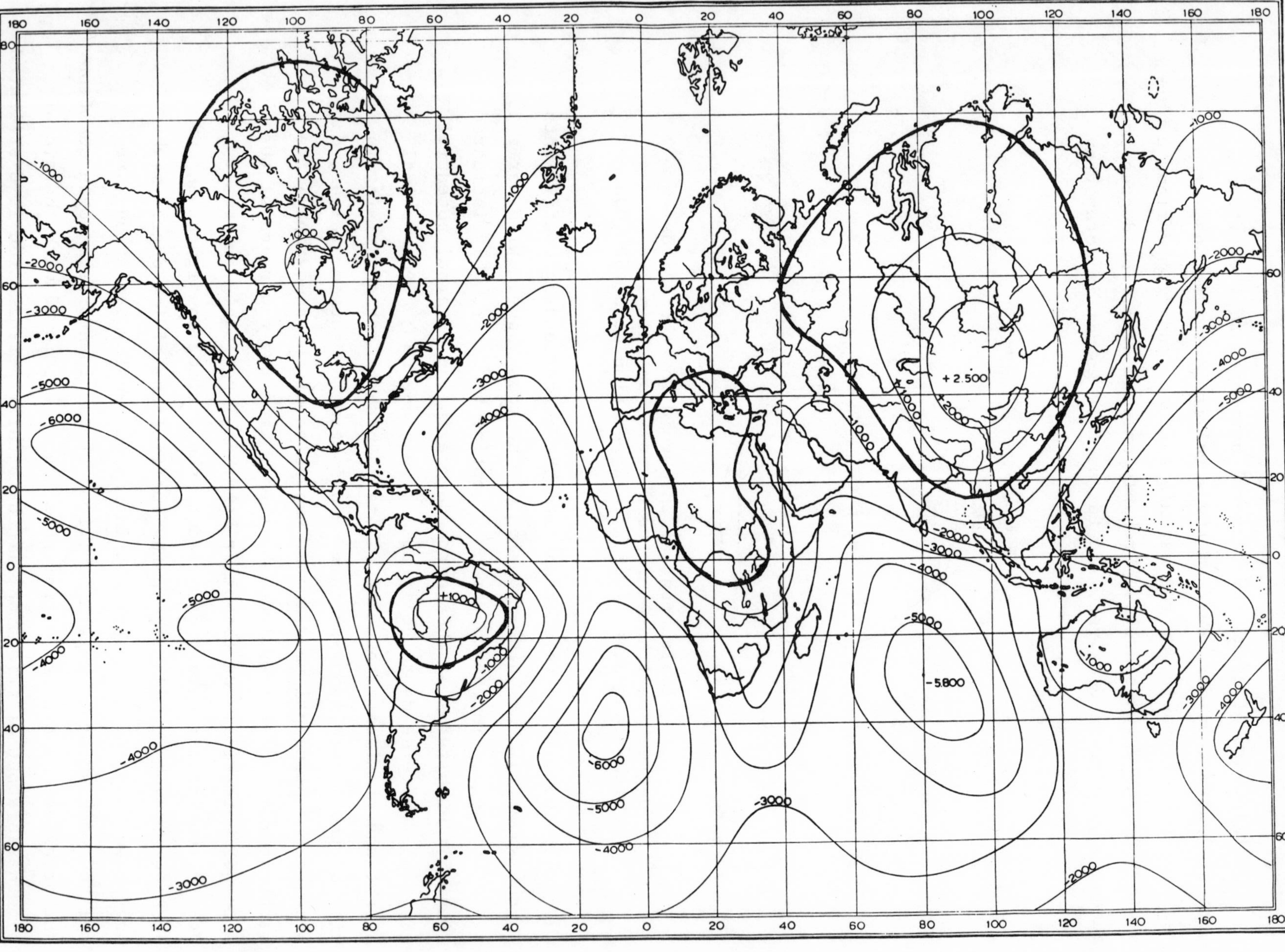

Fig. 5—Topography as expressed by spherical harmonics of the first five degrees (after Vening Meinesz, Ref. 6).

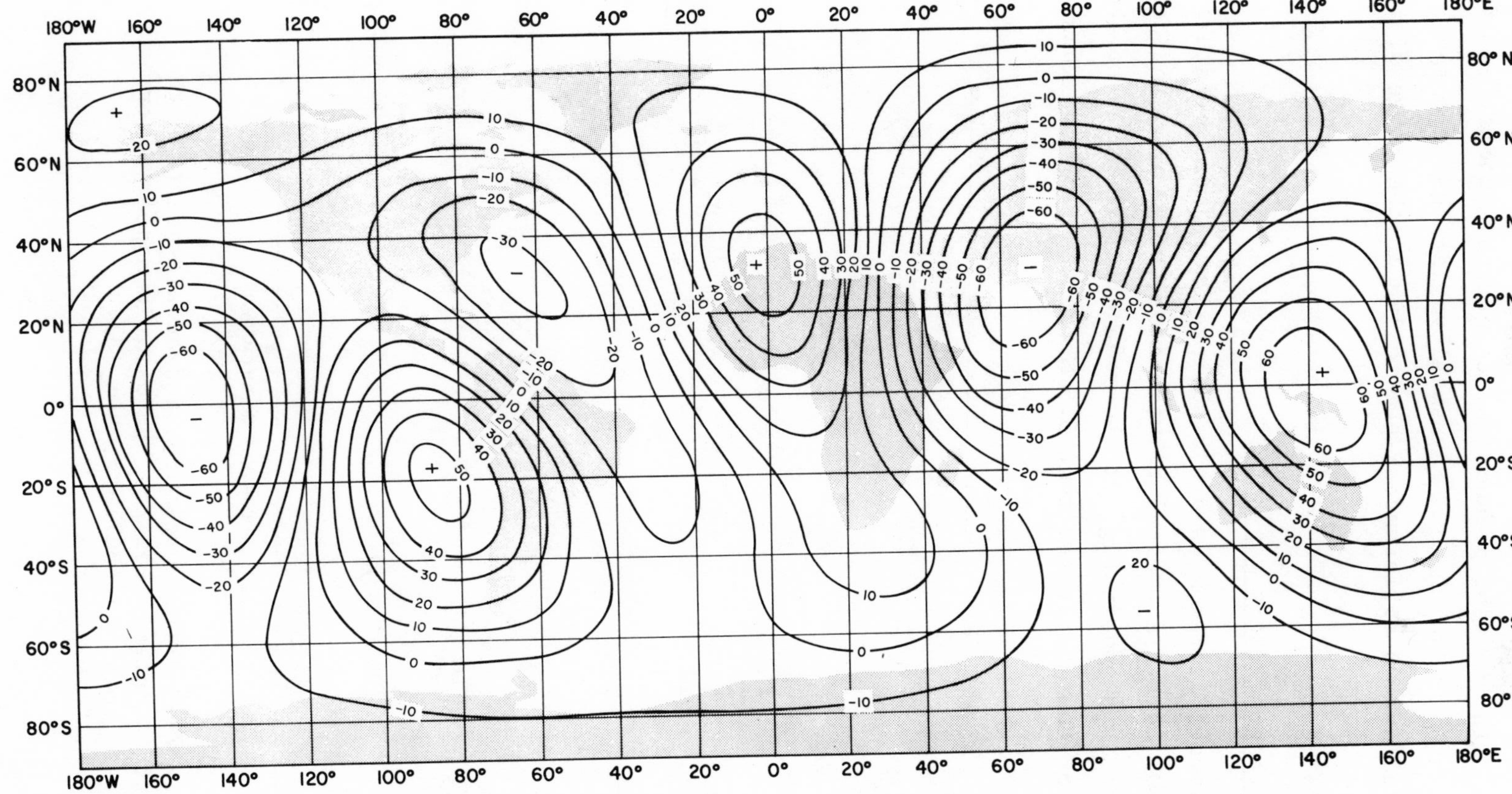

Fig. 6—Undulations of the geoid in meters, as expressed by spherical harmonics of the first four degrees (after Uotila, Ref. 7); the flattening of the reference ellipsoid is 1/298.24.

through the whole mantle, we find anomalous densities of the order of 10^{-4} g/cm^3. This might be produced by temperature anomalies of the order of 10 deg, or, of course, by small variations of composition. The associated stress differences seem to lie in the range of 10 to 100 bars.

Stress Differences within the Earth

These stress differences are associated with mean pressures rising to millions of bars. A typical solution for the density and corresponding mean pressure is shown in Fig. 7; the mean pressure rises to about 1.3 Mb at the base of the mantle and some 3.5 Mb at the center. Thus, the

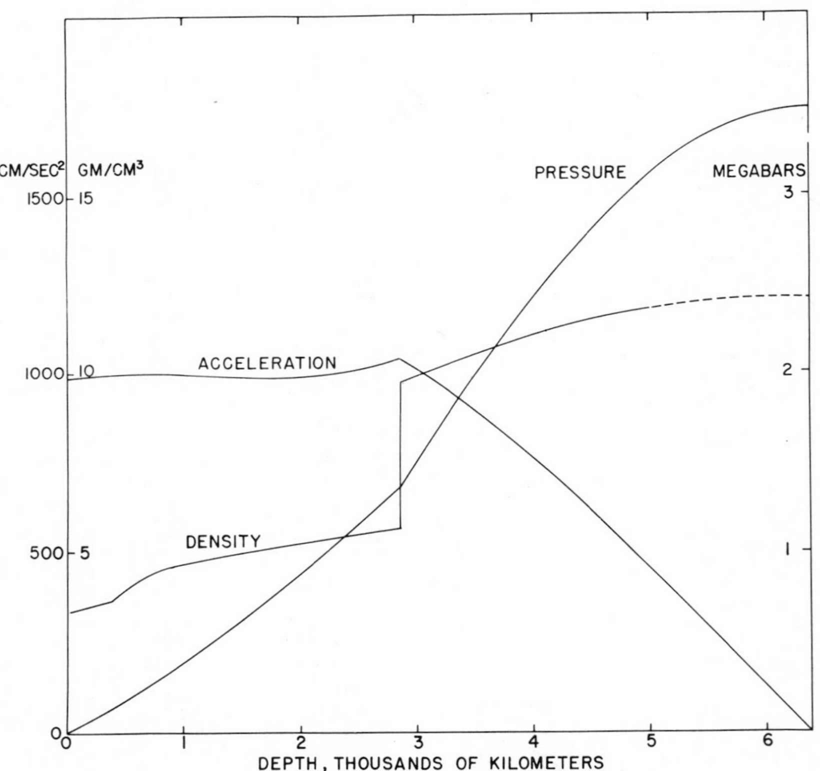

Fig. 7—Density, pressure, and acceleration in the earth's interior on the hydrostatic model (after Bullard, Ref. 8).

approach to hydrostatic equilibrium is very close deep in the mantle; it becomes less close toward the surface, where the stress differences begin to be comparable with the mean pressure. This is especially true within the crust, where we may expect stress differences up to the strength of strong rocks (several kilobars), and the mean pressure cannot much exceed 10 kb. The problem of stresses in the deep interior becomes a search for small departures from large hydrostatic pressures in a medium whose average elastic properties are well known from the seismic travel times, but whose uniformity and anelastic properties are largely conjectural.

The largest stress differences are associated with the support of high mountains and deep ocean trenches. Regardless of how the stresses are distributed in the deep interior, the weight of mountains must be carried on their bases, and there must be compressive stresses at shallow levels of the order of the weight per unit area of the vertical columns, thus of the order of kilobars. Unless the horizontal compressive stresses are of the same order, the maximum shear stresses will also be of the order of kilobars and well within the yield stresses of rocks for short-time laboratory tests, provided all principal stresses are compressive. Whether such stresses produce "creep" over times as long as geological periods is uncertain; the reduction of mountains by erosion is probably much more rapid than their spread by creep at shallow depths. On the other hand, the observations of gravity show that most of the solid surface of the earth is within a few hundred meters of the elevation it would have if the crust were floating in or on a mantle of high density but low strength. This suggests that for stresses of long duration, the yield point must be of the order of 100 bars at no great depth. One of the major problems in the mechanics of the earth is to reconcile this fluid-like behavior with the existence of strong earthquakes at focal depths as great as 700 km and with other paradoxical evidences of long-term strength.

Some Cases in Point

There are pronounced gravity anomalies associated with structures probably formed hundreds of millions of years ago. An example (one of many similar cases) is the gravity maximum over the Green Mountain Anticlinorium in Vermont (Fig. 8), with a relief of some 50 mgal. One interpretation is shown in the figure; the associated stress difference reaches some 500 bars at a depth 10 km or so below the crust. It seems plausible that this feature was formed during the Appalachian orogeny toward the end of the Devonian period. Perhaps

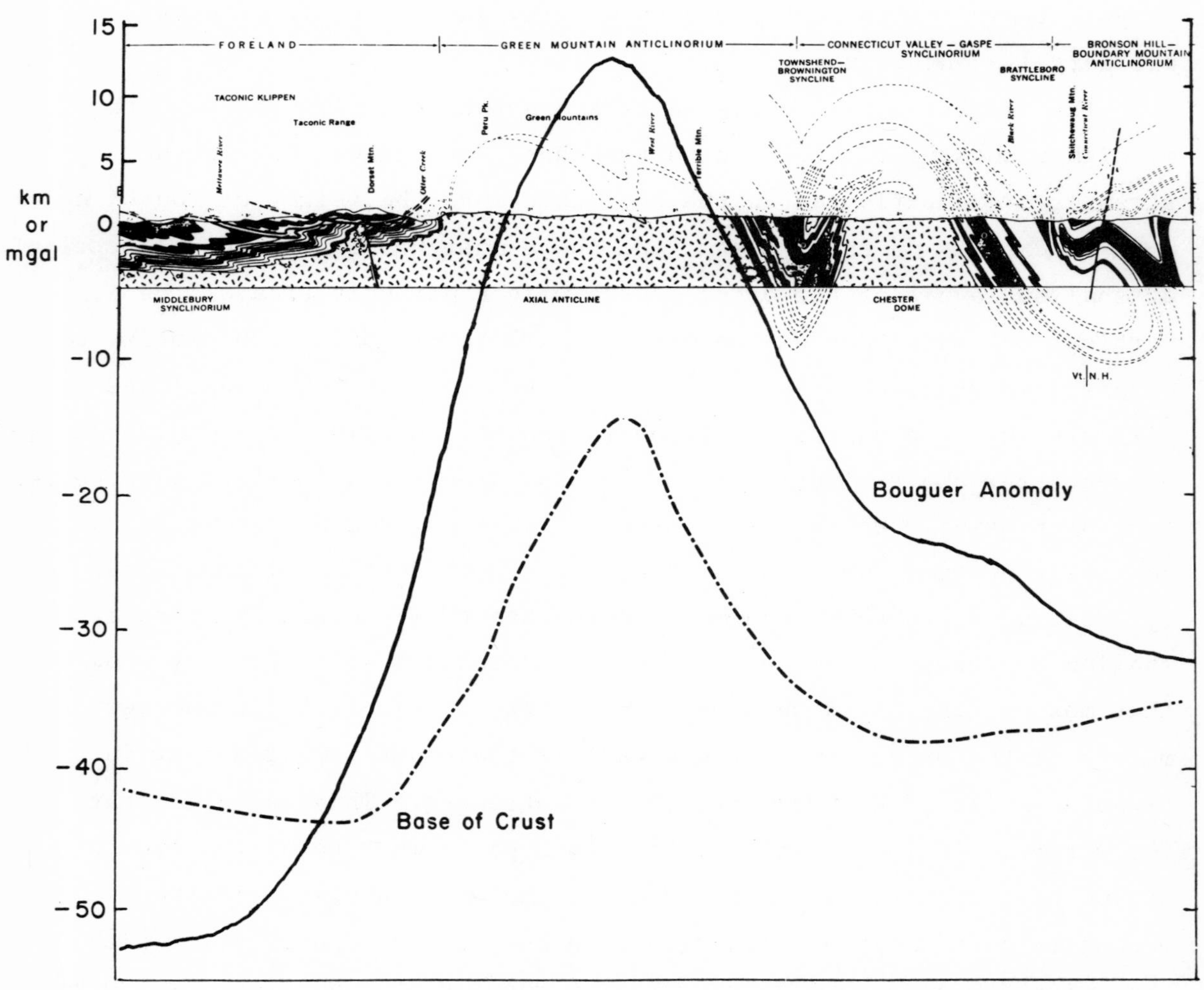

Fig. 8—Geology, Bouguer anomaly, and a possible interpretation for a section across the Green Mountain Anticlinorium, Vermont. Geology from the Geologic Map of Vermont (1963), gravity from Diment (Ref. 9). The structure is shown at true scale.

it was larger originally. At the present time, some of the largest gravity anomalies, such as those of the Indonesian Arc that reach as high as 200 mgal in the negative Meinesz strips, are associated with relatively young geological features, with volcanism and active earthquake zones. Positive anomalies of the same magnitude are found on the Hawaiian volcanoes. The island of Hawaii is an immense mountain rising nearly 10 km from the sea floor, and although it is still growing, much of this mass must have existed for some millions of years. Stress differences of the order of 500 bars are required at depths of 50 to 100 km.

The rise of Fennoscandia after the disappearance of the last glacial icecap has been studied with the object of discovering the time-dependent behavior of the mantle, but the only mechanical model investigated thus far is the viscous one.* The numerical estimates found for the coefficient of viscosity have been applied to many other discussions of flow in the mantle. The viscous model leads to a number of clearly incorrect inferences about the decay rate of anomalies of gravity and probably to misleading conceptions of other internal motions.

On a viscous layer of great depth, the time required for a given approach to floating equilibrium is inversely proportional to the horizontal dimension of the load. For Fennoscandia, with an icecap about 1600 km in diameter, the time for a nearly complete return is given by Haskell as 18,000 years. For Hawaii, with a diameter at the base of about 350 km, the expected time might be about 100,000 years, instead of the probable several million or more. Somewhat similar estimates and discrepancies may be noted for the Indonesian strips; it has been suggested that these are held down by tectonic forces arising from motions of a viscous mantle, but there seems to be no basis for supposing anything of this kind for Hawaii. The departure of the second-degree term in the potential from its hydrostatic value would vanish in a few thousand years on an earth with the viscosity derived from the analysis of the recoil of Fennoscandia.

---

*Editor's note: It may be of interest to study Hast's findings in regard to the stresses in Fennoscandia (see Ref. 7 of W. R. Judd's paper).

Stress Determination

The calculation of stresses produced by surface loads also depends on assumptions concerning mechanical properties. Several examples of distributions under the same normal load are illustrated in Fig. 9.

Vertical load, σ per unit area

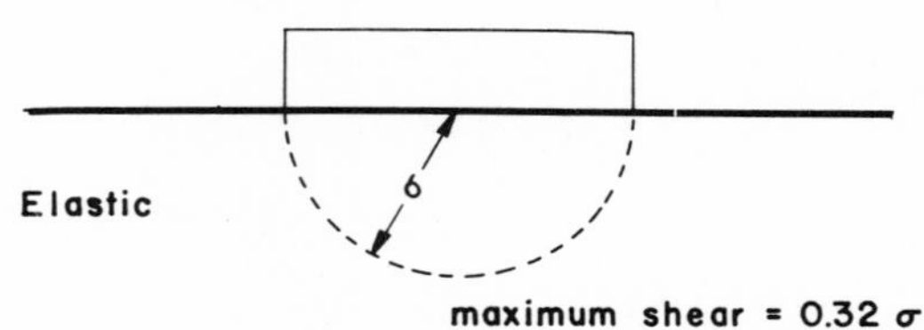

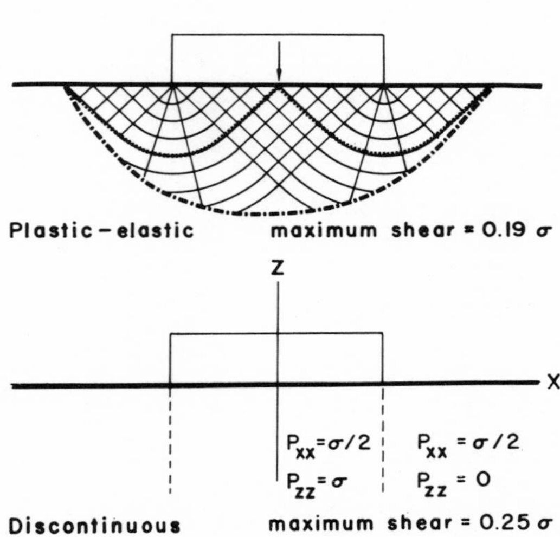

Fig. 9—Solutions for uniformly loaded strips: upper, elastic; center, plastic-elastic (Prandtl, after Hill, Ref. 10); lower, discontinuous stress (Jeffreys, Ref. 1).

These are all two-dimensional solutions for a uniformly loaded strip on a uniform, semi-infinite medium. The load per unit area is $g\sigma$. The elastic solution at the top leads to maximum shear stress of 0.32 $g\sigma$ on the cylindrical surface of radius a (2a is the width of the load). In the center is Prandtl's solution (as modified by Hill[10]) for a plastic-elastic medium under the uniform downward motion of a smooth, rigid punch; the maximum shear is set by the yield condition and corresponds to 0.19 $g\sigma$, the depth of the plastic boundary depending on the width of the load. At the bottom is a stress distribution, suggested by Jeffreys,[(1)] for which the shear stress is everywhere 0.25 $g\sigma$, but the vertical stress is discontinuous at $x = \pm a$.

By way of contrast, Fig. 10 shows Haskell's solution for the displacements beneath a loaded strip on the surface of a semi-infinite viscous medium. More than half the lateral flow takes place below a depth equal to or greater than the width of the load. If the load is 1600 km wide, as in the application to the Fennoscandian icecap, most of the flow takes place at depths greater than 1600 km, and a recalculation for a spherical earth is desirable. In the plastic-elastic

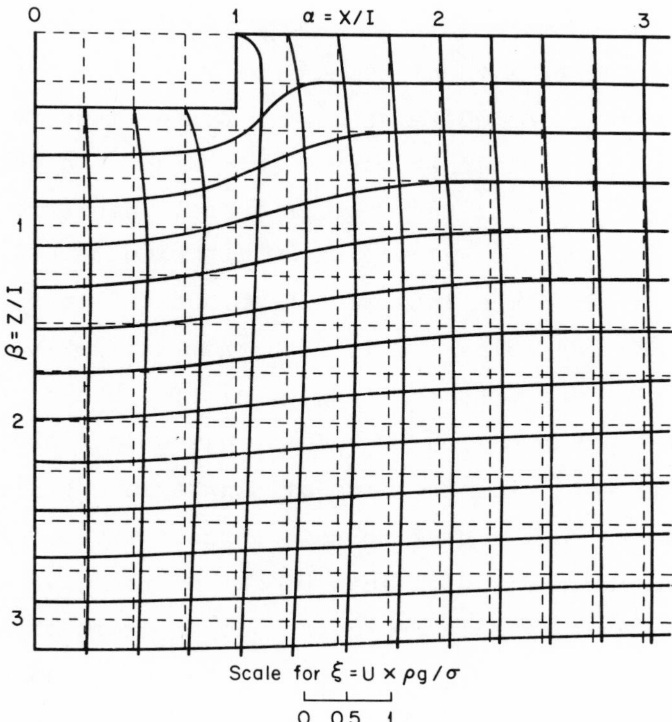

Fig. 10—Asymptotic displacement in semi-infinite viscous medium under strip load (after Haskell, Ref. 11).

solution, the flow takes place near the load, most of the lateral displacement occurring at depths less than half the width of the load. The stress and velocity fields are completely different for the plastic solid and for the viscous fluid, and misleading results will be obtained if the wrong model is chosen.

Structure of the Crust and Mantle

Under surface and near-surface conditions, most rocks are brittle and weak, with tensile strengths of the order of a few hundred bars. Thus, tensile stresses near the surface will be relieved by fracture; even very gentle flexures suffice to cause breaking strains of the order of 10^{-4}. The conception of the crust as a continuous elastic plate is at variance with the limited degree of regionality shown by gravity anomalies. A more suitable image may be a raft of nearly independent blocks with lateral dimensions like the thickness of the crust, perhaps some 30 to 50 km at most. Besides the relatively large blocks, the upper part at least may be fractured on a much finer scale. In their mapping of 9 mi of the Roberts Tunnel, Wahlstrom and Hornback[12] found faults at an average spacing of 20 ft, with several times as many joints. Similar dimensions have been noticed in other tunnels and mines, and one may compare this with the maximum sizes of blocks taken from quarries and with the largest glacial boulders. Curiously, this figure is comparable with the linear dimension, about 15 ft, obtained by Jeffreys[1] for the size of scattering "grains" needed to give the observed damping of seismic surface waves and near-earthquake body waves. Perhaps a division into blocks of this average dimension exists through a considerable depth, the upper crust then resembling a sand pile rather than a solid plate. In the gravitational field, these grains must always be tightly packed together like the stones in an arch, and friction between them may be the controlling factor in deformation rather than plasticity of the intact blocks.

Deeper in the crust and mantle, with all stresses compressive, a plastic model probably becomes appropriate. Nearly ideal plasticity is shown by many rocks under these conditions; an example from the work of Griggs, Turner, and Heard[13] is shown in Fig. 11. These short-time

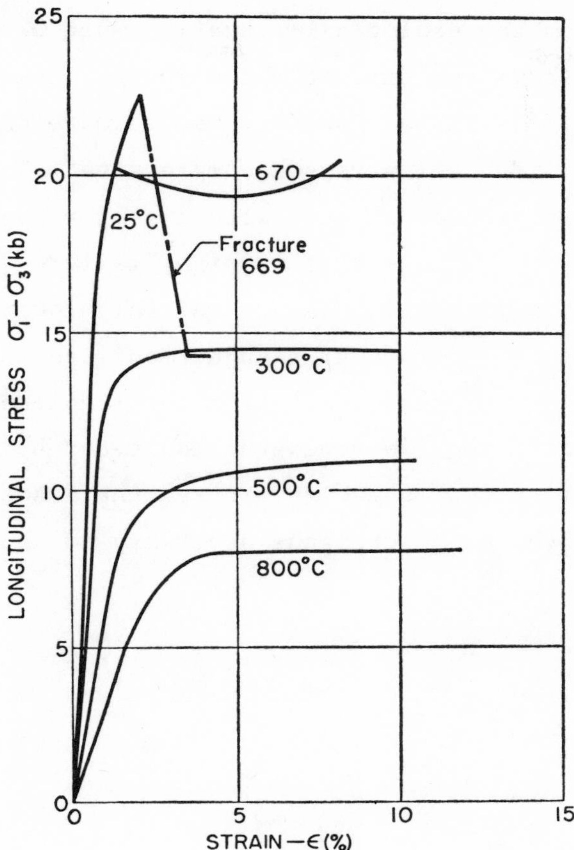

Fig. 11—Stress difference versus strain of Dun Mountain dunite at 5-kb confining pressure in compression (after Griggs *et al.*, Ref. 13).

tests show high values of the shear stress at yield, several kilobars even at 800°C. It seems likely that yield or rupture would take place at lower stresses if they were maintained for long periods. Metallurgists have demonstrated this kind of behavior for various metallic alloys at high temperatures: An example is shown in Fig. 12 where the "creep rupture" stress has been plotted versus time for the nickel alloy Inconel 702 at 1180°C. I have taken the liberty of extending the experimental line, on a log-log plot, to outrageously long periods of time. The remarkable feature is that this does not lead to obviously ridiculous results, but illustrates a kind of behavior that geologists have intuitively attributed to the deeper levels of the earth.

Differences from the geological situation must be noted: The diagram in Fig. 12 shows the time at which rupture takes place under a given steady tensile stress. However, we are more interested in failure under shear, with all stresses compressive. Failure can then mean only a redistribution of stress, with a closer approximation to hydrostatic pressure. If the stress is high, failure takes place quickly with reduction of maximum shear. At this lower level, a longer time must elapse for further readjustment, and so on. A stress can persist for 10^8 or 10^9 years only at a low magnitude. Thus, we might hope to reconcile the low strengths indicated by the gravity anomalies over old features, the higher strengths associated with young features, and the high strengths necessary to account for the energies of earthquakes.

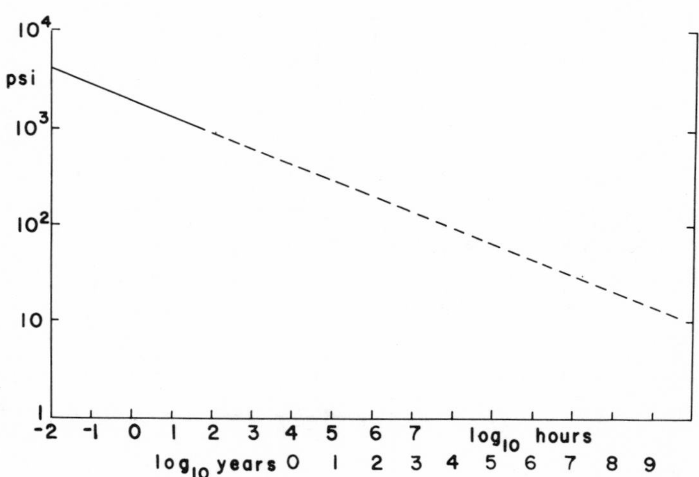

Fig. 12—Creep rupture strength versus time for Inconel 702 at 1180°C.

<u>Earthquake Stresses</u>. The greatest number of earthquakes are shallow, with focal depths less than 60 km; these are in the zone of strong rocks and there is no problem of strength. But earthquakes occur down to some 700 km, in the region named the "asthenosphere" because of its presumably low or vanishing strength. Figure 13 shows the number of earthquakes for 50-km intervals of depth, omitting shallow ones.

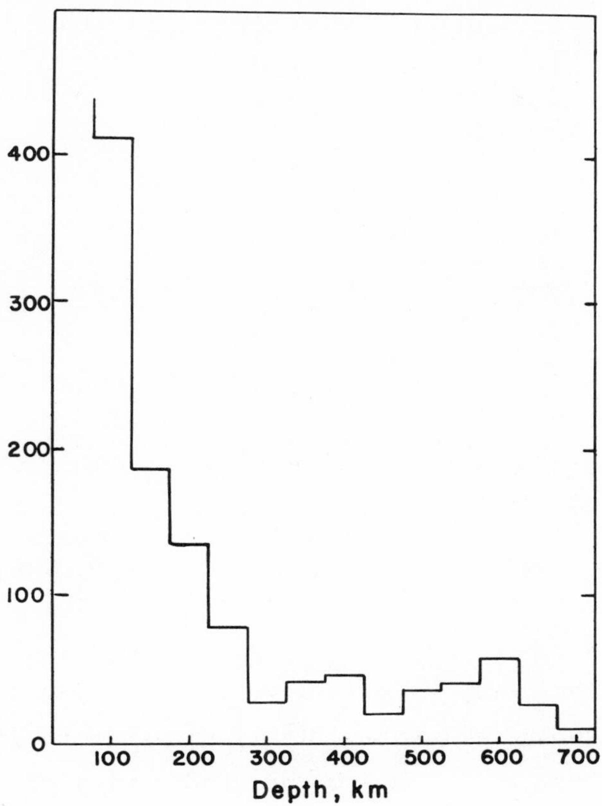

Fig. 13—Number of strong shocks for 50-km intervals of depth in the period (after Gutenberg and Richter, Ref. 14); the number of comparable shallow shocks in the same period is several thousand.

Gutenberg and Richter[14], the authorities for these numbers, notice the slight minimum at 450 km; possibly there is another at 250 km. The average energy for the intermediate shocks (70 to 300 km) is about the same as for the deep shocks (depth greater than 300 km) and about half the average for shallow shocks. Two difficult questions are: How can sufficiently high local stresses be induced in the deep interior; and by what mechanism can they be suddenly released?

The Release Mechanism

A possible suggestion to the mechanism involved in the release of stresses may be obtained from a representation of creep rate or strain rate versus stress given by Orowan[15] (Fig. 14). For low

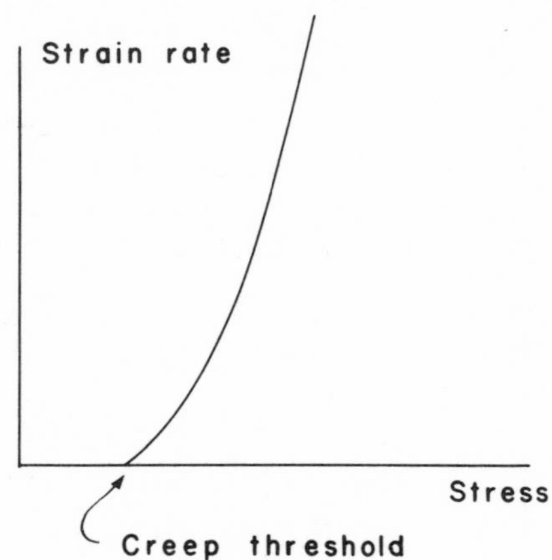

Fig. 14—Possible relation between strain rate and stress (after Orowan, Ref. 15).

values of shear, the creep rate is low, perhaps even zero below a finite creep threshold; evidently, it is not easy to distinguish between a creep rate of zero and one below the sensitivity of measurement. But beyond a certain value of stress, the rate increases rapidly, leading to a type of instability; the amusing term, "catastrophic creep," has been used for this phenomenon. Whether by this mechanism we can escape from the difficulty with sliding friction on deep faults noticed by Orowan, Griggs, and others is not entirely clear. However, it is a mechanism for releasing stress rapidly, converting much of the elastic energy to plastic work in the region of concentrated shear, and possibly producing partial melting. The creep threshold, that is, the stress at which creep begins to accelerate, must be of the order of 100 bars or so. In conjunction with a model such as that of Fig. 12, this means the stresses must develop fairly rapidly.

The distribution of deep shocks, as shown by Gutenberg and Richter,[14] follows in main outline the distribution of shallow shocks; that is, the foci are generally circum-Pacific; only the west coast of North America shows a marked absence of deep shocks. We may speculate that the deep shocks are related to the shallow shocks, failure at shallow levels leading to a rapid redistribution of stress at the deeper levels, large enough to account for the strain energy and to lead to subsequent creep instability at depth. We then have to consider the origin of shallow earthquakes.

Sources of Internal Stresses

Aside from the "passive" stresses produced by surface loads, the principal source of internal stresses must be sought in the thermal regime and in thermally determined density differences. This plunges us into the most debatable questions concerning the interior: the temperatures, the distribution of heat sources, and the mode of transport of heat. I will state briefly what I consider the most acceptable working hypotheses.

Thermal History

For heat sources of adequate staying power and intensity, we have principally the radioactive elements, and the fundamental property of radioactive substances is that they decay. The amounts now present in the earth are substantially smaller than they were 4 or 5 billion years ago, and the amount of heat thus generated is correspondingly less, perhaps one-eighth or one-tenth. Eventually the earth must cool to its surface temperature; the only (and difficult) question is how long a time must elapse before the cooling regime sets in. In the older theories, the earth began hot, and could only grow colder. At present, we commonly envision an initial earth of moderate temperature but undifferentiated, with radioactivity scattered throughout the mass. The temperature then rose until differentiation began, perhaps first as melting of metallic iron at a relatively shallow depth; this pregeological period might have lasted 1 to 2 billion years. Once begun, differentiation proceeded rapidly to form the iron core and the upper mantle, possibly also the nuclei of continents, since rocks have been dated as far back as 3 billion years. The heat sources were concentrated within a few hundred kilometers of the surface and the regime of cooling began. This conclusion is based on the belief that the lower mantle has been thoroughly purged of radioactive elements, a tenent not universally accepted.

Thermal Effects

The mechanical effects of such a thermal history can be looked at in two ways, not necessarily mutually exclusive. We may concentrate on the "thermoelastic" aspect, as in the classical contraction theory; or on motions consequent on density differences in the gravitational field, as in the various convection theories. There are some semantic problems: Convection may be used to denote certain kinds of internal motions that are beyond question, such as the flow associated with gravitational readjustments to erosion and deposition, the obvious motions of liquids in the feeders of volcanoes and lava flows, the rise of salt domes, gneiss domes, and various kinds of intrusions. What is often meant, however, is interchange of material between the upper and lower parts of the mantle, or of the upper mantle. It is my belief that these internal motions are essentially irreversible, in the sense that material that has come from below to the crust or upper mantle never returns to the lower mantle; and that the rising material is differentiated material, either the light liquid fraction formed by partial melting or segregations of intrinsically light solid material. These involve substantial differences of density, and the motion is likely to be confined to narrow zones. With the small density anomalies indicated by the low-degree harmonics of the external potential, it seems probable that even a small creep threshold would suffice to prevent massive overturns.

A number of theoretical papers have been devoted to steady convection and to the initiation of convection in viscous models. A familiar figure (Fig. 15) from the paper by Pekeris[16] shows a circulation according to a second-degree zonal harmonic, in which the whole mantle participates. It is highly desirable that similar studies be undertaken for a plastic solid, with a finite creep threshold and an accelerating creep rate. The pattern of motion will certainly be very different, with flow, if any, confined in relatively narrow slip zones.

The contraction theory, as it has been developed so far, shows us the strains required to maintain continuity in a differentially cooling body. The basic assumption is merely conservation of mass,

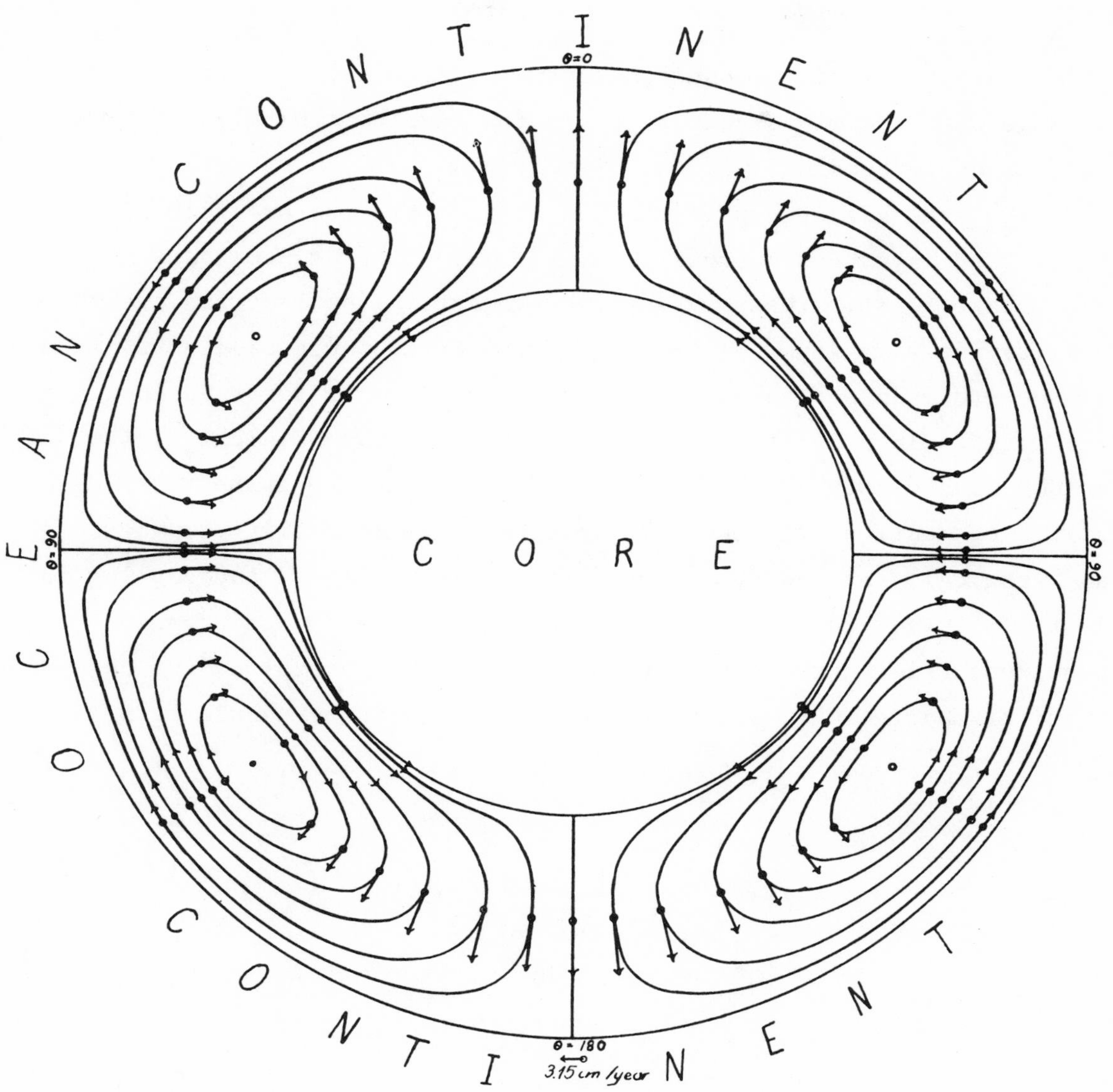

Fig. 15—Flow according to a second-degree zonal harmonic perturbation of temperature in a viscous mantle (according to Pekeris, Ref. 16).

and the only straightforward result is the integral of the normal displacement or velocity over the surface. It tells us nothing directly about tangential movements, as these do not change the volume. It does, however, give a picture of the strains, leaving it to our imaginations to deduce the probable mode of flow or rupture. Jeffreys[(1)] has proposed that the probable form of rupture would be a crumpling of the outer, laterally compressed, shell along two 180-degree arcs at right angles to one another, and that such a distribution "is remarkably like what the Tertiary mountains show." One might also identify such a bulge with the midoceanic rises, which have been heralded as indications of convection.

The stresses set up in the contraction process are much like those that have been postulated by seismologists as necessary at the earthquake focus. The stresses for symmetrical contraction are shown qualitatively in Fig. 16; they are everywhere compressive, but if we remove the mean pressure, we obtain the deviatoric stresses shown on the right. In order to continue to fit in a given conical element of volume with vertex at the center as demanded in the symmetrical case, volume elements above a certain level must be laterally compressed and vertically extended; at the "level of no strain," the cooling is just sufficient to give a fit at the new level and no distortion is required; below this level continuity requires lateral extension and vertical shortening. In the symmetrical case there should be no

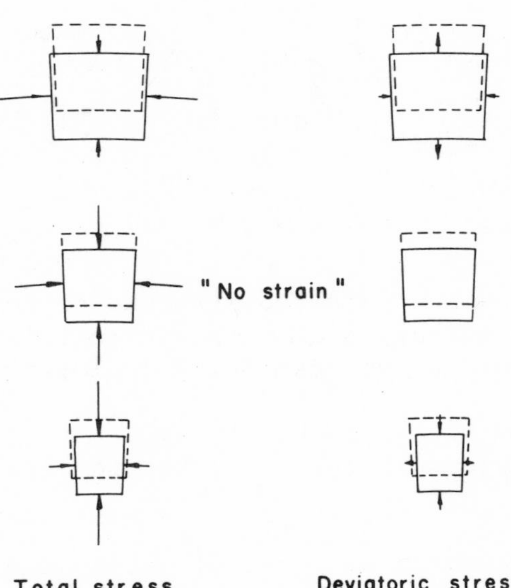

Fig. 16—Qualitative stress distributions associated with symmetrical contraction: upper, above "level of no strain"; center, at "level of no strain"; lower, below "level of no strain." Only the directions are significant, the magnitudes varying with depth.

earthquakes at the "level of no strain." The real earth probably departs from this perfect symmetry and there will be no single "level of no strain"; perhaps its mean position might be identified with one of the minima in the earthquake distribution curve (Fig. 13).

Much of the criticism of the contraction theory has been directed at the amount of radial contraction, which seems small when compared with the horizontal motions indicated by folding, overthrusting, and strike-slip faults. If attention is directed instead to the volume compression of the outer shell, the quantities available for mountain formation and so on become much more impressive; the problem is to determine the nature of the motion. Once the oversimplification of the symmetrical case is left behind, the contraction theory offers as many unsolved problems as does the convection theory. And, again the question is posed as to the behavior of a mass of real material subjected to variations of temperature and density in its own gravitational field. A completely deductive treatment may never be feasible, but it may be possible to make some progress toward reconciling the many conflicting observations now at hand.

REFERENCES

1. Jeffreys, Sir Harold, The Earth, Its Origin, History, and Physical Constitution, Cambridge University Press, New York, 1959, pp. 90, 201, 114.
2. Gutenberg, Beno, "Velocity of Seismic Waves in the Earth's Mantle," Geol. Rundschau, Vol. 45, 1956, pp. 342-353; Trans. Am. Geophys. Union, Vol. 39, 1958, pp. 486-489.
3. Birch, Francis, "Composition of the Earth's Mantle," Geophys. J., Vol. 4, 1961, p. 295.
4. Unpublished communication from Robert G. McQueen, Los Alamos Scientific Laboratory.
5. Wackerle, Jerry, "Shock-Wave Compression of Quartz," J. Appl. Phys., Vol. 33, 1962, pp. 922-937.
6. Vening Meinesz, F. A., Konink. Ned. Akad. Wetenschap. Proc., Ser. B, Vol. 54, 1951, pp. 3-19.

7. Uotila, U. A., "Harmonic Analysis of World-wide Gravity Material," Ann. Acad. Sci. Fennicae, Ser. A, III, 1962, pp. 11-17.

8. Bullard, Sir Edward, "The Density within the Earth," Verhandel. Ned. Geol. Mijnbouwk. Genoot., Vol. 18, 1957, p. 23.

9. Diment, W. H., doctoral thesis, Harvard University, 1954, Fig. 17.

10. Hill, R., The Mathematical Theory of Plasticity, Oxford University Press, New York, 1950, p. 254.

11. Haskell, N. A., "The Motion of a Viscous Fluid under a Surface Load, Part II," J. Appl. Phys., Vol. 7, 1936, pp. 56-61.

12. Wahlstrom, Ernest E., and V. Q. Hornback, "Geology of the Harold D. Roberts Tunnel, Colorado: West Portal to Station 468 & 49," Bull. Geol. Soc. Am., Vol. 73, No. 12, 1962, pp. 1477-1498.

13. Griggs, D. T., F. J. Turner, and H. C. Heard, "Deformation of Rocks at 500° to 800°C," in Rock Deformation, David Griggs and John Handin (eds.), Geol. Soc. Am. Mem. 79, 1960, p. 51.

14. Gutenberg, Beno, and C. F. Richter, Seismicity of the Earth, and Associated Phenomena, Princeton University Press, Princeton, N.J., 1954, pp. 16-18.

15. Orowan, E., "Mechanism of Seismic Faulting," in Rock Deformation, David Griggs and John Handin (eds.), Geol. Soc. Am. Mem. 79, 1960, p. 336.

16. Pekeris, Chain L., "Thermal Convection in the Earth's Interior," Monthly Notices Roy. Astron. Soc., Geophys. Supp., Vol. 3, 1935, pp. 343-367.

DISCUSSION

K. F. DALLMUS (Venezuela):

The amount of flattening at the poles, 70,000 ft, is not only approached by the topographical differences, 63,000 ft, on the crust of earth, but also by the vertical distance, 60,000 ft, through which unconformable surfaces have moved since the beginning of the Paleozoic. The maximum thickness of superimposed sediments also approaches 60,000.

ABSTRACT

Plowshare is the code name for AEC's program on peaceful uses of nuclear explosions--a program that dates back to 1957.

Thermonuclear (fusion) explosions as a method of moving large masses of earth and rock can be relatively inexpensive and also produce a low level of radioactivity as compared to fission reactions. In the early Plowshare studies, very large yields (MT range) were necessary; however, research has now reduced the required yields to the 100-KT range.

Since 1957 thirty-three underground nuclear tests have been performed in tuff, alluvium, granite, and salt. During an underground explosion, vaporization and fusion of the rock adjacent to the shot chamber causes formation of a cavity. The extent of the cavity often is increased by the collapse of its roof; this collapse occasionally forms chimneys, particularly in alluvium. The temperatures drop as the shock wave proceeds outward. There is frequent satisfactory agreement between field measurements and theoretical conservation of energy equations and equations of state.

Nuclear explosions have the following potential peaceful uses: economical block-caving methods in mining; advancement of seismology knowledge by study of the shock waves; and the economical excavation of large canals and highways. For example, it is believed that nuclear explosives were used in Russia for diverting the Kolonga River in the Ural Mountains. The related safety problems in such uses are being studied.

RÉSUMÉ

"Plowshare" est le nom donné en code au programme de l'"Atomic Energy Commission" portant sur les utilisations pacifiques des explosions nucléaires--un programme qui remonte à 1957.

Les explosions thermonucléaires (fusion) utilisées comme moyen de déplacer de grandes masses de terre et de roche peuvent être relativement peu coûteuses et peuvent aussi produire un faible niveau de radioactivité, en comparaison avec les réactions de fission. Lors des premières études de Plowshare, de très grands rendements (de l'ordre de la mégatonne) étaient nécessaires; cependant, les recherches ont maintenant réduit les rendements nécessaires à l'ordre de grandeur de cent kilotonnes.

Depuis 1957, trente-trois essais nucléaires souterrains ont été faits dans des tufs volcaniques, des terrains alluviaux, du granite et du sel. Au cours d'une explosion souterraine, la vaporisation et la fusion de la roche adjacente à la chambre d'explosion provoquent la formation d'une cavité. L'écroulement du plafond de la cavité en augmente souvent les dimensions; cet écroulement forme parfois des cheminées, particulièrement dans les terrains alluviaux. Les températures tombent au fur et à mesure que l'onde de choc va vers l'extérieur. Les résultats des mesures sur les lieux concordent souvent de façon satisfaisante avec les équations théoriques de conservation de l'énergie et les équations d'état.

Les explosions nucléaires peuvent être utilisées à des fins pacifiques pour: creuser économiquement dans les mines, avancer l'état des connaissances sur la sismologie par l'étude des ondes de choc, creuser dans des conditions économiques de grands canaux et de grandes routes. On pense par exemple que des explosifs nucléaires furent utilisés en Russie pour détourner le Kolonga dans les Monts Oural. Les problèmes de sécurité qui s'y rattachent pour de pareilles utilisations sont en cours d'étude.

AUSZUG

"Pflugschar" ist der Deckname für das Programm der Atomic Energy Commission zur friedlichen Verwendung von Kernexplosionen, welches im Jahre 1957 seinen Anfang nahm.

Kern-(Fusions-)Explosionen für das Bewegen von grossen Erd- und Gesteinsmassen können, im Vergleich mit Spaltungsverfahren, verhältnismässig wirtschaftlich sein und auch einen niedrigen Strahlungsausfall ergeben. In den ersten "Pflugschar"-Versuchen waren sehr grosse Ladungen (10^6 Tonnen) notwendig, weitere Forschung hat diese jedoch auf den 10^5 Tonnen Bereich vermindern können.

Seit 1957 sind dreiunddreissig unterirdische Kernversuche in Tuff, Alluvium, Granit und Steinsalz durchgeführt worden. Im Verlaufe einer Untertageexplosion verursachen Verdampfung und Verschmelzung des die Schusskammer umgebenden Gesteines die Bildung einer Höhlung. Diese Höhle wird oft durch das Einbrechen der Firste ausgedehnt, welches häufig Kamine formt, vor allem im Alluvium. Die Temperaturen nehmen mit dem Fortschreiten der Stosswellen nach aussen hin ab. Häufig wird zufriedenstellende Übereinstimmung zwischen den Feldmessungen und den theoretischen Gleichungen der Energieerhaltung und des Spannungszustandes gefunden.

Kernexplosionen haben die folgenden möglichen Anwendungen: wirtschaftliche Abbaumethoden im Bergbau (Block Caving), Förderung der Erkenntnisse auf dem Gebiete der Seismik (durch die Untersuchung von Stosswellen) und wirtschaftliche Erdbewegung für grosse Kanal- und Strassenbauprojekte. Es kann zum Beispiel angenommen werden, dass Kernsprengmittel in Russland für die Umlenkung des Kolonga-Flusses im Ural verwendet wurden. Die mit entsprechenden Anwendungen in Beziehung stehenden Sicherheitsprobleme werden derzeit untersucht.

PLOWSHARE: SCIENTIFIC PROBLEMS

Glenn C. Werth*

Introduction

PLOWSHARE is the code name for the Atomic Energy Commission's program on peaceful uses of nuclear explosions. In this conference you will be informed about studies in rock mechanics, rock stress, and various research tools that have been developed for study in these fields. An underground nuclear explosion in one sense is yet another approach, another tool for studying the properties of rock subjected to violent and large-scale forces. While such a study has not been proposed, it does serve to illustrate that using a nuclear explosive for purely scientific studies is within the Plowshare concept when sufficient need can be demonstrated. But Plowshare is more than that. To us in the program, the controlled use of nuclear explosions in a predictable manner has every possibility of helping mankind in his pursuit of a better life for the peoples of the world--a better life because the energy of the nucleus is put to work constructively to make, for instance, natural resources more readily available, to build water conservation systems, and to improve transportation. These dreams, we realize, will only come through a patient, systematic investigation of the effects of nuclear explosions and the development of accurate prediction methods of these effects. But we in the program are experiencing an excitement in the air, an excitement of anticipation of the first real payoff in terms of industrial or public use of a nuclear explosion experiment.

---

*Deputy Division Leader, Plowshare Division, Lawrence Radiation Laboratory, Livermore, California.

But that is getting ahead of the Plowshare story. Let's go back to the beginning of Plowshare, explore the concept, its advantages and disadvantages, and describe what we have learned about an underground explosion.

History of Plowshare

First of all, the Plowshare concept. With each new invention, the innovator looks around for additional applications. So it was when the first atomic explosion took place in 1946. But military work took precedence and nothing came of such talk. Shortly after World War II, John A. Wheeler and Theodore B. Taylor made separate proposals for using a nuclear explosion to generate isotopes for subsequent recovery. Again no action was forthcoming. It wasn't until 1957 that Dr. Teller succeeded in getting Harold Brown and Gerald Johnson to consider systematically possible peaceful uses of nuclear explosions and to write the initial Plowshare report.

Problems in Uses of Nuclear Energy

Expense

Now to the basic advantages of nuclear explosions. In large quantity the energy from a nuclear explosion, on a per unit energy basis, is cheap. And it is particularly cheap if the energy comes from a fusion rather than a fission reaction. But therein lies the first dilemma of Plowshare. The cheap fusion (thermonuclear) reactions are inherently very large scale, in the megaton range. How do you safely control such a large explosion and put it to useful purpose? So, in part, real progress in Plowshare has had to wait for advances in weapon technology. But progress has been made, and it is now possible to announce that the Plowshare experiment called Sedan, which had a yield of 100 KT, was more than 70 per cent fusion. So now we are in a position to realize the cheapness of thermonuclear reactions in units as low as 100 KT.

Safety

The radioactivity produced by a nuclear explosion is dangerous, and peaceful applications must be accomplished in complete safety. But here we are fortunate, inasmuch as a nuclear explosive like Sedan, where the majority of the energy comes from fusion, also produces a level of radioactivity much reduced from that which would be produced in a fission reaction of equal yield. So progress has been made in weapon technology to produce more manageable yields that still give the cheapness and reduced activity of the thermonuclear reaction. These steps have moved Plowshare significantly nearer to its goal. More work, of course, can be done along these lines.

Experimental Results

In seeking applications of nuclear explosions, we are first and foremost dealing with an explosion--and what are explosives fundamentally used for? To break rock. And if buried deep enough in rock, the radioactivity is also contained. Therefore, the pioneering experiments in the Plowshare effort were underground. Rainier was fired in 1957. It was 1.7 KT buried 899 ft deep in volcanic tuff at the Nevada test site. Four million cubic feet of rock were broken by the explosion (Fig. 1). In 1959 Logan and Blanca were fired. The 19.2-KT Blanca produced 66 million cubic feet of broken rock (Fig. 2). Postshot exploration work and the beginnings of a theory of underground explosions made possible a description of the explosive process. The results were reported by Johnson, Higgins, and Violet in 1959.[(1)] But in talking with industries interested in producing broken rock, their reply was rightly: "Your experiments are fine, but we don't want to break volcanic tuff. Do you know what would happen in other rock types?" And our answer was that we thought something similar would happen, but we had no data to prove it. The moratorium on weapons tests came along, and Plowshare marked time for three years, unable to follow the pioneering experiments with further ones to answer these very legitimate questions.

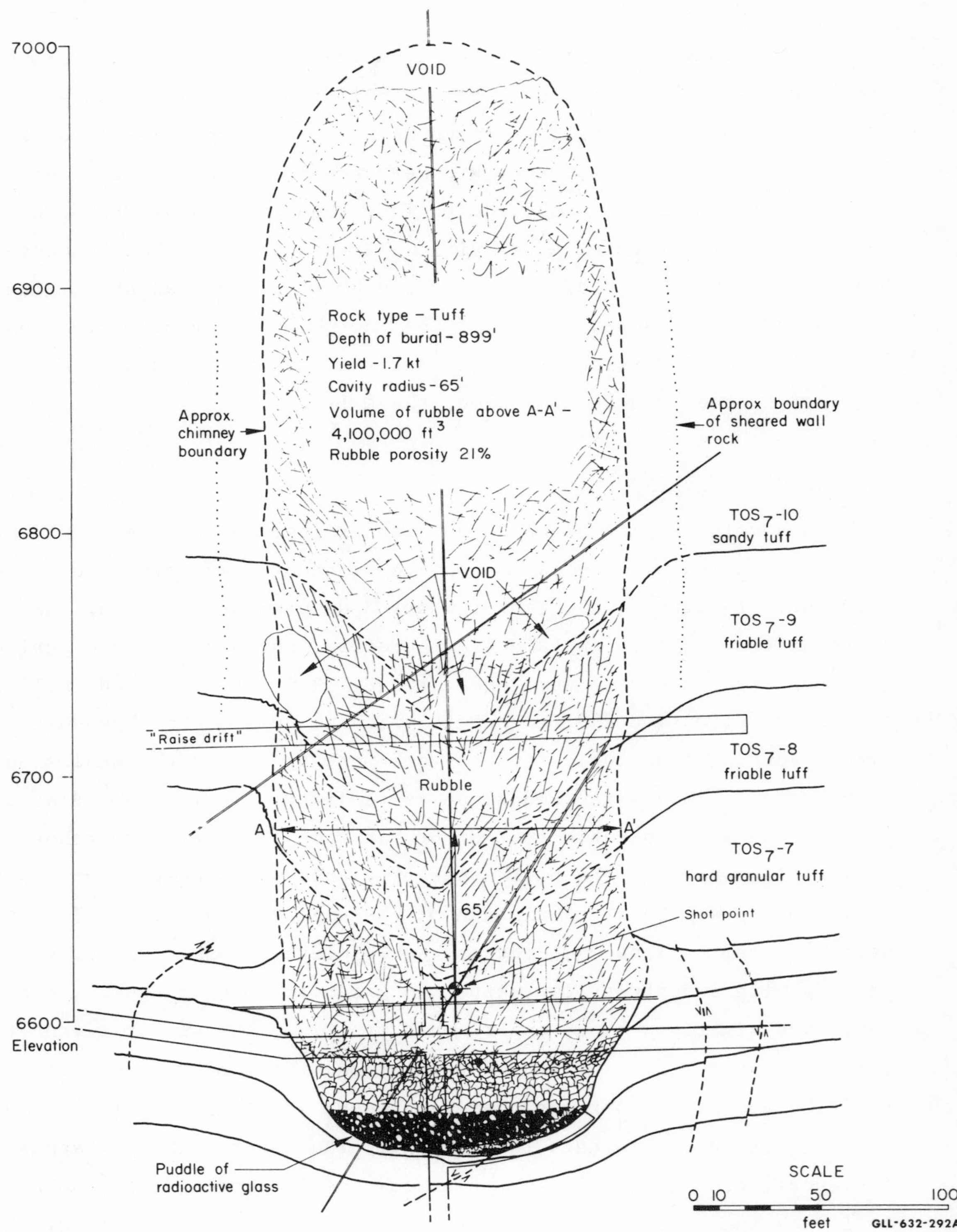

Fig. 1—Rainier schematic cross section.

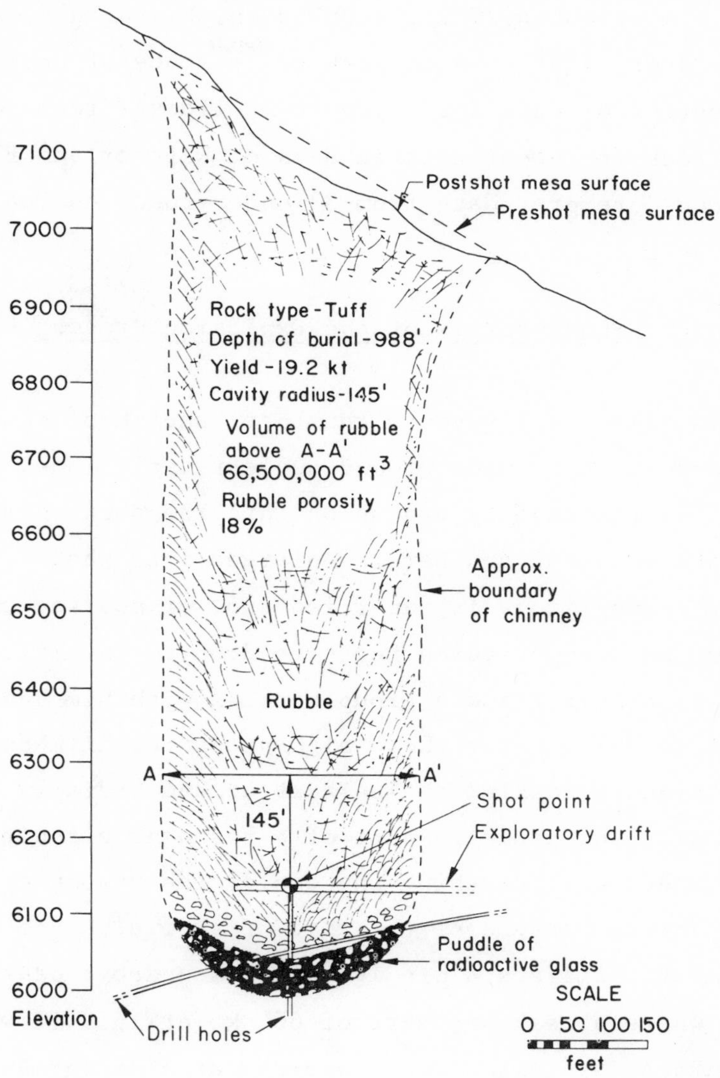

Fig. 2—Blanca schematic cross section.

In September 1961 testing was resumed and Plowshare could proceed. The testing in Nevada was underground in tuff and in alluvium. We have learned how to place the nuclear explosive in tunnels and down drill holes. We have learned what to do and what not to do to ensure containment of radioactivity. Furthermore, these experiences in tuff and alluvium, rounded out with the Department of Defense shot in granite and the Plowshare shot in salt near Carlsbad, New Mexico, have made possible what we believe is a rather comprehensive understanding of at least the basic properties of underground detonations. We have a new report coming out by Charles Boardman, David Rabb, and Dick McArthur[2] of our laboratory that is a comprehensive study of underground explosions and a collection of data far beyond the wildest dreams of a Plowshare enthusiast back in the pioneering days of Rainier and Blanca. They summarize and interpret data from 33 underground nuclear explosions.

Fundamentals of the Explosion Process

I would like to give you a thumbnail sketch of several of the more important processes in underground nuclear explosions.

Energy is released by explosion into the shot chamber. The rock vaporizes and the shot chamber is enlarged into what is called the cavity. Additional rock is vaporized and, of course, the shock wave proceeds out and temperatures begin to drop to the melting point of the rock. Fred Holzer and a group working with him have invented measurement techniques for these high-pressure, high-temperature regions. In one technique a coaxial cable is stretched radially outward from the shot room. The cable is part of a resonant circuit and as the shock wave moves out, crushing the cable, the resonance changes and gives an accurate measure of the time of arrival. Another technique by Dave Lombard utilizes a pin method to measure a peak pressure. This work, where a peak pressure of 664 kb at 15.9 ft was measured for the 5-KT Hardhat explosion, was reported at a recent Physical Society meeting.[3]

These measurements are compared with theoretical calculations. Initial work was done by John Nuckolls, but now Fred Seidl of our laboratory is developing the technique further. The method, of course, utilizes high-speed digital computers. The basic theoretical approach was first used, I believe, by von Neumann and is sometimes referred to as the brute-force method of integrating the conservation equations. Spherical symmetry is assumed and radial zones are set up. The conservation equations of mass, momentum, and energy are coded into the computer program to be sequentially applied zone by zone. The initial energy is thus allowed to propagate into adjacent zones and the motion is followed sequentially in time. The conservation of energy equation or equation of state is so written that, depending on ranges of temperatures and pressures, the material behaves like a vapor, a liquid, a crushed solid, a plastic solid, and finally like an elastic solid.

Correlation Problems

Getting an agreement between theory and experiment is a bit like pulling oneself up by one's own boot straps. You are not sure of the equation of state--although you have some independent measurement of the Hugoniot, but not over complete range. Now that we have measurements on a number of explosions in different media, it is possible to state that for earlier portions of the explosion history we get quite satisfactory agreement in tuff, alluvium, and granite. Our agreement in salt is not as satisfactory. It has taken years of experimental and theoretical work to reach this point, and we are not satisfied that we have gone far enough, but when it does come out in published form, I believe that you will find that it is a first-class piece of scientific work in a difficult field.

We have a two-dimensional version of this computing technique that we have applied to more complicated problems.[4] You will be hearing more about calculations of this general type from Dr. Brode of RAND in tomorrow's session.

Empirical Results

We have mined back through the region of the Rainier explosion and can define the cavity (Fig. 3). Directly below the shot point is found a 1- to 2-ft-thick puddle of frozen rock melt. The outer boundary of this melt can be traced up the sides of the spherical cavity where it thins out to 6 to 8 in. in thickness. Ninety per cent of the radioactivity is trapped in this melt in fused glass--insoluble to ground water penetrating through it. In some explosions we have drilled down rather than mined to locate this radioactive melt and thus establish maximum cavity radii. We find that the cavity expands adiabatically

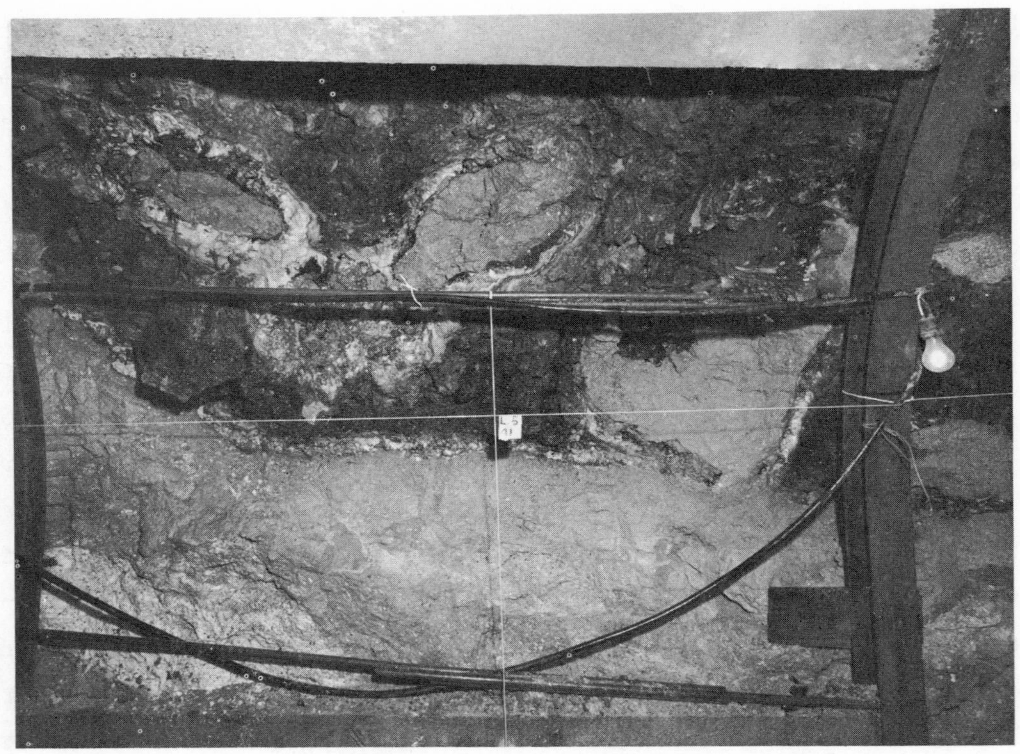

Fig. 3—Edge of the Rainier cavity showing the dark rock that was melted by the explosion in contrast with the lighter colored rock that was not melted (Lawrence Radiation Laboratory photo).

until the volume is such that the pressure in the cavity is equal to the weight of rock overhead. For tuff, where we have about a dozen measurements, the cavity radii are in agreement with the simple theory to within 8 per cent, which is quite a satisfactory prediction for any practical application. Radii depend on rock type, of course, but not as strongly as one might think. So salt, granite, tuff, and alluvium radii only differ by ± 20 per cent, and the amount of water is perhaps the important variable.

When the cavity has grown to its maximum extent, the roof falls in. The cavity of the Gnome shot, in salt, grew to about 55 ft. The roof fell in filling up about half of the cavity and yielding a radius to the ceiling of 85 ft (see Frontispiece). The collapse of the roof continued upward until the rock was strong enough to arch. I climbed into this cavity six months after the detonation. The radioactivity had been buried by the roof fall in essentially a reverse fall-out shelter. The temperature was 135° F--caused, of course, by the heat deposited in the rock by the explosion. The coloration of the rock is magnificent--the white of the undamaged salt, the blue of the salt exposed to radiation, and the reds of the polyhalite impurities. I really think Plowshare should set up a tourist concession and go into competition with Carlsbad Caverns.

In the case of the tuff and granite, the roof fall continues upward until the resulting configuration is a cylinder of rubble with a diameter equal to that of the cavity and a height of the cylinder about five times the diameter. The rock that falls occupies a larger volume and is more porous. The collapse of the cavity roof and the resultant column of broken rock is called the chimney. We have mined back into the chimney in Rainier and in Hardhat, the 5-KT shot in granite. The edge of the chimney is very distinct--the preshot rock on one side and broken rock on the other. Visual estimates of the size of the rock in the chimney are 50 per cent of the rock less than a half a foot, 40 per cent between a half a foot and one foot, and 10 per cent greater than one foot.

The shots in alluvium produce a chimney too, but since the material shows only a small tendency to bulk up when it slides into the cavity, the chimney generally grows to the surface and forms a subsidence crater (in some cases) with a volume about equal to the volume of the original cavity. There is little hope of practical application of these subsidence craters because they occur in dry, porous alluvium and are quite small compared to explosion-produced craters.

Application to Mining

The mining industry has been watching our work in tuff and particularly in granite with special interest. In the block-caving method of mining, tunnels are driven into the mountain below the ore body. From these tunnels, or drifts, raises are made that flare out into what are essentially funnels. Above the funnels a room is mined out 9 ft high with pillars left in place to support the ceiling. The pillars are then blasted out, the roof falls 9 ft, and rock is drawn down through the funnels and out of the mountain.

It has occurred to many mining people that a nuclear explosion may make this block-caving process a lot more efficient because the broken rock is produced not by dropping the ceiling a mere 9 ft, but by throwing rock up, say, 100 ft, and letting it fall. Thus, there is a more effective crumbling of the rock and hence far easier removal and processing.

It is the job of Plowshare to design and conduct experiments to present and interpret data. If application is to come, industry must take the initiative not only in this country but throughout the world to come forward and say: "We have a problem. Perhaps it can best be solved by use of a nuclear explosive. Let's look into the problem and explore the feasibility."

Application to Seismology

Let us turn from the possible industrial uses of underground nuclear explosions to scientific uses. The earthquake seismologist interested in studying the crust of the earth has always been handicapped because his generator of seismic waves, an earthquake, is unpredictable in location and time. Thus, his data are obtained only on a catch-as-catch-can basis. The underground nuclear explosion has revolutionized this science. Controlled experiments can now be planned. The seismic results from Gnome have been called astounding.

Seismic amplitudes at 1000 mi in one direction were found to be 150 times larger than amplitudes in another direction. The high amplitudes and associated high velocities to the northeast from the Gnome site were unpredicted; this points up the new information that can be learned by controlled experimentation in seismology, particularly when the sources of seismic waves are not limited to the earthquake regions of the world.*

---

*These Gnome seismic results are published in a special issue on the Gnome Symposium in the Bull. Seismological Soc. Amer., Vol. 52, No. 5, December, 1962, by Romney, Herrin, and others.

The nuclear explosion lifts the study of the earth's crust out of what can be deduced from time measurements and relative amplitude to one of permitting use of absolute amplitude in the interpretation. We have done a great deal of work on this subject at Livermore. The Sandia Corporation and the Stanford Research Institute have developed techniques for measurement of the motion in the first 1000 ft from nuclear explosions. From these measurements it is possible to deduce an equivalent seismic source function. We have been able to take these close-in source measurements and relate them theoretically to the amplitude of the seismic waves from 250 to 700 km from sources in tuff, alluvium, granite, and salt, to an accuracy of about 20 per cent.[5]

In the future we will use nuclear explosions in controlled experiments on structural damage to houses and buildings caused by seismic waves. Such research relates to building better earthquake-proof buildings and should take that engineering area out of the catch-as-catch-can field and put it into the field of controlled experimentation.

Nuclear Excavation

Let us move on to one final subject, and a very important subject for Plowshare, that of nuclear excavation. There are interesting statements in the Russian literature which I would like to quote.

At the time of the first Russian explosion in 1949, Vishinsky claimed that the Soviet effort was to serve constructive aims:

> Right now we are utilizing atomic energy for our economic needs in our own economic interest. We are raising mountains; we are irrigating deserts; we are cutting through the jungle and the tundra; we are spreading life, happiness, prosperity, and welfare in places wherein the human footsteps have not been seen for a thousand years.*

There are indications that the Russians have followed up this early proposal. In 1956 explosions of many kilotons were used in China for the purpose of uncovering ore fields. In 1957 the Kolonga

---

*Statement to the United Nations, November, 1949.

River in the Ural Mountains was deflected into a new bed by 30 explosions of 100 tons each, which produced overlapping craters forming a new river bed. We have been most interested in these and similar experiments, and we asked the Russians to show us the explosion sites and discuss with us these constructive possibilities. Unfortunately, the Russians refused any such cooperation and did not permit inspection of the sites. They claim that the explosions were performed by conventional high explosives, a statement that we could not verify since we could not enter the sites. What we do know is that explosions similar to those performed by the Russians could have been performed with nuclear explosives, that such explosions even when performed by conventional means are needed as a preparation for bigger explosions in which nuclear power is utilized, and that for really big enterprises nuclear explosions are certainly more economical than the application of chemical high explosives. Thus, the early words and the later deeds of the Russians lent some plausibility to the assumption that Russian nuclear power is indeed being used in this important field.

(At this point in his address, Dr. Werth showed two motion pictures and made the following remarks.)

The first is known as our "shoot 'em up" film because we show five underground detonations and because it represents what we knew prior to the Sedan explosion of last summer. The second film shows Sedan. You can judge how well we did in predicting the Sedan results at 100 KT from the previous work shown in the first film. Milo Nordyke has been leading our work in the cratering area[6] (Figs. 4 and 5).

Conclusion

Plowshare is going ahead with plans for further cratering experiments. We are planning to repeat the Sedan experiment in hard rock and to start a series of experiments on the simultaneous detonation of a row of nuclear explosions in hard rock to produce a channel that would be needed for a canal or a highway cut. The theory of cratering is being developed and the various safety problems are being studied.

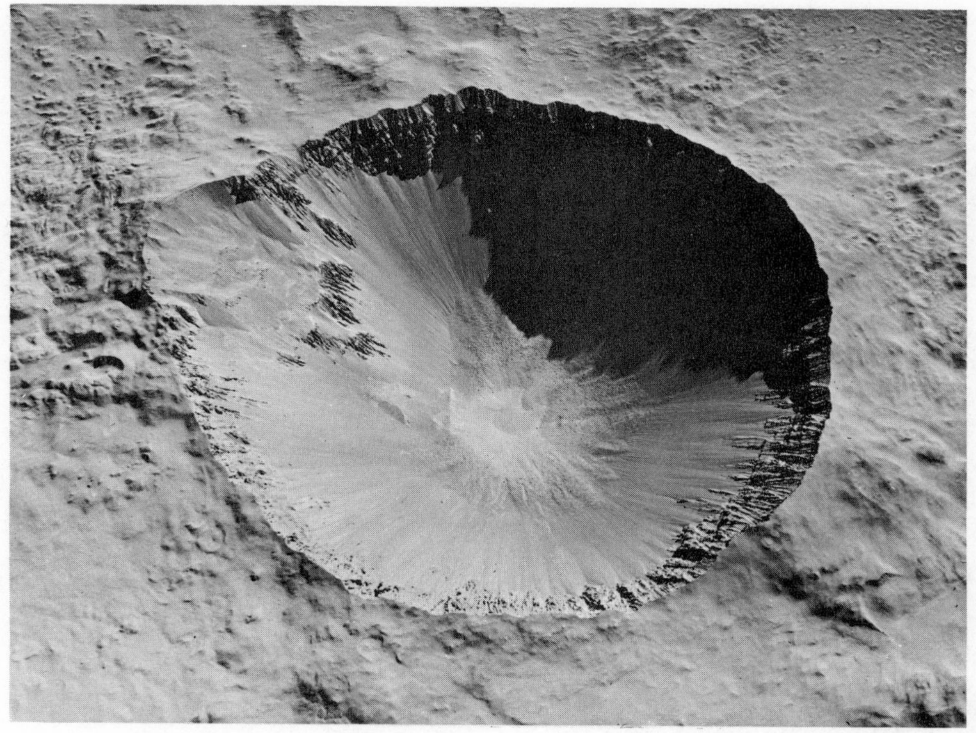

Fig. 4—Aerial view of the Sedan crater. The crater is 320 ft deep, 1200 ft in diameter, and was produced by a 100-KT explosion (Lawrence Radiation Laboratory photo).

Plowshare is being carried on at the Lawrence Radiation Laboratory, where the nuclear explosion and the resultant radioactivity have been witnessed again and again, and I can assure you that familiarity breeds caution. It is for that reason we are insistent that in all of our Plowshare work we develop the technical understanding of nuclear explosions to the point where we can predict with reliability what will happen and what safety precautions are required. We have this main framework of knowledge in contained underground detonations. We are developing it for cratering, and when we are ready, and only then, will we strike out into actual application.

Fig. 5—View of the wall of the Sedan crater taken from the crater floor (note the man in the picture) (Lawrence Radiation Laboratory photo).

REFERENCES

1. Johnson, G. W., G. H. Higgins, and C. E. Violet, "Underground Nuclear Detonations," J. Geophys. Res., Vol. 64, 1959, pp. 1457-1470.
2. Boardman, C. R., D. D. Rabb, and R. D. McArthur, Characteristic Effects of Contained Nuclear Explosions for Evaluation of Mining Applications, Lawrence Radiation Laboratory Report UCRL-7350, 1963.
3. Lombard, D. B., and D. V. Power, Hydrodynamic Pressure Measurement in Granite Near an Underground Nuclear Explosion, Lawrence Radiation Laboratory Report UCRL-7221, 1963.
4. Maenchen, G., and S. Sack, The TENSOR Code, Lawrence Radiation Laboratory Report UCRL-7316, 1963.
5. Werth, G. C., and R. F. Herbst, "Comparison of Seismic Waves from Nuclear Explosions in Four Mediums," J. Geophys. Res., Vol. 68, 1963, pp. 1463-1475.
6. Nordyke, M. D., "Nuclear Cratering and Preliminary Theory of the Mechanics of Explosive Crater Formation," J. Geophys. Res., Vol. 66, 1961.

PART II--FUNDAMENTALS

John W. Handin, Chairman

Consideration of current empirical knowledge of the rheological properties of rocks and of recent advances in our mechanistic understanding of fracture, nonlinear viscosity, and stored strain energy--factors of importance in estimating stress in naturally deformed crustal materials.

PRINCIPES FONDAMENTAUX

Examen des connaissances empiriques actuelles des propriétés rhéologiques des roches et des progrès récents en ce qui concerne notre compréhension analytique de la fissuration, de la viscosité non-linéaire et de l'énergie potentielle des contraintes--facteurs essentiels pour l'estimation des contraintes dans les matériaux naturellement déformés de l'écorce terrestre.

GRUNDLAGEN

Betrachtung der gegenwärtigen empirischen Kenntnisse über die rheologischen Eigenschaften von Gesteinen und neuere Fortschritte im Verständnis der Vorgänge beim Bruch, der nichtlinearen Viskosität und der Formänderungsenergie--wichtige Umstände beim Abschätzen von Spannungen im natürlich verformten Material der Erdkruste.

CONTENTS

INTRODUCTION

John W. Handin*

IN ALL ITS RAMIFICATIONS of current engineering, geophysical, and geological interest, rock mechanics is by any definition a field of bewildering complexity. The range in order of magnitude of the variables is enormous--stresses from a small fraction of a bar to 10 megabars, temperatures from near zero to a million degrees, and strain rates from 10^6 to perhaps 10^{-15} per second. The materials are structurally and chemically complicated, heterogeneous, and anisotropic. At very high stress levels the hydrodynamic model seems justified. For low stresses of short duration, the assumption of linear elasticity yields good results. But over the broad spectrum between, deformations are rarely even approximated by the classical idealized models.

Our as yet meager understanding of rock deformation is hardly surprising, and we should not suppose that our comprehension of the behavior of relatively simple materials is at all adequate, either. Recently Shockley(1) had to say "in no case is a definitely established conceptual picture available for the processes that occur when a metal deforms or fractures under stress." Furthermore, active interest in rock mechanics by any but an obstinate few is quite recent though certainly welcome.

Even if we limit our consideration to the natural state of stress in the earth's crust, the problems remain formidable. Suppose we are asked to calculate this stress at any time and place from first principles. If we fully understood the energetics,

*Exploration and Production Research Division, Shell Development Company, Houston, Texas.

we could determine the causative surface and body forces and the boundary conditions upon them in terms of stresses and/or displacements and their rates. And if we knew the proper rheological properties, we could work out the stress field and its associated deformations in space and time, provided, of course, the mathematics was tractable.

In the geological literature the first attempt to predict large-scale stress distributions semiquantitatively seems to be that of Anderson[2] who supposed that in flat country one principal stress would be nearly vertical. The relative magnitudes of the vertical and horizontal stresses could then be determined from fault patterns, provided faulting was governed by the Coulomb condition. Hafner[3] and Sanford[4] made more sophisticated elastic analyses to show that principal stress trajectories, and hence faults, could be curved. Odé[5] worked out the elastic stress distribution associated with an igneous intrusion and predicted the related dike pattern by assuming that the dikes were filled extension fractures everywhere normal to the least principal pressure. Varnes[6] investigated the faulting around another intrusion from the standpoint of ideal plasticity.

Although all this work has much enhanced our knowledge of natural states of stress, the origins of natural forces are, in fact, usually conjectural, and boundary conditions are seldom well defined. More seriously still, the real stress-strain-time relations are largely unknown. These depend on the composition and fabric of the rock; the natural environment--confining pressure, pore pressure, temperature, and appropriate thermodynamic potentials; and the state of stress itself. These relations are not subject to direct measurement and must be learned through controlled laboratory experiments in which the natural environment is simulated as realistically as possible.

In practice, our determination of stress begins with what we can observe in the field--namely, the general state of strain at an instant in geologic time. Our purely geometric description can sometimes be combined with sparse data on absolute age. Even then our picture of the deformation is at best kinematic. Any inference about the state of stress again demands knowledge of the rheological properties for appropriate conditions. These may be essentially atmospheric for the recoverable viscoelastic deformations recorded during strain-relief operations. But the permanent deformations delineated by petrofabric methods, fractures and finite strains, may have occurred in regions of high pressure, high temperature, and chemical activity.

Petrofabric studies of experimentally deformed rocks and their natural counterparts provide useful information about the mechanisms of deformation--fracture, flow by cataclasis, intracrystalline slip, or recrystallization. The petrofabric data often provide clues to the directions of the principal stresses at the time of deformation. They may also suggest the most probable rheological behavior, since this is intimately related to the deformation mechanism. Our estimates of the absolute magnitudes of the stresses, however, are still merely intelligent guesses. Empirical knowledge of the rheological properties of rocks has grown rapidly during the last 10 years, but is very far from complete. We need, in particular, more realistic theories to allow meaningful extrapolations of laboratory data beyond the limits of the test.

This part deals with recent advances in our mechanistic, rather than merely phenomenological, understanding--especially of fracture in the brittle state, of nonlinear viscosity, and of stored strain energy, all of which are important in estimating the state of stress in deformed crustal materials. Authors and panelists are all eminently qualified by virtue of their own active research in rock mechanics.

REFERENCES

1. Shockley, W. (ed.), Imperfections in Nearly Perfect Crystals, John Wiley & Sons, Inc., New York, 1952, p. vii.
2. Anderson, E. M., The Dynamics of Faulting, Oliver and Boyd, London, 1942.

3. Hafner, W., "Stress Distributions and Faulting," Bull. Geol. Soc. Am., Vol. 62, April, 1951, pp. 373-398.
4. Sanford, A. R., "Analytical and Experimental Study of Simple Geologic Structures," Bull. Geol. Soc. Am., Vol. 70, January, 1959, pp. 19-52.
5. Odé, H., "Mechanical Analysis of the Dike Pattern of the Spanish Peaks Area, Colorado," Bull. Geol. Soc. Am., Vol. 68, May, 1957, pp. 567-575.
6. Varnes, D. J., "Analysis of Plastic Deformation According to Von Mises' Theory with Application to the South Silverton Area, San Juan County, Colorado," U.S. Geol. Surv. Prof. Paper 378B, 1962.

DISCUSSION

J. L. SERAFIM (Portugal):

The engineers who have studied hydraulic structures made of such porous materials as concrete and rock know that there is an important problem when there are differences in interstitial or pore pressures in the material. The forces created by such differences can bring about high stresses and rupture the structure. The value of these forces depends on certain properties of the material.

I think I can offer something for consideration to this meeting that concerns the determination of those properties of porous materials.

We determined experimentally such properties for concrete[*,**] and then used the same method to determine the same properties for a few samples of some rocks.[***] When we have in a material a

---

[*] J. L. Serafim, "A Subpressas nas Barragens," Publication 59, L.N.E.C., Lisbon, 1954.

[**] J. L. Serafim, "Discussion of 'Permeability, Pore Pressure and Uplift in Gravity Dams,' by Roy W. Carlson," Trans. Am. Soc. Civil Engrs., Vol. 122, 1957, pp. 603-608.

[***] J. L. Serafim, "Discussion of 'Hydraulic Pressures in Concrete,' by T. C. Powers," Proc. Am. Soc. Civil Engrs., Paper 904, February, 1956, pp. 55-57.

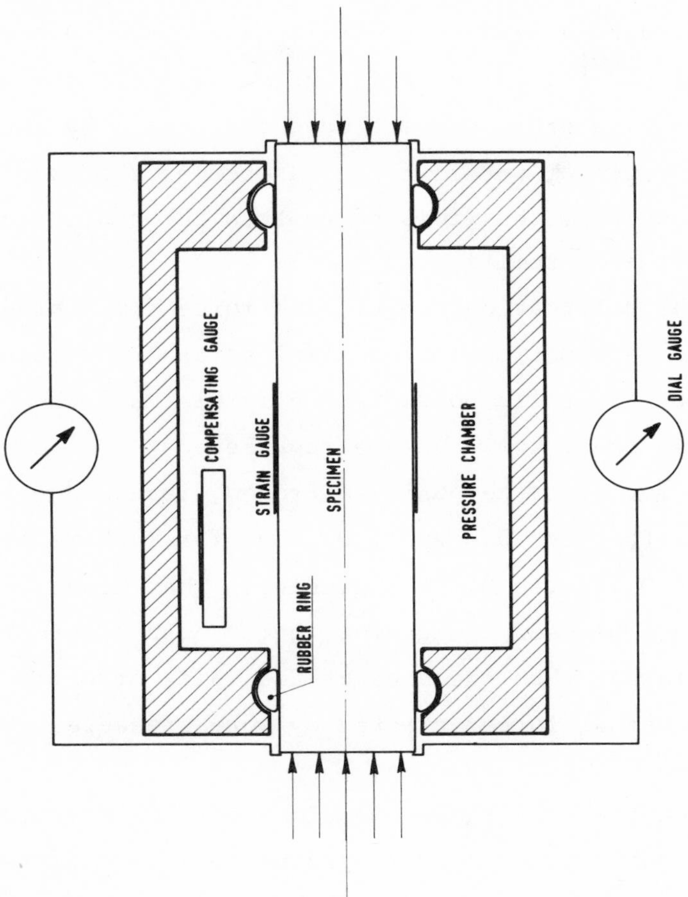

Fig. 1—Triaxial chamber.

difference in interstitial pressure from one point to another at a distance of dx, the force dF due to a difference in pressure dp is $dF = \eta dp$, η being called boundary porosity and defined as the percentage of the area of the bonds in the projected total area of the surface of rupture by tension.

In previous tests conducted up to rupture by other research workers, the value of the boundary porosity was found to be close to unity for concrete. To determine this property we used the triaxial chamber shown in Fig. 1.

The specimen is sealed against the pressure chamber by means of rubber rings under pressure. Axial external loads can be applied direct to the specimen and its deformations can be obtained either by dial gauges or by strain gauges, having a compensating gauge inside the chamber.

This compensating gauge can also indicate the volumetric compressibility of the solid matter in the material if its readings are compensated by another gauge placed on steel. By applying the pressure of nitrogen or water in the chamber as the specimen is completely unjacketed, the fluid percolates through it and creates a force toward the extremities.

Now, if the deformation due to this force for a pressure P is equal to the deformation for an axial stress P, we can conclude that the area of the solid bonds is negligible, and the boundary porosity is also equal to 1, before rupture.

The tests for concrete, using nitrogen, revealed values of η very close to 1 (between 0.8 and 1.0), but for rock we obtained lower values of 0.4, even 0.3 for granite. That means that the area of the bonds between the grains in the case of granite is quite large or even larger than the area of the voids or of the cracks, and that the internal forces created by pore pressures are lower than in the case of concrete.

Really, I think that these questions of area of voids, pore pressures, and solid bonds between grains or crystals present basic problems to be understood and solved when dealing with rock in nature. The fact that rocks are in an internal state of compression, and that the grains are held together by bonds that we can break at a certain time, can account for much phenomena that we observe in rock masses. For instance, when we extract samples we can destroy some of those bonds which, of course, will produce an expansion of the rock, and so the samples will not show the same properties that the rock had when it was in the interior of the mass. But I wish to re-emphasize that if in our rock masses we have a liquid inside the voids, we must consider the stresses in the solid material due to the pressures in the liquid. Such stresses depend on the value of the coefficient, η, which is a characteristic of our rock mass.

J. W. HANDIN in reply:

I certainly agree with Dr. Serafim that the water content of rocks is of enormous importance; in fact, as much so as that of soils. We probably accept this as more or less axiomatic. I have serious doubts about the concept of boundary porosity, which almost always turns out to be unity. We have also done a number of experiments on rocks in which pore and confining pressure were applied independently at rather high values to 30,000 psi.*

One problem with the concept he has proposed is that of attributing the magnitude of the boundary porosity to the relative area occupied by the bonds between grains. This implies that failure is intergranular and through the bonding material when, in fact, the boundary porosity of a sandstone with a bulk porosity of the order of 15 per cent is still 1 in our experience, even though the fractures are propagated through the grains, not through the bonds.

D. L. BIGGS (USA):

Has the effect of oriented dislocation arrays on deformation or vice versa been studied? If so, what is the status of the art?

J. W. HANDIN in reply:

As far as I know the only significant study to date of the effects of dislocations on rock deformation is that of Carter, Christie, and Griggs at the University of California, Los Angeles. This important work should appear soon in _The Journal of Geology_.

---

*J. W. Handin _et al._, "Experimental Deformation of Sedimentary Rocks under Confining Pressure: Pore Pressure Tests," _Bull. Am. Assoc. Petrol. Geologists_, Vol. 47, May, 1963, pp. 717-755.

ABSTRACT

Some recent experimental studies of brittle fracture are described, with discussion of the general applicability of Mohr and Griffith theories of fracture to rocks.

Five brittle rocks (two dolomites, a diabase, a granite, and a quartzite) that were relatively isotropic, homogeneous, and of negligible porosity were fractured at room temperature. Owing to a novel shape of sample used, a wide range of stress states could be applied from nearly pure axial tension through axial compression with superposed confining pressure. The sample shape, a cylinder with central section of reduced diameter, appeared to eliminate many of the "end effects" of the conventional compression test. Special loading techniques permitted behavior just before fracture to be observed; partially fractured material could sometimes be recovered to reconstruct the process of fracture growth. Brittle fractures apparently start at grain boundaries, which loosen and become partially detached as fracture is approached. Straight sections of grain boundaries may act as "Griffith cracks." Tensile strength and grain-size dependence of strength agree roughly with that predicted by Griffith theory if crack length is taken as maximum grain diameter.

Behavior of the silicate rocks closely follows the McClintock-Walsh modification of Griffith theory for a coefficient of friction of 0.9 to 1.5. The Mohr envelopes of the silicate rocks are straight lines in approximately the field of compression and roughly parabolic at the intersection with the σ axis.

A gradual transition exists between extension fracturing and faulting. The angle of inclination of the fracture to the direction of maximum compression was observed to vary gradually from 0° to about 30°, as stresses were altered from tensile to compressive. Thus, extension fractures and faults may not be two different types of fracture, as is widely believed, but rather members of one continuous series.

RÉSUMÉ

Quelques études expérimentales récentes de cassure sont décrites, avec une discussion des possibilités générales d'application des théories de Mohr et de Griffith sur les fractures des roches.

Cinq roches cassantes (deux dolomites, une diabase, un granit, et un quartzite) qui étaient relativement isotropes, homogènes, et de porosité négligeable furent fracturées à température ambiante. Grâce à la forme nouvelle de l'échantillon utilisé, on put employer une grande variété d'états de contraintes, depuis une traction presque purement axiale jusqu'à une compression axiale avec pression confinant superposée. La forme de l'échantillon, cylindre à section centrale de diamètre réduit, sembla éliminer beaucoup des "effets d'extrémité" de l'essai de compression conventionnel. Des techniques de chargement spéciales permirent d'observer le comportement juste avant la fracture. On put parfois récupérer des matériaux partiellement fracturés pour reconstituer le processus de développement de la fracture. Les cassures commencent apparemment aux limites des grains qui se desserrent et se détachent partiellement au fur et à mesure qu'on approche de la fracture. Les parties droites des limites des grains peuvent se comporter comme des fissures de Griffith. La résistance à la traction et la relation de cette résistance à la grosseur du grain concordent en gros avec celles prévues par la théorie de Griffith si on prend comme longueur de la fissure le diamètre maximum du grain.

Le comportement des roches siliceuses suit de près la modification par McClintock-Walsh de la théorie de Griffith pour un coefficient de friction de 0.9 à 1.5. Les enveloppes de Mohr des roches siliceuses sont des lignes droites dans la zone de compression approximativement, et en gros paraboliques à l'intersection avec l'axe des σ.

Une transition graduelle existe entre la fracture par extension et la faille. On a remarqué que l'angle d'inclinaison de la fracture avec la direction de compression maximum variait graduellement de 0° à 30° environ, lorsque l'on passait des efforts de tension à ceux de compression. Ainsi les fractures par extension et les failles pourraient ne pas être deux types différents de fractures comme on le croit généralement, mais appartenir à une série continue.

AUSZUG

Einige neuere experimentelle Untersuchungen über das spröde Brechen werden hier beschrieben, wobei auf die allgemeine Anwendbarkeit der Bruchtheorien von Mohr und Griffith auf Gesteine eingegangen wird.

Fünf spröde Gesteine, zwei Dolomite, ein Diabas, ein Granit und ein Quarzit, die verhältnismässig isotrop, homogen und von vernachlässigbarer Porosität waren, wurden bei Zimmertemperatur zum Bruch gebracht. Mit Hilfe einer neuartigen Form des Versuchsstückes war es möglich, eine grosse Vielfalt von Spannungszuständen aufzubringen, von fast reinem achsialen Zug bis zum achsialen Druck mit überlagerter Umschliessung. Die Form des Versuchsstückes, ein Zylinder mit im Mittelbereich verkleinertem Durchmesser, schloss offensichtlich viele der "Endeffekte" von gewöhnlichen Kompressionsversuchen aus. Besondere Belastungsmethoden machten es möglich, das Verhalten der Proben im Augenblick vor dem Bruch zu beobachten. Teilweise gebrochenes Material konnte manchmal wiedergewonnen werden und somit der Verlauf des Bruchvorganges rekonstruiert werden. Spröde Brüche beginnen offensichtlich an den Korngrenzen, welche erst gelockert und dann teilweise schon vor Erreichen des Bruches gelöst werden. Gerade Teilbereiche der Korngrenzen können hierbei als "Griffith Risse" wirken. Die Abhängigkeit zwischen Zugfestigkeit und Korngrösse stimmt ungefähr mit dem durch die Griffith Theorie vorausgesagten Verhältnis überein, wenn die Risslänge gleich dem maximalen Korndurchmesser angenommen wird.

Das Verhalten der Silikatgesteine folgt für Reibungsbeiwerte von 0.9 bis 1.5 recht genau der Modifizierung der Griffith Theorie durch McClintock-Walsh. Die Mohr'schen Hüllinien der Silikatgesteine verlaufen im Druckbereich ungefähr geradlinig und beim Schnitt mit der σ -Achse annähernd parabolisch.

Ein allmählicher Übergang zwischen Zug- und Scherbrüchen wurde beobachtet. Mit Änderung der Beanspruchung von Zug auf Druck vergrösserte sich der Winkel zwischen der Bruchlinie und der Richtung grösster Spannung allmählich von 0 bis 30°. Es ist daher möglich, dass Zug- und Scherbrüche nicht verschiedene Bruchart sind, was weitgehend angenommen wird, sondern Glieder einer kontinuierlichen Reihe.

BRITTLE FRACTURE OF ROCKS

William F. Brace*

Introduction

MECHANICAL PROPERTIES of rock under conditions comparable to those in the earth's crust have been extensively studied in recent years.[1-14] In spite of a great many observations, however, particularly of failure of rock under stress, we can not begin to answer even rather basic questions, important both to the geologist trying to understand how rocks have become folded and faulted and to the engineer trying to create or support large openings underground. For example, what determines the absolute strength of a particular rock? Intuitively, hardness of the minerals seems to be most important. However, there must be other factors, in view of the observation that limestone may be stronger than granite[13] or dolomite stronger than quartzite.[15] Or, take the question of the strength of large masses of rock. In the laboratory, one studies behavior of small samples which are as nearly free from cracks as possible. How does the strength of such samples compare with that of a cubic meter or kilometer of rock with all the joints, faults, and other natural planes of weakness? Are there several fundamentally different kinds of fractures in rock, analogous perhaps to the large variety observed in metals?[16] And finally, to what extent can one reconstruct the stress state which produced a given naturally occurring fracture?

Many of these questions could be answered if a theory of strength of rocks were available. One important element of a theory of strength is a failure condition. Search for a generally applicable failure

---

*Professor, Massachusetts Institute of Technology, Cambridge, Mass.

condition has been intensive starting with the first, classic study of von Kármán.[17] After trying all possible failure or yield conditions for metals, as well as failure conditions originally conceived for brittle material, most investigators have concluded that no existing failure law holds for rocks in general, or even for a single rock under different conditions of loading. It has been generally found[4,11] that the best approximation for room temperature tests is given by the Mohr criterion[18,19] but even this failure law does not predict correctly relative behavior under different types of loading—for example, under bending and compression. Search, then, for a failure law in particular, and a theory of strength of rocks in general has not been too successful.

Part of the difficulty may be traced to two sources: the complicated microscopic character of failure, and uncertainty as to stress applied in certain types of tests.

Failure or ductility, which are macroscopic phenomena, are the result of several microscopic processes: true brittle fracture—by definition, a process which produces no permanent change in material other than separation into parts; dissipative processes such as gliding or viscous flow; and processes such as rotation of grains or frictional sliding of grains about one another. Although one or another of these processes may predominate in typically brittle or ductile macroscopic behavior, in general they probably operate together. In a lot of the previous work the rocks studied and pressures and temperatures of the experiments were such that more than one of these processes probably did operate. By their very nature, microscopic processes are hard to identify so that it has often been unclear just how fracture or flow did occur in particular experiments.

In many investigations, failure stress is compared from one type of test to another. Conclusions are often clouded by uncertainties in the stresses imposed in some of these tests. For example, "end effects" in compression testing of short cylinders introduce large unknown errors,[20-24] particularly at small confining pressure. In any test in which stresses are calculated from elastic theory (bending of bars, collapse of hollow cylinders, indentation), errors may be introduced because of the rather common nonlinear elasticity of rocks at small confining pressure.[9,25-29] The errors involved here are probably at least as large as, say, the difference in strength which is supposed to characterize compression and extension of short cylinders.

In view of these difficulties, it appeared worth while to continue search for a general failure condition by choosing material and testing conditions which would (a) enable macroscopic behavior to be studied while only one microscopic process operated, and (b) permit accurate determination of stresses applied to samples. In several recent studies* it has been suggested that Griffith theory of fracture might apply to rocks when behavior on a microscopic scale was purely brittle. It was decided to explore this possibility further, for it seemed an easy matter to find geologic material and testing conditions which would give purely brittle behavior. The anticipated result would be a fracture theory, applicable to purely brittle behavior. In order to develop a theory of failure or a strength theory of general applicability, it would be necessary to consider individually the other microscopic processes and to investigate transitional behavior. This final step would not be the same as a study of transitional macroscopic behavior.[6]

In the present study, fracture of several brittle, isotropic, crystalline rocks was produced under a wide range of stress conditions. Experiments were so designed that stress at fracture would be known accurately, and also, that partially fractured material could be obtained, from which stages in the microscopic development of a fracture could be observed. Stresses at fracture were compared with prediction of Mohr and Griffith theories.

The first part of this report is a description of rocks, apparatus, and experimental procedure. Microscopic observations are summarized, to be given in detail elsewhere. Those who are primarily interested in conclusions and implications might turn immediately to my section entitled "Discussion," page 147.

---

*See Refs. 10, 15, 30, and 31.

Rocks Studied

Rocks were selected which would fracture under conditions which included moderate confining pressure, without first undergoing gliding or viscous flow: a quartzite, a granite, a diabase, and two dolomites. Although gliding occurs in the mineral dolomite, experimental deformation of a coarse-grained dolomite rock[32] produced almost no twinning at room temperature at pressures up to 5 kb. Robertson[13] showed that a fine-grained dolomite rock, the "Blair dolomite," behaved in an apparently brittle fashion at room temperature at confining pressures up to 25 kb.

The rocks were selected for homogeneity, isotropy, and low porosity. Bulk density, mineral density, and porosity (Table 1) were determined from the total volume and the dry and saturated weights of precisely ground right circular cylinders. Dry weight was obtained after heating for 48 hours in a vacuum at 230°F, and saturated weight after the sample was boiled in water under a vacuum for 8 hours. Probably not all pore space was reached in this way for rocks of this low porosity, but the values of porosity shown are probably not in error by more than 20 per cent. This is indicated by measurements of two samples of Blair dolomite having much different total volumes. One pair of samples of total volume 8 cc gave a porosity of 0.0022 and 0.0017. The other pair of volume 35 cc gave a porosity of 0.0023 and 0.0033.

The Frederick diabase,[33] Cheshire quartzite,[13] and Westerly granite[3,33] have been described elsewhere. The Westerly granite used in the present study was collected by the writer at Westerly, Conn., near the source of the so-called G-1 granite.[34] The Cheshire quartzite contains, aside from quartz, about 5 per cent partly altered feldspar and a trace of ankerite.

The Dunham dolomite (physically similar to the "Rutland dolomite" of Ref. 33) was collected from Route 7 a short distance west of Williamstown, Mass., and contains, aside from dolomite, about 1 per cent quartz. It is massive and light gray in color. The Webatuck dolomite was collected from the quarry of the White Marble Company, Webatuck, Dutchess County, N.Y., and contains, aside from dolomite, a fraction of

Table 1

ROCK PROPERTIES

Rock	Density		Porosity	Grain dia		Velocity (km/sec)
	Bulk (g/cc)	Mineral (g/cc)		Max (10^{-3} mm)	Av (10^{-3} mm)	
Solenhofen limestone	2.544	2.670	0.047	14.5	8	(5.5, 5.5, 5.6)[a]
Marble	2.697	2.715	0.007	480	200	(4.7, 4.8, 5.9)[a]
Blair dolomite	2.841	2.849	0.002	74	45	7.10, 7.02, 7.11
Webatuck dolomite	2.851	2.867	0.005	660	450	4.4, 5.0, 4.7
Dunham dolomite	2.840	2.864	0.008	250	75	(5.6, 5.8, 5.4)[a]
Westerly granite	2.621	2.646	0.009	750	500	3.6, 4.1, 4.3
Cheshire quartzite	2.626	2.646	0.007	640	300	4.3, 4.8, 5.1
Frederick diabase	(3.012)[a]	3.020	0.003	700	175	(6.76, 6.76, 6.78)[a]

[a]Ref. 33, Tables 5 and 6.

a per cent of phlogopite and magnetite. It is white and of medium grain size. The Blair dolomite (from the same formation as the "Blair dolomite" of Refs. 4 and 13) is a very fine-grained, dark gray dolomite; one sample which was chemically analyzed[35] contained 85 per cent dolomite, 6 per cent calcite, and 9 per cent insoluble residue.

Two crystalline limestones are included in Table 1: the Solenhofen limestone,[6,13,33] a typical buff, somewhat porous lithographic limestone, and a white marble (called here simply "marble"), probably from West Rutland, Vt., physically similar to the Danby marble of Ref. 33.

All the rocks except the dolomites appeared to be homogeneous and free of compositional banding and dimensional grain orientation. Preferred orientation of grain axes in the quartzite and marble is extremely weak.

The Blair dolomite possesses a faint wavy compositional banding, perhaps stylolitic in origin, which can be observed on a weathered or polished face. This inhomogeneity seems to have little influence on the elastic isotropy of this rock (Table 1). The Webatuck dolomite shows a perceptible dimensional orientation of grains in thin section, although the rock is structurally featureless in hand specimen.

Maximum and average grain diameter of the rocks (Table 1) were estimated visually from thin or polished sections. Elastic anisotropy of the rocks is indicated by measurements of compressional wave velocity at room pressure in three mutually perpendicular directions (the three numbers given in the last column of Table 1). Velocity was measured using methods and equipment developed by Prof. F. Birch.[33]

Important information on the structural character of porosity of a rock is given by variation of compressional wave velocity with confining pressure.[25] Variation of Young's modulus with pressure has the same significance. Static Young's modulus, which was measured at different pressures as described below, is given for the dolomites and silicate rocks as a function of pressure in Fig. 1. The theoretical average for an aggregate of quartz grains (the mean of the Voigt and Reuss values[25]) is shown for comparison with the data for quartzite.

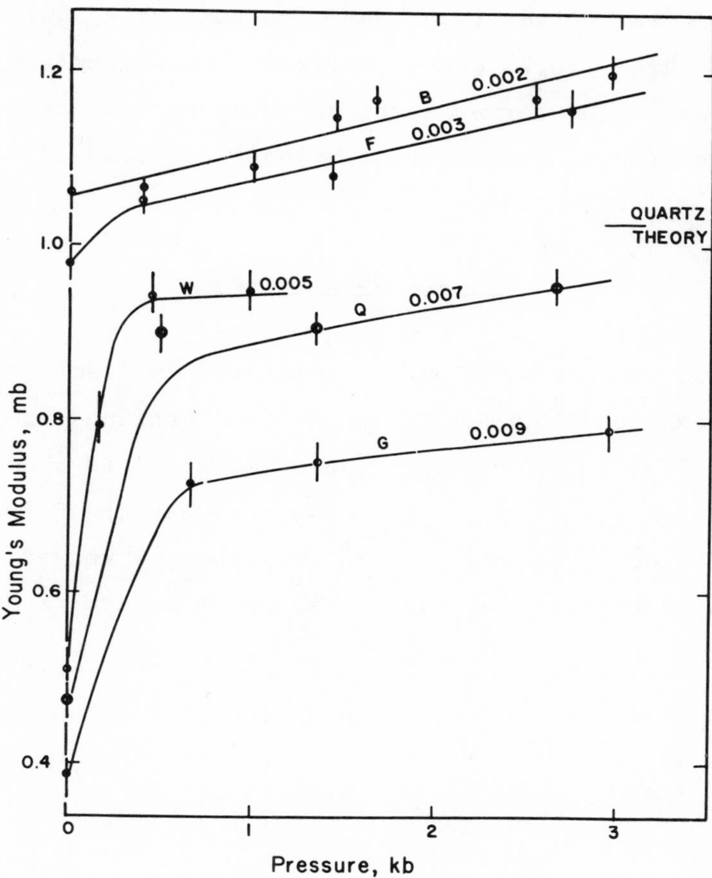

Fig. 1—Static Young's modulus as a function of pressure.

In Fig. 1, B is Blair dolomite, F is diabase, W is Webatuck dolomite, Q is quartzite, and G is granite. The numbers shown with each curve are porosity.

For the Blair dolomite and the diabase, Young's modulus increases gradually with pressure throughout. For the other three rocks it increases very rapidly with pressure up to about 0.5 kb and then increases at about the same rate as for the other two rocks. This sort of rapid initial increase has been generally attributed to a closing of pore space. In crystalline rocks such as studied here the porosity which can be reduced in this way probably exists as thin wedgelike openings between grains.[25] For the granite, quartzite, and Webatuck dolomite it may be inferred, therefore, that an appreciable amount of the relatively high porosity of these rocks (Table 1) occurs as grain boundary openings, rather than, say, isolated round holes. On this basis, these three rocks can be differentiated from the Blair dolomite and the diabase. The former might be termed loose structured, the latter compact structured, or simply loose and compact rocks, respectively.

Experimental Method

Apparatus was designed which would produce fracture under a wide range of stress conditions. The conditions attained are shown diagrammatically in Fig. 2, in which a cube of material is shown under a homogeneous state of stress. C is axial stress and P is confining pressure. In the experiments to be described, C ranged from tension through zero to compression, and P ranged from zero to compression.

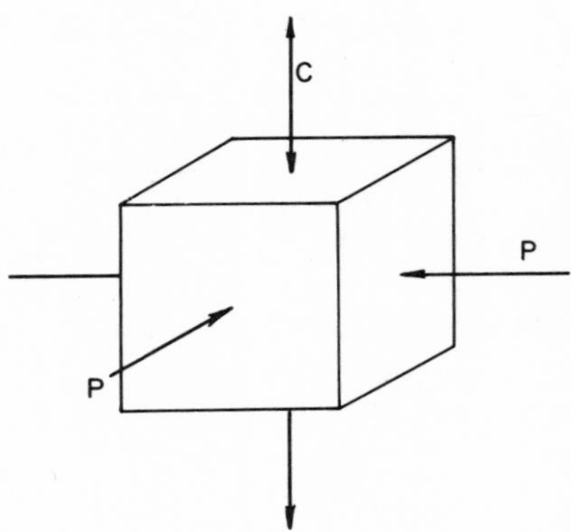

Fig. 2—Diagrammatic sketch of stresses applied.

Shape and Preparation of Specimen

A new feature of the experiments was the shape of sample used. In nearly all previous studies of compression or extension of rocks, specimens had the shape of a right circular cylinder; the ratio of length to diameter varied from 1 to 3. Several characteristics of the failure of such specimens suggested that improvement had to be made in the shape if significance were to be attached to the absolute value of breaking strength. For one thing, barreling, formation of conical fragments, and the association of fractures with the ends of the specimens[24] suggested that the stress state in short cylindrical samples may be far from uniform. The influence of the length-diameter ratio[36,37] and the large discrepancy between static and dynamic elastic constants[38] was further disturbing in this respect. The large variation typical of, say, room pressure, room temperature, and compressive strength of short cylindrical samples[37,38] suggested that this sort of test was very sensitive to minor differences between samples. This lack of reproducibility does not seem to extend to tests carried out at fairly high confining pressure and at elevated temperature. Reproducibility is often better than 5 per cent and agreement excellent between laboratories for these conditions.[35]

Irregularities such as these have usually been traced to frictional forces which arise because of elastic mismatch between the ends of the specimen and the testing machine.[20-24]

One type of sample, in which most of this effect is eliminated, has been used in the present study (Fig. 3). It is a long cylinder containing a central section of reduced diameter. The transition between this section (the throat) and the two ends (the heads) is accomplished by a large radius fillet. The area of the throat is one-fourth that of the heads; therefore, axial stress due to an axial load applied to the ends of the sample will always be four times as great in the throat as in the heads. Any disturbance in the stress field near the contact of the head and the testing machine will have little influence on behavior of material in the throat.

Two complications arise for the sample shape shown in Fig. 3 which do not exist for the short cylindrical sample: the bending tendency under axial load, which is discussed in the next section, and the stress concentration due to the presence of the fillet. The effect of the fillet is to increase locally the apparent axial stress in the throat. The maximum increase occurs near the point where the fillet joins the straight part of the throat. The amount of increase has been determined[39] by photoelastic studies and theoretical examination of the stresses around notches of hyperbolic cross section. For the dimensions of the samples used (Fig. 3) the maximum stress concentration due to

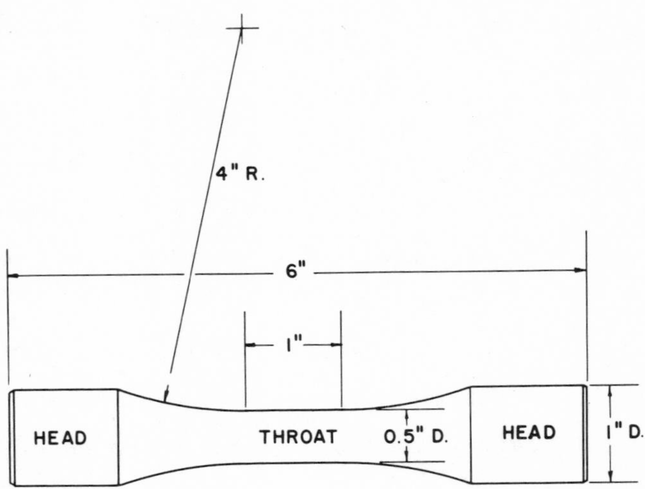

Fig. 3—Shape of sample.

the presence of the fillet is about 2 per cent with an uncertainty of one-half per cent.

Samples of roughly this shape were first used by Voigt[(40)] and later by Bridgman.[(41)] The experiments differed in that the entire sample was subjected to hydrostatic pressure and the reduced center section was used simply for localizing axial stress.

To prepare a sample, a cylinder was first cored from a block. This cylinder was mounted between centers in a cylindrical grinder and surface ground to nearly the final diameter of the head. The throat region was then shaped by a grinder especially built for the purpose. A grinding wheel turning about an axis parallel to the cylinder axis was mounted in such a way that it could also be rotated to cut out the fillet shown in Fig. 3. Even for the hardest rocks the grinder produced a final surface in the throat region which was extremely smooth and precisely concentric with the surface of the heads. The technique was first applied to the softer dolomites and, after experience was gained, to the harder silicate rocks. The time required for coring and grinding was about 3 to 4 hours per sample.

Following surface grinding, the sample was usually etched in an appropriate solvent (HCl or HF) to remove fractured and loosened surface material. Only a fraction of 0.001 in. was removed in this process.

Bending in the Specimen; Strain Gauges

The other disadvantage of the specimen shape used is the tendency to bend under an axial load, although as it turned out this rarely happened. Of some sixty samples tested only one failed by buckling, as shown by a tension failure on one side of the throat. To evaluate the bending, wire resistance strain gauges were mounted on the surface of the throat. A typical arrangement is shown in Fig. 4(a). As shown in Appendix A, either three or four gauges mounted axially enable the increase in axial stress due to bending to be determined. The one assumption that must be made is that the column of rock at the throat behaves as an ideally elastic body in bending. The use of four axial gauges in the throat enabled this assumption to be verified. Four gauges were used in a number of runs in which it was found that departure from ideal elastic behavior just before fracture of the sample was 1 per cent or less of the average strain and therefore negligible.

Constantan foil gauges mounted on epoxy film were used (Baldwin Lima Hamilton Company).* They were cemented directly to the rock using araldite epoxy cement (CIBA Company). The independent effect of hydrostatic pressure acting on the gauges was evaluated so that they could be used to measure strain even when the rock was jacketed and subjected to hydrostatic pressure.

The Pressure Vessel

The essential elements of the pressure vessel used are shown in Fig. 5. The sample sits in the central cavity of the vessel. An axial load is applied by the piston, and fluid pressure is applied to the throat region by a fluid in the central cavity. The heads of the specimen project out through O-ring seals so that the fluid pressure is not applied to the ends of the sample. If F is force on the piston, P is fluid pressure, and A_h and A_t are head and throat cross-sectional areas, respectively, then total axial stress, C, in the throat region is given by

$$C = \frac{F}{A_t} + \frac{P(A_h - A_t)}{A_t} ,$$

---

*Details of strain gauges and mounting procedure will be published in a future paper.

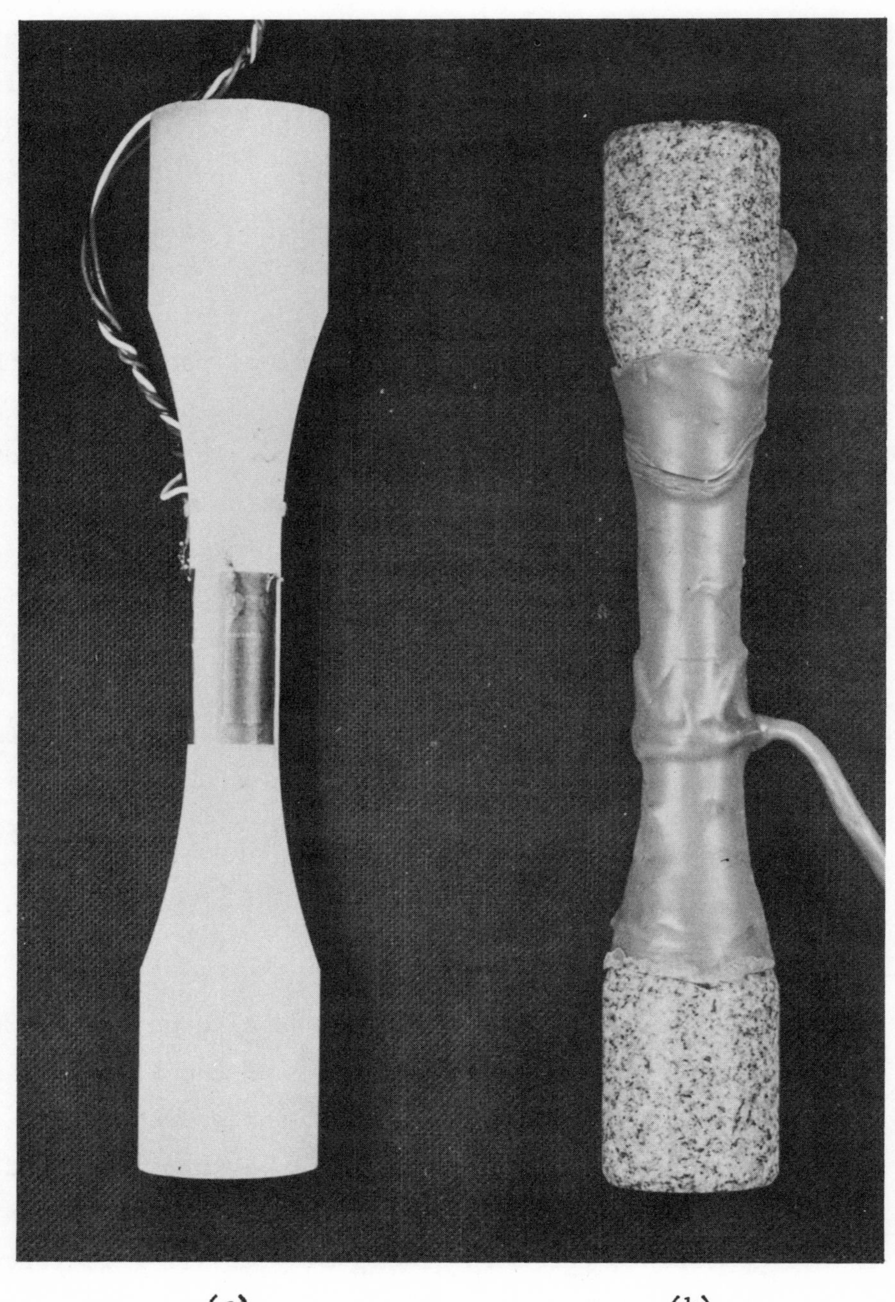

(a) (b)

Fig. 4—Photographs of sample.

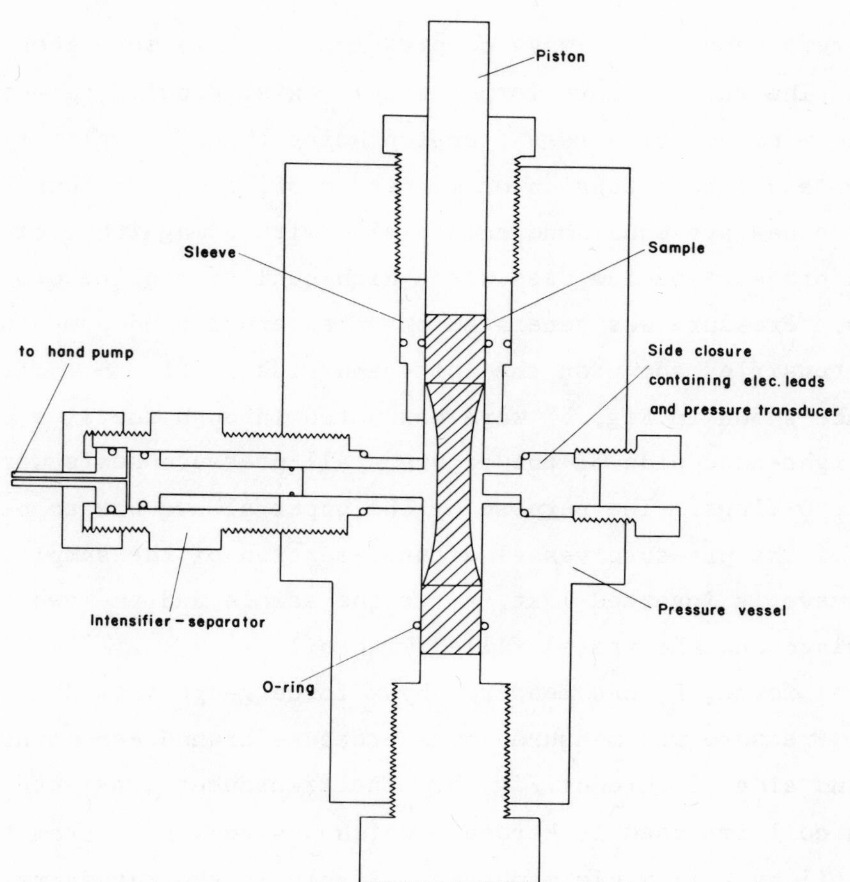

Fig. 5—Apparatus.

in which tension is positive, pressure is positive, and negative F acts toward the specimen. For the sample dimensions used, A_h is about $4A_t$, so that

$$C = \frac{F}{A_t} + 3P \ . \tag{1}$$

If P is zero then C is always compressive; if F is zero then C is tensile. The stress state is not simple axial tension, however. The maximum principal stress (tension being figured positive) is axial and equal to a tension of magnitude 3P, and the other two principal stresses are equal and compressive with a magnitude of P.

The pressure medium used was a high-quality machine oil free of solvents. Pressure was generated by an external hand pump in the small intensifier shown on the left-hand side of Fig. 5. Strain gauge leads (not shown in Fig. 5) were conducted through the side closure on the right-hand side of the figure. All pressure seals were effected by rubber O-rings. The purpose of the upper sleeve was to permit filling of the pressure vessel after insertion of the sample. The upper sleeve is inserted last, after the sample and the two side closures are in place and the vessel filled with oil.

Axial force, F, was measured by a force gauge outside the pressure vessel. Pressure was measured by a pressure transducer mounted in the right-hand side closure of Fig. 5. The transducer consisted of a manganin coil immersed in kerosene which was separated from the heavy machine oil by a flexible membrane. Strain in the throat region of the specimen was measured by the strain gauges. Under load, the cross-sectional area changes by a small amount. Stress, C, was corrected for this change in area.

Axial force was generated by a hand pump working into a piston-cylinder system external to Fig. 5. Although the capacity of this system was about 100 tons, the hand pump was so constructed that load could be applied in increments of a few tens of pounds if desired.

Friction at the O-rings bearing on the sample was determined precisely as follows. The sample was replaced by a short plug of the same rock type and having the exact dimensions of the head of the sample. With the central cavity filled with oil, force on one side of the O-ring could be compared with pressure on the other side. The difference was found to be a constant fraction of the force applied by the pressure medium, and was practically independent of the type of rubber O-ring used, its hardness, and whether it was badly scored from previous experiments. Friction amounted to 1.50 (± 0.25) per cent of the force applied by the pressure medium throughout the range of pressures used in the experiments (3.5 kb).

A typical run required 1 to 2 hours, of which the greater part was taken up with measurements. Insertion of samples, soldering leads, filling with oil, removal of sample, and cleaning up required 30 to 45 minutes.

Termination of an Experiment Just Prior to Fracture

Most of the rocks in this study broke explosively, even under confining pressure. This destroyed many of the subtle features formed during early stages of fracture growth. To understand how fractures originated, therefore, it was necessary to develop a technique for stopping an experiment an instant before failure of the rock.

Several factors contributed to the success of this operation. First, breaking stress was found to be quite reproducible (Table 2 below), and after the first one or two samples were fractured the fracture of the next could be approached carefully. Second, fracture was almost always preceded by flattening of the force-strain curve. This change in slope was particularly easy to detect when force and strain were continuously recorded on an XY recorder. Third, the press used to apply axial load was kept as rigid as possible; this reduced the elastic energy which would be released when the specimen began to fail. Finally, a small microphone with amplifier was used during the experiments to detect subaudible noises. A rapid buildup of these noises is known to precede fracture, and in one or two cases this gave a signal far enough ahead so that a partially fractured specimen could be obtained.

Jacketing of Specimen

The sample had to be jacketed to prevent penetration of the pressure transmitting fluid. Owing to the fact that the O-ring seals of the pressure vessel bear directly on the heads of the specimen and because of the tapered shape, jacketing was a rather complicated procedure. In general, the heads were sealed against the pressure in a different way than the throat region.

It was found that rocks of less than 1 per cent porosity could be effectively jacketed against fluid penetration during the period of an experiment if (a) the fluid was fairly viscous and (b) the outer 1 mm or so of surface material was impregnated with a tough cement-like material. The material used was the same epoxy cement used for mounting the strain gauges. At a temperature of 100°C or so the cement penetrates some distance into the rock, whereupon it hardens and seals the surface of the rock against fluid penetration. Excess is wiped from the surface before hardening so that the outside dimensions of the body are unaltered.

The heads of a sample were impregnated with the epoxy cement in this way. The throat was not impregnated as this might alter the properties of the rock in this critical region. The throat region of a compression specimen, together with strain gauges and gauge leads, was encased in a rubbery compound (Flexofix, Pyroil Corporation) which extended some distance toward the heads and overlapped the impregnated region. This compound had the desirable quality that it could be applied as a fluid. The samples were dipped in the fluid, which was then allowed to harden and encase the samples. A finished compression specimen ready for testing is shown in Fig. 4(b).

As discussed below, special care was taken with the jacketing of extension samples of the compact rocks. Instead of jackets made of Flexofix, various metal jackets were used to cover the throat region. Strain gauges were not mounted on most extension samples so that the metal was in direct contact with the rock surface. Jackets were prepared of various thicknesses of annealed copper and of type 302 stainless steel. Such a jacket was made as follows. A strip of the sheet metal was cut to the form shown in Fig. 6(b). It was then rolled

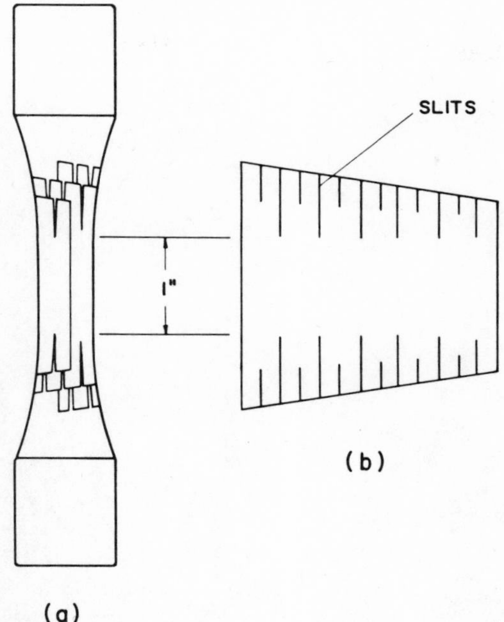

Fig. 6—Jacketing of an extension sample.

tightly over the throat area, as shown in Fig. 6(a). Care was taken that all seams, wrinkles, and other irregularities in the jacket were nearly parallel to the cylinder axis, because the planes along which an extension sample fractures are nearly perpendicular to this axis. After a little experience was gained, the metal jackets (even the full hard 0.002-in. steel) could be made to lie smoothly over the throat.

A copper jacket could be made by collapsing a soldered tube down onto the throat section; again care had to be taken that folds in the tube ran parallel with the cylinder axis. A typical jacket after being subjected to pressure is shown in Fig. 7. The opening at the end is sealed as described below. Scale in the figure is given by the small diameter of sample, which is 1/2 in.

A Flexofix jacket was next cast over the metal jacket which completely sealed the sample against fluid penetration. If a fairly thick jacket (1/16 in.) was cast, considerable relative movement of metal and rock could take place without leakage.

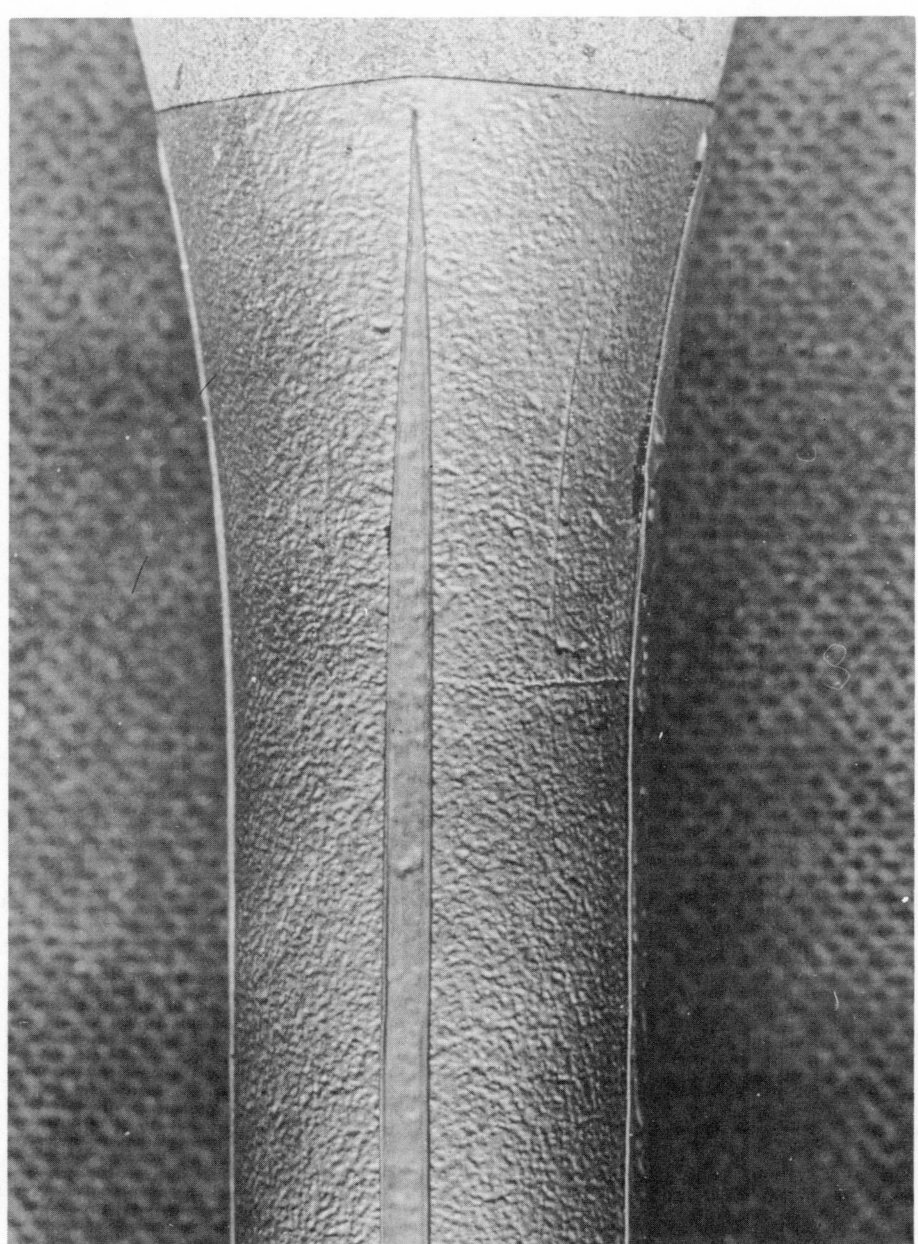

Fig. 7—Partially fractured extension sample.

Accuracy of Measurements and Calculated Stresses

Output of the force gauge, pressure gauge, and strain gauges was measured directly with a bridge or displayed on an XY recorder.

Several force gauges were used to cover different ranges of load. Each gauge was calibrated by a Moorehouse proving ring. Over the range of loads used, accuracy was better than 1 per cent. If the uncertainty in friction at the O-rings is included, accuracy of force measurements is 1.5 per cent. Sensitivity is much greater than this.

Pressure measurements at pressures greater than 1.5 kb had an accuracy of about 1 per cent. This included errors due to measurement, to instrument drift, and to calibration of the recorder. Sensitivity was about 0.01 to 0.02 kb so that uncertainty was greater for pressure less than 1.5 kb.

The accuracy of strain measurements was about 2 per cent, although sensitivity is much greater. Part of the error was due to uncertainties in the bending analysis. Using four gauges and Eq. (A-2) (Appendix A), departures of 1 per cent were observed from ideal elastic bending. The rest of the error arose because the strain gauges, being of finite width, indicate an average strain over their surface rather than strain at a point. For the gauges used, actual strain could deviate by as much as 1 per cent from the indicated strain.

Taking these factors into account and allowing a 1 per cent uncertainty in the strain concentration factor, the probable error[(42)] of the indicated axial forces and stresses for the experiments in compression is 5 per cent.

In extension the uncertainties are much greater because the two quantities on the right-hand side of Eq. (1) are nearly equal in magnitude. For pressures greater than 1 kb the probable error in the axial stress reported is about 20 per cent. For lower pressures the limit of sensitivity of the pressure measurement is approached, and again the values of axial stress reported in these extension experiments have an uncertainty of about 20 per cent.

The effect of rubber jackets on the axial stresses in either compression or extension experiments was negligible. However, the strength of the metal jackets used must be taken into account. This is done (Appendix B) by assuming that the metal jacket and the rock deform elastically as a unit. This gives a correction to the axial stress which amounts to about 10 per cent, with a small uncertainty due to the elastic moduli of metal and rock used. A much greater error would exist if rock and metal did not behave as assumed. This assumption seems to be justified for two reasons. First, it was observed that the steel jacket sometimes ruptured normal to the axis of the cylinder after the rock had broken in several places. This rupture could only form if large frictional forces held rock and jacket together, as the jacket is not attached in any way to the rock or to the testing machine. Second, two samples of the same material were tested in extension at the same pressure; one had a copper jacket, the other a steel jacket. Applying the appropriate corrections for the two jackets, the two results agreed fairly well. This would not have been the case had the above assumption and the computed correction been grossly in error.

Observations

Microscopic Observations and the Stress-Strain Curve

An idealized stress-strain curve (Fig. 8) shows typical behavior of rocks in compression tests both at room pressure and under confining pressure. Regions of this characteristic curve can be correlated with microscopic behavior. This behavior is summarized here.*

In regions I and II behavior is elastic; nearly all strain is recoverable. The degree of curvature in region I (the difference between solid and dotted lines) varies for different rocks. In general, the compact rocks have a straight stress-strain curve in this region, whereas the loose rocks show rather pronounced curvature. Hast,[26] Nishihara,[27] and Matsushima[9] give stress-strain curves for granite, marble, and sandstone which have the same pronounced initial curvature as the loose rocks of this study. If stress is applied and then removed

*A more detailed presentation is being prepared at the present time.

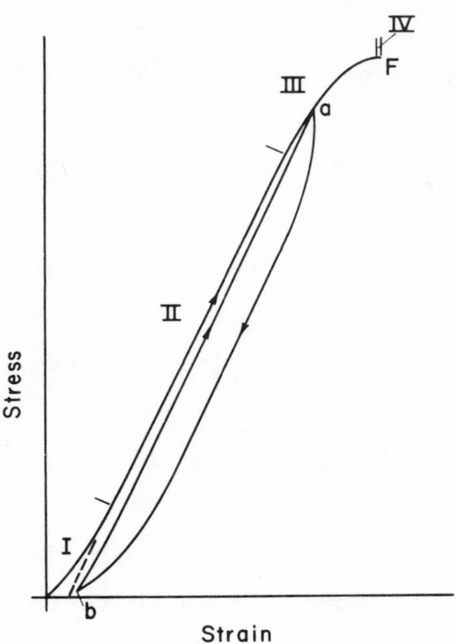

Fig. 8—Idealized stress-strain curve.

from the loose rocks in regions I and II, hysteresis loops are traced which have a complex and not very consistent pattern. With the application of a few tenths of a kilobar of hydrostatic pressure, the curved region I of a loose rock becomes straight. If initial slope of the stress-strain curve (initial Young's modulus) is plotted against pressure, a curve such as that shown in Fig. 1 is obtained.

Starting with region III, important permanent changes in the microscopic character of both loose and compact rock occur. These changes, which accompany a gradual flattening of the stress-strain curve, are quite subtle and can best be detected in a polished section under dark field illumination. The rock takes on a somewhat lighter color, which can be traced at high magnification to reflection of light at grain boundary surfaces. Apparently, grains are becoming detached at their boundaries; when this happens the boundary surface becomes totally reflecting and therefore easily visible. These reflecting surfaces become more numerous as fracture is approached.

Another significant change occurs in region III. By placing strain gauges perpendicular and parallel to the axis of the specimen, lateral as well as longitudinal strain can be observed. The ratio of these strains is nearly constant in regions I and II. In region III the lateral strain (an extension for a compression test) begins to increase more rapidly. In region IV it often increases very rapidly; Poisson's ratio may become 0.5 or even 1.0 (also noted by Matsushima[9]). This can probably be interpreted as due to an increase in the volume of the sample as deformation proceeds (see also Bridgman[43]). Although somewhat less pronounced, increase in Poisson's ratio and, therefore, volume increase in region III are also observed in tests carried out under confining pressure.

If the load is reduced in region III, a hysteresis loop such as that drawn between points a and b (Fig. 8) is traced out. Usually a small amount of permanent strain remains if load is removed entirely.

A great deal happens very rapidly in region IV, which is shown as a very small region of nearly constant stress. Based on the study of a number of specimens with partially formed fractures, that is, fractures which do not extend completely through the specimen, two stages can be differentiated. In the first stage, cracks begin to grow out of grain boundaries at many sites throughout the rock; these are boundaries that became detached in region III. In a second stage, large throughgoing fractures form out of systems of these cracks. If the fracture is of the fault type, in which the principal relative motion is one of shear, then the fracture almost always grows from a system of _en échelon_ cracks. If the final fracture is an extension fracture, in which the principal motion is normal to the plane of the fracture, then the fracture appears to grow out of a single crack.

As a fault grows out of a system of _en échelon_ cracks, it must break through and bridge the material between individual cracks. Therefore, a great deal of pulverized material is found at a fault; if there has been appreciable relative motion of the sides of the fault, then this debris is smeared out into slickensides. The surfaces of extension fractures never have these quantities of debris. Although both types of surface are somewhat rough, the surface of faults is usually more nearly planar, whereas the surface of extension fractures is broken by minute bumps and depressions.

No evidence of slip or gliding was detected in the silicate rocks, or even in the limestone and dolomite samples which were caught before total destruction, although the method used (estimation of intensity of gliding in thin and polished sections) was not very sensitive. It is quite possible that a small amount of gliding flow occurred. Appreciable local twinning is produced when a specimen is allowed to fracture completely; this occurs close to fractures where there has been pronounced shearing motion.

Numerical Data

Five rocks were studied throughout the entire range of stress conditions possible: the Blair and Webatuck dolomites, the Westerly granite, the Cheshire quartzite, and the Frederick diabase. Room pressure data from three other rocks—the Dunham dolomite, the Solenhofen limestone, and the marble—were also obtained. The data are in three parts: room pressure tests, confined compression tests, and extension tests under confining pressure. In a compression test the axial compression is greater; in an extension test it is less than the confining pressure.

Data from a typical compression test, as recorded on an XY recorder, are shown in Fig. 9. The ordinate is axial force, the abscissa axial strain. Confining pressure was first increased from zero to about 0.5 kb along the path OO'. This operation produces a zigzag path because fluid pressure and axial load must be raised independently to produce a state of hydrostatic pressure in the throat of the specimen. Each of these operations creates its own trace with a particular slope. From point O' axial load only was increased and the force-strain curve of the rock was traced out. As dimensions of the specimen change by a negligible amount, this is equivalent to a stress-strain curve. Strain from all three gauges mounted on the rock was recorded at intervals. Here, the curve shown was recorded using gauge 1 so that the pairs of points alongside the curve at intervals belong to gauges 2 and 3. Load was dropped twice to observe hysteresis and then finally increased to the point of fracture, F.

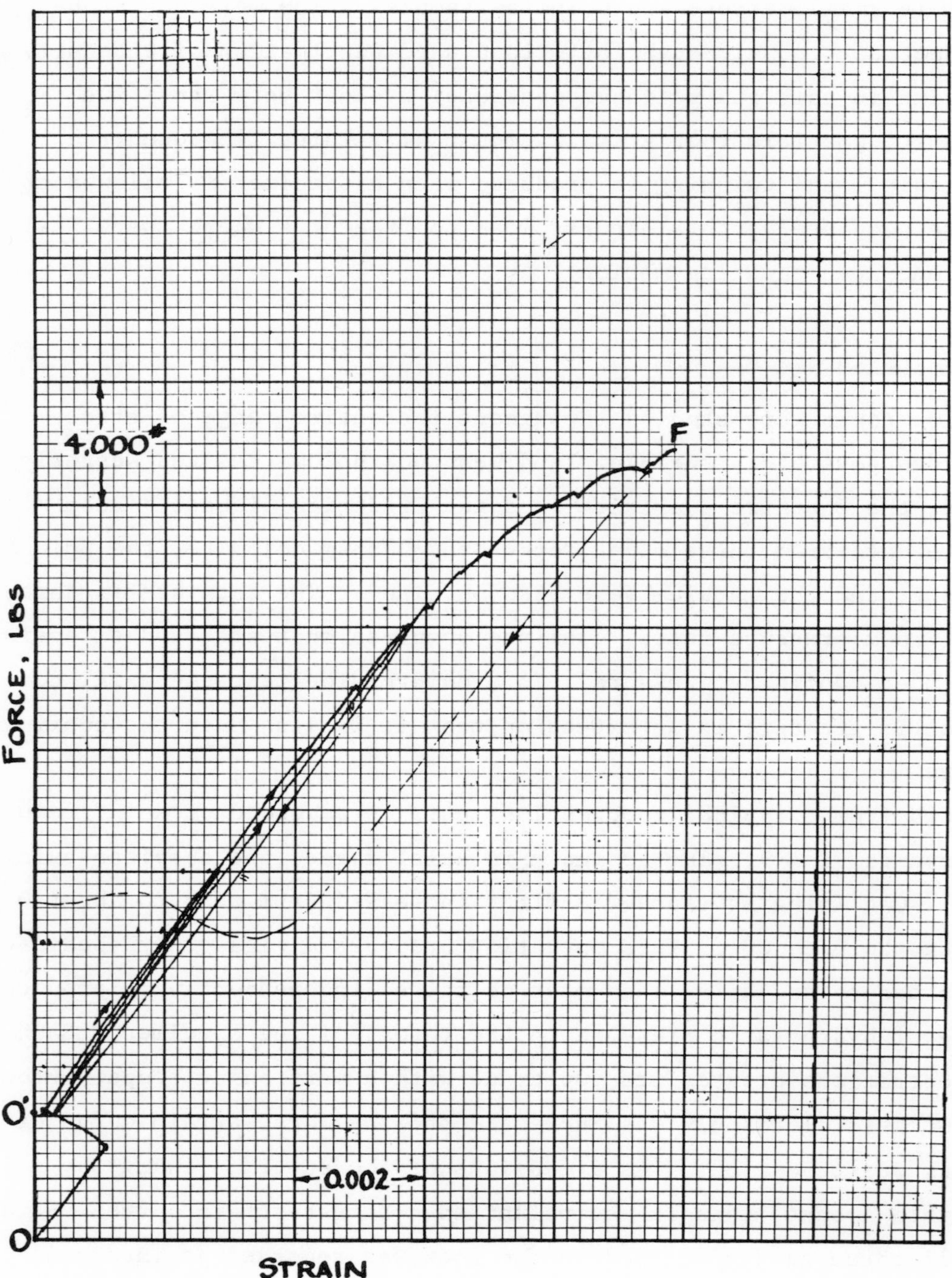

Fig. 9—Typical compression test data.

In a typical extension experiment, shown in Fig. 10, axial force and pressure are recorded. Axial force is again nearly equivalent to axial stress because change of area is negligible. At all points along the line OH (Fig. 10), the throat region of the specimen is under hydrostatic pressure. In this experiment fluid pressure and axial load were raised (along the upper zigzag path to the left of OH) so that at point O' the throat of the sample was under hydrostatic pressure of about 2.5 kb. Then axial load was reduced, along the path O'F until fracture occurred at F. Pressure and axial load were then reduced to zero along the second zigzag path.

Data from tests such as these are given in Tables 2, 3, and 4. The quantity F is the maximum axial load applied. C is the calculated axial tensile or compressive stress. In most tests the specimen broke at this stress; in a few tests load was dropped as the force-strain curve was observed to level off, and the specimen was removed intact or only partially fractured. C_o is the value of C for room pressure compression tests.

The C shown has been corrected for changes in lateral dimensions of the sample, for stress concentration in the sample due to its shape, and for local increase in the average axial stress due to bending. C is therefore the extreme axial stress anywhere in the throat region. To allow for bending, C has been multiplied by the factor $(1 + e'_{max})$ where e'_{max} is the maximum strain increase due to bending, assuming that bending is linear in the cross section, and that stress is linearly proportional to strain (Appendix A). The quantity $\bar{e}_{max}$ is the average strain at fracture, obtained from the average of the axial gauge mounted on the rock. Extension is positive, contraction negative.

The quantity θ is the average angle of inclination, measured from the cylinder axis, of fracture surfaces. It is a rough figure, for inclination of a fracture may vary locally by as much as 10 degrees. Typical fractures in compression and extension are shown in Figs. 11 and 12, respectively.

From left to right in Fig. 11: Solenhofen limestone, P = 0, θ = 20 degrees; Blair dolomite (B-7), P = 0.23 C_o, θ = 19 degrees; Blair dolomite (B-19), P = 0.36 C_o, θ = 26 degrees; granite (G-5), P = C_o, θ = 29 degrees.

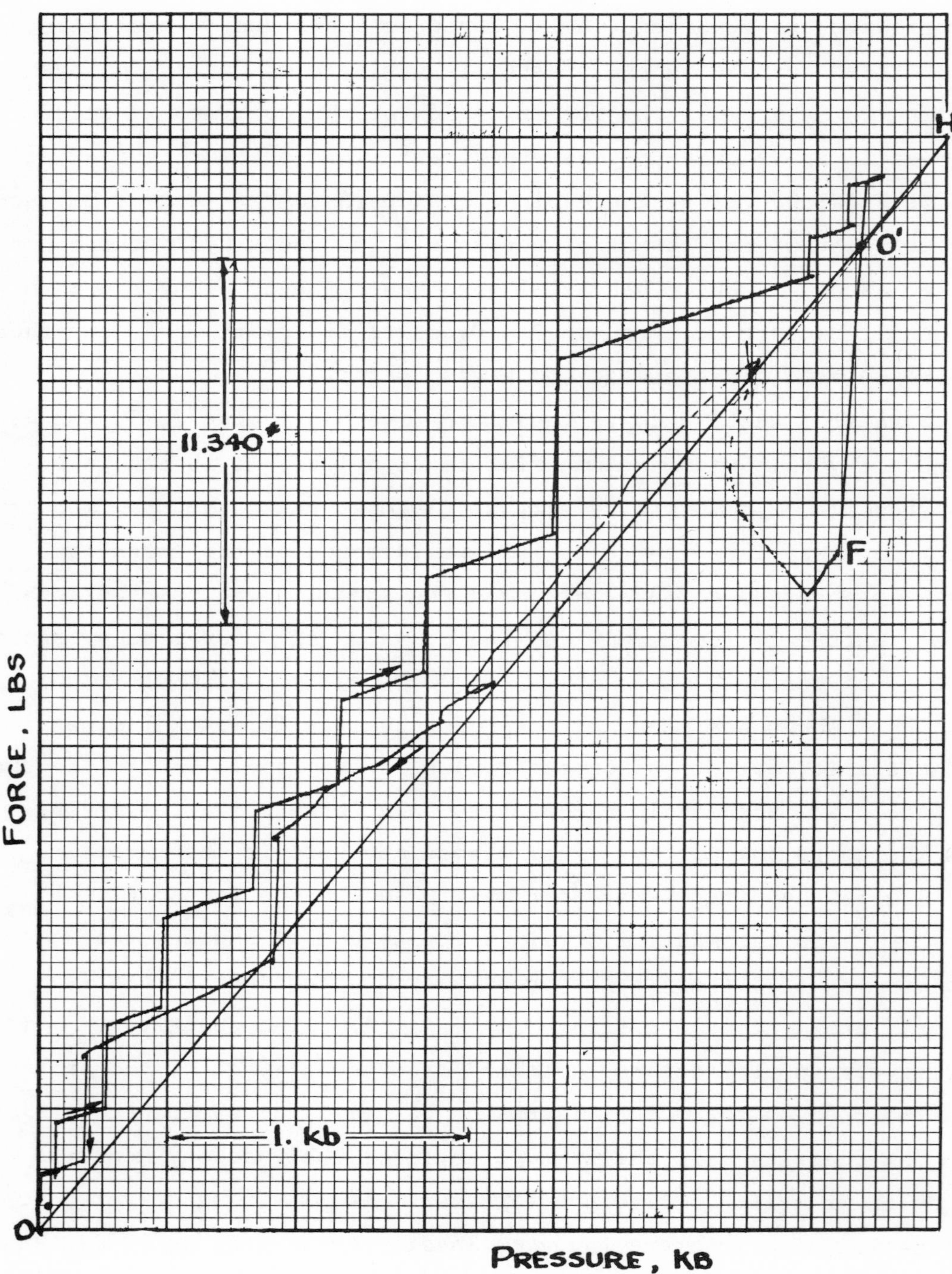

Fig. 10—Typical extension test data.

Table 2

ROOM PRESSURE COMPRESSION TESTS

Rock	Test No.	- F (lbs)	$\lvert C_o \rvert$ (kb)	$\bar{e}_{max}$ (%)	e'_{max} (%)	θ (deg)	E_o (Mb)	E_{max} (Mb)	E_{dyn} (Mb)
Blair	2	13,900	5.14	-0.52	3.2	23	1.05	1.05	
	3	13,800	5.00	-0.53	0.8	i	1.04	1.04	
	6	14,000	5.51	-0.50	9.9	23	1.10	1.10	
	Average		5.07[a]				1.06		1.09[b]
Webatuck	1	3,820	1.49	-0.27	5.7	i	0.54	0.70	
	2	3,770	1.48	-0.26	6.4	22	0.48	0.73	
	Average		1.48				0.51		0.52
Dunham	1	5,870	2.28	-0.41	5.7	20	0.72	0.72	
	2	6,150	2.32	-0.38	2.2	21	?	0.68	
	Average		2.30						--
Solenhofen	1	8,760	3.22	-0.66	2.5	19	0.568	0.568	
	2	7,870	2.86	-0.60	1.3	20	0.542	0.542	
	Average		3.04				0.555		0.588[b]
Marble	1	1,380	0.55	-0.25	11.0	18	---	0.24	
	2	1,380	0.51	-0.25	3.6	24	---	0.24	
	3	1,400	0.54	-0.23	8.0	23	---	0.24	
	Average		0.54						0.25[b]
Diabase		13,370	4.87	-0.55	1.5	14	0.979	0.979	1.015[b]
Quartzite		12,700	4.61	-0.60	1.1	i	0.48	0.79	0.66
Granite		6,300	2.29	-0.44	1.2	22	0.39	0.56	0.41

NOTES: kb and Mb mean kilobars (10^9 dynes/cm^2) and megabars (10^{12} dynes/cm^2), respectively; i means sample was recovered intact.

[a]The average shown is the mean of samples 2 and 3. One of the gauges during test number 6 was not functioning properly so that the e'_{max} shown is high and therefore the C_o calculated is also too high.

[b]Modulus determined from measurements of the resonant frequency of bars.

Table 3

COMPRESSION TESTS UNDER CONFINING PRESSURE

Rock	Test No.	P (kb)	- F (lbs)	C (kb)	$\bar{e}_{max}$ (%)	e'_{max} (%)	θ (deg)
Webatuck	6	0.16	7,330	- 2.26	-0.36	5.	i
	5	0.48	12,500	- 3.20	-0.48	5.	i
	4	1.07	21,500	- 4.82	-0.45	6.0	17
Blair	8	0.46	25,800	- 8.26	-0.90	6.5	13
	7'	0.94	35,200	-10.17	-1.30	5.	19
	7	1.15	38,800	-10.55	-1.65	3.	i
	4	1.57	44,200	-11.25	-1.31	3.	i
	9	1.84	47,100	-12.41	-1.30	8.5	26
	10'	2.73	58,500	-13.73	-1.30	6.7	i
	11	3.49	66,600	-14.71	-3.0	9.0	i
Diabase	10	0.49	27,000	- 8.06	-1.65	2.1	i
	4'	1.60	50,200	-13.10	-1.37	3.3	i
	3	3.18	80,570	-20.65	-2.2	6.8	26
Granite	4	0.83	32,500	- 9.45	-1.6	3.0	24
	5	1.50	47,900	-14.51	-2.2	1.1	29
Quartzite	7	0.61	39,200	-11.85	-1.28	0.6	i
	3[a]	1.62	74,000	-21.65	-2.70	2.4	24
	4[a]	3.08	107,500	-28.45	-3.40	0.9	26

NOTE: i means the sample was recovered intact.

[a]The axial force measurement in this test is suspect.

Table 4

EXTENSION TESTS UNDER CONFINING PRESSURE

Rock	Test No.	P (kb)	- F (lbs)	Jacket type	C (kb)	θ (deg)
Webatuck	8	0.24	1,780	F	+0.08	0
	9	0.93	7,750	F	+0.02	6
	7	2.39	20,300	F	-0.10	18
Blair	10	0.11	0	F	+0.34	0
	16	0.19	470	S	+0.36	0
	15'	2.23	17,350	S	+0.43	0
Diabase	9	0.25	826	S	+0.43	0
	8	0.50	3,000	S	+0.39	0
	7	1.46	11,000	S	+0.40	0
	7'	2.65	26,300	None	-1.47	Intr.
	5	2.77	25,400	S'	-0.77	Intr.
	4	2.53	21,400	C	-0.13	Intr.
	6	2.65	21,200	S	+0.30	5
Quartzite	8	0.15	470	S	+0.24	0
	2	0.60	4,080	S	+0.31	0
	5	2.61	21,150	S	+0.22	0
Granite	3	0.30	1,810	S	+0.21	0
	2	1.51	12,180	S	+0.12	0
	6	1.59	12,950	C	+0.12	4
	7	2.39	20,100	C	-0.05	14

NOTES: Intr. means jacket material appears to have penetrated rock and caused fracture. S means 0.002-in. steel, S' means 0.0005-in. steel, C means 0.002-in. copper, and F means rubber.

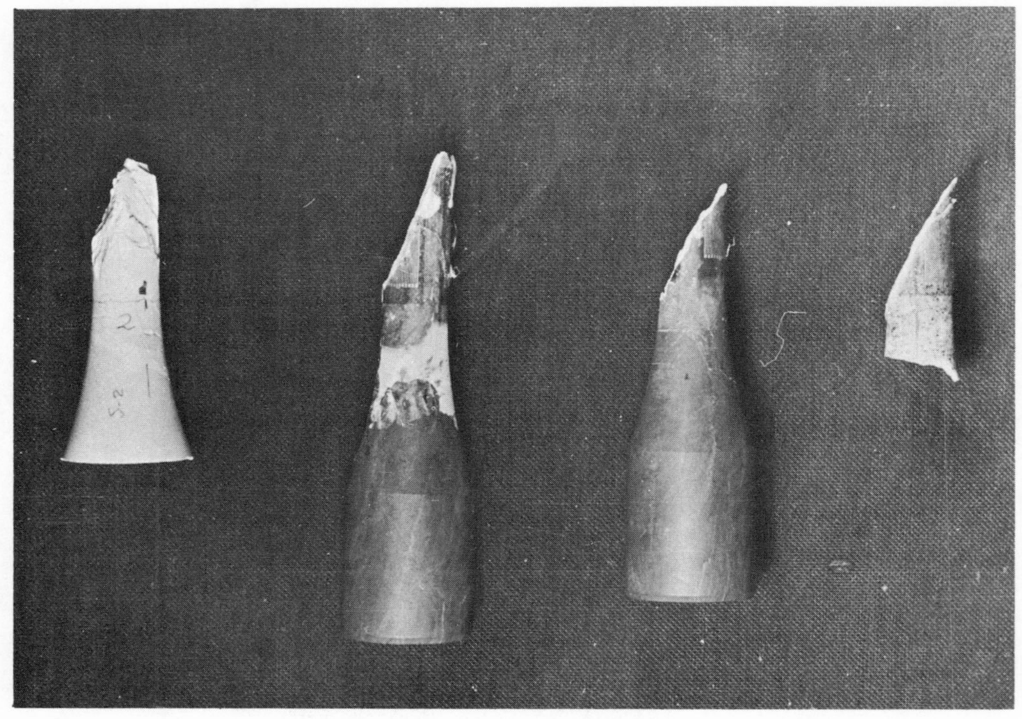

Fig. 11—Typical fractures in compression test.

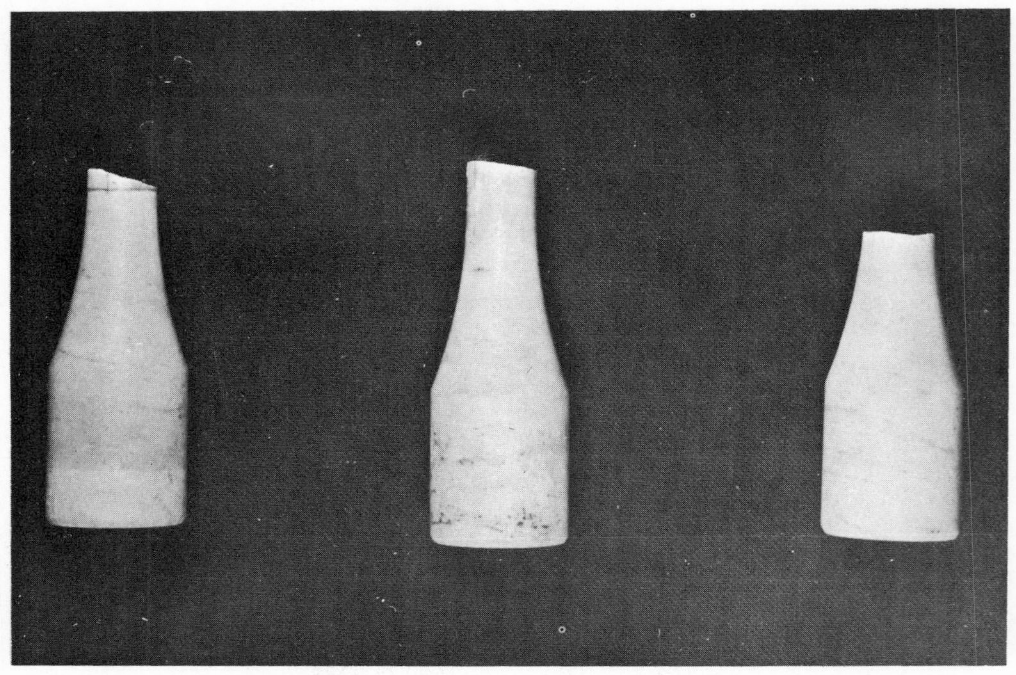

Fig. 12—Typical fractures in extension test.

In Fig. 12,* from left to right: (W-7), $P = 2.0\ C_o$, $\theta = 18$ degrees; (W-9), $P = 0.63\ C_o$, $\theta = 6$ degrees; (W-8), $P = 0.16\ C_o$, $\theta = 0$ degree.

Initial (E_o) and maximum (E_{max}) Young's modulus were measured directly from the recorded data sheets in compression tests. Young's modulus was not measured in extension tests. E_{dyn} is the dynamic Young's modulus determined from compressional wave velocity[33] and static Poisson's ratio.

A correction for the strength of the jacket (Appendix B) has been made in the calculated extreme axial stresses in the extension tests (Table 4). Strain is not reported for extension tests, nor is bending. Several measurements of bending carried out in early extension tests showed that increase in axial stress due to bending here was less than 1 per cent and could be neglected.

The maximum stresses applied to the five rocks in the above experiments are plotted in Figs. 13 through 17. These are shown as plots of σ_1 versus σ_3, where σ_1 is the maximum and σ_3 the minimum principal stress in the throat at fracture. The probable error in these numbers is shown by size of the line at each point. Open and full circles are extension and compression tests, respectively. Numbers in parentheses are the values of θ when specimen fractured.

Intrusion Fracturing

In a number of extension experiments, behavior observed was similar to that described by Bridgman as the "pinch-off" effect. In the extension of cylindrical samples Bridgman found[41,44,45] that the fracture of glass, Solenhofen limestone, and cast iron was affected by surface conditions of the specimen and by the nature of material in contact with the surface. For example, unjacketed cast iron or Solenhofen limestone broke normal to the cylinder axis, whereas the same material jacketed with copper foil either broke along a fault or flowed "plastically." The stresses necessary to rupture glass in extension

---

*All samples in Fig. 12 are of Webatuck dolomite.

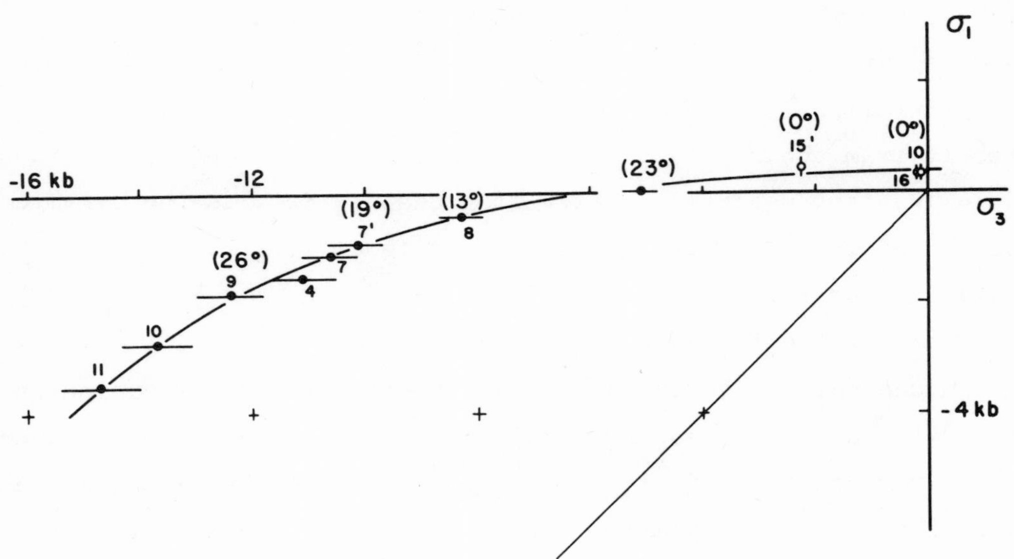

Fig. 13—Failure conditions of Blair dolomite.

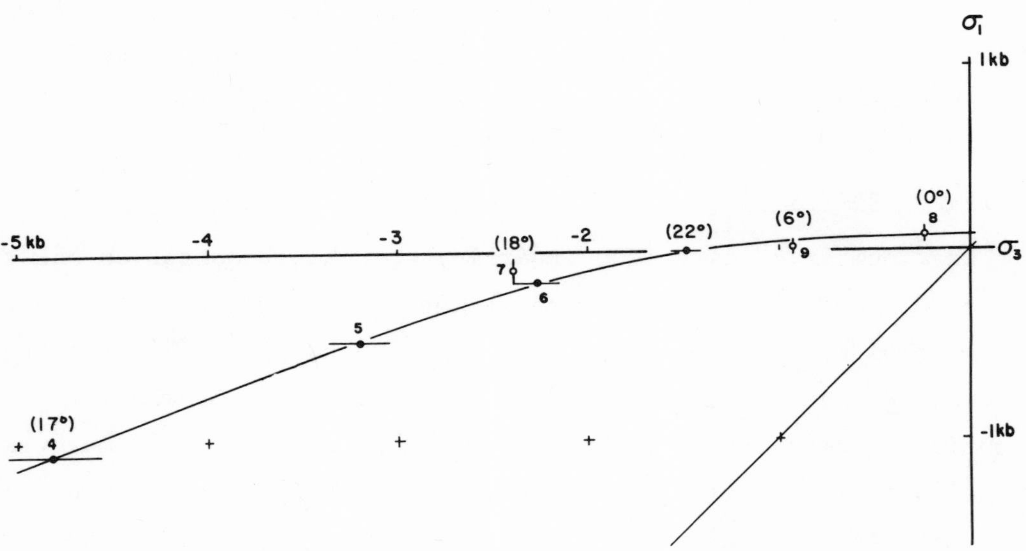

Fig. 14—Failure conditions of Webatuck dolomite.

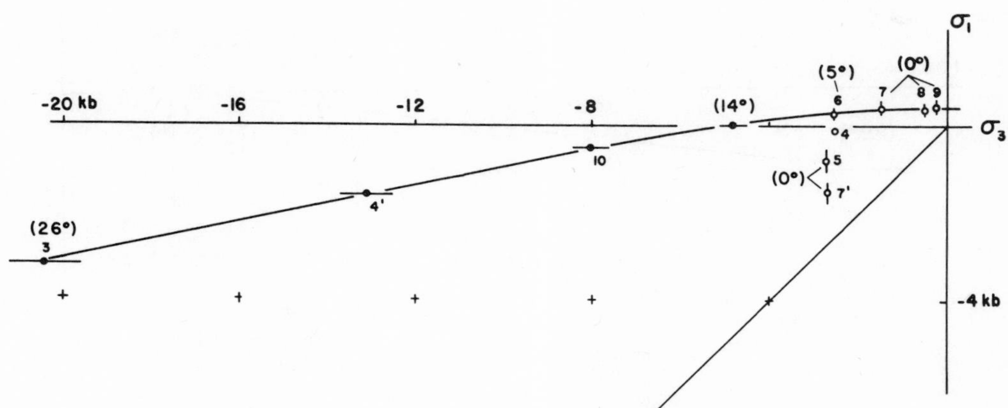

Fig. 15—Failure conditions of diabase.

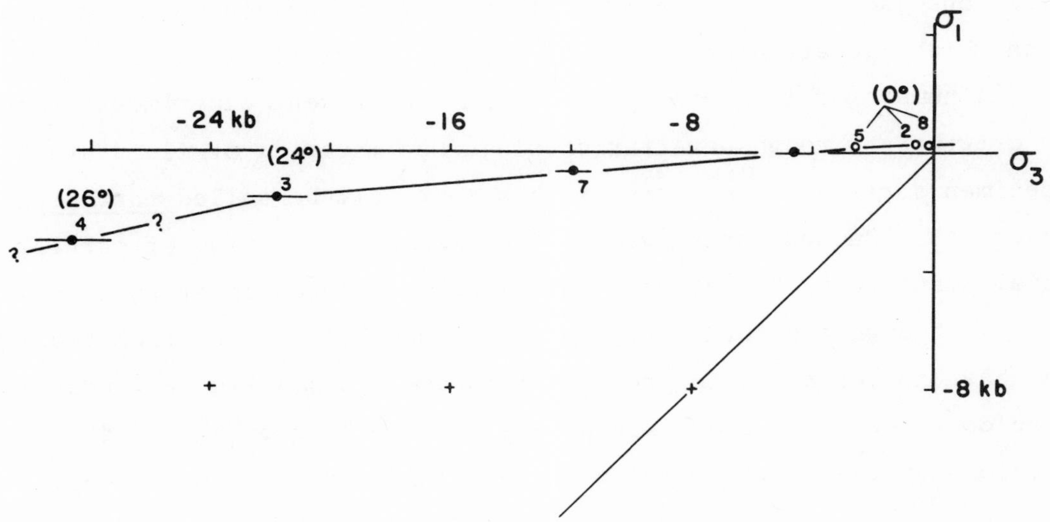

Fig. 16—Failure conditions of quartzite.

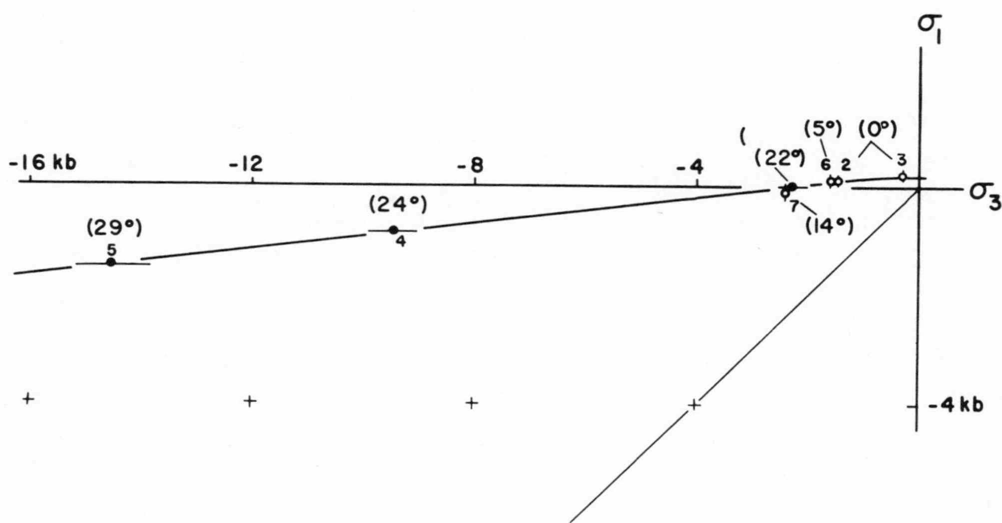

Fig. 17—Failure conditions of granite.

depended on whether a rubber, lead, or copper jacket was used, and on how the glass specimen had been prepared. The fractures normal to the specimen axis, which resembled tension fracture even though all stresses were compressive, were thought to be due to penetration, either of pressure medium or of the jacket material itself. The difference between one jacket and another was thought to be due to relative ease with which jacket material could penetrate surface cracks.[46]

A number of fractures produced in the present experiments appeared to originate through penetration of jacket material or fluid at the specimen surface. Fractures of this sort will be called <u>intrusion fractures</u>. One was preserved in the diabase in a stage of partial development (Fig. 7). Here, a crack has formed which extends only partly through the throat region of the sample. Copper from the jacket has been forced by fluid pressure into the crack. Had the experiment been continued, the crack would have presumably been extended across the specimen by the wedging action of the fluid and the ductile copper jacket.

Intrusion fracturing only occurred with the compact rocks. A series of experiments was performed with one of these, the diabase, to determine the effect of using different jacket materials, including one which was not very ductile. Diabase samples 7', 5, 4, and 6 were all fractured in extension at a pressure close to 2.5 kb (Table 4; Fig. 15). Sample 7' was unjacketed; sample 5 had a 0.0005-in. stainless steel, type 302, annealed jacket; sample 4 had a 0.002-in. annealed copper jacket; and sample 6 had a 0.002-in. stainless steel, type 302, full hard jacket. The stress difference required to rupture these samples varied from about 1 kb to 3 kb. A more significant difference in behavior, however, was the mode of fracture. Both copper and thin stainless steel jackets appeared to penetrate the diabase; fractures normal to the specimen axis were produced. The thick stainless steel jacket did not penetrate the rock, and it failed in a shear fracture. Although the shear fracture, or fault, had an inclination of only 6 degrees, it was clearly a shear fracture because of the shear offset left imprinted on the jacket.

In a number of other experiments in which faults rather than extension fractures were produced in extension tests, it was found that the fractures often appeared to start at the surface, as in the sample shown in Fig. 7, whereas faults formed anywhere in the sample.

Pressure Dependence of Strength in Compression and Extension

Data from compression and extension tests have been plotted in Figs. 13 through 17. Features common to the behavior of all five rocks include the following. First, a smooth curve can be fitted to points representing stresses at fracture, for a given rock, if one excludes tests in which intrusion fracturing was observed. Points representing intrusion fracturing fall widely off the trend of the other data. Second, for the silicate rocks and probably also for the Webatuck dolomite, curves through the data points for the confined compression tests approach straight lines. Third, fracture has occurred at about the same stresses in compression and extension experiments. Thus, for example, the data point for test 7 of the granite (Fig. 17) nearly coincides with the point representing the room pressure compression test. The extreme principal stresses were nearly the same in these two tests, although the intermediate principal stress was different; in extension it had the value of the maximum compression, in compression it had the value of the minimum compression. The coincidence of the two results for granite and for Webatuck dolomite suggests that the value of the intermediate principal stress has little influence in fracture of these rocks. Third, the angle of inclination of fracture, θ, increases monotonically and gradually from 0 degree near the σ_1 axis to around 30 degrees for compression tests under relatively high confining pressure. (See also Figs. 11 and 12.) Finally, the tendency of data points from extension tests to fall irregularly off the best smooth curve reflects the rather high probable error (20 per cent) of data from these tests. Projection of the curves to the σ_1 axis gives an intercept which is the conventional tensile strength, T_o.

Evaluation of the New Sample Shape

Several observations suggest that the shape of sample used here yielded somewhat more reliable data than are usually obtained in compression and extension tests. First, petrographic examination of samples deformed to total strains of 1 or 2 per cent (total permanent strain was 10^{-4} to 10^{-3}) showed that deformation such as grain boundary loosening was distributed uniformly throughout a sample. Measurements of permanent change of shape showed no barreling. Second, fractures form without any preference as to location or orientation in the throat region (Figs. 11 and 12); fractures never form in the heads, at the contact with the piston. Third, room pressure compressive strength is more reproducible than in the conventional compression test (Table 2); and, finally, static and dynamic Young's modulus agree for the present test about as well as can be expected for room pressure measurements (Table 2).

There is probably not much point in comparing coefficient of variation of compressive strength measured in two types of tests, without having at least ten measurements from each test. However, some indication of the difference in this variation is given by comparing results from Table 2 with the breaking strengths of four circular cylinders of diameter 0.5 in. and length 1 in. For the same Solenhofen limestone tested in this study, strengths were 2.60, 2.28, 2.37, and 2.14 kb, with an average of 2.35 kb and a scatter of ± 10 per cent. In this study, C_o (Table 2), uncorrected for bending, was 3.04 kb ± 6 per cent. For the Blair dolomite, conventional strengths were 5.34, 4.25, 4.95, and 4.77 kb, with an average of 4.85 kb ± 12 per cent. Using the long cylinders of this study, C_o, uncorrected for bending, was 4.96 kb ± 0.5 per cent. In both cases the scatter was smaller using the new sample shape than using a circular cylinder of length-diameter ratio of 2.

A very significant difference between the two types of compression tests was the way the rock fractured at room pressure. Fractures in the small circular cylinders were largely parallel with the axis of the cylinder. Fractures in the present experiments are almost never parallel with the axis. The reason for this is not known, but presumably it has to do with "end effects." Tensile stress on radial planes can be produced in a number of ways by end constraint in short compression specimens. (See, for example, Refs. 21 and 22.) Radial fractures in the short circular cylinders were presumably caused by this tension; removal of end constraint in the throat region of the long cylinders of this study apparently suppressed radial fracturing in the throat region.

Discussion

Internal and Intrusion Fracturing

To aid in discussion, certain results are summarized diagrammatically (Fig. 18). All fractures that were produced in the samples are divided into two groups. Intrusion fractures formed when pressure medium or jacket material was forced into the rock; intrusion fractures appear to start at the surface of a sample and are invariably perpendicular to the direction of the least compression. Internal fractures form when intrusion of foreign material is prevented; internal fractures form anywhere in the sample. Small fractures have been found which are completely contained in the interior of a sample. The stress required for intrusion fracturing depends upon properties of both rock and intruding substance. The stress required for internal fracturing depends solely on properties of the rock.

A typical failure curve, which is the locus of stress conditions at fracture, is shown for intrusion and internal fracturing (Fig. 18). Coordinates are the extreme principal stresses, here designated P and Q,

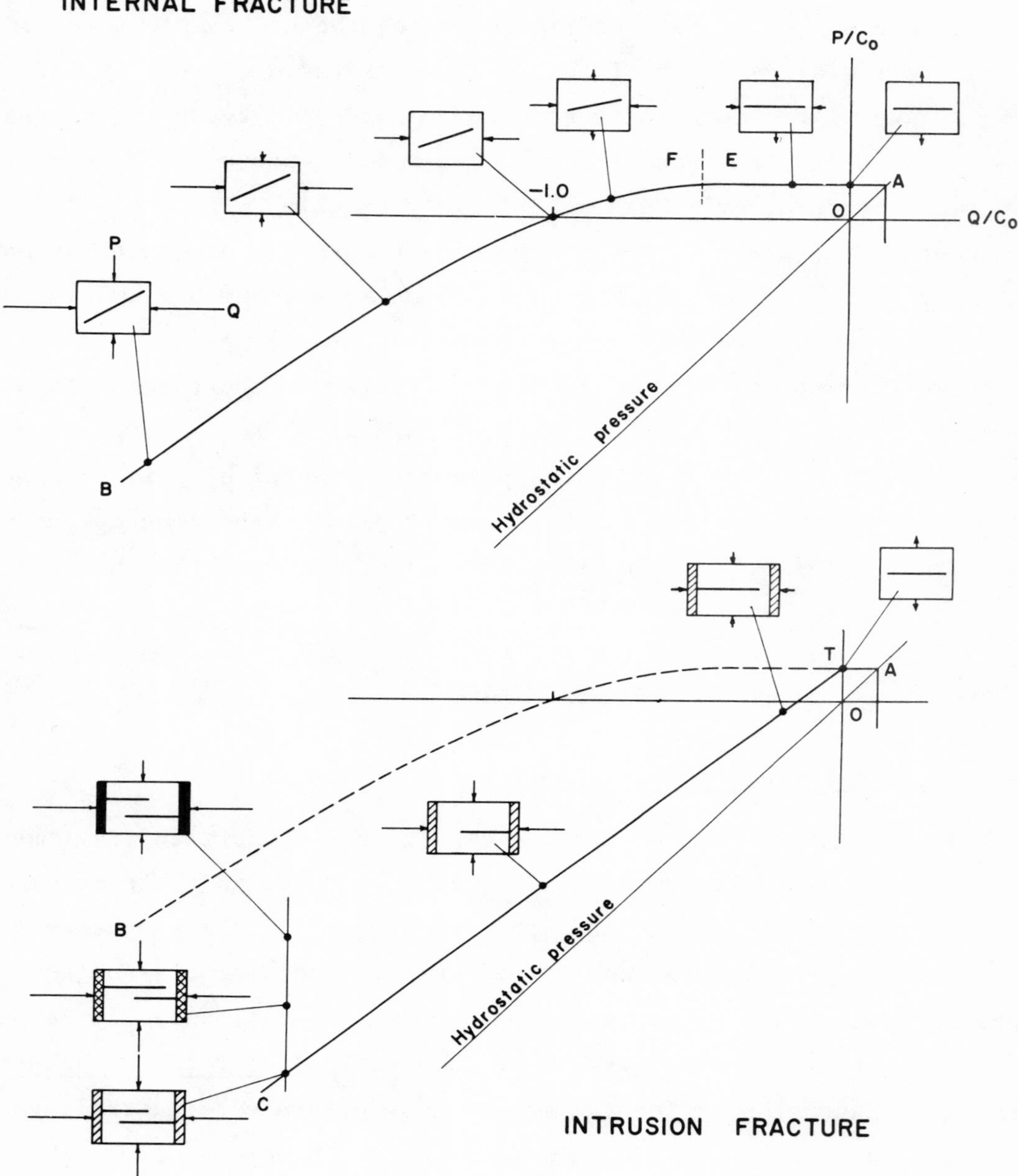

Fig. 18—Diagrammatic sketch of results.

divided by room pressure compressive strength, C_o. Compressive stress is negative, tensile stress positive. A block of homogeneously stressed material is shown at several points along the failure curve (AB in the upper diagram, ATC in the lower) with arrows which represent the failure stresses in their correct relative size. The heavy line within each block shows the approximate orientation of fractures. The vertical dotted line in the upper diagram is the boundary between conditions for _extension fracturing_ and _faulting_. In the former, motion of the two parts of the fracture is normal to the fracture surface; in the latter, parallel with the surface. Both types of fracture form as a result of internal fracturing; only extension fractures form by intrusion fracturing.

In the lower diagram (Fig. 18) stresses at fracture are shown applied to blocks of material through an intermediate foreign substance, drawn crosshatched or full black. For one particular foreign substance—say, rubber or copper—a failure curve such as ATC is traced out. Failure states for other foreign substances will not lie on this curve. In the lower left, failure states for three different foreign substances are shown. As the strength of the foreign substance increases, and therefore its ability to intrude the rock decreases, the failure stresses for the rock approach those for internal fracture. If the material cannot intrude, then the point representing these stresses falls on AB shown dotted in the lower diagram.

A Transition between Faulting and Extension Fracturing

Griggs and Handin(47) summarized previous knowledge of rock fracture which was gained through experiments and which relates to geologic problems. They suggest a two-fold classification of fractures as _faults_ or _extension fractures_, and further suggest a transition between the two as pressure and temperature of experiments increase, and therefore ductility of the rock increases. Their classification is retained here, with somewhat more emphasis on difference in relative motion rather than on difference with respect to stress axes. The transition between the two fracture types observed here differs only in detail from the transition which they proposed.

First, the transition observed in this study probably has nothing to do with ductility. Thus, it takes place for granite, for example, at pressures so low that behavior by any set of standards is wholly brittle. Second, Griggs and Handin's transition between extension fracturing and faulting has an intermediate stage during which extension fracturing is associated with macroscopic wedges of shear. Such wedges were not observed here but Griggs and Handin suggest the wedges could be microscopic, and hence invisible for extremely brittle material. Thus their intermediate stage might not be observed for extremely brittle material. Their analysis of formation of extension fractures contains the assumption inherent in the Griffith theory: that fractures grow, even when macroscopic stress is compressive, due to local tension at the tips of cracks.

A final difference is that their extension fractures form not only when tensile stress is present, but also when stress is wholly compressive. Here, they form, for internal fracturing, only when one component of stress is tensile (out to dotted line between F and E in Fig. 18). For intrusion fracturing, they may form when all stress is compressive. The reason for this difference is probably that Griggs and Handin include both internal and intrusion fractures in their analysis.

Comparison of Observations and Theory

Of the various possibilities only the Mohr and Griffith theories[18,19] might reasonably be expected to apply to the internal fracturing of present experiments. Griffith theory has the great advantage that it is based on a microscopic model which has been found to be approximately correct for other brittle material, such as glass.[16,48] The basis of the Mohr criterion, on the other hand, and of the specialization of it most commonly used, the Navier-Coulomb law,[49] has been seriously questioned,[30,49,50] although the Mohr criterion holds empirically for many rocks. However, although both theories start with different premises, they predict macroscopic behavior which is similar.[10,51]

The Griffith Model of the Fracture Process. Griffith[48] assumed that brittle material contains sharp flaws or cracks which are present before the material is subjected to stress. Fracture is supposed to occur when, due to applied stress, certain of these flaws enlarge and spread through the material. The Griffith theory predicts, for a given material, the stress conditions which will make flaws of a given size enlarge. In a given material, with flaws of constant length, the theory predicts fracture strength for a general state of stress, in terms of tensile strength. The theory does not predict the path a growing crack will take except in simple tension or what will happen when two or more growing cracks approach one another.

Griffith postulated a material, then, which has certain structural characteristics before fracture. It is of interest to see if rocks have these characteristics, before attempting to apply Griffith theory.

Rocks obviously contain many features which might be called flaws: joints, cleavage, parting parallel with bedding planes, faults, and the like. If these larger scale features are eliminated from discussion, features on the scale of the grain size or smaller remain. Certain observations(15) suggest that the largest and therefore the most critical of these are grain boundaries. A number of the microscopic observations made in this study support this idea. Grain boundaries at many sites in a rock became loosened as stress was increased prior to fracture (region III, Fig. 8). At the instant before fracture a rock was filled with loosened sections of grain boundaries which had various lengths and orientation. Growth of cracks out of these sections of grain boundaries apparently led to fracture.

Thus, if sections of grain boundaries are taken as Griffith cracks, the crystalline rocks studied here appear to come fairly close to the ideal material postulated by Griffith, if one considers these rocks at the instant before fracture (the end of region III, Fig. 8).

McClintock and Walsh(31) modified the Griffith theory to include the closing of Griffith cracks and the development of frictional forces across crack surfaces; this is a likely situation for material fractured by systems of stresses that are wholly compressive. Several observations suggest that in compression tests made here grain boundaries, although loosened, are in frictional contact at fracture of the rock. Grain boundaries of the loose rocks may not be in full contact before application of stress. However, the pressure effect on elastic constants suggests, as discussed above, that initial grain boundary openings are largely reduced by application of a few hundred bars of pressure. Hence, grain boundaries must be at least locally in contact above these pressures. For both types of rock, hysteresis in the force-strain curve becomes marked if load is cycled in region III (Fig. 8). One explanation for this hysteresis is that under stress, crack surfaces within a rock slide by one another by a small amount and that motion is partly resisted by frictional forces. Then at a given stress, strain for decreasing load would be greater than strain for increasing load. This is observed, for example, in Figs. 8 and 9.

Thus, indirect evidence suggests that the "Griffith cracks" here make frictional contact in compression tests. The McClintock-Walsh modification of the Griffith theory should therefore predict fracture stresses for rocks more nearly correctly than the original theory, which does not include the effect of frictional forces at crack surfaces.

Tensile Strength and Grain Size Dependence of Strength. A number of quantitative tests can now be made of predictions of Griffith theory. For certain of the rocks, predicted and observed tensile strength can be compared. The two-dimensional theory[52] predicts a tensile strength which will be approximately valid for three dimensions.[16] That is,

$$T_o = \sqrt{\frac{2E\gamma}{\pi c}} \, , \qquad (2)$$

where T_o is tensile strength, E and γ are Young's modulus and specific surface energy of the material, respectively, and c is one-half the length of the Griffith crack. γ has been recently measured for calcite[53] and for quartz and feldspar.[54]

Because the grain boundary flaw enlarges by cutting into new grains, E of mineral rather than E of rock is used. The E used for calcite and quartz is the average of the Voigt and Reuss values for an aggregate.[25] E of feldspar is calculated from measurements of linear compressibility.[55] Measured T_o is found by projecting a smooth curve through data points for extension tests (Figs. 13 through 17) to intersect the σ_1 axis.

T_o for marble, for which these data are not available, was assumed to be equal to $C_o/10$. Crack length, 2c, is taken as the longest straight section of grain boundary. For quartzite 2c is much smaller than average or maximum grain diameter because grain boundaries are quite irregular. Grain boundaries of the other rocks are quite straight so that 2c is set equal to maximum grain diameter.

Calculated and observed T_o are compared in Table 5. Agreement is no better than about a factor of 2. This is partly the influence of large uncertainties[(15)] in E and γ. Also, it is hard to know exactly what to use for maximum grain diameter, as grain size invariably has a rather broad frequency distribution with an upper limit that varies from sample to sample.

Table 5

TENSILE STRENGTH

Rock	γ (erg/cm^2)	E (mb)	c (10^{-3} mm)	T_o (bars) Calculated	Observed
Marble	230	0.85	480	72	54
Quartzite	500	1.03	75	300	280
Granite	8000	1.2	750	400	210
Diabase	8000	1.2	700	420	400

Uncertainties in E and γ are avoided by applying Eq. (2) to suites of rocks of the same composition. Fracture stress should be proportional to inverse square root of maximum grain diameter, if Griffith cracks are assumed to be sections of grain boundaries. This relation held approximately for indentation hardness of quartzites and diabases[(15)] but not for calcite marbles and dolomites. The present data for carbonate rocks are plotted in Fig. 19. The slopes of -0.51 for calcite marble and -0.56 for dolomite are in good agreement with a predicted slope of -0.50. The lack of agreement for the hardness measurements[(15)] was probably due to plastic flow of these materials during indentation.

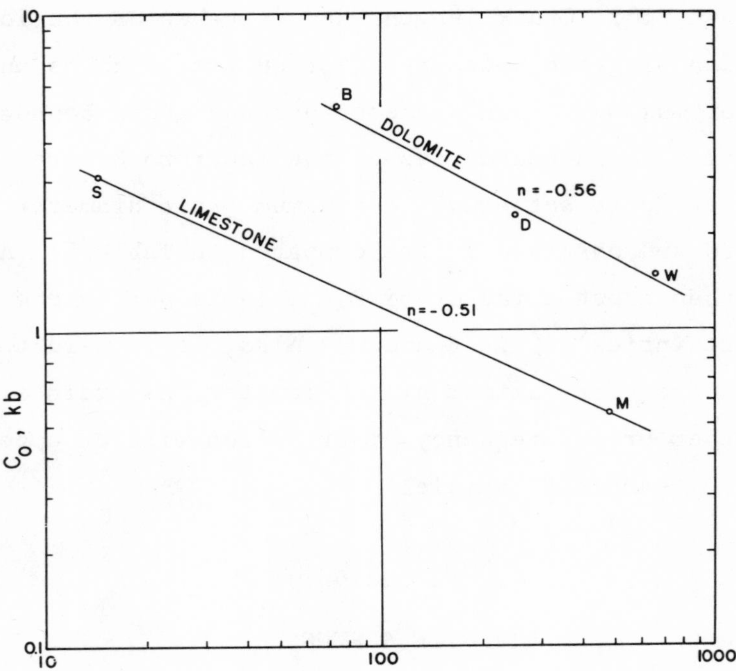

Fig. 19—Plot of C_o versus maximum grain size.

Pressure Dependence of Internal Fracture Stress. Prediction and observation are most easily compared in a plot of σ_1/C_o versus σ_3/C_o, where these are the extreme principal stresses divided by room pressure compressive strength (Fig. 20). In this plot the normal Griffith theory predicts fracture of a material with strength C_o when stresses reach the curve marked GT. For example, failure under simple tension occurs at a point such as a; failure under simple compression occurs at (-1.0, 0). In the present experiments, stress states to the left of the line OB can be reached. For present sample dimensions (Fig. 3), OB has a slope of -6 in Fig. 20.

In the McClintock-Walsh modification of the Griffith theory,[31] failure occurs at the following condition:

$$\mu(\sigma_3 + \sigma_1 - 2\sigma_c) + (\sigma_1 - \sigma_3)(1 + \mu^2)^{\frac{1}{2}} = 4T_o(1 - \sigma_c/T_o)^{\frac{1}{2}} , \qquad (3)$$

where σ_1 and σ_3 are the extreme principal stresses, T_o is tensile strength, μ is the coefficient of friction at crack surfaces, and σ_c is the stress

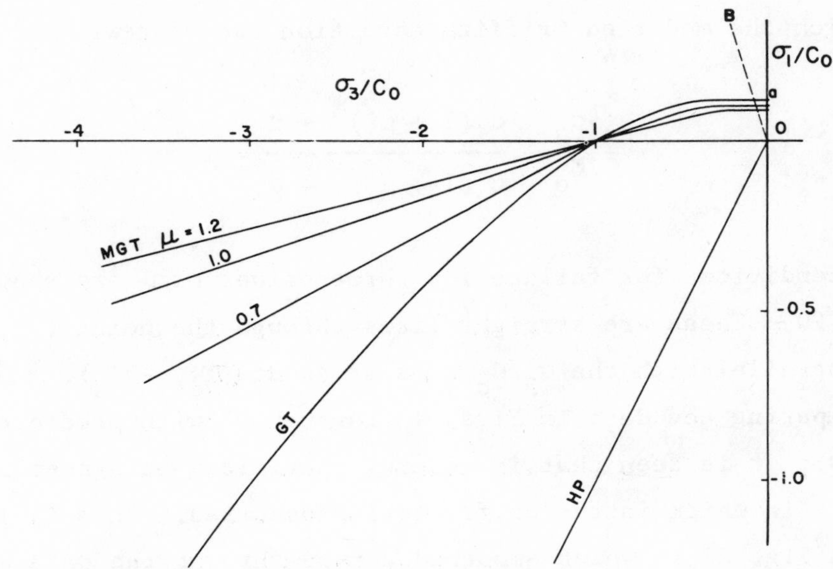

Fig. 20—Predicted failure conditions.

normal to the crack required to close it. If the Griffith cracks are taken to be segments of grain boundaries, then these are probably in contact for both compact and loose rocks for all of the compression tests carried out under confining pressure. A rough measure of σ_c for the loose rocks is given by the pressure required to eliminate the initial pressure effect for Young's modulus (Fig. 1). This amounts to less than -2 times the tensile strength of the rocks, and is small enough that σ_c can probably be set equal to zero in the above equation.

With σ_c equal to zero, the modified Griffith condition becomes

$$\mu(\sigma_3 + \sigma_1) + (\sigma_1 - \sigma_3)(1 + \mu^2)^{\frac{1}{2}} = 4T_o \ . \tag{4}$$

The strength in simple compression is C_o. Setting $C_o = -\sigma_3$, and $\sigma_1 = 0$, the ratio of compressive to tensile strength is

$$\frac{C_o}{T_o} = \frac{4}{(1 + \mu^2)^{\frac{1}{2}} - \mu} \ , \tag{5}$$

from which the modified Griffith condition can be rewitten as

$$-\frac{\sigma_3}{C_o} + \frac{\sigma_1(1+\mu^2)^{\frac{1}{2}} + \mu}{C_o(1+\mu^2)^{\frac{1}{2}} - \mu} = 1 \ . \qquad (6)$$

Stress conditions for failure for three values of μ are shown as MGT in Fig. 20. These are straight lines through the point (-1.0, 0). They become parallel with the σ_3/C_o axis at about $(T_o, -3T_o)$.

Comparing the data in Figs. 13 through 17 with predicted behavior (Fig. 20), it is seen that in general there is good agreement, except for tests in which intrusion fracturing occurred. This is further compared in Fig. 21 in which smoothed curves through the data are replotted in principal stress coordinates normalized by dividing by C_o. Only data from compression tests are considered; further consideration of data from extension tests seems unjustified until some of the large probable error in these tests can be removed. The curves for the three silicate rocks, which are very nearly straight lines, coincide with the modified Griffith conditions when μ has a value of 0.9 to 1.5. Curves for both dolomites are concave downward; the curve for Webatuck dolomite coincided with the modified Griffith condition at $\mu = 0.9$ for about one-half its length. The curve for the Blair is strongly concave throughout.

Very few measurements have been made of coefficient of friction of rocks and minerals; therefore, it is not certain whether a value of 0.7 to 1.5 is a reasonable one. Also it is not certain whether the coefficient to be used in the modified Griffith theory is that which pertains to (a) large bodies in apparent contact or (b) small bodies making nearly complete contact on a microscopic scale. The former is measured when two pieces of rock, such as the fractured cylinders in Jaeger's experiments,(56) slide by one another; the coefficient of friction here ranges from about 0.6 to 1.2. (See also Refs. 3 and 57.)

Several observations suggest that the coefficient of friction to be used here is that for large bodies, although in fact the frictional forces of interest probably form between pairs of grains. First of all, contact between grains, just before fracture, is probably not complete on a

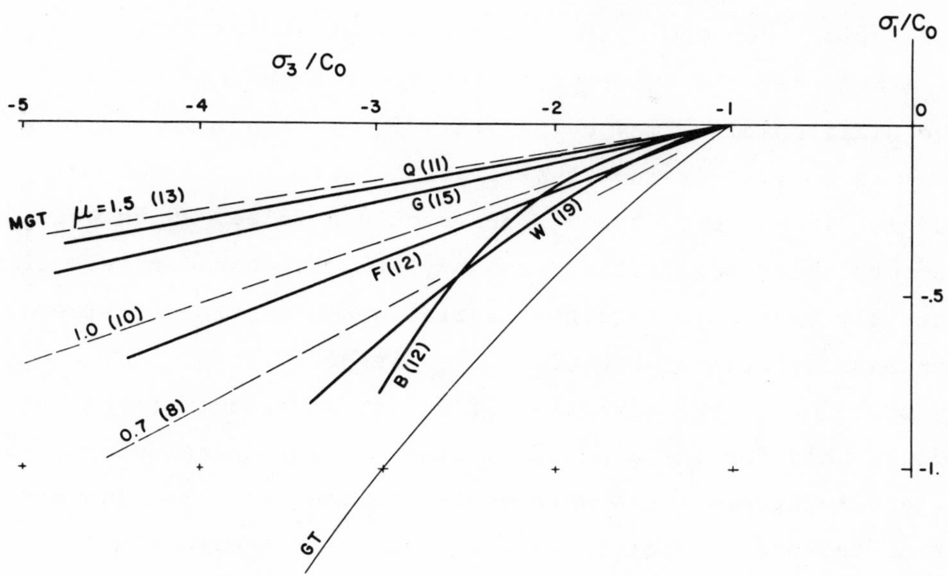

Fig. 21—Observed failure conditions.

microscopic scale. A minute opening up of grain boundaries is indicated by the microscopic observations noted above as characterizing region III (Fig. 8). Also, the rapid increase of volume of the samples just prior to fracture is most likely due to enlargement of grain boundary openings. Therefore, contact across "Griffith cracks" is probably not complete just prior to fracture, and the situation probably resembles that at the contact of large bodies.[58] A value of around 1.0 is thus probably not unreasonable in the modified Griffith theory.

Again, agreement is fairly good between observed fracture stresses in compression and failure conditions predicted by the McClintock-Walsh modification of the Griffith theory. Qualitative agreement was suggested previously[30,31] on the basis of data taken from the literature.

Theoretical ratio of room pressure compressive strength to tensile strength (Eq. (5)) is compared with observed; this ratio is given in parentheses at each of the curves in Fig. 21. Agreement is rather poor, due in part to the uncertainty in the measured value of T_o.

Disagreement between observation and prediction for the dolomites may be due to several causes. Gliding may be taking place, particularly in the tests at two or more kb confining pressure, although none was detected in the samples. The two curves (Fig. 21) are bending in the same direction they would if gliding flow of grains were taking place. Another cause for the discrepancy between observation and prediction for the Blair dolomite may be the relatively high impurity content. The rock is so fine-grained that it is hard to locate this material, but judging from one or two of Robertson's samples, which were even more impure than usual, it is probably at grain boundaries. If this is generally true, the foreign material would exert a considerable influence on frictional behavior at grain boundaries. The probable effect would be in the direction of reducing the coefficient of friction that would hold for the contact of clean grains, particularly at the higher confining pressures when grains in the rock must support greater stress differences. This possibility is in line with the trend of the curve for the Blair dolomite in Fig. 21.

Previous Tests of Griffith Theory. Clausing,[51] Murrell,[10] and Odé[59] first pointed out the similarity of the Griffith theory to the Mohr criterion: Griffith theory predicts a parabolic Mohr envelope for fracture. When Griffith theory was later modified to include closing of cracks in compression,[31] the prediction turned out to be identical to the Navier-Coulomb law in the field of compression.[30] This immediately suggested that the Griffith criterion was approximately valid for rocks.

However, when examined in detail, the failure criterion of the original Griffith theory was not found to hold.[31] Murrell did find that a number of rocks have quasi-parabolic Mohr envelopes[60,61] (in agreement with the normal Griffith theory), but detailed studies[4,13,17] have shown that rocks of this kind are typically quite ductile and often flow before fracturing. Typically brittle rocks such as granite and quartzite (for which the Griffith criterion is most likely valid) have a straight Mohr envelope in compression, in disagreement with the normal Griffith criterion.

Murrell[61] recently extended the Griffith theory to three-dimensional stress states. By analogy with the Griffith theory in two dimensions, the failure surface in principal stress space was assumed to have the form of a paraboloid in the field of compression.

This gives sections in two dimensions which are approximately, but not exactly, the same as the Griffith criterion. The same disagreement exists, therefore, as noted above for behavior of typically brittle rocks. In addition, Murrell's theory predicts that room pressure strength in the compression test (minimum and intermediate principal compressions equal) will be one-half that in the extension test (maximum and intermediate compressions equal). A difference is often observed, but this is of the order of 10 per cent,[6,13,62] and in preliminary tests here, the difference was zero.

Comparison has occasionally been made of the angle of fracture with the angle of the critical Griffith crack which is given by the theory.[7,51] This is pointless, however, for stress fields other than purely tensile. A crack in a field of compression does not propagate in its own plane[63] and therefore cannot form a macroscopic fracture having the same angle of inclination.

Robertson[13] determined the failure stresses of Solenhofen limestone in four types of tests: three-point bending, crushing of hollow and solid cylinders, and punching of disks. The constant, T_o, computed for these stresses agreed for all but the punching tests to within about 10 per cent. This is probably within the limits with which the failure stresses were known in the bending and hollow cylinder tests. There is very poor agreement with T_o computed for the punching tests, for unknown reasons; agreement is also poor for the solid cylinder tests carried out at high confining pressure, due probably to gliding flow before fracture (the samples are strained up to 7 per cent before fracture). Heard[6] found that the Griffith theory failed to correlate results for compression and extension of Solenhofen limestone under confining pressures up to 5 kb and temperatures up to 500°C. However, agreement of calculated T_o at room temperature is within 15 per cent. This is in line with the difference in ultimate strength in compression and extension for Carrara marble[62] and concrete.[64] The magnitude of the difference for material like quartzite has never been determined outside of this study.

Handin and Hager[4] found that ultimate maximum shear stress and the mean stress have an approximately linear relationship for a wide variety of rocks at failure. The slope ranges from 0.7 for limestones to 1.0 for the very brittle rocks such as quartzite. It is of interest to compare this with prediction of the modified Griffith theory (chosen rather than the normal Griffith theory because tests were carried out at high confining pressure).

Calling the maximum shearing stress at failure τ_m, then

$$\tau_m = \frac{\sigma_3 - \sigma_1}{2} ,$$

and the mean stress, σ_m, is

$$\sigma_m = \frac{\sigma_1 + \sigma_2 + \sigma_3}{3} .$$

Substituting in Eq. (4), we have

$$\sigma_m + \tau_m \left(\frac{2(1 + \mu^2)^{\frac{1}{2}}}{3\mu}\right) = \frac{4T_o}{3\mu} + \frac{\sigma_1}{3} . \qquad (7)$$

Under high confining pressure, the terms on the right will be very small. The equation gives, then, a straight line which nearly goes through the origin. The slope ranges from 0.8 to 1.0 for a coefficient of friction, μ, of 0.6 to 1.0. This is a reasonable value of μ here, so that the prediction of the modified Griffith theory is in good agreement with Handin and Hager's results. Robertson[13] also found that fracture stresses, when plotted as τ_m versus σ_m, showed a linear relationship with a slope of 1.0, again in good agreement with the modified Griffith theory.

Internal fracturing of this study probably corresponds with the "pinch-off effect" of Bridgman.[41] This behavior was regarded for some time as a refutation of stress theories of fracture in general and of Griffith theory in particular. However, Gurney and Rowe[65,66] by experiments with glass showed that this behavior is quantitatively predictable if one considers the effect of intrusion of the high-pressure fluid into surface cracks. A predicted failure curve for intrusion fracturing would be TC in Fig. 18; TC is nearly parallel with the line representing hydrostatic pressure in this plot, and at a distance about equal to the tensile strength, T_o, away from it. This is approximately what was observed here.

Areas of Uncertainty in Application of Griffith Theory. Several observations appear to disagree with either the microscopic model postulated in the Griffith theory, or with the actual fracture criterion derived from the theory. There are also several fairly fundamental questions regarding derivation of the fracture criterion.

Use of elasticity theory for the extremely high stresses near a crack tip has been questioned by Bridgman.[(45)] Gurney[(67)] partially disposed of this objection by showing that the average stress concentration over finite distances the order of the atomic spacing in natural material is still very close to the theoretical stress concentration at a point.

Another uncertain procedure is the use of isotropic elastic equations to derive the fracture criterion for a general two-dimensional stress field, as done in Griffith's second paper.[(52)] In the derivation one goes, in effect, from the crack tip to the stresses at the boundary of the material, and traverses material which, in the case of rock, may be elastically anisotropic. Many rocks, though initially isotropic, become decidedly anisotropic under stress.[(9,68)] This is apparent from the nonlinear stress-strain curve, such as that shown in Fig. 8, which is typical of many rocks.[(26,27)] This could cause marked departure from theoretical behavior. In the present study this is being evaluated by comparing behavior of rocks that do not become anisotropic under stress (such as the diabase) with the behavior of those that do (such as the granite).

Many investigators report a dependence of failure stress on the intermediate principal stress.[*] Griffith did not treat three-dimensional stress states, but consideration of three-dimensional cracks[(31,69)] suggests that the influence of σ_2 is negligible. Provisional results here

---

[*]See Refs. 6, 32, 62, and 64.

support this view. Disagreement of the other observations could be explained by the effect of gliding flow (two crystalline limestones show this effect: the Carrara marble[62] and the Solenhofen limestone[6]), by systematic errors in determination of stress caused by end effects, or by anisotropy of starting material.[32]

It has been recently shown experimentally that a critical Griffith crack, for other than purely tensile stress fields, does not propagate in its own plane.[63] It grows in a direction approaching the direction of the maximum compression, and for constant applied stress, stops propagating after a distance of a few crack lengths. A macroscopic shear fracture or fault apparently grows, in a manner not yet clearly understood, from *en échelon* sets of such cracks. As Griffith theory gives stresses at the instant a critical crack starts to grow, and if actual cracks in compression grow but then stop, it could be that the stress to cause macroscopic fracture is different from the stress to start crack growth. In this case, Griffith theory would not predict compressive fracture strength and could not serve as the basis of a theory of strength in compression. This possibility is currently being investigated along with the relation of the angle of critical Griffith crack to the angle of the macroscopic shear fracture.

Mohr Failure Criterion. It is of interest to put typical data from this study in the form of a Mohr envelope for comparison with the Mohr theory of failure. Data for the Frederick diabase are shown in Fig. 22. Mohr circles have been drawn for fracture stresses from compression and extension tests and an envelope constructed which comes closest to being tangent to all the circles. The angle of inclination of fractures, θ, as predicted by the Mohr strength theory for this envelope, is compared with the observed angle.

The envelope is approximately parabolic out to its intersection with the circle representing the room pressure compressive strength, C_o. Beyond that it is a straight line inclined to the σ axis at 40 degrees. The observed angles of fractures for tests in compression agree with predicted angles according to Mohr strength theory to within 4 degrees. Angles for extension tests differ by large amounts from predicted angles, but do agree in trend. That is, both observed and

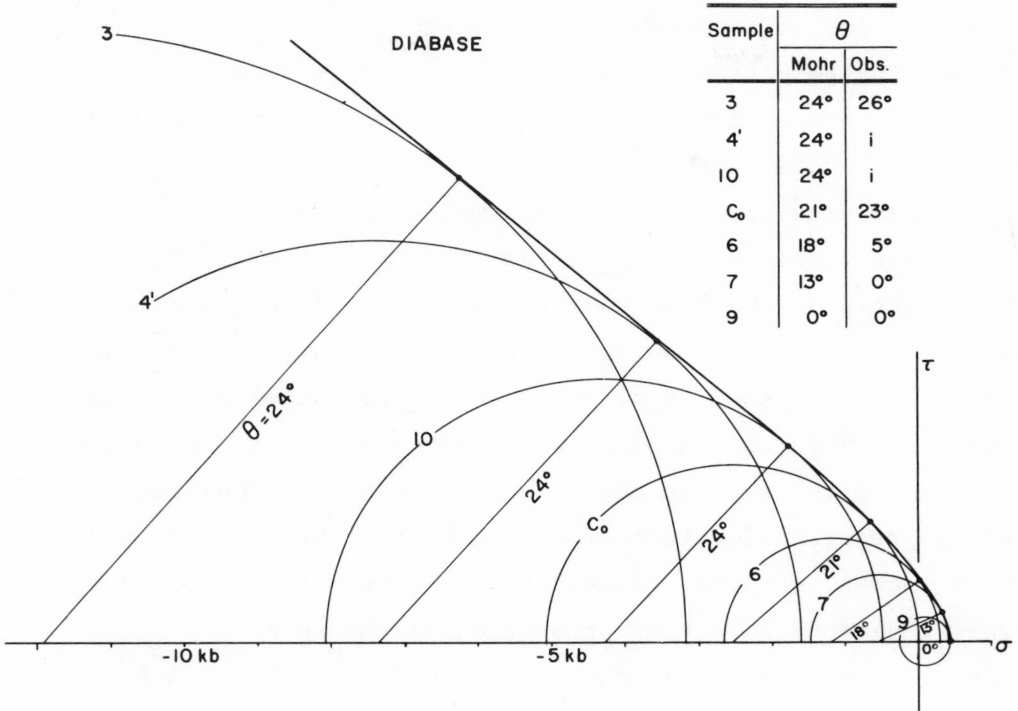

Sample	θ	
	Mohr	Obs.
3	24°	26°
4'	24°	i
10	24°	i
C_0	21°	23°
6	18°	5°
7	13°	0°
9	0°	0°

Fig. 22—Mohr envelope of diabase.

predicted angles decrease monotonically from about 20 degrees to 0 degree as maximum stress difference at fracture decreases.

The envelope to the left of the origin is approximately a straight line, in agreement with experimental results of many workers for brittle rocks, and the same as that given by the McClintock-Walsh modification of the Griffith theory.(30) The approximately parabolic portion near the origin agrees with general theoretical predictions made for brittle material.*

Conclusions

A quartzite, a granite, a diabase, and two dolomites were fractured in tests designed to permit accurate determination of stress. The first three rocks remained brittle on a microscopic scale in all tests, whereas the dolomites may have undergone gliding flow, or viscous flow at

*See Refs. 60, 61, and 70-72.

grain boundaries in some tests. A cylindrical sample with a reduced center section permitted, in a single type of test, stresses to be applied which ranged from nearly uniaxial tension to tension or compression under confining pressure.

The fractures produced in the samples fall into two groups: intrusion fractures and internal fractures. Intrusion fractures, which produce both the longitudinal splitting of the uniaxial compression test and the "pinch-off" fracturing in extension tests, form when material such as pressure medium or jacket penetrates surface cracks. This only happens when such material is in contact with the surface of the rock in the plane of the maximum compressive stress. The stress required for intrusion fracturing depends on properties of rock and jacket, or pressure medium; the stress required for internal fracturing depends solely on properties of the rock.

Continuous transition was observed between extension fracturing and faulting as applied stress was varied from tensile to compressive. The angle of inclination of fractures to the direction of σ_3 varies gradually from 0 degree to about 30 degrees. Thus, faulting and extension fracturing may not be two discrete types of fracture in rocks, as is widely believed, but may be simply members of a continuous series.

There is good qualitative agreement between fracture stress of the granite, quartzite, and diabase and prediction of the McClintock-Walsh modification of the Griffith theory, taking the coefficient of friction to be 0.9 to 1.5. Absolute value of tensile strength of calcite marble, quartzite, granite, and diabase agreed within better than a factor of two with tensile strength predicted by Griffith theory. In the calculation, Griffith crack length was taken as maximum grain diameter. Support for this assumption was given by microscopic observations of partially fractured material. Fractures appear to start at grain boundaries, and "Griffith cracks" are probably more or less straight sections of grain boundaries. Additional indirect support was provided by the observed relation between maximum grain diameter and fracture stress for suites of limestone and dolomite; the relationship agreed closely with predictions of Griffith theory.

Excluding the dolomites, agreement of observations with predictions of Griffith theory of fracture is therefore fairly good. Experimentally, the dependence of fracture stress on σ_2 needs to be studied in more detail for these brittle rocks; preliminary results show negligible dependence. Other published results of experimental rock deformation show approximate agreement with Griffith theory, when those experiments are selected for which it is reasonably certain that the only microscopic process of deformation was brittle fracture.

At present, the biggest areas of uncertainty in application of Griffith theory to rocks are (1) the validity of using isotropic elastic theory (necessary for derivation of the general failure criterion) for rocks at small confining pressure and (2) the possibility that the fracture stress in compression is not the same as the stress necessary to cause cracks to grow (which is the stress given by the Griffith theory).

In the Mohr diagram, results of these tests give an envelope for fracture which is straight in approximately the region of compression and very nearly parabolic at the intersection with the σ axis. The fracture angle predicted by the Mohr theory from this envelope does not agree with the observed angle, particularly near the origin of the Mohr diagram.

Acknowledgments

The author is indebted to the Committee on Experimental Geology and Geophysics of Harvard University for his appointment as research fellow in geophysics, and to Prof. F. Birch for unrestricted use of the Dunbar Laboratory. The tedious preparation of samples, the quality of which influenced the entire course of the study, was done by Mr. Arthur Ames. Professor Birch was a constant source of ideas and criticism, and Mr. Harold Ames' aid in design and construction of the apparatus was invaluable. The design of the sample grew out of conversations with Prof. F. A. McClintock and members of Foster-Miller Associates. Dr. J. B. Walsh made a number of key suggestions during the course of the work. Drs. E. C. Robertson and John Handin read an early version of the manuscript and offered suggestions which led to important changes in the text. The manuscript was prepared while the writer was a Guggenheim fellow and guest of the Geological Institute of the University of Vienna.

Appendix A

The amount of bending in an elastic cylinder can be found if axial strains are measured at three or more points. In Fig. 23 three gauges are mounted 120 degrees apart at points a, b, and c on the surface of a cylinder shown in section normal to the axis. The cylinder is bent about axis 00 so that the extreme axial strains, e_1 and e_3, are at 90 degrees to this axis, at d and f. Assuming that the strain due to bending varies linearly across the section, average axial strain is

$$\bar{e} = \frac{e_1 + e_3}{2} .$$

The maximum axial strain due to bending is

$$e' = \frac{e_1 - e_3}{2} .$$

In the present problem axial strain is measured at three points, giving values e_a, e_b, and e_c. The quantities e' and $\bar{e}$ are to be found from these.

The axial strain at a is obviously

$$e_a = \frac{e_1 + e_3}{2} + \frac{e_1 - e_3}{2} \cos \alpha ,$$

in which α is the angle between the radii to a and d. Similar expressions for the strains at b and c are

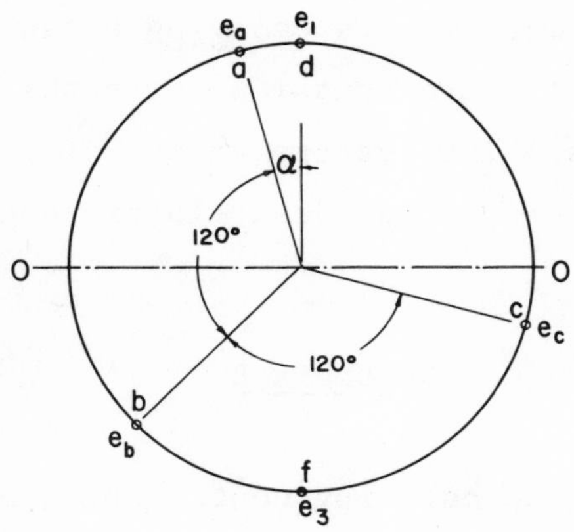

Fig. 23—Bending of a cylinder.

$$e_b = \frac{e_1 + e_3}{2} + \frac{e_1 - e_3}{2} \cos(\alpha + 120^\circ) ,$$

$$e_c = \frac{e_1 + e_3}{2} + \frac{e_1 - e_3}{2} \cos(\alpha + 240^\circ) .$$

But these equations are analogous to those giving the strain in an equiangular strain rosette[73] for which the desired quantities are

$$\bar{e} = \frac{(e_a + e_b + e_c)}{3} ,$$

$$e' = \frac{2}{3} \{(e_a - e_b)^2 + (e_a - e_c)^2 + (e_b - e_c)^2\}^{\frac{1}{2}} . \qquad \text{(A-1)}$$

By similar analysis of an array of four axial strain gauges giving strains e_a, e_b, e_c, and e_d at points 90 degrees apart, in which the equations are analogous to those of a 45-degree strain rosette, one obtains

$$\bar{e} = \frac{e_a + e_c}{2} = \frac{e_b + e_d}{2} ,$$

$$\text{(A-2)}$$

$$e' = \frac{1}{2} \{(e_a - e_c)^2 + (e_b - e_d)^2\}^{\frac{1}{2}} .$$

To determine whether a material is behaving as an ideal elastic body, one can use the first of these equations with the measured strains from four strain gauges. The average strain of diametrically opposed gauge pairs should agree for the ideal elastic body.

Appendix B

To determine the load carrying contribution of the jacket in the extension tests it is assumed that jacket and rock remain in contact and do not slide by one another. Then the axial strains in each will be the same, and the sum of the loads carried by each equals the total load applied to the rock. These relations give the following expression for a correction factor to be applied to the axial stress in the rock:

$$\frac{\sigma}{\sigma_a} = \frac{1}{1 + E_j A_j / E_r A_r} , \qquad \text{(B-1)}$$

where σ is the true stress in the jacketed rock, σ_a is the stress figure as though there were no jacket, E_r and E_j are Young's modulus of rock and jacket, and A_r and A_j are cross-sectional areas of rock and jacket, respectively. The ratio A_j/A_r here was 0.04 for steel and 0.02 for copper, and using for steel E_j = 1.93 mb and for copper E_j = 1.25 mb, correction factors can be obtained corresponding to values of Young's modulus at strains near the points of fracture (Table 6).

Table 6

CORRECTION FACTORS FOR METAL JACKETS

Rock	E_r (mb)	Correction factor		
		0.002-in. steel	0.0005-in. steel	0.002-in. copper
Diabase	0.98	0.928	0.980	0.975
Blair	1.07	0.933	-	-
Quartzite	0.48	0.860	-	-
Granite	0.39	0.835	-	-
Webatuck	0.94	-	-	0.974

REFERENCES

1. Bredthauer, R. O., "Strength Characteristics of Rock Samples under Hydrostatic Pressure," Trans. ASME, Vol. 79, 1957, pp. 695-708.
2. Dreyer, W., "Über das Festigkeitsverhalten sehr verschiedenartiger Gesteine," Bergbauwissenschaften, Vol. 2, No. 7, 1955, pp. 183-191.
3. Griggs, D. T., F. J. Turner, and H. C. Heard, "Deformation of Rocks at 500° to 800°C," Geol. Soc. Am. Mem. 79, 1960, pp. 39-104.
4. Handin, J., and R. V. Hager, Jr., "Experimental Deformation of Sedimentary Rocks under Confining Pressure: Tests at Room Temperature on Dry Samples," Bull. Am. Assoc. Petrol. Geologists, Vol. 41, 1957, pp. 1-50.
5. Handin, J., and R. V. Hager, Jr., "Experimental Deformation of Sedimentary Rocks under Confining Pressure: Tests at High Temperature," Bull. Am. Assoc. Petrol. Geologists, Vol. 42, 1958, pp. 2892-2934.
6. Heard, H. C., "Transition from Brittle Fracture to Ductile Flow in Solenhofen Limestone as a Function of Temperature, Confining Pressure and Interstitial Fluid Pressure," Geol. Soc. Am. Mem. 79, 1960, pp. 193-226.
7. Hobbs, D. W., "The Strength and Stress-Strain Characteristics of Oakdale Coal under Triaxial Compression," Geol. Mag., Vol. 97, 1960, pp. 422-435.
8. Jaeger, J. C., "Punching Tests on Disks of Rock under Hydrostatic Pressure," J. Geophys. Res., Vol. 67, 1962, pp. 369-373.
9. Matsushima, S., "On the Flow and Fracture of Igneous Rocks," Disaster Prevent. Res. Inst., Kyoto Univ., Bull., No. 36, 1960, pp. 1-9.
10. Murrell, S. A. F., "The Strength of Coal under Triaxial Compression," in Mechanical Properties of Nonmetallic Brittle Solids, W. H. Walton (ed.), Interscience, New York, 1958, pp. 123-145.
11. Paterson, M. S., "Experimental Deformation and Faulting in Wombeyan Marble," Bull. Geol. Soc. Am., Vol. 69, 1958, pp. 465-476.
12. Price, N. J., "The Strength of Coal-measure Rocks in Triaxial Compression," Natl. Coal Bd. MRE Rpt. No. 2159, May, 1960.

13. Robertson, E. C., "Experimental Study of the Strength of Rocks," Bull. Geol. Soc. Am., Vol. 66, 1955, pp. 1275-1314.

14. Robinson, L. H., "The Mechanics of Rock Failure," Quart. Colo. School Mines, Vol. 54, 1959, pp. 177-200.

15. Brace, W. F., "Dependence of Fracture Strength of Rocks on Grain Size," Penn. State Univ. Mineral Ind. Expt. Sta. Bull., No. 76, 1961, pp. 99-103.

16. Orowan, E., "Fracture and Strength of Solids," Rpt. Progr. Phys., Vol. 12, 1949, pp. 185-232.

17. von Kármán, T., "Festigsversuche under allseitigem Druck," Z. Ver. Deut. Ing., Vol. 55, 1911, pp. 1749-1757.

18. Jaeger, J. C., Elasticity, Fracture and Flow, John Wiley & Sons, Inc., New York, 1962.

19. Nadai, A., Theory of Flow and Fracture of Solids, McGraw-Hill Book Company, Inc., New York, 1950.

20. Bauschinger, J., "Experimentelle Untersuchunger über die Gesetzen der Druckfestigkeit," Mitt. aus dem Mech.-techn. Lab. München, No. 7, 1876, p. 188.

21. Filon, L. N. G., "On the Elastic Equilibrium of Circular Cylinders under Certain Practical Systems of Loads," Phil. Trans. Roy. Soc. London, A, Vol. 198, 1902, pp. 147-233.

22. Lisowski, A., "Failure of Rock Cubic Specimens in the Light of the Theory of Elasticity," Acad. Polonaise Sci. Bull., Ser. Sci. Tech., Vol. 7, No. 5, 1959, pp. 341-351.

23. Salmassy, O. K., W. H. Duckworth, and A. D. Schwope, Behavior of Brittle-state Materials, U.S. Air Force WADC Tech. Rpt. No. 53-50, Part I, 1955.

24. Seldenrath, T. R., and J. Gramberg, "Stress-Strain Relations and Breakage of Rocks," in Mechanical Properties of Nonmetallic Brittle Solids, W. H. Walton (ed.), Interscience, New York, 1958, pp. 79-102.

25. Birch, F., "The Velocity of Compressional Waves in Rocks to 10 Kilobars, Part II," J. Geophys. Res., Vol. 66, 1961, pp. 2199-2224.

26. Hast, Nils, "The Measurement of Rock Pressure in Mines," Sveriges Geol. Undersokn. Årsbok, Vol. 52, 1958, pp. 1-183.

27. Nishihara, M., "Stress-Strain Relation of Rocks," Doshisha Eng. Rev., Vol. 8, No. 2, 1957, pp. 32-55.

28. Reynolds, H. R., Rock Mechanics, Crosby, Lockwood & Son Ltd., London, 1961.

29. Zisman, W. A., "Young's Modulus and Poisson's Ratio with Reference to Geophysical Application," Proc. Natl. Acad. Sci., U.S., Vol. 19, 1933, pp. 653-686.

30. Brace, W. F., "An Extension of the Griffith Theory of Fracture of Rocks," J. Geophys. Res., Vol. 65, 1960, pp. 3477-3480.

31. McClintock, F. A., and J. Walsh, "Friction on Griffith Cracks in Rocks under Pressure," U.S. Natl. Congr. Appl. Mech., Berkeley, 1962.

32. Handin, J., and H. W. Fairbairn, "Experimental Deformation of Hasmark Dolomite," Bull. Geol. Soc. Am., Vol. 66, 1955, pp. 1257-1274.

33. Birch, F., "The Velocity of Compressional Waves in Rocks to 10 Kilobars, Part I," J. Geophys. Res., Vol. 65, 1960, pp. 1083-1102.

34. Fairbairn, H. W., et al., "A Cooperative Investigation of Precision and Accuracy in Chemical, Spectrochemical and Modal Analysis of Silicate Rocks," U.S. Geol. Surv. Bull. 980, 1951.

35. Handin, John, personal communication, 1962.

36. Dreyer, W., "Über die Bruchfestigkeit mono- und polykristallinen Gesteins in Abhängigkeit von Prüfkörperform, Belastungsgeschwindigkeit und Art der Einspannung," Bergbauwissenschaften, Vol. 5, No. 1, 1958, pp. 15-22.

37. Obert, L., S. L. Windes, and W. I. Duvall, "Standardized Tests for Determining the Physical Properties of Mine Rock," U.S. Bur. Mines Rep. Invest., No. 3891, 1946, pp. 1-67.

38. Wuerker, R. G., "Annotated Tables of Strength of Rock," AIME Petroleum Branch Paper No. 663-G, 1956.

39. Peterson, R. E., Stress Concentration Design Factors, John Wiley & Sons, Inc., New York, 1953.

40. Voigt, W., "Beobachtungen über Festigkeit homogener Deformation," Ann. Phys. und Chem., Vol. 67, 1899, pp. 452-458.

41. Bridgman, P. W., "Breaking Tests under Hydrostatic Pressure and Conditions of Rupture," Phil. Mag., Vol. 24, 1937, pp. 63-80.

42. Topping, J., Errors of Observation and Their Treatment, Chapman and Hall, London, 1955.

43. Bridgman, P. W., "Volume Changes in the Plastic Stages of Simple Compression," J. Appl. Phys., Vol. 20, 1949, pp. 1241-1251.

44. Bridgman, P. W., "Plastic Flow and Rupture of Rocks: Discussion of a Paper by C. Zener and J. H. Holomon," Trans. Am. Soc. Metals, Vol. 33, 1944, pp. 226-228.

45. Bridgman, P. W., Studies in Large Plastic Flow and Fracture, McGraw-Hill Book Company Inc., New York, 1952.

46. Bridgman, P. W., "Fracture and Hydrostatic Pressure," in Fracturing of Metals, American Society for Metals, Cleveland, 1947, pp. 246-261.

47. Griggs, D., and J. W. Handin, "Observations on Fracture and a Hypothesis of Earthquakes," Geol. Soc. Am. Mem. 79, 1960, pp. 347-364.

48. Griffith, A. A., "The Phenomena of Flow and Rupture and Flow in Solids," Phil. Trans. Roy. Soc. London, Vol. 221, 1921, p. 163.

49. Hubbert, M. K., and W. W. Rubey, "Role of Fluid Pressure in Mechanics of Overthrust Faulting," Bull. Geol. Soc. Am., Vol. 70, 1959, pp. 115-206.

50. Voigt, W., "Zur Festigkeitslehre," Ann. Physik, Vol. 4, 1901, pp. 567-591.

51. Clausing, D. P., "Comparison of Griffith's Theory and Mohr's Failure Criterion," Quart. Colo. School Mines, Vol. 54, 1959, pp. 285-296.

52. Griffith, A. A., "Theory of Rupture," in First Internatl. Congr. Appl. Mech., Delft, 1924, pp. 55-63.

53. Gilman, J. J., "Direct Measurements of the Surface Energy of Crystals," J. Appl. Phys., Vol. 31, 1960, pp. 2208-2218.

54. Brace, W. F., and J. B. Walsh, "Some Direct Measurements of the Surface Energy of Quartz and Orthoclase," Am. Mineralogist, Vol. 47, 1962, pp. 1111-1122.

55. Birch, F., et al., "Handbook of Physical Constants," Geol. Soc. Am., Spec. Papers No. 36, 1942.

56. Jaeger, J. C., "The Frictional Properties of Joints in Rocks," Geofis. Pura Appl., Vol. 43, 1959, pp. 148-158.

57. Orowan, E., "Mechanism of Seismic Faulting," Geol. Soc. Am. Mem. 79, 1960, pp. 323-345.

58. Bowden, F. P., and D. Tabor, The Friction and Lubrication of Solids, Clarendon Press, Oxford, 1954.

59. Odé, H., "Faulting as a Velocity Discontinuity in Plastic Deformation," Geol. Soc. Am. Mem. 79, 1960, pp. 293-321.

60. Murrell, S. A. F., "The Effect of High Pressure on Brittle Fracture," in Proc. Conference on Phys. and Chem. of High Pressure, London, June 27, 1962.

61. Murrell, S. A. F., "A Criterion for Brittle Fracture of Rocks and Concrete under Triaxial Stresses, and the Effect of Pore Pressure on the Criterion," Proc. 5th Symposium Rock Mech., 1963 (to be published).

62. Böker, H., "Die Mechanik der bleibenden Formänderung in kristallinisch aufgebauten Körpern," Ver. Deut. Ing. Mitt. Forsch., Vol. 175, 1915, pp. 1-51.

63. Bombolakis, E. G., and W. F. Brace, "A Note on Brittle Crack Growth in Compression," J. Geophys. Res., Vol. 68, 1963, pp. 3709-3713.

64. Richart, F. E., A. Brandtzaeg, and R. L. Brown, "A Study of Failure of Concrete under Combined Compressive Stresses," Univ. Illinois Bull. Eng. Expt. Sta., Vol. 185, 1928.

65. Gurney, C., and P. W. Rowe, "The Effect of Radial Pressure on the Flow and Fracture of Reinforced Plastic Rods," RAE Rpt. No. Mat. 5, Rpts. and Mem. No. 2283, May, 1945.

66. Gurney, C., and P. W. Rowe, "Fracture of Glass Rods in Bending and under Radial Pressure," RAE Rpt. No. Mat. 8, Rpts. and Mem. No. 2284, November, 1945.

67. Gurney, C., "The Effective Stress Concentration at the End of a Crack Having Regard to the Atomic Constitution of Materials," RAE Rpt. No. Mat. 10, Rpts. and Mem. No. 2285, December, 1945.

68. Tocher, Don, "Anisotropy in Rocks under Simple Compression," Trans. Am. Geophys. Union, Vol. 38, 1957, pp. 89-94.

69. Sack, R. A., "Extension of Griffith's Theory of Rupture to Three Dimensions," Proc. Phys. Soc. (London), Vol. 58, 1946, pp. 729-736.

70. Leon, A., "Über die Scherfestigkeit des Betons," Beton und Eisen, Vol. 34, 1935, pp. 130-135.

71. Mohr, O., Abhandlungen aus dem Gebiete der technischen Mechanik, Ernst & Sohn, Berlin, 1906.

72. Muehlberger, W. R., "Conjugate Joint Sets of Small Dihedral Angle," J. Geol., Vol. 69, 1961, pp. 211-219.

73. Hetenyi, M., Handbook of Experimental Stress Analysis, John Wiley & Sons, Inc., New York, 1950.

74. Hobbs, D. W., "The Strength and Stress-Strain Characteristics of the M.R.E. Representative Coals under Triaxial Compression," Natl. Coal Bd. MRE Rpt. No. 2198, October, 1961.

75. Hubbert, M. K., "Mechanical Basis for Certain Familiar Geologic Structures," Bull. Geol. Soc. Am., Vol. 62, 1951, pp. 355-372.

76. Matsushima, S., "Variation of the Elastic Wave Velocities of Rocks in the Process of Deformation and Fracture under High Pressure," Disaster Prevent. Res. Inst., Kyoto Univ., Bull., No. 32, 1960, pp. 1-8.

77. Terzaghi, K., "Stress Conditions for the Failure of Saturated Concrete and Rock," Am. Soc. Testing Mater. Proc., Vol. 45, 1945, pp. 777-801.

78. Weibull, W., "A Statistical Theory of the Strength of Materials," Ing. Svetenskaps Acad., Hand. No. 151, 1939.

DISCUSSION

J. A. TALOBRE (France):

A significant contribution has been made by Mr. Brace concerning the question of rock fracture. His study of intrusion fracturing elucidates many abnormalities experienced during confined compression tests, and, consequently, is of prime interest. But I feel that many more tests would still be necessary to allow for an undebatable conclusion about the validity of the Griffith theory.

I wish first to insist on the fact that the Mohr criterion must not be considered as a theory. A criterion explains nothing. It is a process by the help of which different test results can be compared. The Mohr criterion is such a process. We know that it is far from perfect. Anyhow, more than once, it has proved to be helpful.

To my knowledge, the Griffith theory was never used as a criterion. Through this old theory, Griffith tried to explain by the existence of cracks some knotty points concerning the scatter of test results. The explanation seems simple, too simple to be accepted. A strong reason to suspect its validity is that this theory has been claimed verified, even when obviously it did not apply. It is supposed to hold for single crystals even in which no crack could ever be detected. It is supposed to hold for rocks with granular structure, even when the intergranular contacts are particularly tenacious, and consequently when fractures cut across the grains.

In fact, the Griffith theory does not elucidate the real reasons of fracture. In the two-dimensional formula that Mr. Brace recalled, the length of the cracks has a much smaller influence than the product of the modulus of elasticity and of the specific surface energy. We know that a correlation exists between these characteristics and the strength of the material, and it is not surprising if there is a satisfactory fit between all these parameters, whatever the length, nature, and influence of the cracks and flaws.

Another point worthy of notice is that the real Mohr envelope is not constituted, as surmised by Griffith, by one parabola only. A particular envelope corresponds to each type of failure. Hence, there are at least two different intersecting envelopes for each material. We must also consider that ruptures of different types may occur in the same sample at the same time. This may largely alter the envelope. In fact, sometimes the Mohr envelope is found to be polygonal. However, this shape, although common, is not prevalent, and we must admit that the dispersion inherent in the tests does not allow very precise statements.

In conclusion, a large amount of experimental work is still necessary for the origination of a scientific theory of failure. I foresee that the experiments to come will show that rupture is essentially a molecular phenomenon. Let us in the meantime thank Mr. Brace for his excellent and useful work.

W. F. BRACE in reply:

Regarding the importance of crack length, we have examined this here and elsewhere (Ref. 15 of my paper) in a preliminary way. The observed relation between fracture strength and crack length came out fairly close to the relation predicted by Griffith theory for a suite of rocks of the same composition and texture, but of varying grain diameter, and therefore, of varying Griffith crack length. In such a suite of materials, modulus and surface energy are, in effect, held constant while crack length is allowed to vary.

It is certainly true that fracture involves breakage of bonds and other phenomena at a molecular level. But it is also clear that one must postulate flaws to obtain strengths which are of the magnitude of those observed. Calculated strength assuming only bond breakage is absurdly high (Ref. 15), and one is led, much as in the case of metals, back to the necessity of flaws. In the case of metals, the flaws are dislocations and in the case of brittle material, Griffith cracks.

S. D. WOODRUFF (USA):

My question concerns the transverse tangential stresses which exist at the ends of openings in specimens of elastic materials which are subjected to uniaxial compressive stress.

Assuming that the long axis of the opening is parallel to the direction of the applied compressive stress: Does the transverse tensile stress at the ends of the opening decrease as the length-width ratio increases; that is, would the tensile stresses still be of the same magnitude if the opening were a crack with an infinite length-width ratio?

W. F. BRACE in reply:

As near as we know at present, this stress concentration is independent of crack length. It is similar to the situation around circular holes, which gives a stress concentration that is independent of hole diameter. Dr. J. B. Walsh has recently analyzed cracks in compression and found that the elastic energy associated with a crack parallel with the direction of compression is extremely small. Growth of such a crack does not release elastic energy, and such a crack does not propagate the way a crack in tension does.

If one is concerned with pure compression, then such tensile stress at the tips of these cracks can not account for the splitting often observed in a compression test.

J. C. JAEGER (Australia):

I would like to discuss the question of whether intrusion fractures can be caused by penetration of the metal jackets around the specimens into the specimens themselves. For unjacketed specimens, it seems pretty clear that intrusion fracturing occurs very rapidly when confining fluid at high pressures penetrates into Griffith cracks at the surface. With jacketed specimens, I have never seen any evidence of penetration of jacket metal at the surfaces of failure, and have attributed failure to puncturing of the jacket (at its soldered joint) which has allowed confining fluid access to the specimen. I have also observed cases in which the jacket over a crack has remained intact. I find it difficult to visualize penetration of a metal jacket into a Griffith crack of the conventional dimensions, and feel that if intrusion fractures can be initiated without rupture of the jacket, the process may consist of forcing some weaker, or suitably oriented, portion of the rock inward as a wedge, which would constitute a different and important mechanism of failure.

I would also like to ask Professor Brace if he can say anything more about the distribution over the volume of the rock of the microfractures which he observed in his stage three. In particular, is it uniform over the volume of the specimen or is it a roughly planar distribution? Dr. Cook and I have made a number of experiments on Rand quartzite which develops a milky appearance at high stress, suggesting a uniform distribution of grain boundary fractures. If there is a random distribution of microfractures throughout the rock, what is the mechanism by which a major fracture develops?

W. F. BRACE in reply:

I feel that there is essentially no difference between the intrusion of a ductile metal such as copper and the intrusion of the fluid pressure medium, particularly when one is dealing with rocks such as granite and quartzite, which can support enormous stress differences.

Evidence such as that shown in Fig. 7 seems fairly convincing to us, at least, that we are dealing with fractures which start at the surface. We sectioned this sample and found that the fracture went a short way into the rock and stopped; copper from the jacket lined the fracture.

As to the distribution of microfractures, this is a matter currently being studied. We have no specific information on the very important question of which grain boundaries are activated first. This is also being studied at present.

The development of _en échelon_ arrays of cracks is also being currently investigated. Dr. E. G. Bombolakis (now at Boston College) studied this problem photoelastically and found that certain arrays of cracks might start to enlarge at much lower stresses than for an isolated crack. The problem of how the fracture develops from the array, how it breaks through from crack to crack, is also being studied.

ABSTRACT

Most of the experiments on creep of rocks covered in this review were made at room temperature and pressure, under constant stress, and in a dry, chemically unchanged environment. A binary classification according to strain rate and to total strain makes qualitative comparison easy between types of experimental and of natural rock deformation. The creep tests on rocks and ice were performed under constant maximum stress differences, specific strain rates and maximum strains. Experimental data on creep of rocks at room temperature and pressure are usually fitted by empirical equations combining elastic and inelastic moduli. A general creep equation is

$$\epsilon = \epsilon_e + \epsilon(t) + At + \epsilon_T(t) .$$

Several forms have been proposed for $\epsilon(t)$, from the simple relation, $\epsilon(t) = B \ln (t)$, to the more complex Kelvin-Voigt $\epsilon(t) = C[1 - \exp(-t/t_k)]$.

Viscoelasticity theory is useful in determining relaxation mechanisms in dynamic measurements of internal friction and in impact loading, but the theory does not apply to creep and flow of rocks. In particular, the coefficient of viscosity $\eta = \sigma/\dot{\epsilon}$ is not constant; this is shown by room temperature creep results and by the equation used to express high temperature and pressure results:

$$\dot{\epsilon}/\dot{\epsilon}_o = \exp(-Q/RT) \sinh(\sigma/\sigma_o) .$$

An empirical, sigmoidal curve of σ versus log $\dot{\epsilon}$ probably fits the data for rocks better than that of the hyperbolic sine function; evidence for the existence of a flat, low stress part of the curve is from analogy with creep of aluminum at high temperature and from geophysical evidence of the distribution of mass in the earth's mantle.

RÉSUMÉ

La plupart des expériences mentionnées dans cette récapitulation portant sur le fluage des roches furent faites à température et pression ambiantes, sous contrainte constante, et dans un milieu sec et chimiquement inaltéré. Une classification binaire suivant le taux de contrainte et suivant la contrainte totale facilite la comparaison qualitative entre les types de déformation expérimentaux et naturels. Les essais de fluage sur roche et glace furent faits sous de conditions de différences de contrainte maximum constantes, et avec des valeurs déterminées de taux de tension et de contraintes maxima. Les résultats expérimentaux sur le fluage des roches à température et pression ambiantes vérifient habituellement des équations empiriques combinant les modules élastiques et inélastiques. Une équation générale de fluage s'écrit:

$$\epsilon = \epsilon_e + \epsilon(t) + At + \epsilon_T(t) .$$

Plusieurs formes ont été proposées pour $\epsilon(t)$, depuis la relation simple $\epsilon(t) = B \ln (t)$ jusqu'à celle, plus complexe, de Kelvin-Voigt $\epsilon(t) = C[1 - \exp(-t/t_k)]$.

La théorie de la viscoélasticité est très utile pour la détermination des mécanismes de relaxation dans les mesures dynamiques de friction interne et dans le cas du chargement dynamique, mais la théorie ne s'applique pas au fluage et à l'écoulement des roches. En particulier, le coefficient de viscosité $\eta = \sigma/\dot{\epsilon}$ n'est pas constant; ceci est démontré par les résultats de fluage à température ordinaire et par l'équation utilisée a exprimer les résultats à haute température et pression:

$$\dot{\epsilon}/\dot{\epsilon}_o = \exp(-Q/RT) \sinh(\sigma/\sigma_o) .$$

Une courbe empirique sigmoïde représentant σ en fonction de Log $\dot{\epsilon}$ s'accorde probablement mieux aux résultats sur les roches qu'une courbe en sinus hyperbolique. Les preuves de l'existence d'une partie plate à faible contrainte dans la courbe proviennent de l'analogie avec le fluage de l'aluminium à haute température et de données géophysiques sur la distribution de la masse dans l'écorce terrestre.

AUSZUG

Die meisten in diesem Überblick behandelten Kriechversuche an Gesteinen wurden unter Raumtemperaturen und -drücken, unter gleichmässiger Belastung, und in trockener, chemisch unveränderter Umgebung durchgeführt. Eine binäre Einteilung nach Kriechgeschwindigkeit und nach Gesamtformänderung erleichtert den qualitativen Vergleich zwischen den Arten der versuchsmässigen und natürlichen Verformung von Fels. Die hier untersuchten Kriechversuche, wurden unter gleichbleibenden maximalen Spannungsunterschieden, spezifischen Formänderungsgeschwindigkeiten $\dot{\epsilon} = 10^{-12}$ bis 10^{-1} sec^{-1} und die maximalen Formänderungen durchgeführt.

Versuchsdaten über das Kriechen von Gesteinen bei Zimmertemperaturen und -drücken werden gewöhnlich in empirischen Gleichungen ausgedrückt, welche die elastischen und nichtelastischen Anteile zusammenfassen. Eine allgemeine Kriechgleichung ist $\epsilon = \epsilon_e + \epsilon(t) + At + \epsilon_T(t)$. Verschiedene Formen wurden für $\epsilon(t)$ vorgeschlagen, von der einfachen Abhängigkeit $\epsilon(t) = B \ln (t)$ zur mehr verwickelten Form $\epsilon(t) = C[1 - \exp(-t/t_k)]$ für den visko-elastischen Körper nach Kelvin-Voigt.

Die visko-elastische Theorie ist brauchbar um den Entspannungsvorgang bei dynamischen Messungen von innerer Reibung und bei Stossbelastungen zu bestimmen. Diese Theorie ist jedoch nicht auf das Kriechen und Fliessen von Gesteinen anwendbar. Der Beiwert der Viskosität $\eta = \sigma/\dot{\epsilon}$ ist vor allem nicht konstant. Dies ist durch die Ergebnisse über das Kriechen unter Raumtemperatur bewiesen, ferner durch die Gleichung, die die Ergebnisse unter hohen Temperaturen und Drücken ausdrückt:

$$\dot{\epsilon}/\dot{\epsilon}_o = \exp(-Q/RT) \sinh(\sigma/\sigma_o) .$$

Eine empirische, sigmaförmige Kurve von σ zu log $\dot{\epsilon}$ passt wahrscheinlich besser zu Gesteinen als die hyperbolische Sinusfunktion. Der Beweis für das Vorhandensein einer flachen Kurvenstrecke bei niedere Spannung ist durch die Ähnlichkeit mit dem Kriechen von Aluminium bei hoher Temperatur und durch geophysikalische Beweisführung über die Massenverteilung im Erdmantel gegeben.

VISCOELASTICITY OF ROCKS

Eugene C. Robertson*

Introduction

IN CLASSICAL MECHANICS, the elastic solid of Hooke and the viscous liquid of Newton are two idealizations of the behavior of materials commonly made; in both, the relations between the relevant mechanical variables are linear. These relations are limiting cases for ideal materials and are treated in the theories of elasticity and hydrodynamics. The behavior of real materials is sometimes represented fairly closely by viscoelasticity theory, in which theories for ideal elastic and viscous materials are combined. Although elastic, plastic, and viscous models are very useful as mathematical approximations, empirical equations often better describe real mechanical behavior; in fact, empirical data for time dependent properties of real materials will be emphasized in this paper.

The terminology of deformation deserves attention here in order to avoid ambiguity. The word deformation itself refers to all the qualitative characteristics of mechanics of materials, such as rupture, yielding, and flow and fracture between and in grains. A general term for the nonlinear, time dependent behavior of real materials is rheology; under rheology, large as well as small strains and the effects of heat and hydrostatic pressure are considered as a function of time. In contrast with rheology, under viscoelasticity only small strains at low temperatures and pressures are fitted into phenomenological equations with elasticity and viscosity coefficients. The term elastic describes a linear stress-strain behavior with recovery of strain on unloading and without regard to time effects; in

*U. S. Geological Survey, Silver Spring, Maryland.

practical use, full recovery of strain is not important. In the behavior of an inelastic material, strains are nonlinear with stresses, and strain hardening may occur; but as in elasticity, strain rates are not considered.

The term plastic describes an ideal behavior in which the initial deformation is elastic, and then at some yield stress, deformation continues without limit, that is, without strain hardening. No account is taken of time effects, but a linear relation is assumed between the vectors of stress and strain. Some ductile metals and some rocks under high temperature and at slow strain rates are nearly plastic, although the yield stresses are not independent of temperature and strain rate.

The effect of time is considered under anelasticity, a word introduced by Zener[1] for the inelastic relaxation of very small deformations, usually studied as forced vibrations. The term flow has connotations of liquid behavior and is commonly used for inelastic deformation at nearly constant strain rate under any load. Creep refers to inelastic deformation under a constant load, usually below the yield stress, and at any low strain rate.

Experimental studies of rocks will be the primary concern here, although certain results of experiments on metals, helpful to interpretation of rock deformation, will be mentioned. Unless otherwise stated, the rocks were deformed in a dry, chemically unchanged environment, and no regard has been taken of their previous strain history. Except for studies on single crystals of minerals and on certain rocks, the rock samples are considered for mathematical simplicity to have been homogeneous and isotropic. The assumption is that there is a statistically random orientation and a relatively small size of the grains of the inherently anisotropic mineral constituents. Geologic model experiments can give helpful qualitative information if dimensionally correct, but the emphasis will be on results obtained from rock samples.

Comprehensive reviews of experimental results on the rheology of rocks and rocklike materials are given by Nadai,[2,3] Schmid and Boas,[4] Walton,[5] and in Griggs and Handin;[6] time dependent behavior is discussed by Freudenthal,[7] Cottrell,[8] Schoeck,[9] Reiner,[10] and Murrell and Misra.[11]

LeComte[12] and Kendall[13] have very kindly permitted use of material from their unpublished doctoral and master's theses. The assistance of Louis Peselnick in clarifying concepts in anelasticity is gratefully acknowledged.

Classification of Rock Deformation

To clarify the relationships among the various types of deformation experiments on rocks, classification according to rate of strain and total strain is presented in Table 1. Typical geologic processes of rock deformation are also placed in the classification to show at least qualitatively the applicability of the laboratory results. In addition, temperature and pressure effects can be superimposed without confusion.

The magnitude of the stress and the rate at which the stress is applied are the causal factors in the rheology of solids, but total strain and strain rate are used in this classification because it is their values that are always measured, and the stresses are inferred. Whether strain or stress units are used is not greatly important in Table 1 because, in general, strain waves closely follow and are proportional to the stress waves. The categories are qualitative, although quantitative limits are suggested. In the intermediate groups, the limits of strain and strain rate are not given because they vary widely with rock type and environmental conditions.

The term recovery, referring in Table 1 to the result of small deformation, is used to indicate that most if not all the strain may be recovered; generally, some recovery of strain occurs in intermediate flow and fracture. After large strain by flow or fracture, failure by loss of cohesion occurs. Unquestionably, there is overlap of recovery, fracture, flow, and failure in rock deformation, but only the predominant effects need be considered for the classification.

Table 1

QUALITATIVE CLASSIFICATION OF THE DEFORMATION OF ROCKS IN NATURAL PHENOMENA AND IN TYPICAL TESTS

Total Strain (ϵ)	General result of deformation	Strain Rate ($\dot{\epsilon}$): Small		Intermediate		Large	
		Monotonic ($\dot{\epsilon} < 10^{-5}$ sec^{-1})	Oscillatory ($f < 10^{-4}$ cps)	Monotonic	Oscillatory	Monotonic ($\dot{\epsilon} > 10^{2}$ sec^{-1})	Oscillatory ($f > 10^{2}$ cps)
Small ($\epsilon < 10^{-4}$)	Recovery	Creep tests; continental glacier re-bound	Torsion pendulum tests; tidal stresses and axial wobble of the earth	Elastic de-formation tests	Free oscillations of the earth; earthquake waves	Unconfined impact tests	Ultrasonic pulse and resonance tests
Intermediate	Prefailure flow or fracturing	Creep and flow tests; glacier flow; surface subsidence over mine opening; solifluction	Volcano reservoir pulsations	Inelastic deformation tests; landslides	Fatigue tests	Unconfined impact tests; meteorite impacts	Deformation by under-ground ex-plosions
Large ($\epsilon > 10^{-1}$)	Failure	Strength tests; jointing and faulting; geo-synclinal down-warping; conti-nental drift; mantle convec-tion currents	Epeirogenic pulsations	Strength tests; faulting; rock bursts in mines	Crushing and grinding	Unconfined impact tests; meteorite impacts	Deformation by under-ground ex-plosions

It will not be necessary to discuss every box in Table 1, as many of the results are self-evident. Impact and dynamic testing, in the large strain rate column, are important enough to warrant separate treatment. Creep experiments, which are discussed in the next section, lie in the small strain rate, monotonic column of Table 1 because at low stresses the strain rates usually are below 10^{-5} sec^{-1}. In fact, unless the temperature is raised to about half the melting point in Kelvin degrees, higher strain rates cannot be attained at high, constant stress due to the onset of failure; this is true of metals,[9,14] and is indicated for ice in Fig. 5A, page 197. If the temperature of silicate and carbonate rocks is raised, the hydrostatic pressure also must be raised to avoid loss of intergranular cohesion by differential thermal expansion or by mineral dissociation. Temperature plus pressure effects on creep have been fairly well studied for ice; as for other rocks, there are the recent, very important experiments of Heard on Yule marble[15] and of LeComte on halite.[12]

Creep

Experimental Results

A creep test is the simple deformation of a rock under a constant load, usually below the elastic limit, in which the time dependence of strain is observed. Research on the creep of rocks at room temperature and pressure has received much attention, as can be seen in Table 2. Explanations of the table are given in its footnotes; some of the experiments will be discussed in detail below. Creep studies of concrete and cement are not listed because their behavior differs considerably from that of rocks due to the presence of relatively plastic aluminates and hydrous silicates.

More than 60 years ago, Nagaoka[30] observed the continued creep strain of a sandstone specimen under a constant load. He found that he had to stay far below the elastic limit in order to determine usable moduli of elasticity for the eighty rocks he studied. In Fig. 1, the creep observations of Phillips[31] are shown; strain-time

Table 2

COMPILATION OF CREEP EXPERIMENTS ON VARIOUS ROCKS AT ROOM TEMPERATURE[a] AND ON ICE AT TEMPERATURES NEAR THE FREEZING POINT

Reference and Rock Type	Number of Tests	Test Duration (days)	Maximum Stress σ (bars)	Maximum Strain ϵ (cm/cm)	Average Strain Rate[b] $\dot{\epsilon}$ (sec^{-1})	Creep Stages[c] 1 2 3	Stress Type, Temperature, and Pressure[d]
Rocks and Minerals							
Evans (Ref. 16)							
Marble, slate, granite, sandstone	8	0.1	10^2	10^{-3}	10^{-7}	X	Comp
Evans and Wood (Ref. 17)							
Slate, granite, marble	42	0.01	10^3	10^{-3}	10^{-8}	X	Comp
Goguel (Ref. 18)							
Limestone, marl, talc	5	0.01	10^3	10^{-2}	10^{-5}	X X	Comp; P = 0.4 to 2 kb
Griggs (Ref. 19)							
Limestone	4	0.1	10^4	10^{-1}	10^{-5}	X X X	Comp; P = 10 kb
Griggs (Ref. 20)							
Limestone	1	550	10^3	10^{-4}	10^{-12}	X	Comp
Halite	1	42	10^2	10^{-2}	10^{-9}	X X	Comp
Shale	1	144	10_2	10^{-2}	10^{-9}	X X	Comp
Alabaster	2	20	10^2	10^{-1}	10^{-9}	X X	Comp dry; in water
Griggs (Ref. 21)							
Alabaster	1	40	10^2	10^{-3}	10^{-10}	X	Comp dry
Alabaster	7	100	10^2	10^{-2}	10^{-9}	X X X	Comp in water
Alabaster	3	3	10^2	10^{-2}	10^{-8}	X X X	Comp in water; P = 1 kb

Gunter and Parker (Ref. 22)							
Halite	6	0.1	10^{2}	10^{-2}	10^{-6}	X X	Comp
Hardy (Ref. 23)							
Sandstone	2	18	10^{3}	10^{-4}	10^{-10}	X X	Comp
Hematite ore	3	1	10^{2}	10^{-3}	10^{-6}	X X	Comp
Haskell (Ref. 24)							
Earth's crust	—	10^{6}	—	10^{-4}	10^{-15}	X	Fennoscandian glacier rebound
Iida, Wada, Aida, and Shichi (Ref. 25)							
Serpentine	3	13	10^{2}	10^{-3}	10^{-10}	X X	
Basalt	10	6	10^{2}	10^{-4}	10^{-10}	X X	
Rhyolite	4	100	10^{2}	10^{-4}	10^{-11}	X X	
Andesite	4	14	10^{2}	10^{-5}	10^{-9}	X X	
Granodiorite	2	6	10^{2}	10^{-5}	10^{-10}	X X	
Inouye and Tani (Ref. 26)							
Coal	39	10^{-3}	10^{2}	—	10^{-2}	X	Beam
Kendall (Ref. 13)							
Limestone	14	1	10^{3}	10^{-2}	10^{-8}	X X X	Comp
Halite	14	1	10^{2}	10^{-1}	10^{-7}	X X X	P = 1 kb
Kuznetsov (Ref. 27)							
Sandstone	1	1	10^{2}			X X	Tension
Rock salt	1	1	10			X X	Tension
LeComte (Ref. 12)							
Halite	17	10	10^{2}	10^{-2}	10^{-8}	X X	Comp; $T = 30^{\circ}$ to $200^{\circ}C$; P = 0 to 1 kb
Lomnitz (Ref. 28)							
Granodiorite	9	3	10^{2}	10^{-4}	10^{-10}	X X X	Torsion
Gabbro	5	3	10^{2}	10^{-5}	10^{-10}	X X X	Torsion
Matsushima (Ref. 29)							
Granite	6	2	10^{3}	10^{-4}	10^{-9}	X	Comp
Granite	2	12	10^{3}	10^{-4}	10^{-10}	X	Comp

Table 2—continued

Reference and Rock Type	Number of Tests	Test Duration (days)	Maximum Stress σ (bars)	Maximum Strain ϵ (cm/cm)	Average[b] Strain Rate $\dot{\epsilon}$ (sec^{-1})	Creep[c] Stages 1 2 3	Stress Type,[d] Temperature, and Pressure
Rocks and Minerals (continued)							
Nagaoka (Ref. 30) Sandstone	1	0.01	1	10^{-2}	10^{-5}	X	Torsion
Phillips (Ref. 31) Shale	15	1	10^{2}	10^{-3}	10^{-8}	X	Cantilever
Pomeroy (Ref. 32) Coal	1	50	10	10^{-1}	10^{-9}	X X	Cantilever
Robertson (Ref. 33) Limestone	50	0.1	10^{4}	10^{-1}	10^{-6}	X	Comp; P = 0.3 to 4 kb
Roux and Denkhous (Ref. 34) Quartzite	1	20	10^{2}	10^{-3}	10^{-10}	X X	Comp
Ice							
Glen (Ref. 35) Ice	13	6	1	10^{-3}	10^{-8}	X X	Comp
Ice	5	6	10	10^{-2}	10^{-5}	X X	T = -1.5°C
Glen (Ref. 36) Ice	36	2	10	10^{-1}	10^{-7}	X X X	Comp; T = 0° to -13°C
Glen (Ref. 37) Ice	—	11	1	10^{-1}	10^{-7}	X	Glacier observations
Glen and Perutz (Ref. 38) Ice, single crystal	9	2	1	10^{-1}	10^{-7}	X	Tension

Gold (Ref. 39) Ice	12	0.2	10	10^{-2}	10^{-6}	X X X	Comp; T = -10°C
Griggs and Coles (Ref. 40) Ice, single crystal	41	0.2	10	10^{-1}	10^{-5}	X	Comp; T = -1 to -18°C
Jellinek and Brill (Ref. 41) Ice	23	0.2	1	10^{-4}	10^{-8}	X X	Tension; T = -5° to -15°C
Ice, single crystal	4	0.2	1	10^{-4}	10^{-8}	X	Tension; T = -5°C
Meier (Ref. 42) Ice	—	10^{3}	1	10^{-2}	10^{-10}	X	Glacier observations
Rigsby (Ref. 43) Ice	20	1	1	1	10^{-4}	X X	Shear; P = 1 to 300 bars; T = 5° to -20°C
Steinemann (Ref. 44) Ice, single crystal	51	1	1	1	10^{-5}	X	Shear; T = -2.3°C
Steinemann (Ref. 45)							
Ice	2	5	1	10^{-2}	10^{-9}	X X X	Tension; T = -1.9°C
Ice	16	5	10	10^{-2}	10^{-7}	X X X	Comp; T = -1.9°C
Ice	15	5	1	10^{-2}	10^{-9}	X X X	Tension; T = -4.8°C
Ice	20	5	10	10^{-2}	10^{-7}	X X X	Comp; T = -4.8°C
Ice	2	5	1	10^{-2}	10^{-9}	X X X	Tension; T = -11.5°C
Ice	9	5	10	10^{-2}	10^{-7}	X X X	Comp; T = -11.5°C
Ice	4	5	10	10^{-1}	10^{-8}	X X X	Comp; T = -22°C
Ice	6	10	1	10^{-1}	10^{-7}	X X X	Torsion; T = -1.9°C

NOTE: See next page for footnotes.

FOOTNOTES TO TABLE 2

[a]In most investigations listed, more than one creep test was made on each rock type, either on different samples or on the same sample. Orders of magnitude are used in order to encompass the data from several tests with one number. In a few instances, sample dimensions had to be guessed in order to calculate strains.

[b]The strain rate is a simple over-all average, the maximum strain divided by the total time; typical values are given if the range is large.

[c]Creep stages observed during tests are numbered 1 for the primary stage, 2 for the secondary stage, and 3 for the tertiary stage.

[d]Room temperature and pressure conditions prevailed unless otherwise specified. Experiments on ice were necessarily made at temperatures below $0^{\circ}C$ as marked. Abbreviations are: kb, kilobar; P, hydrostatic pressure; T, temperature; Comp, compressive stress; Torsion, torsion stress; Tension, tensile stress; Cantilever, cantilever beam in bending; Beam, midpoint bending.

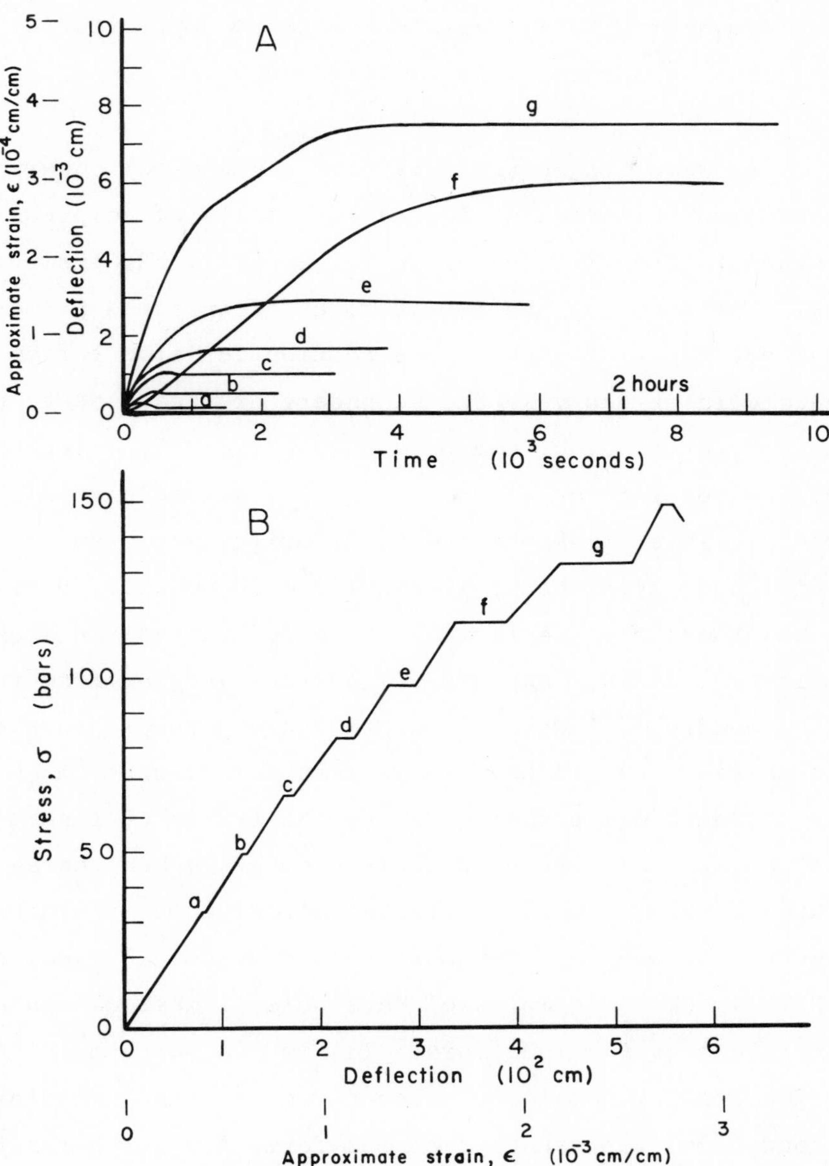

Fig. 1—A: Creep curves of shale loaded as a cantilever beam; letters refer to constant stress lines in B. B: Stress-strain curve of sample in A. (After Phillips, Ref. 31, Figs. 4 and 3.)

data are plotted for seven constant loads on a shale beam under room conditions; the change of strain decays in typical fashion, taking longer for higher stresses. The jogs in the stress-strain line are like those observed by Nagaoka, and are well-known phenomena. The jogs illustrate the importance of time on the stress-strain ratio;

that is, they demonstrate that Hooke's law is an approximation in the rheology of rocks, apropos only for continuously and rapidly applied stress.

In a similar way, Griggs (Ref. 19, Figs. 9 and 10) observed the creep of Solenhofen limestone at increasing levels of compressive stress, as shown in Fig. 2. The creep curves (Fig. 2B) exemplify the three stages, familiar from studies of metal creep: the primary, or transient, or decelerating strain rate portion for time t from $0 < t < 2000$ sec in all curves; the secondary or steady state portion, $t > 2000$ sec for curves a, b, and c, and $2000 < t < 5000$ sec for curve d; and the tertiary or accelerating strain rate portion, $t > 5000$ sec for curve d. This limestone sample of Griggs was open to kerosene pressure fluid under hydrostatic pressure P = 10,000 bars during the experiment, and therefore the form of curve d, with strain accelerating to rupture, is considerably expanded. In fact, judging from recent experiments of Handin,[46] Griggs' sample under pore pressure equal to confining pressure should have shown tertiary creep at much lower stress. Its strength was probably due to the low initial permeability of the limestone, so that the pore fluid did not weaken the sample until fractures developed at high strain, allowing the fluid to penetrate and weaken the sample. Kendall's creep curves[13] show the same three stages for jacketed samples of Solenhofen limestone and halite, which he compressed at P = 1000 bars. In Fig. 2A, with much different strain and time scales, Lomnitz's creep curve[28] for granodiorite under room conditions represents a common form of a curve leading to rupture. Most creep curves for silicate rocks are of this latter type.

Upon release of the load in a creep test, partial recovery of the strain is observed, the amount depending on the load and the rock type. The creep and recovery curves of three rocks in Fig. 3 from Evans[16] show the differences in behavior and also their similarities; the similar curvature for a given rock upon application and upon release of load was also observed by Michelson (Ref. 47, p. 22).

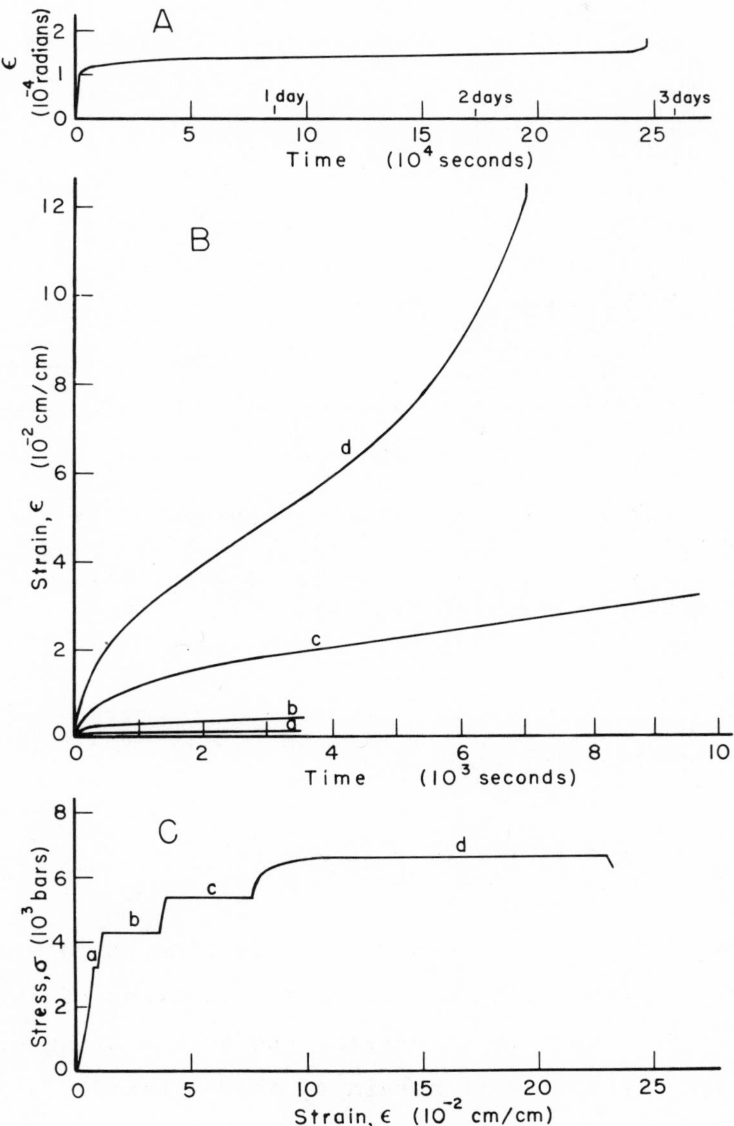

Fig. 2—A: Creep curve of granodiorite under constant torque stress of 140 bars (after Lomnitz, Ref. 28, Fig. 4). B: Creep curve of Solenhofen limestone in compression, P = 10,000 bars; letters refer to constant stress lines in C (after Griggs, Ref. 19, Fig. 10). C: Stress-strain curve of sample in B (after Griggs, Ref. 19, Fig. 9).

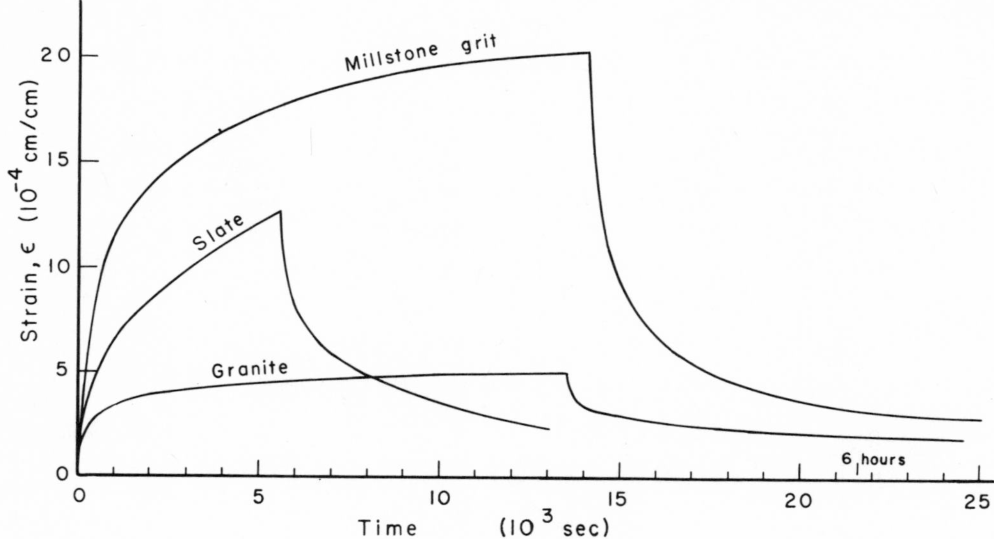

Fig. 3—Creep and creep recovery curves under constant stress; millstone grit, $\sigma = 85$ bars; slate, $\sigma = 85$ bars; and granite, $\sigma = 165$ bars (after Evans, Ref. 16, Fig. 4).

Creep Equations

A general equation for creep is

$$\epsilon = \epsilon_e + \epsilon(t) + At + \epsilon_T(t) , \qquad (1)$$

where ϵ is strain, ϵ_e is elastic strain, t is time, A is a constant, $\epsilon_T(t)$ is a function expressing the accelerating creep of the tertiary stage, and $\epsilon(t)$ is a function expressing the decelerating creep of the primary stage. The elastic strain ϵ_e occurs immediately after loading as in the plots of Figs. 1 and 2; the steady state term, At, accounts for strain linear with time, which occurs after the transient creep stage. No simple expression can be given for $\epsilon_T(t)$, and it will not be considered further. Many forms for transient creep $\epsilon(t)$ have been proposed.

Transient creep is often represented by a logarithmic relation:*

$$\epsilon(t) = B \ln t . \qquad (2)$$

---

*See Refs. 13, 18, 28, 29, 32, and 33.

A logarithmic plot of Griggs'[20] long-term observations of the creep of a Solenhofen limestone specimen in compression is shown in Fig. 4. Experience in metals[9] shows that whereas Eq. (2) applies to creep at low temperatures, the well-known power law of E. N. da C. Andrade,

$$\epsilon(t) = Bt^{1/3} , \tag{3}$$

fits creep data better for higher temperatures. So little work has been done on rocks at elevated temperatures that Andrade's law has not been evaluated for them, although Glen[36] found it to fit his creep data for ice near its melting point.

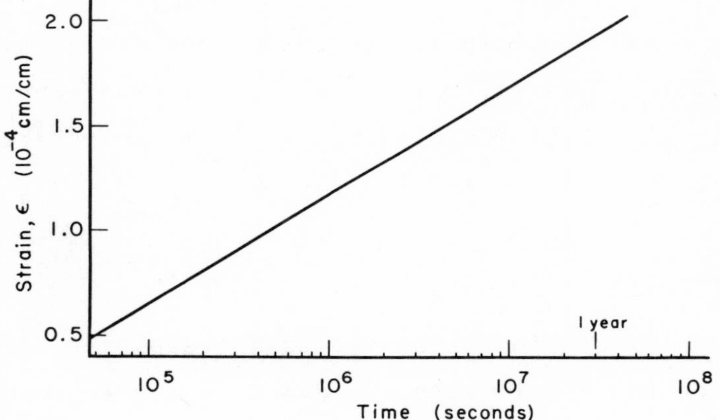

Fig. 4—Creep of Solenhofen limestone under constant stress; σ = 1400 bars (after Griggs, Ref. 20, Fig. 4).

The viscoelastic model of Kelvin and Voigt for constant stress provides another relation for transient creep,

$$e(t) = B[1 - \exp(-t/t_k)] , \tag{4}$$

in which t_k is the retardation time constant for transient creep; at the time $t = t_k$, the strain reaches $(1 - 1/e)B = 0.632B$, where B is the maximum, asymptotic value of strain at infinite time, and e is the base of natural logarithms. The viscoelastic model of Burger adds the expression for a Maxwell material to that of the Kelvin-Voigt material to account for steady state creep at constant stress:

$$\epsilon(t) + At = B[1 - \exp(-t/t_k)] + C(1 + t/t_m) \ , \qquad (5)$$

where C is a constant involving the stress, and t_m is the relaxation time constant; at the time $t = t_m$, stress decreases by 1/e at constant ϵ.

It is difficult to fit creep data for rocks to viscoelastic Eqs. (4) and (5) because both the two time constants and the finite limiting strain must be determined. As shown in Fig. 4, creep of limestone exhibited the ϵ - log t relation even after 550 days. Such continued strain precludes estimating the asymptotic value of strain and obtaining a satisfactory fit to the exponential curves of Eqs. (4) and (5). However, for $t < 1$ day, Jellinek and Brill (Ref. 41, pp. 1204-1205) found Eq. (5) to fit their data fairly well on the creep of ice to $\epsilon = 2 \times 10^{-4}$.

Effect of Stress on Creep Rate

The amount of strain ϵ observed in a creep test after a given time depends on the strain rate $\dot{\epsilon} = d\epsilon/dt$, which in turn depends on the amount of stress σ applied. (The symbol σ is used for maximum stress difference and is referred to as stress.) Figures 1 and 2 show this for limestone and shale, and Fig. 5B for ice. An empirical equation commonly used to relate the variables is

$$\dot{\epsilon} = A\sigma^n \ , \qquad (6a)$$

and approximately

$$\epsilon = At\sigma^n \ , \qquad (6b)$$

where A and n are constants. A tabulation of values of n is given in Table 3; they were calculated from transient creep data taken from several studies on rocks listed in Table 2. Figures for ϵ at some

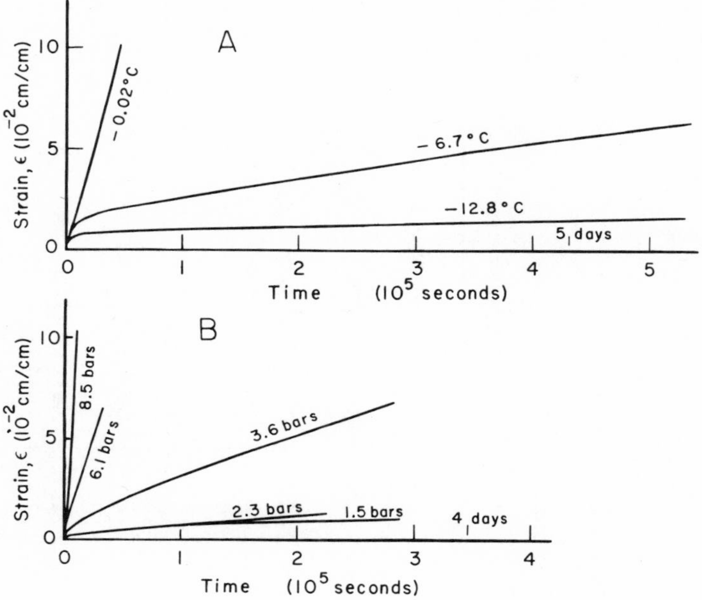

Fig. 5—A: Creep of polycrystalline ice at various temperatures, at the same stress, $\sigma = 6$ bars. B: Creep of polycrystalline ice at various stresses, at $T = -0.02°C$. (After Glen, Ref. 36, Figs. 4 and 3.)

convenient but arbitrary t were plotted against σ on log-log graphs to find n (Eq. (6b)). For very small strains at low stresses, $n = 1$, as found by Lomnitz; for larger strains, say, $\epsilon > 10^{-3}$, n varies between 2 and 3. If experimental results permit, data from steady state creep are ordinarily used in Eq. (6), but significant analysis can be obtained from primary creep data also.

Michelson[47,48] found an exponential increase of strain at high stress. Michelson's results on marble, limestone, talc, slate, and shale are given only as constants to a complicated equation; therefore, making comparisons with other work is difficult. A typical result is that for limestone:

$$\epsilon = 600\sigma e^{0.10\sigma} + 20\sigma e^{0.3\sigma}(1 - e^{-1.2\sqrt{t}}) + 9\sigma e^{0.3\sigma} t^{0.3} + 11\sigma e^{0.8\sigma} .$$

Similarly, Dorn[14] in his investigation of creep of metals at high temperature found that at high stresses

Table 3

VALUES OF THE EXPONENT n IN THE EQUATION $\dot{\epsilon} = A\sigma^n$ IN THE CREEP OF VARIOUS ROCKS AT ROOM TEMPERATURE

Rock Type	Maximum Strain ϵ (10^{-3} cm/cm)	Maximum Stress σ (10^3 bars)	Hydrostatic Pressure P (10^3 bars)	Exponent n	Reference
Slate	1	3.6	0	1.8	Evans and Wood (Ref. 17)
Granite	1	3.5	0	3.3	
Limestone	50	6.7	10	5.0	Griggs (Ref. 19)
Alabaster (in water)	5	0.2	0	2.0	Griggs (Ref. 20)
Limestone	7	1.4	0	1.7	Kendall (Ref. 13)
Limestone	5	3.5	0.7	3.0	
Limestone	4	4.1	1	7.9	
Halite	20	0.3	0	1.9	
Gabbro	0.01	0.1	0	1.2	Lomnitz (Ref. 28)
Granodiorite	0.2	0.1	0	1.0	
Granite	0.08	1.0	0	2.1	Matsushima (Ref. 29)
Shale	3	0.1	0	2.7	Phillips (Ref. 31)
Limestone	300	5.8	2	3.7	Robertson (Ref. 33)
Limestone	180	6.3	3	1.8	
Limestone	100	5.6	4	1.8	

$$\dot{\epsilon} = Ae^{B\sigma} . \tag{7}$$

At low stresses, Dorn used Eq. (6a).

Studies of steady state creep of ice attest to the same generalization on n in Eq. (6) as for rocks: At low stresses for very small strains, n = 1, and at higher stresses for larger strains, n > 1. Jellinek and Brill(41) found n = 1, for $\sigma < 1.5$ bars, and n = 1.4, for $2.4 < \sigma < 1.5$ bars. In another study (Fig. 5B) of polycrystalline ice, Glen(36) found n = 3.17, for $1.5 < \sigma < 10$ bars; in an earlier study, Glen(35) found n = 4 for the same stress range. Steinemann(45) found n = 1.85 at σ = 1 bar and n = 4.16 at σ = 15 bars. Two creep studies were made on single crystals of ice, which deformed in what seems to be almost entirely in the tertiary stage;(40,44) plots of the results indicate a range of n = 1.5 to 4, depending on the elapsed time chosen (Eq. (6b)).

A comprehensive plot of data from eight investigations (including laboratory experiments and studies of glacier flow) by Meier (Ref. 42, Fig. 40) is given in Fig. 6; some duplicate points are not shown. For the solid curve in Fig. 6, Meier found that the relation between

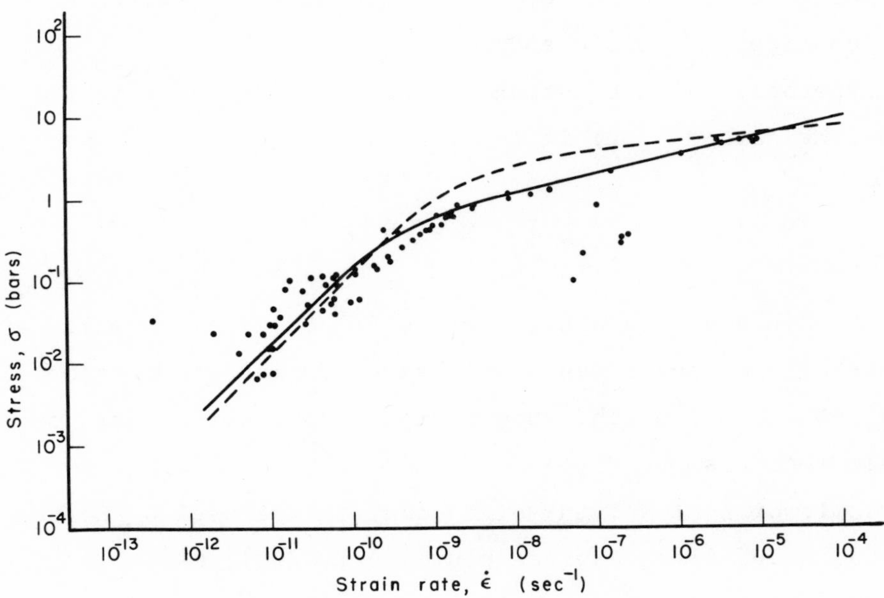

Fig. 6—Creep and flow of ice, stress versus strain rate (after Meier, Ref. 42, Fig. 40).

octahedral shear stress and octahedral shear strain rate (steady state creep) could be expressed by

$$\dot{\epsilon} = A\sigma + B\sigma^{4.5} , \qquad (8)$$

with $A = 5.7 \times 10^{-10}$ bar^{-1} sec^{-1} and $B = 4.1 \times 10^{-9}$ bar^{-1} sec^{-1}. There is some scatter to the points in Fig. 6, but the change in slope at low stresses is obvious. The six low points at about $\dot{\epsilon} = 10^{-7}$ sec^{-1} are probably erroneous (Ref. 42, p. 44). As indicated in Table 3 and the accompanying discussion relating to Eq. (6), $n = 1$ for low σ and $n > 1$ at higher σ, as in Eq. (8).

H. Hughes (1962, oral communication) pointed out the importance of measuring the strain rate in the range of stress of actual interest, citing his finding that rods of uranium metal creep at a rate linear with stress at very low stresses at moderate temperatures. Data for uranium, obtained at higher stress, extrapolated by the exponential law of Eq. (6), for $n = 3$, gave a much too low creep rate. This important observation may be applied to ice in Fig. 6, as an example. If the $\dot{\epsilon}$ of ice (Fig. 6) observed in the range $\sigma = 1$ to 10 bars were extrapolated to $\sigma = 10^{-2}$ bars by the equation $\dot{\epsilon} = B\sigma^{4.5}$, the calculated value would be $\dot{\epsilon} = 10^{-18}$ sec^{-1}, instead of the observed $\dot{\epsilon} = 10^{-11}$ sec^{-1}, a very considerable difference.

A hyperbolic sine equation can be fitted roughly to the data in Fig. 6 (the dashed line) in the form

$$\dot{\epsilon} = M \sinh N\sigma , \qquad (9)$$

where M and N are constants. Griggs[21] evolved a similar expression from his data on the creep of alabaster in water; Goranson[49] derived Griggs' equation from the theory of thermodynamics based on a change of phase mechanism.

A semiempirical Kelvin-Voigt equation was used by Ruppeneit (Ref. 50, Chap. 6) to express his experimental results on the transient creep of sandstone, sandy shale, and argillaceous shale:

$$(\sigma/E)^n = t_k\dot{\epsilon} + B .$$

The following are approximate values of the constants for the three rocks: The elastic modulus E is about 2 x 10^5 bars, the retardation time t_k is about 10^5 sec, the constant n is 1.2, and the constant B is about 0.1.

Temperature and Pressure Effects

Many studies of the effect of heat and confining pressure on the strength of rocks have been made, usually at moderate strain rates, which fall in the small and intermediate $\dot{\epsilon}$ and intermediate and large ϵ categories of Table 1; for example, in Griggs, Turner, and Heard(51) and in Heard,(52) deformation of marble and limestone was at $\dot{\epsilon} = 10^{-3}$ to 10^{-5} sec^{-1}, temperature T = 25° to 800°C, and hydrostatic pressure P = 1 to 5000 bars. A good bibliography of similar research is available in Griggs and Handin.(6) In general, however, the variation of $\dot{\epsilon}$ in these studies at high temperature and pressure has been too small, or insufficient data have been provided to allow conclusions on time dependence of creep to be drawn.

In four recent experimental programs, strain rate has been given particular emphasis. LeComte(12) studied polycrystalline salt (NaCl) at rates $\dot{\epsilon} = 10^{-7}$ to 10^{-9} sec^{-1}, under constant stress σ, at T = 25° to 200°C, and P = 1 to 10^3 bars. A few qualitative conclusions may be drawn from LeComte's results: The $\dot{\epsilon}$ in salt is increased by increase of temperature or of stress difference, and increase of confining pressure or of grain size causes a relatively small decrease in $\dot{\epsilon}$. Robertson(33) observed in limestone a many-fold decrease in $\dot{\epsilon}$ with increase of confining pressure from 1 to 4000 bars. Serdengecti and Boozer(53) studied sandstone, limestone, and gabbro under constant conditions, in the ranges $\dot{\epsilon}$ = 1 to 10^{-5} sec^{-1}, T = 25° to 150°C, and P = 1 to 1400 bars. They observed an increase in the ultimate strength of the rocks with an increase in confining pressure or of strain rate, and a decrease in strength with increase of temperature, although the changes in strength were small, about 10 per cent for the maximum change of conditions.

Heard(15) has made an important and comprehensive investigation of Yule marble, in the ranges $\dot{\epsilon} = 10^{-1}$ to 10^{-8} sec^{-1}, T = 25° to 500°C, and P = 5000 bars; all tests were made at constant $\dot{\epsilon}$, T, and P. An example of his observations is given in Fig. 7, in which the importance of differences in $\dot{\epsilon}$ is very clearly shown. Heard's results and those at 600° and 800°C quoted by him are summarized by the solid line segments in Fig. 8; the shape of the curves is like the one for ice in Fig. 6, a much larger increment of σ being needed at low σ to obtain a unit increase of $\dot{\epsilon}$ than at high σ.

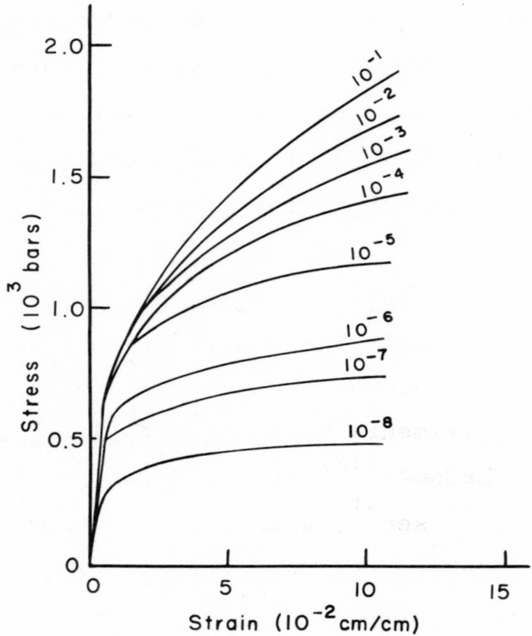

Fig. 7—Stress-strain curves for one cylinder of Yule marble tested at various strain rates under hydrostatic pressure P = 5000 bars, at T = 500°C (after Heard, Ref. 15, Fig. 11).

As a contribution to the problem of the strength of the upper mantle of the earth, it is tempting to suggest that the low $\dot{\epsilon}$ end of the σ-log $\dot{\epsilon}$ curves in Figs. 6 and 8 change form and flatten. Extension of each curve to the left to give it a sigmoid shape is warranted by only bits of evidence. One enlightening piece of information is the analogy between the shapes of the curves of σ-log $\dot{\epsilon}$ at various temperatures for pure aluminum in Fig. 9 with Heard's curves for Yule marble in Fig. 8; the data for Fig. 9 were taken from Zhurkov

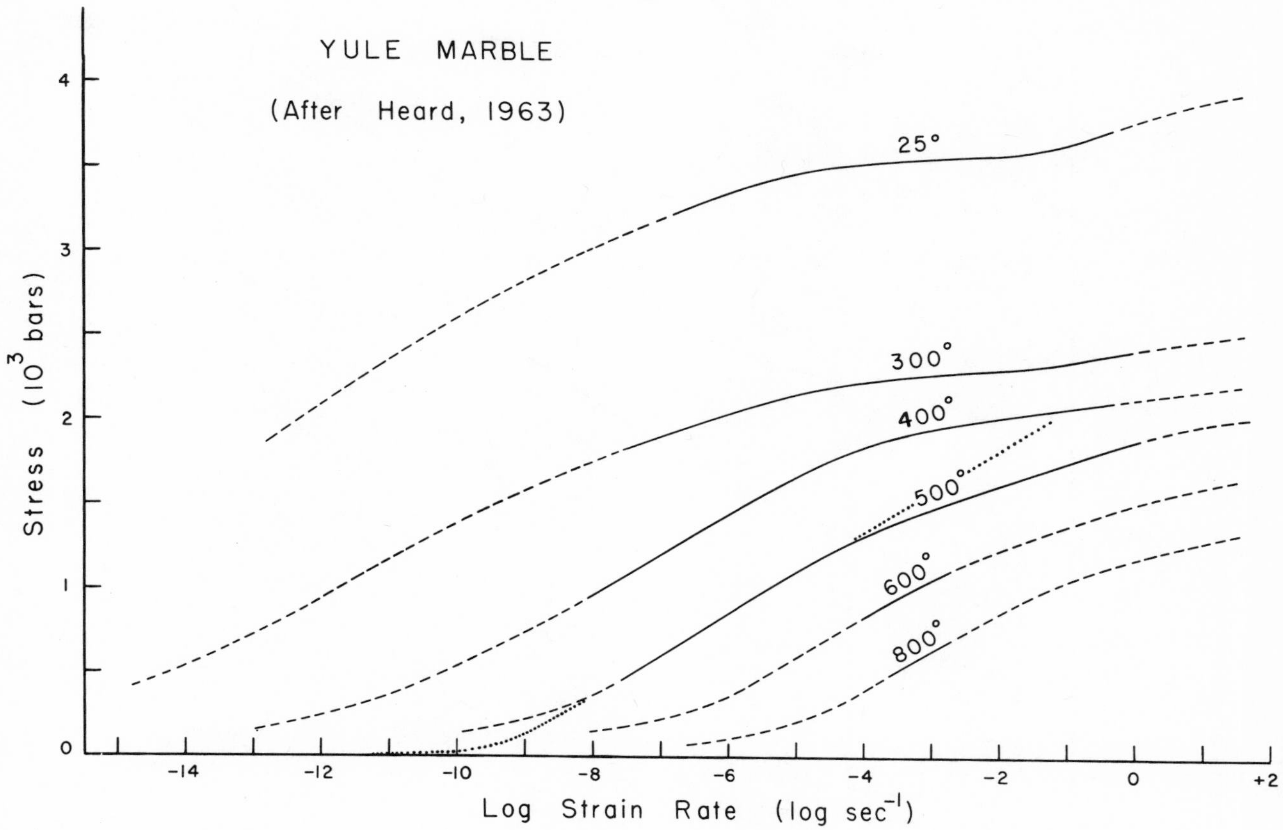

Fig. 8—Stress-strain rate curves for one cylinder of Yule marble at 10 per cent strain; solid lines cover experimental points; dotted line is a hyperbolic sine extrapolation at 500°C (after Heard, Ref. 15, Fig. 12).

and Sanfirova(54) under constant conditions of $\dot{\epsilon}$ and T for each test. The observed data in Fig. 8 for marble have been extended as dashed lines to lower and higher $\dot{\epsilon}$, similar to the observed curves for aluminum in Fig. 9.

Birch, in his paper in this symposium, "Megageological Considerations in Rock Mechanics," has reviewed the evidence from the earth satellites on the gravitational potential and has determined that stress differences of about 10^2 bars, his "creep threshold," must be supported in the upper mantle of the earth to account for the observed anomalies. Assuming that the curves for silicate rocks would be like those for marble in Fig. 8, $\sigma = 10^2$ bars is not implausible for discernible creep, $\dot{\epsilon} = 10^{-13}$ sec^{-1}, for the conditions

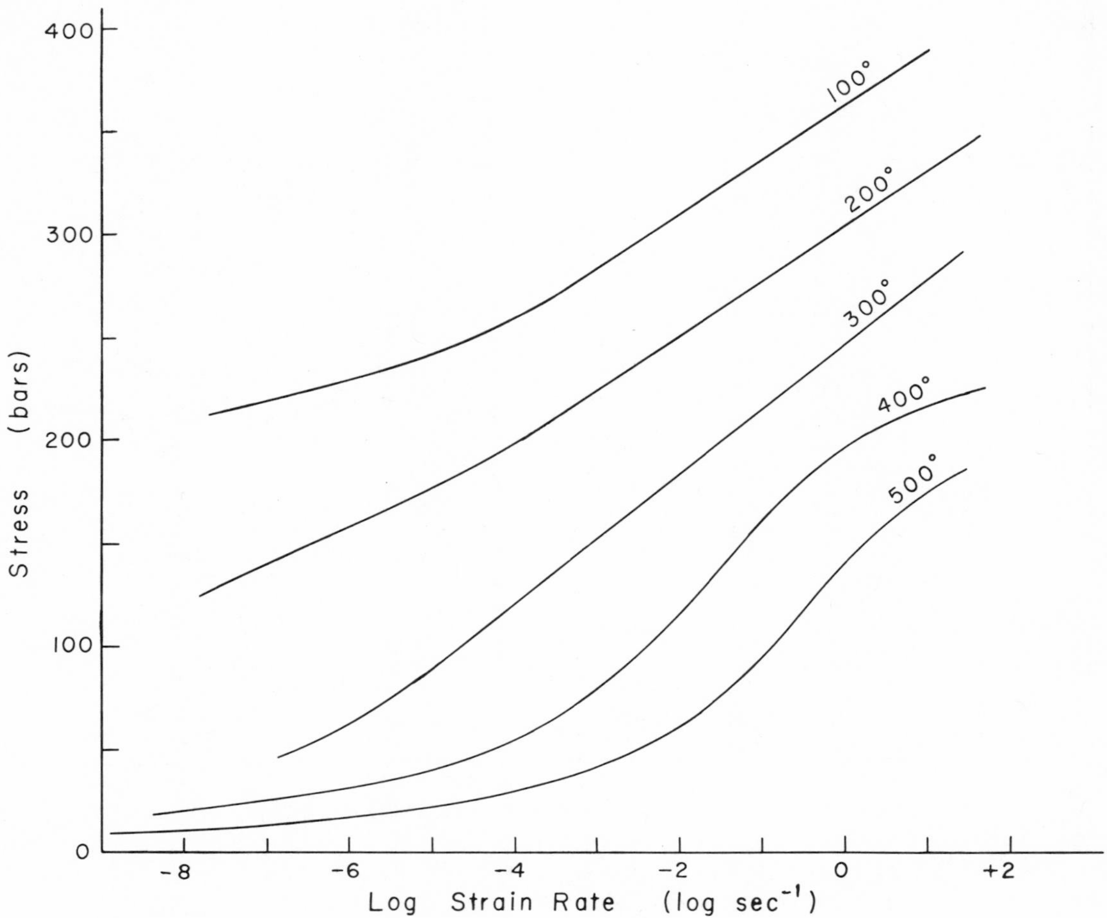

Fig. 9—Stress-log strain rate curves for pure, polycrystalline aluminum at various temperatures (after Zhurkov and Sanfirova, Ref. 54, Fig. 3).

of T and P for the upper mantle. G. J. F. MacDonald (1963, oral communication) has calculated a strength for the upper mantle of the same order from another analysis of satellite data.

Glass behaves mechanically like a silicate rock, and so a lower limit for the σ to cause creep at room temperature and pressure can be deduced from the following observations. Morey[(55)] describes the rigidity of a 6-ft diameter, 1-ton glass mirror; no creep within "a quarter of a wavelength of light" was observed after standing on edge for 25 years. Thus at σ = 5 bars, a maximum limit of $\dot{\epsilon} = 10^{-16}$ sec^{-1} is indicated, implying that a higher σ is required to attain creep at that $\dot{\epsilon}$, and furthermore implying that the σ-log $\dot{\epsilon}$ curve flattens at low $\dot{\epsilon}$.

Finally, there is the geophysical argument (Ref. 56, p. 199) that small isostatic anomalies persist near the earth's surface which indicate that a minimum load of 10^2 bars is required before perceptible adjustment will occur. This also implies a flattening of the σ-log $\dot{\epsilon}$ curve for silicate rocks.

By appropriate analysis, Heard[15] has been able to fit his data to an equation developed by Eyring from irreversible thermodynamics:

$$\dot{\epsilon}/\dot{\epsilon}_0 = \exp\ (-Q/RT)\ \sinh\ (\sigma/\sigma_0)\ , \tag{10}$$

where $\dot{\epsilon}_0$ and σ_0 are constants, Q is the activation energy of self-diffusion, R is the gas constant, and T is absolute temperature.

(The shape of the curve on a σ-log $\dot{\epsilon}$ plot is shown by the dotted line in Fig. 8.) LeComte[12] and Jellinek and Brill[41] have also used Eq. (10) to fit their data into; Murrell and Misra,[11] Stacey,[57] and others have supported the use of the equation in creep and flow of rocks. Weertman[58] has used σ^n, with n a constant, in place of the stress term in Eq. (10). However, the use of sinh (σ/σ_0) or of σ^n to extrapolate to the low geologic strain rate (for example, 10^{-13} sec^{-1}) in the earth's crust or mantle may lead to large errors, in view of the previous finding of a flattening in slope of the curve of σ-log $\dot{\epsilon}$. The effects of high pressure and temperature do not alter the argument, as only the position not the shape of the curve would be changed. It is well to remember the admonition of Hughes that in order to be certain, the $\dot{\epsilon}$ should be measured at the actual σ under consideration.

An apparent conclusion from this review of creep equations is that all are essentially empirical, principally because the phenomena involved are complex and not easily expressed in terms of strain rate, stress difference, temperature, and pressure. The mechanisms of creep are not easily identifiable in the equations, although Heard's (Ref. 15, p. 189) microscopic observations of recrystallization in his deformed samples support the diffusion mechanism on which Eq. (10) is based.

Mechanisms of Creep

Fracturing (including cataclastic flow) is an important mechanism in the creep of rocks. Matsushima[29] found in creep studies of granite that the lateral strain increased much more rapidly with each higher level of stress than the longitudinal strain (parallel to the stress); he concluded that the samples were being split apart. Evans and Wood[17] similarly observed greater lateral than longitudinal strains in the creep of slate, granite, and marble. Their results and those of Matsushima, if analyzed as for Table 3, give exponents n (Eq. (6)) greater for lateral than for longitudinal creep. Robertson[33] concluded that fracturing is important in the creep of Solenhofen limestone; for example, the density was decreased 11 per cent in one sample which was shortened by 27 per cent, under 2000 bars hydrostatic pressure. Bridgman,[59] in measuring the linear compressibilities of soapstone, marble, and diabase, observed an increase of volume with increase of load, and a recovery on unloading, which he concluded to be a recovery from fracturing. Stroh[60] calculated that 80,000 cracks/cm^2, each 13μ by 0.04μ, would account for an observed 0.045 per cent decrease in the density of metallic nickel, twisted to a strain of 2.34. Orowan[61,62] has long advocated the importance of fracturing in the deformation of metals and of rocks under pressure.

The details of the mechanisms of twin and translation gliding in minerals are known primarily from studies of plasticity in single crystals, in which the effects of stress difference, hydrostatic pressure, and temperature were considered but not time dependence. Recrystallization is another important mechanism, especially at high temperature. It is not necessary to review all the work done, although the following may be mentioned. Thorough investigations were made on calcite by Turner, Griggs, and Heard[63] and on dolomite crystals by Higgs and Handin;[64] a tabulation by Buerger[65] of glide planes and directions has been brought up to date by John Handin (1962, unpublished tables). A helpful review of gliding and plasticity in single crystals of metals is given by Paterson.[66] A review of deformation mechanisms of rock salt is in Schmid and Boas.[4] Dislocation mechanisms have been much studied lately on another ionic crystal, lithium fluoride.[67]

None of the mechanisms of creep can easily be correlated with the stages of creep or with the terms in creep equations, and this must be done before a good analytical theory can be developed. Fracturing within mineral grains is difficult to distinguish from intragranular gliding, Griffith cracks, or dislocations, especially if the discontinuities heal on release of stress. It has been hypothesized that the matrix of certain aggregated rocks, like sandstone or shale, behaves like a viscoelastic substance, and this model has been used successfully. Emery[68] has deduced viscoelastic creep and recovery equations from just such a model and applied them to residual strain observed in many rock samples. Biot[69] has evolved a useful theory of folding by buckling of viscoelastic layers, which is like that observed in sedimentary rocks. Phillips[70] studied the effect of the matrix minerals in his shale samples; he determined by a sizing method the content of quartz and clay, and by correlating these determinations with his creep tests, he discovered that the maximum strain varied directly with clay content.

Viscoelasticity

General

The theory of viscoelasticity provides a theoretical basis for analyzing data of time dependent strain and for extrapolation beyond the range of experiment. In particular, analysis of problems of creep and flow of rocks in the earth requires some such basis, and in this role viscoelasticity can be helpful, as long as its limitations are understood. The preceding experimental results afford data for its evaluation and for modifications to make it more applicable to rocks. Certain fundamentals can be considered, but no attempt will be made here to review the theory of viscoelasticity. In this section, experimental results in the small strain column of Table 1 will be covered, especially in the large strain rate, monotonic and oscillatory categories.

In linear viscoelasticity, some terms are in velocities and other terms are in units of strain and of stress, with all terms to the first power. Sometimes tandem and parallel arrangements of dashpots and springs, simulating viscous and elastic elements, are used as props to visualize the mechanics of the equations. The equations from these and from other tacitly phenomenological models in one dimension are treated at length by several authors in Eirich,(71) by Nadai,(3) by Freudenthal,(7) and in a clear exposition by Reiner.(10) Bland(72) develops a three-dimensional theory by generalizing from a one-dimensional mechanical model; Biot(73) deduces his three-dimensional theory by postulating certain potential and dissipation functions taken from irreversible thermodynamics.

Standard Linear Solid

The relaxation time for a solid, expressed in viscoelasticity theory as the ratio of the viscosity η to some elastic modulus M, is a parameter often used in viscoelasticity for the time of decay to 1/e of stress or strain. Relaxation under constant stress or constant strain is an important property of rocks, but only for small strains, before irreversible effects occur. The equation for the standard linear solid will help set the stage with usable relations between viscoelastic constants.

Equation (5) is the integrated equation at constant stress of Burger's model, which can be simplified and given in the more easily manipulated form of the standard linear solid (Ref. 1, p. 43):

$$\sigma + t_m \dot{\sigma} = M_s(\epsilon + t_k \dot{\epsilon}) \quad . \qquad (11)$$

The coefficient t_m is the relaxation (Maxwell) time at constant strain, t_k is the retardation (Kelvin) time at constant stress, and M_s is the static (or relaxed) elastic modulus. For clarity, the two time constants should be distinguished, but in anelasticity they are often lumped into a mean relaxation time t_r.

The creep and creep recovery curves in Fig. 3 represent the retardation of strain at constant stress, after loading and upon removal of load. (The distinction between constant load and constant stress can be neglected.) If an asymptotic limiting strain is estimated, values of the time constant t_k for each curve can be calculated and for these short-term tests, t_k would be about 10^3 sec; if the tests were longer, t_k would be longer. If the rock samples were held at constant strain, and the stress allowed to relax out, a plot of creep and recovery curves of σ-t at constant strain would have much the same appearance as the ϵ-t curves in Fig. 3; also, the relaxation time t_m would have a similar significance for a plot such as that of t_k in Fig. 3. Often in experiments on rocks (and metals) both stress and strain recover simultaneously, so that only a combination relaxation time t_r can be determined.

It is important to make clear at once that relaxation and retardation times are not constant; they vary with temperature and pressure, and probably are not significant for large strains. This does not detract from their usefulness as concepts in creep, flow, and high-strain-rate rock deformation, however. Similarly, the parameters of viscosity and elasticity are not constant over a wide strain-rate-frequency range, but nevertheless are indispensable concepts, that is, as operational tools.

To gain a better understanding of relaxation and the rheological properties of rocks in general, impact and anelastic dynamic tests should be considered. These topics will be discussed next.

Impact Tests

The initial loading in a creep experiment is often considered to be instantaneous and elastic, but actually there is a lag of strain behind stress. This lag is observed clearly in impact tests in which the effect of high stress rate is to compress the strain in time and produce a high strain rate, although strain never quite catches up with stress. Figure 10 shows three results from investigations of impact loading of steel,[74] concrete,[75] and rocks;[76] these curves demonstrate the time relations of stress and strain for very fast loading. Table 4 gives a condensed summary of their results and the results from shock loading of several rocks using a high-explosive method.[77]

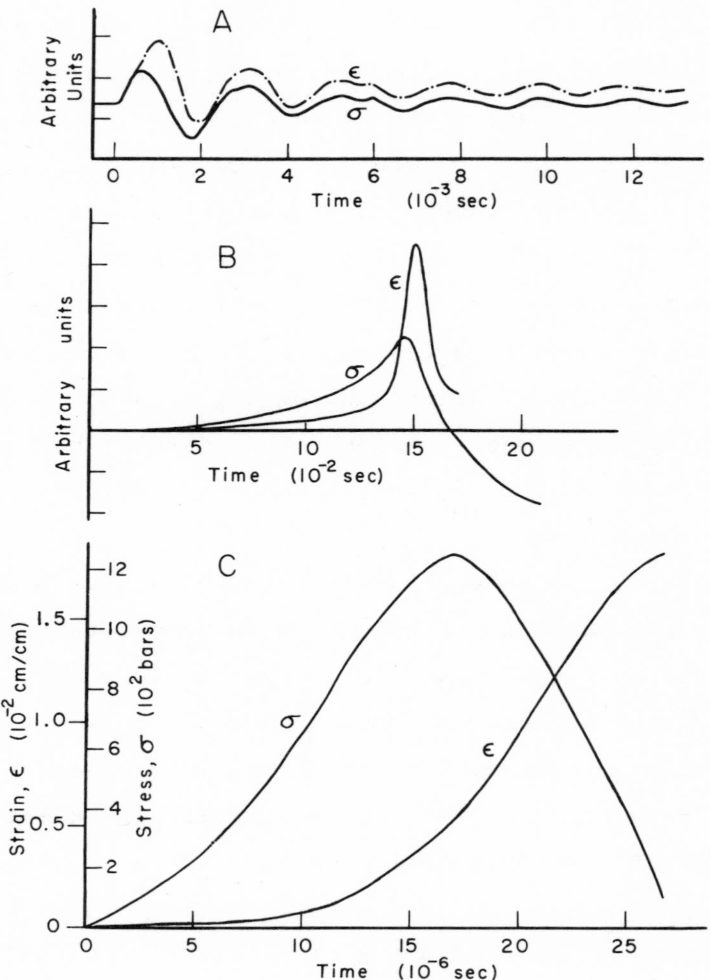

Fig. 10—A: Stress and strain waves in an impact test on 4340 steel (after Smith et al., Ref. 74, Fig. 7). B: Stress and strain waves in an impact test on concrete (after Watstein, Ref. 75, Fig. 7). C: Stress and strain waves in an impact test on Darley Dale sandstone (after Attewell, Ref. 76, Figs. 7 and 8).

The wave trains of stress and strain in A of Fig. 10 illustrate dynamic effects in steel, and the single cycle of strain and stress waves in pulse loading of concrete and of sandstone are shown in B and C of Fig. 10.

The curves in Fig. 10 make it easier perhaps to visualize how the inherent mechanical properties of a material control its behavior, especially the lag of strain behind stress. The maximum stress and

Table 4

IMPACT TESTS ON VARIOUS ROCKS AT ROOM TEMPERATURE

Specimen Type	Number of Tests	Stress		Strain		Reference
		Average Rate $\dot{\sigma}$ (bars/sec)	Maximum σ (10^3 bars)	Average Rate $\dot{\epsilon}$ (sec^{-1})	Maximum ϵ (cm/cm)	
Darley Dale sandstone	1	10^8	1	10^3	10^{-2}	Attewell (Ref. 76)
Granodiorite	3	10^8	2	10^2	10^{-3}	
Carrara marble	3	10^8	2	10^3	10^{-2}	
Tunstead limestone	1	10^8	2	10^3	10^{-2}	
Dolomite	2	10^8	2	10^2	10^{-3}	
Weak concrete	3	1	0.2	10^{-6}	10^{-3}	Watstein (Ref. 75)
	1	10^2	0.2	10^{-3}	10^{-3}	
	1	10^5	0.2	1	10^{-3}	
	1	10^6	0.3	10	10^{-3}	
Strong concrete	3	1	0.4	10^{-6}	10^{-3}	
	1	10^3	0.5	10^{-3}	10^{-3}	
	1	10^6	0.6	1	10^{-3}	
	1	10^6	0.9	10	10^{-3}	
Steel, SAE 4340, hardened	1	10^8	10	10	10^{-2}	Smith, Pardue, and Vigness (Ref. 74)
Rock salt	14		10^2		0.3[a]	Lombard (Ref. 77)
	6		10^3		0.5	
Granite	11		10^2		0.3	
	2		10^3		0.5	
Rhyolite tuff	16		10^2		0.4	
Dolomite	2		10^2		0.2	
Limestone	4		10^2		0.3	
	2		10^3		0.4	
Andesite	3		10^2		0.2	
Basalt	4		10^3		0.3	
Marble	6		10^2		0.2	
Taconite	4		10^2		0.3	
	3		10^3		0.5	
Oil sand	3		10^2		0.4	
	3		10^3		0.5	
Oil shale	19		10^2		0.3	

[a]In these shock experiments, the strain is the change in volume calculated from particle and shock velocities.

strain for the steel sample (Fig. 10A) were $\sigma = 16 \times 10^3$ bars, $\epsilon = 0.10$; it was stretched by a method in which the initial input energy was absorbed by plastic deformation of the sample during the decay of oscillations of the high inertia components coupled to the sample. For the concrete specimen (Fig. 10B), the rise time of stress due to a dropping weight was more gradual; after a critical stress was reached, the rise time of strain was sharp, possibly due to the plasticity of the concrete; the specimen failed by rupture. Despite the high stresses and high strains induced in the sandstone (Fig. 10C) and the other four rocks of Attewell,[76] the deformation was fully recovered, showing no residual strain when measured within 2 minutes after the tests. The relaxation time for these rocks is very short compared with slowly strained rocks; relaxation times may vary with strain rate, but more likely, the mechanisms of relaxation probably differ at different strain rates (Ref. 1, Frontispiece).

Internal Friction Anomalies

In his basic work on anelasticity, Zener (Ref. 1, Chap. 5) by integration of Eq. (11) for the standard linear solid obtains a relation between the relaxation times τ_m and τ_k and the unrelaxed and relaxed elastic moduli (which can also be called the dynamic and static moduli, respectively):

$$\frac{M_u}{M_r} = \frac{\tau_k}{\tau_m} . \tag{12}$$

From Eq. (12) and the expression for the logarithmic decrement δ (lag of ϵ behind σ) in dynamic strain at frequency f, Zener derives another equation:

$$\tan \delta = \frac{M_u - M_r}{M} \cdot \frac{ft_r}{1 + (ft_r)^2} , \tag{13}$$

where $M = \sqrt{M_u M_r}$, and $t_r = \sqrt{t_m t_k}$. With this relation it can be shown that at some frequency dependent on the relaxation properties of a material, a peak of δ is centered at the break in slope between the values of M_u and M_r. Zener points out that such a peak in δ can be correlated with a mechanism of relaxation in the solid.

Kê[78] in an important set of experiments on aluminum found a relaxation peak in δ, and a change of 30 per cent in M as a function of temperature, as in A and B of Fig. 11. He used aluminum wire in a torsion pendulum, oscillating at f = 0.8 cps. Kê concluded that the phenomenon was a relaxation at the grain boundaries of the polycrystalline aluminum primarily because it was not observed in single crystal aluminum. Other studies have corroborated Kê's findings, and the usefulness in metallurgy of studies of internal friction is apparent from the very great interest in the subject.[79]

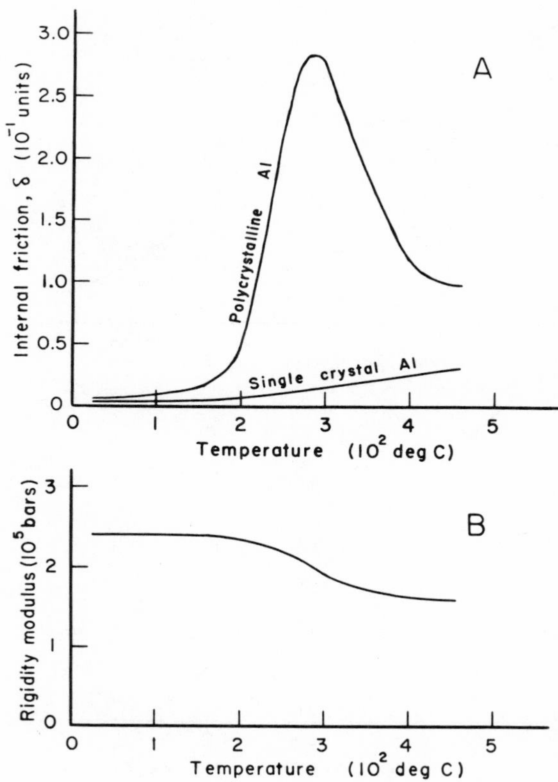

Fig. 11—A: Variation of internal friction with temperature. B: Relaxation of rigidity in polycrystalline aluminum. (After Kê, Ref. 78, Figs. 2 and 4.)

Peselnick and Zietz[80] compared the δ of limestone with δ of a calcite crystal and concluded that the internal friction probably occurs at the grain boundaries of the limestone. Peselnick and Outerbridge[81] measured δ in limestone at f = 5 cps, 10^4 cps, and 10^7 cps and found no evidence of a relaxation mechanism at room temperature and pressure; Peselnick (1963, oral communication) remarks that in no other study of internal friction in rocks has a peak in δ been found either.

A relaxation study of single crystals of ice was made by Kneser <u>et al.</u>,[82] using a resonance technique. Torsional oscillations at ten different frequencies, f = 0.69 x 10^3 cps to 81 x 10^3 cps, and over a temperature range from 0° to -25°C showed the dependence of a peak in δ on frequency, shown in Fig. 12A. In addition, the dependence of maximum damping frequency on temperature was found, as shown in Fig. 12B. The obvious relaxation mechanism is translation gliding on

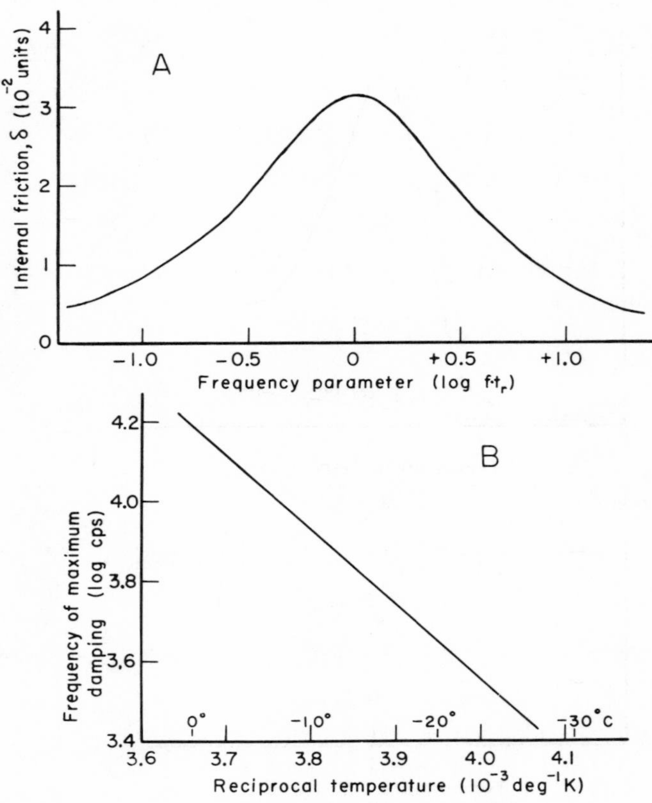

Fig. 12—A: Maximum of internal friction. B: Maximum damping frequency as a function of temperature. (After Kneser <u>et al.</u>, Ref. 82, Figs. 2 and 1.)

the {0001} plane because the axis of the rods of ice in torsion was perpendicular to {0001}, thus permitting shear on that plane, the only glide plane verified in ice as yet.[83]

In Fig. 12A the relaxed modulus M_r for the lower frequencies on the left is 6 per cent smaller than the unrelaxed modulus M_u at the higher frequencies on the right. As Zener (Ref. 1, Frontispiece) suggests, there might be as many as six such anomalies in a frequency spectrum of 10^{20} cps, due to as many mechanisms of relaxation. The differences in M over the spectrum could add up to a large over-all change.

The effect of temperature is similar to that of frequency, but inverse. In a temperature range in which the relaxation mechanism operates at the frequency employed, there is an internal friction maximum and a drop from the unrelaxed to relaxed M (Fig. 11). Furthermore, increase of temperature increases the frequency of maximum damping (Fig. 12B). Experimentally, it is easier to locate internal friction maxima as a function of temperature than of frequency.

In their study of B_2O_3 glass by a resonance method, Birch and Bancroft (Ref. 84, Fig. 3) found the beginning of a peak in δ and a decrease in M with temperature. These features may represent an incipient relaxation, not revealed perhaps because of the high frequency, $f = 10^4$ cps; but as they occur just before melting, they may also be attributes of the melting process.

Experiments in anelasticity are significant because they reveal the mechanisms of attenuation of energy in a solid. Such studies are much needed for rocks and minerals.

Viscosity Parameter

There is some uncertainty about applying the theory of viscoelasticity, particularly the viscosity coefficient, to rocks (Refs. 84 and 85; Ref. 86, Chap. 10). It seems better to analyze the available experimental results empirically, rather than to try to fit stress and time dependent strain data into hydrodynamic theory. In fact, the use of the term viscosity itself for rocks seems unfortunate; as

Goranson[87] pointed out, viscosity has a definite meaning in hydrodynamics and is used loosely when applied to solids, because for solids the coefficient varies with stress, stress rate, and time (except for steady state creep). Goranson suggested using the word <u>mobility</u> instead for the vastly more complicated phenomenon in solids. Also, he suggested that mobility be expressed by the inverse of viscosity units, that is, the reciprocal poise. The suggestion is a good one, although it is not followed in this paper because of long usage and pusillanimity.

The viscosity, $\eta = \sigma/\dot{\epsilon}$, has been determined for rocks from monotonic creep tests, made at room temperature in a narrow stress range, $\sigma = 10^2$ to 10^3 bars, and in this range η does not change much. The averaged values for η in poises range from 10^{15} to 10^{17} for halite,[12,20] 10^{16} to 10^{17} for coal,[32] 10^{16} to 10^{18} for alabaster,[21] 10^{21} for Fennoscandian crustal rebound,[24] and 10^{22} for limestone.[20] From his experiments, Heard[15] gives $\eta = 10^{18}$ to 10^{22} poises for Yule marble in the ranges $\dot{\epsilon} = 10^{-10}$ to 10^{-14} sec^{-1} and T = 25° to 400°C.

Disregarding the fact that the mechanisms of creep may not be the same over the whole range of $\dot{\epsilon}$, let us calculate the viscosity η for the high, middle, and low $\dot{\epsilon}$ parts of the 400°C curve of Heard, as extended in Fig. 8. At σ = 2000 bars, $\eta = 10^{11}$ poises; at σ = 1000 bars, $\eta = 10^{17}$ poises; and at σ = 200 bars, $\eta = 10^{21}$ poises. The value of η will obviously increase rapidly at lower stresses.

It is fairly certain that η is a greatly changing parameter over the range of σ and of $\dot{\epsilon}$ of interest. Instead, we can define $\sigma/\log \dot{\epsilon}$ as a new parameter, and call it logarithmic viscosity η^*; it has the advantage that its value will reflect the sigmoidal shape of the curve of σ-log $\dot{\epsilon}$, as in Fig. 8. However, η^* is only an empirical parameter, and is not very useful in the theory of viscoelasticity. We may conclude from our analysis of the creep studies that η is not a constant, and that therefore linear viscoelasticity is not applicable in general for rock deformation.

<u>Logarithmic Viscosity and Internal Friction</u>

Comparisons of the σ-log $\dot{\epsilon}$ curve in Fig. 8 with the M and δ curves in Figs. 11 and 12, and the M and δ versus f curves of Zener

(Ref. 1, Fig. 21) suggest a similarity; these curves are shown diagrammatically in Fig. 13. Analogy of logarithmic viscosity η^* with internal friction δ is tempting in view of the similarity of the dimensional units, M/f in dynes cm^{-2} sec, and $\sigma/\log\dot{\epsilon}$ in dynes cm^{-2} log sec. If a relation can be found between them, measurement of internal friction would offer a simple experimental approach to determining values of η^* and of mechanisms of creep and relaxation.

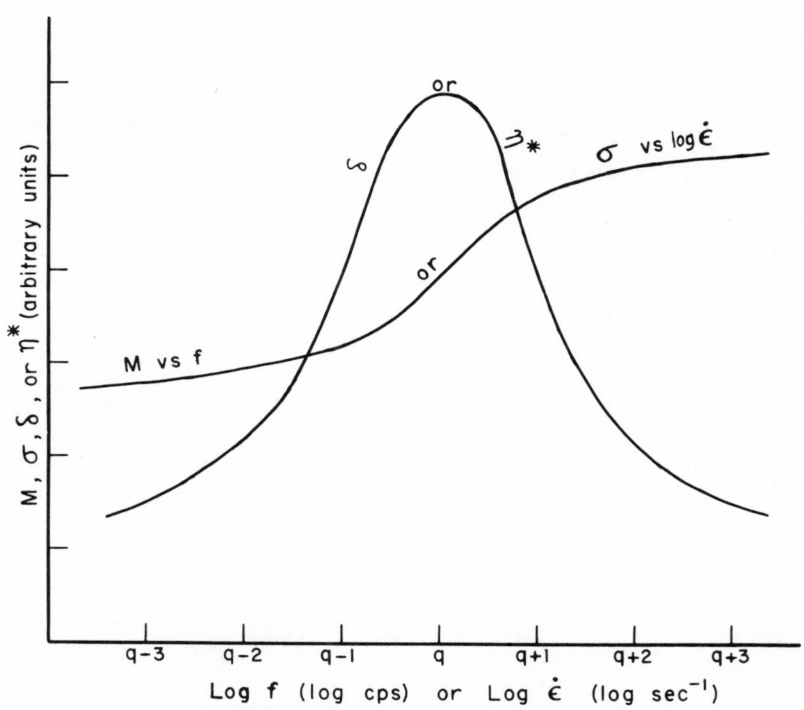

Fig. 13—Elastic modulus M and internal friction δ as a function of frequency f (after Zener, Ref. 1, Fig. 21), and stress σ and logarithmic viscosity η^* as a function of strain rate $\dot{\epsilon}$. (Diagrammatic.)

The great difference in total ϵ between the two techniques, very low ϵ for internal friction and relatively large ϵ for creep, suggests that it would be fortuitous if the dissipation mechanism were the same as the creep mechanism. However, there is pertinent evidence from work on aluminum. The peak in δ found by Kê (Ref. 78, Fig. 11) occurred at 300°C at $\dot{\epsilon} = 10^{-5}$ sec^{-1}; the point at $\dot{\epsilon} = 10^{-5}$ sec^{-1} on the σ-log $\dot{\epsilon}$ creep curve at 300°C for aluminum was observed metallographically by Servi and Grant[(88)] to mark the boundary between plastic flow between the grains and the onset of separations between the grains, that is, at incipient fracture. It is not obvious that the grain boundary attenuation mechanism at maximum $\epsilon = 10^{-5}$ is the same as the creep fracturing mechanism at maximum $\epsilon = 10^{-1}$, but the correlation is very interesting. Without more experimental evidence, perhaps the only valid parallel to be noted is that activation of anomalous behavior seems to occur at an inflection of the σ-log $\dot{\epsilon}$ curve just as it occurs at the inflection of an M-f curve. At low σ the mechanism of creep can be twin and translation gliding, self-diffusion, impurity diffusion, or dislocation mechanisms, and at high σ the deformation seems to occur by fracturing,[33,62] or at high temperature by recrystallization.[15] Therefore, the inflection of the σ-log $\dot{\epsilon}$ curve may occur because an activation σ is required to initiate fracturing or recrystallization.

REFERENCES

1. Zener, C., Elasticity and Anelasticity of Metals, University of Chicago Press, Chicago, 1948.
2. Nadai, A., Theory of Flow and Fracture of Solids, Vol. 1, McGraw-Hill Book Company, Inc., New York, 1950.
3. Nadai, A., Theory of Flow and Fracture of Solids, Vol. 2, McGraw-Hill Book Company, Inc., New York, 1963.
4. Schmid, E., and W. Boas, Plasticity of Crystals (a translation of Kristallplastizitat), F. A. Hughes and Company, London, 1950.
5. Walton, W. H. (ed.), Mechanical Properties of Nonmetallic Brittle Materials, Interscience Publishers, New York, 1958.
6. Griggs, D., and J. Handin (eds.), Rock Deformation, Geol. Soc. Am. Mem. 79, 1960.
7. Freudenthal, A. M., The Inelastic Behavior of Engineering Materials and Structures, John Wiley & Sons, Inc., New York, 1950.
8. Cottrell, A. H., "The Time Laws of Creep," J. Mech. Phys. Solids, Vol. 1, 1952, pp. 53-63.
9. Schoeck, G., "Theory of Creep," in Creep and Recovery, American Society for Metals, Cleveland, 1957, pp. 199-226.

10. Reiner, M., Deformation, Strain, and Flow, Interscience Publishers, Inc., New York, 1960.

11. Murrell, S. A. F., and A. K. Misra, "Time-dependent Strain or Creep in Rocks and Similar Nonmetallic Materials," Bull. Inst. Mining Met., Vol. 71, Pt. 7, 1962, pp. 353-378.

12. LeComte, P., "Creep and Internal Friction in Rock Salt," doctoral dissertation, Harvard University, 1960.

13. Kendall, H. A., "An Investigation of the Creep Phenomena Exhibited by Solenhofen Limestone, Halite, and Cement under Medium Confining Pressures," Master of Science thesis, Texas Agricultural Mining College, 1958.

14. Dorn, J. E., "Some Fundamental Experiments on High Temperature Creep," J. Mech. Phys. Solids, Vol. 3, 1954, pp. 85-116.

15. Heard, H. C., "The Effect of Large Changes in Strain Rate in the Experimental Deformation of Rocks," J. Geol., Vol. 71, No. 2, 1963, pp. 162-195.

16. Evans, R. H., "The Elasticity and Plasticity of Rocks and Artificial Stone," Proc. Leeds Phil. Lit. Soc., Vol. 3, Pt. 3, 1936, pp. 145-158.

17. Evans, R. H., and R. H. Wood, "Transverse Elasticity of Natural Stones," Proc. Leeds Phil. Lit. Soc., Vol. 3, Pt. 5, 1937, pp. 340-352.

18. Goguel, J., Introduction à l'étude méchanique des déformations de l'écorce terrestre, Imprimerie Nationale, Paris, 1943.

19. Griggs, D. T., "Deformation of Rocks under High Confining Pressures," J. Geol., Vol. 44, No. 5, 1936, pp. 541-577.

20. Griggs, D. T., "Creep of Rocks," J. Geol., Vol. 47, No. 3, 1939, pp. 225-251.

21. Griggs, D. T., "Experimental Flow of Rocks under Conditions Favoring Recrystallization," Bull. Geol. Soc. Am., Vol. 51, 1940, pp. 1001-1034.

22. Gunter, B. D., and F. L. Parker, The Physical Properties of Rock Salt as Influenced by Gamma Rays, Oak Ridge National Laboratory Report ORNL-3027, 1961.

23. Hardy, H. R., "Time-dependent Deformation and Failure of Geologic Materials," Quart. Colo. School Mines, Vol. 54, No. 3, 1959, pp. 134-175.

24. Haskell, N. A., "The Viscosity of the Asthenosphere," Am. J. Sci., Ser. 5, Vol. 33, No. 193, 1937, pp. 22-28.

25. Iida, K., T. Wada, Y. Aida, and R. Shichi, "Measurements of Creep in Igneous Rocks," J. Earth Sci., Nagoya Univ., Vol. 8, No. 1, 1960, pp. 1-16.

26. Inouye, K., and H. Tani, "Rheology of Coal, IV, Strength and Fracture of Coal," Bull. Chem. Soc. Japan, Vol. 26, No. 4, 1953, pp. 200-204.

27. Kuznetsov, G. N., Mekhanicheskiye svoystva gornykh porod (Mechanical Properties of Rocks), Ugletekhizdat, Moscow, 1947.

28. Lomnitz, C., "Creep Measurements in Igneous Rocks," J. Geol., Vol. 64, No. 5, 1956, pp. 473-479.

29. Matsushima, S., "On the Flow and Fracture of Igneous Rocks," Disaster Prevent. Res. Inst., Kyoto Univ., Bull. 36, 1960, pp. 1-9.

30. Nagaoka, H., "Elastic Constants of Rocks and the Velocity of Seismic Waves," Phil. Mag., Vol. 50, No. 302, 1900, pp. 53-68.

31. Phillips, D. W., "Further Investigation of the Physical Properties of Coal-Measure Rocks and Experimental Work on the Development of Fractures," Trans. Inst. Mining Engr., Vol. 82, Pt. 5, 1932, pp. 432-450.

32. Pomeroy, C. D., "Creep in Coal at Room Temperature," Nature, Vol. 178, No. 4527, 1956, p. 279.

33. Robertson, E. C., "Creep of Solenhofen Limestone under Moderate Hydrostatic Pressure," Geol. Soc. Am. Mem. 79, 1960, pp. 227-244.

34. Roux, A. J. A., and H. G. Denkhous, "An Analysis of the Problem of Rock Bursts in Deep Level Mining, Part 2," Chem. Met. Mining Soc. South Africa J., Vol. 55, 1954, pp. 103-124.

35. Glen, J. W., "Experiments on the Deformation of Ice," J. Glaciology, Vol. 2, No. 12, 1952, pp. 111-114.

36. Glen, J. W., "The Creep of Polycrystalline Ice," Proc. Roy. Soc. (London), Ser. A., Vol. 228, 1955, pp. 519-538.

37. Glen, J. W., "Measurement of the Deformation of Ice in a Tunnel at the Foot of an Ice Fall," J. Glaciology, Vol. 2, No. 20, 1956, pp. 735-746.

38. Glen, J. W., and M. F. Perutz, "The Growth and Deformation of Ice Crystals," J. Glaciology, Vol. 2, No. 16, 1954, pp. 397-402.

39. Gold, L. W., "The Cracking Activity in Ice during Creep," Can. J. Phys., Vol. 38, No. 9, 1960, pp. 1137-1148.

40. Griggs, D. T., and N. E. Coles, "Creep of Single Crystals of Ice," U. S. Army Snow, Ice, and Permafrost Research Establishment, SIPRE Rept. 11, 1954.

41. Jellinek, H. H. G., and R. Brill, "Viscoelastic Properties of Ice," J. Appl. Phys., Vol. 27, No. 10, 1956, pp. 1198-1209.

42. Meier, M. F., "Mode of Flow of Saskatchewan Glacier, Alberta, Canada," U. S. Geol. Surv. Profess. Papers, No. 351, 1960.

43. Rigsby, G. P., "Effect of Hydrostatic Pressure on Velocity of Shear Deformation of Single Crystals of Ice," U. S. Army Snow, Ice, and Permafrost Research Establishment, SIPRE Rept. 32, 1957.

44. Steinemann, S., "Results of Preliminary Experiments on the Plasticity of Ice Crystals," J. Glaciology, Vol. 2, No. 16, 1954, pp. 404-412.

45. Steinemann, S., "Experimentelle Untersuchungen zur Plastizitat von Eis" (Experimental Investigation of the Plasticity of Ice), Beitr. Geol. Schweiz, Hydrologie, No. 10, 1958.

46. Handin, J., R. V. Hager, Jr., M. Friedman, and J. N. Feather, "Experimental Deformation of Sedimentary Rocks under Confining Pressure: Pore Pressure Tests," Bull. Am. Assoc. Petrol. Geologists, Vol. 47, No. 5, 1963, pp. 717-755.

47. Michelson, A. A., "The Laws of Elastico-viscous Flow," J. Geol., Vol. 28, No. 1, 1920, pp. 18-24.

48. Michelson, A. A., "The Laws of Elastico-viscous Flow, I," J. Geol., Vol. 25, No. 5, 1917, pp. 405-410.

49. Goranson, R. W., "'Flow' in Stressed Solids: An Interpretation," Bull. Geol. Soc. Am., Vol. 51, 1940, pp. 1023-1033.

50. Ruppeneit, K. V., Davlenie i smeshchenie gornykh porod v lavakh pologopadayushchikh plastov (Pressure and Displacement of Rocks in Sloping Lava Beds), Ugletekhezdat, Moscow, 1957.

51. Griggs, D. T., F. J. Turner, and H. C. Heard, "Deformation of Rocks at 500° to 800°C," Geol. Soc. Am. Mem. 79, 1960, pp. 39-104.

52. Heard, H. C., "Transition from Brittle to Ductile Flow in Solenhofen Limestone as a Function of Temperature, Confining Pressure, and Interstitial Fluid Pressure," Geol. Soc. Am. Mem. 79, 1960, pp. 193-226.

53. Serdengecti, S., and G. D. Boozer, "The Effects of Strain Rate and Temperature on the Behavior of Rocks Subjected to Triaxial Compression," Penn. State Univ., Mineral Ind. Expt. Sta. Bull. 76, 1961, pp. 83-97.

54. Zhurkov, S. N., and T. P. Sanfirova, "Svyaz mezhdu prochnostyu i polzuchestyu metallov v splavov" (Relation Between Strength and Creep of Metals and Alloys), Zh. tekhnicheskoi fiziki, Akad. Nauk, SSSR, Vol. 28, No. 8, 1958, pp. 1719-1726.

55. Morey, G. W , "The Flow of Glass at Room Temperature," J. Op. Soc. Am., Vol. 42, No. 11, 1952, pp. 856-857.

56. Jeffreys, H., The Earth, Its Origin, History, and Physical Constitution, Cambridge University Press, New York, 1952.

57. Stacey, F. D., "The Theory of Creep in Rocks and the Problem of Connection in the Earth's Mantle," Icarus, Vol. 1, No. 4, 1963, pp. 304-313.

58. Weertman, J., "Mechanism for Continental Drift," J. Geophys. Res., Vol. 67, No. 3, 1962, pp. 1133-1139.

59. Bridgman, P. W., "Volume Changes in the Plastic Stages of Simple Compression," J. Appl. Phys., Vol. 20, 1949, pp. 1241-1251.

60. Stroh, A. N., "The Existence of Microcracks after Cold-Work," Phil. Mag., Ser. 8, Vol. 2, 1957, pp. 1-4.

61. Orowan, E., "Fracture and Strength of Solids," Rept. Progr. Phys., Vol. 12, 1949, pp. 185-232.

62. Orowan, E., "Mechanism of Seismic Faulting," Geol. Soc. Am. Mem. 79, 1960, pp. 323-345.

63. Turner, F. J., D. T. Griggs, and H. Heard, "Experimental Deformation of Calcite Crystals," Bull. Geol. Soc. Am., Vol. 65, 1954, pp. 883-934.

64. Higgs, D. V., and J. Handin, "Experimental Deformation of Dolomite Single Crystals," Bull. Geol. Soc. Am., Vol. 70, 1959, pp. 245-278.

65. Buerger, M. S., "Translation-gliding in Crystals," Am. Mineralogist, Vol. 15, 1930, pp. 45-64.

66. Paterson, M. S., "The Plastic Deformation of Single Crystals," J. Australian Inst. Metals, Vol. 1, No. 2, 1956, pp. 112-124.

67. Johnston, W. G., and J. J. Gilman, "Dislocation Velocities, Dislocation Densities, and Plastic Flow in Lithium Fluoride Crystal," J. Appl. Phys., Vol. 30, 1959, pp. 129-143.

68. Emery, C. L., "Testing Rock in Compression," Mine Quarry Eng., Vol. 26, No. 4, 1960, pp. 164-168, and No. 5, 1960, pp. 196-202.

69. Biot, M. A., "Theory of Folding of Stratified Viscoelastic Media and Its Implications in Tectonics and Orogenesis," Bull. Geol. Soc. Am., Vol. 72, No. 11, 1961, pp. 1595-1620.

70. Phillips, D. W., "The Nature and Physical Properties of Some Coal-measure Strata," Trans. Inst. Mining Engr., Vol. 80, Pt. 4, 1931, pp. 212-242.

71. Eirich, F. R. (ed.), Rheology, Theory and Applications, Vol. 1, Academic Press, Inc., New York, 1954.

72. Bland, D. R., The Theory of Linear Viscoelasticity, Permagon Press, New York, 1960.

73. Biot, M. A., "Linear Thermodynamics and the Mechanics of Solids," in Proc. 3d U. S. Natl. Congr. Appl. Mech., American Society of Mechanical Engineers, 1958, pp. 1-18.

74. Smith, R. C., T. E. Pardue, and I. Vigness, The Effect of Axial Dynamic Loads on the Mechanical Properties of Certain Steels, Naval Research Laboratory Report 4468, 1954.

75. Watstein, D., "Effect of Straining Rate on the Compressive Strength and Elastic Properties of Concrete," J. Am. Concrete Inst., Vol. 24, No. 8, 1953, pp. 729-744.

76. Attewell, P. B., "Response of Rocks to High Velocity Impact," Bull. Inst. Mining Met., Vol. 71, No. 670, 1962, pp. 705-724.

77. Lombard, D. B., "The Hugoniot Equation of State of Rocks," Penn. State Univ., Mineral Ind. Expt. Sta. Bull. 76, 1961, pp. 143-152.

78. Kê, T. S., "Experimental Evidence of the Viscous Behavior of Grain Boundaries in Metals," Phys. Rev., Vol. 71, No. 8, 1947, pp. 533-546.

79. Newick, A. S., H. S. Sack, and P. J. Leurgans (eds.), "Conference on Internal Friction due to Crystal Lattice Imperfections," Acta Met., Vol. 10, No. 4, 1962, pp. 271-500.

80. Peselnick, L., and I. Zietz, "Internal Friction of Fine-grained Limestones at Ultrasonic Frequencies," Geophysics, Vol. 24, No. 2, 1959, pp. 285-296.

81. Peselnick, L., and W. F. Outerbridge, "Internal Friction in Shear and Shear Modulus of Solenhofen Limestone over a Frequency Range of 10^7 Cycles per Second," J. Geophys. Res., Vol. 66, No. 2, 1961, pp. 581-588.

82. Kneser, H. O., S. Magun, and G. Ziegler, "Mechanische Relaxation von einkristallinem Eis" (Mechanical Relaxation in Single Crystals of Ice), Naturwissenschaften, Vol. 42, No. 15, 1955, p. 437.

83. Glen, J. W., "The Mechanical Properties of Ice: I. The Plastic Properties of Ice," Advan. Phys., Vol. 7, No. 26, 1958, pp. 254-265.

84. Birch, F., and D. Bancroft, "The Elasticity of Glass at High Temperatures, and the Vitreous Basaltic Substratum," Am. J. Sci., Vol. 240, 1942, pp. 457-490.

85. Knopoff, L., and G. J. F. MacDonald, "Attenuation of Small Amplitude Stress Waves in Solids," Rev. Mod. Phys., Vol. 30, No. 4, 1958, pp. 1178-1192.

86. Munk, W. H., and G. J. F. MacDonald, The Rotation of the Earth, Cambridge University Press, London, 1960.

87. Goranson, R. W., "Fracture and Flow in Stressed Solids," Trans. Am. Geophys. Union, Vol. 21, 1940, pp. 698-700.

88. Servi, I. S., and N. J. Grant, "Creep and Stress Rupture Behavior of Aluminum as a Function of Purity," Am. Inst. Mining Met. Petrol. Engrs., J. Metals, Vol. 3, No. 10, 1951, pp. 909-916.

89. Griggs, D. T., M. S. Paterson, H. C. Heard, and F. J. Turner, "Annealing Recrystallization in Calcite Crystals and Aggregates," Geol. Soc. Am. Mem. 79, 1960, pp. 21-37.

DISCUSSION

R. G. COLEMAN (USA):

Can we with our present knowledge of experimental rock strength demonstrate that "tectonic overpressures" obtain during certain types of metamorphism? I am in particular referring to the formation of such high-pressure minerals as jadeite, aragonite, and kyanite. These minerals demand such high pressures it seems unlikely that their stable formation could be accomplished by depth of burial alone within the earth's crust.

E. C. ROBERTSON in reply:

It is a matter of having not only a knowledge of the strength of rocks, but also something of the geochemistry of formation of these high-pressure minerals, aragonite, jadeite, and kyanite. Dr. Coleman is referring, I believe, to the fact that at low pressure and temperature these minerals are metastable, and that they would be metastable in the shallow depths of the earth's crust, where they are probably formed. The question is, then, why do the metastable minerals form in the crust rather than their alternative polymorphs which *are* stable at low pressure?

Let us consider aragonite first. As a matter of fact, we know that aragonite can originate at essentially zero pressure; it forms under 10 ft of water in the banks of the Bahama Islands, and that is not a very high pressure. The chances are that the formation of aragonite rather than the stable polymorph, calcite, is a geochemical problem and has to do with the metastability induced by the chemistry of the environment.

Jadeite as found in nature is not pure; it is always a mixture of diopside and jadeite, and as such might have a stability field which would come fairly close to the existing pressures near the surface of the crust. Impure jadeite, as it is actually found in rocks, might very well form at a shallow depth because the pressures there would be adequate.

As to kyanite, the problem is somewhat more difficult because the geochemistry of the formation of kyanite, and its polymorphs, sillimanite and andalusite, is not known. As indicated by Sidney Clark, tectonic "overpressure," creating stress differences up to the rock strength, can exist. This "overpressure" is a lateral stress considerably greater than that induced by the superincumbent load of rock, and is presumably due to a tectonic process, wherever that comes from. It is possible that a stress difference can develop which is twice as large as that to be expected from the lithostatic pressure, and the mean stress might be adequate to form kyanite at not too great depths.

The catalytic action of shearing stresses may be important in producing kyanite at low temperatures. It has been found in the laboratory, for example, that quartz inverts to coesite, a high-pressure form of silica, at 100°C, if a shearing stress is applied to the sample at the same time that it is compressed and heated. Otherwise, if overpressure and shearing stress are not adequate, I can only suggest that kyanite forms metastably because of the geochemical environment, as does aragonite.

K.-H. HÖFER (Germany):*

Mr. Robertson demonstrates in his very interesting paper the importance of realizing that the relaxation time is not constant but may vary with temperature and stress. We can support this statement from our own experiences in testing of rock both in the laboratory and *in situ*.

The dependence of the strain rate on the stressing, as shown in Fig. 14,** was proved by *in situ* measurement on pillars in potash salt mines. My collaborator, geophysicist F. Schuppe, studied the creep relations, as obtained *in situ*, and derived a

---

**In absentia*. Contribution translated from the German by K. W. John. The mineralogic terms used by Dr. Höfer are translated literally; as chemical equivalents were not available, it was not possible to put these words in mineralogic terms more familiar (or possibly more accurate) to the American reader.

**K.-H. Höfer, "Beitrag zur Frage der Standfestigkeit von Bergfesten im Kalibergbau," *Freiberger Forschungsh*., A 100, 1958.

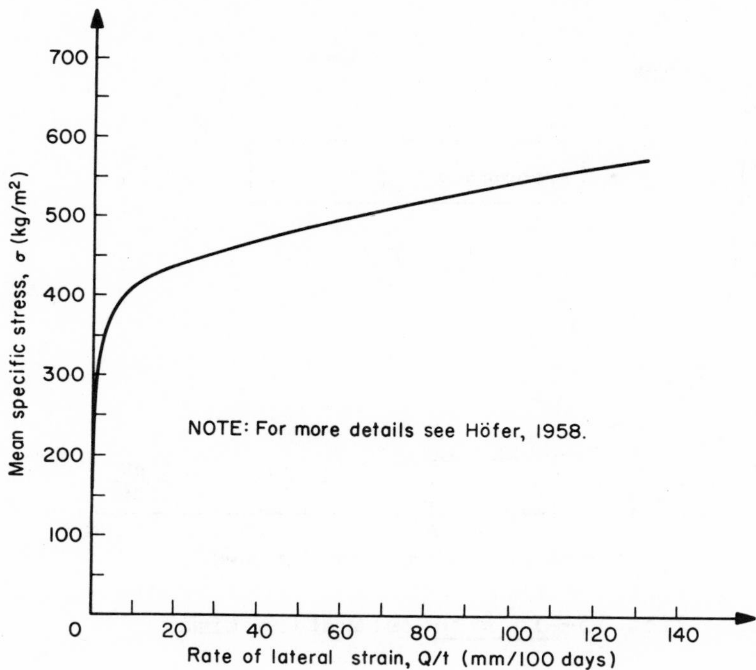

Fig. 14—Rate of lateral strain in pillar as function of mean specific stress (after Höfer, 1958).

relationship between relaxation time τ and rate of deformation $\dot{\epsilon}$, which is shown in Fig. 15.

The laboratory investigations show that both relaxation time and modulus of elasticity, as determined from the creep strain, have a dependence on stress σ. Figure 16 gives the results on rock salt, Fig. 17 those on hard salt, and Fig. 18 those on carnallite.

It is particularly interesting to make reference to the dependence of the modulus of elasticity on the stress. The considerable increase of the modulus of elasticity in the first part and its abrupt decrease at higher stresses are remarkable. Similar results were also obtained by Zetsche and Hauser* for copper, both not exposed and exposed to neutron radiation (see Fig. 19). The stressing at the maximum value of the modulus of elasticity can be considered to represent effective creep limits, since at these stresses the

*"Messungen des E-Moduls nach Kaltverformungen an bestrahltem und unbestrahltem Kupfer," Physica Status Solidi 2, K31-K35, Akademie-Verlag, Berlin, 1962.

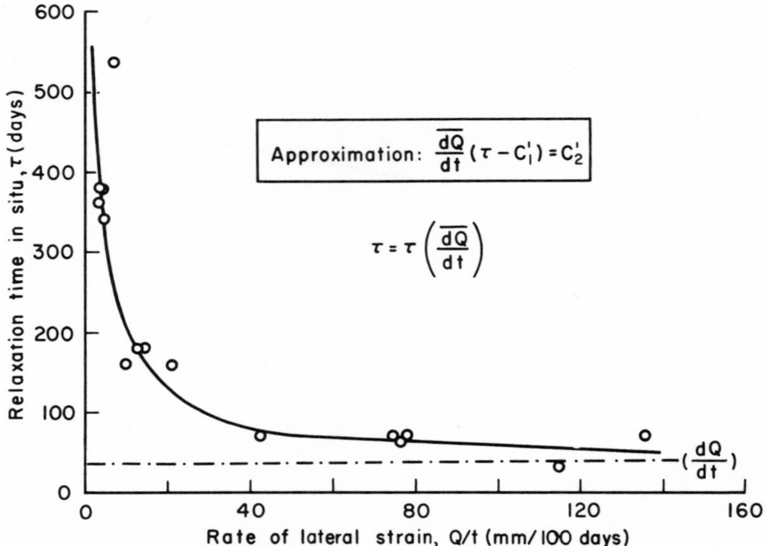

Fig. 15—*In situ* relaxation time.

material yields considerably under the external loads, which results in a considerable increase in strain rate. From the investigation it can further be concluded that the rheological model, by Kelvin and Newton, which is apparently satisfactory for rock salt, is not sufficient. The model proposed by F. Schuppe,* which is stress dependent, provides a better representation (see Fig. 20).

E. C. ROBERTSON in reply:

I have looked over this contribution. To try to go through it in detail would take too much time as there are several diagrams. I will just briefly summarize.

In his remarks, Dr. Höfer is adding observational information to my statement that relaxation times vary with stress. He describes his measurements of creep in pillars in several potash salt mines in Germany. He studied the relaxation time and the modulus of elasticity of the rock salt, as affected by the stress developed in the pillars. This stress, I think, he simply computed from the

*"Physikalische Kennziffern zur Beurteilung zeitabhängiger Verformungsprozesse der Salzgesteine," *Bergakademie*, 1963; "Ein rheologisches Modell für das Salzgesbirge," *Bergakademie*, 1963.

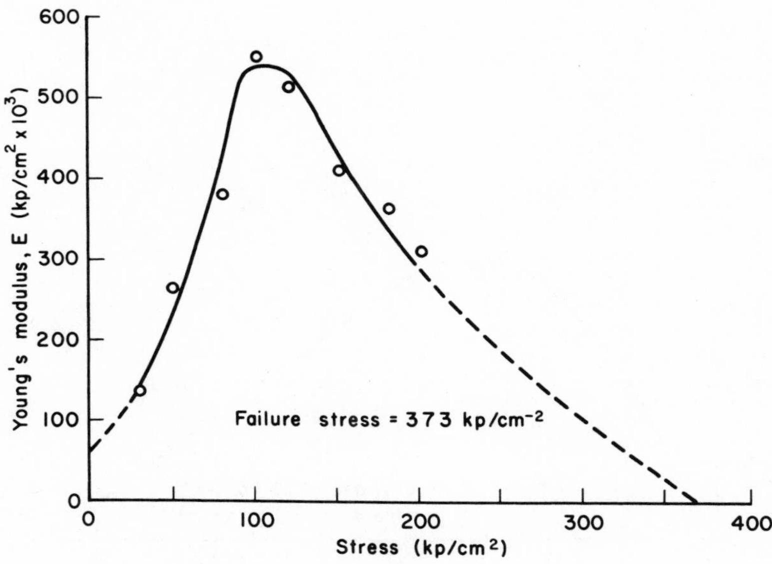

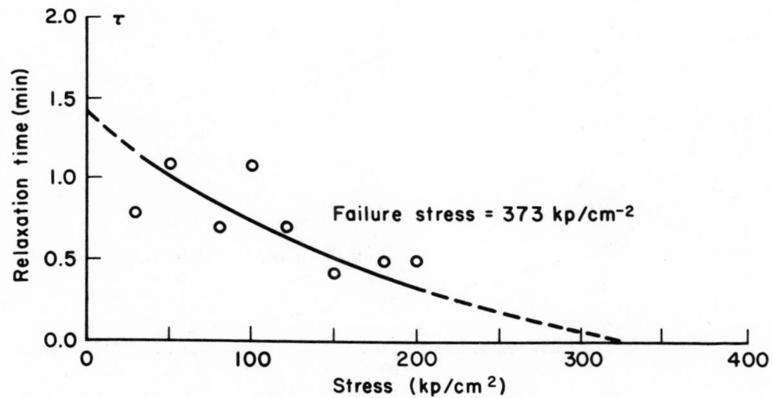

Fig. 16—Relaxation time and modulus of elasticity, rock salt (68, Werra).

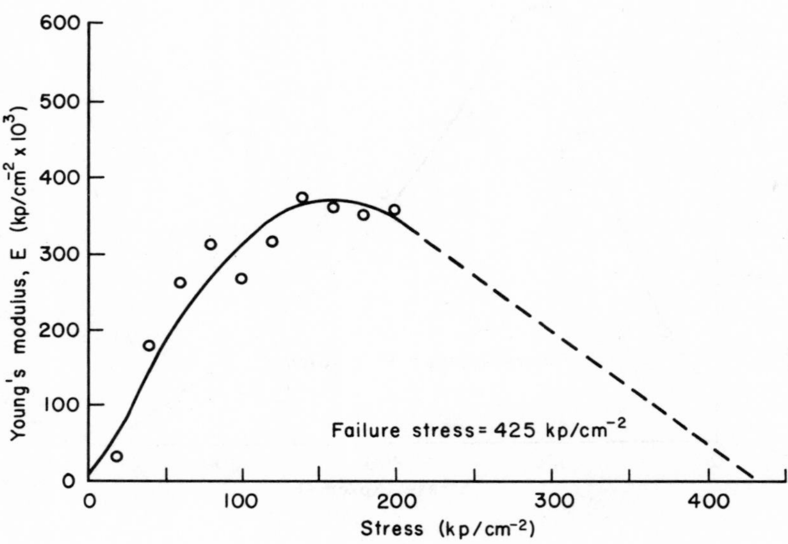

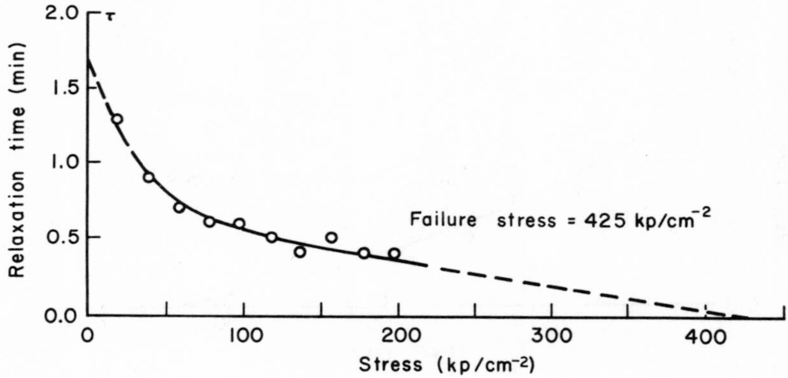

Fig. 17—Relaxation time and modulus of elasticity, hard rock salt (4.1).

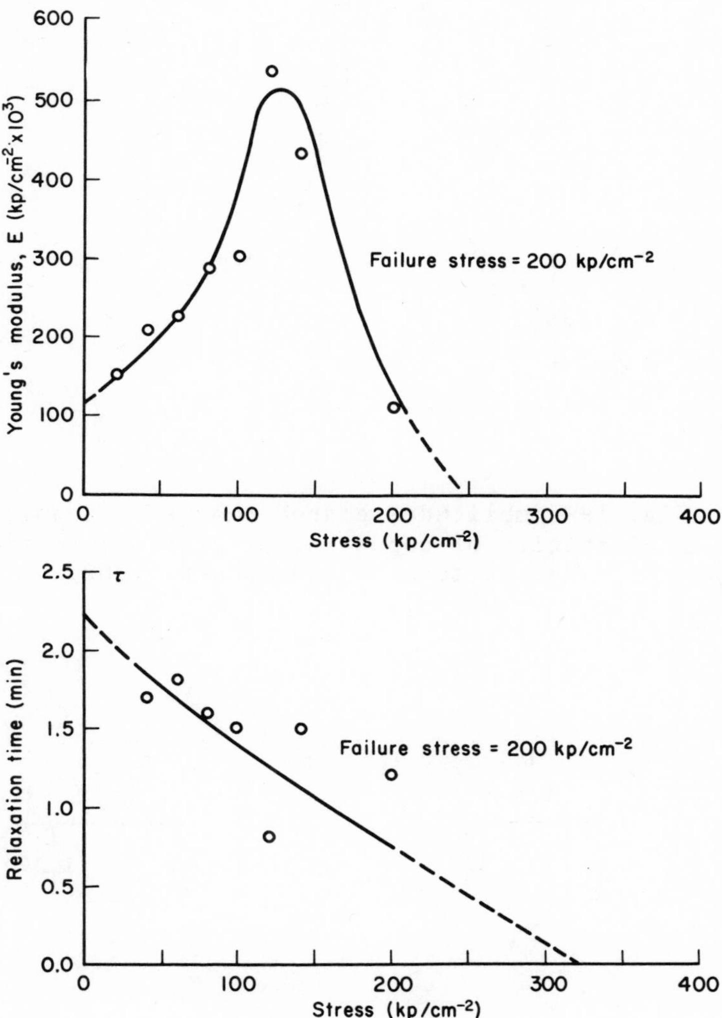

Fig. 18—Relaxation time and modulus of elasticity, carnallite (F. 1, Bernburg).

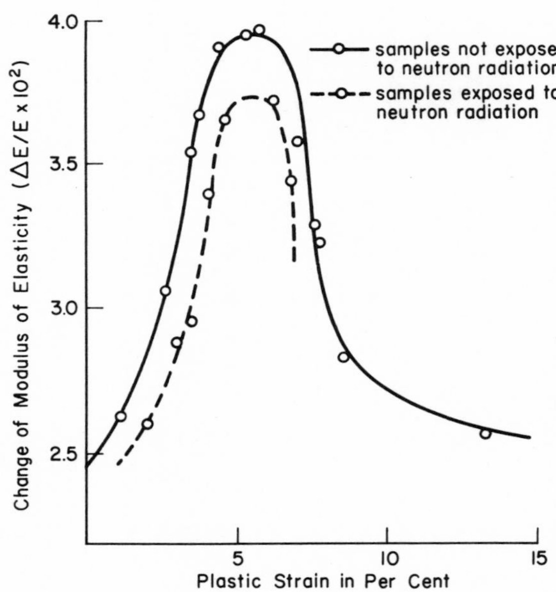

Fig. 19—Amplitude-related change of modulus of elasticity of copper after cold deformation (after Zetsche and Hauser, 1962).

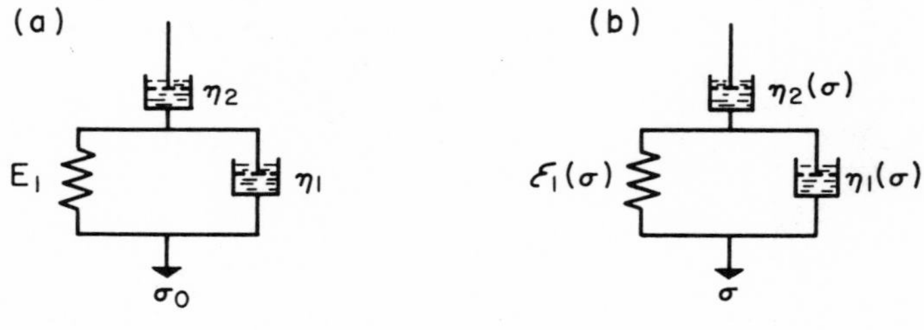

$$\varepsilon = c_2 \left\{ (1 - e^{-t/\tau_1}) + \frac{t}{\tau_2} \right\}$$

whereas $c_2 = \frac{\sigma_0}{E_1}$, σ_0=constant

$$\tau_1 = \frac{\eta_1}{E_1}, \quad \tau_2 = \frac{\eta_2}{E_1}$$

$$\varepsilon = \frac{c_2}{c_1} \left\{ -\int_{B_1} e^{-t/\tau_1}\, dt + \int_{B_2} \frac{c_1}{\tau_2}\, dt \right\}$$

whereas $\tau'_{1,2} = \tau'_{1,2}(\sigma)$

Fig. 20—Models for rock salt: (a) Kelvin-Newton model; (b) after investigations by Schuppe.

lithostatic pressure of the overlying rock. He found a considerable variation in the modulus of elasticity and in the relaxation time with stress in the pillars.

Dr. Höfer also quotes from the work of a colleague, Schuppe, who has developed a mathematical model for the creep results. This differs from the rheological model of Kelvin in that the viscosity and the relaxation time enter as functions of stress.

Dr. Höfer's results are very interesting, but I have reservations about some of his results. For example, the curve showing a maximum value for the modulus of elasticity at some intermediate stress might be misleading. It is probable that the modulus of elasticity does increase initially with increasing stress, as Dr. Brace shows in Fig. 1 of his paper, but Höfer's curve has a bell shape. This indicates that inelastic as well as elastic phenomena are involved. That the modulus drops off sharply at higher stresses is probably due to the development of fracturing with loss of cohesion, and so it is not a measure of elasticity in this range. Dr. Höfer has done a lot of work, and he has done it in the place of greatest interest, namely, in rock underground.

ABSTRACT

The paper deals with some arguments based on observed and measured data concerning the way in which rock reacts to loads and stored strain energy. A mechanism is advanced to support source bed theory and to explain, in part, how high lateral stresses can occur in sedimentary beds through simple uplift.

Rock is discussed as a heterogeneous, aeolotropic medium, a product of its own history. Under examination it is apparent that any predictions on the reactions of a rock mass to a change in load will have to be made in the light of the present transient state of strain of the rock mass.

RÉSUMÉ

Cette communication traite de quelques arguments qui se fondent sur les données observées et mesurées concernant la manière dont la roche réagit aux charges et à l'accumulation d'énergie de déformation. On propose un mécanisme pour étayer la théorie de couches de source et pour expliquer en partie comment il peut se produire des efforts latéraux élevés dans les couches sédimentaires par simple soulèvement.

On parle de la roche comme d'un milieu hétérogène et aelotropique, produit de sa propre histoire. Il apparaît à l'examen que toute prédiction sur les réactions d'une masse rocheuse vis-à-vis d'un changement de charge devra se faire en tenant compte de l'état actuel transitoire de la déformation de la masse rocheuse.

AUSZUG

Diese Arbeit behandelt einige auf Beobachtungen und Messungen beruhende Beweisführungen über die Reaktion des Gebirges gegenüber Belastungen und aufgespeicherter Formänderungsenergie. Ein Mechanismus wird beschrieben, der einmal die Theorie des Schichtungsursprunges (source-bed) unterstützt und ferner teilweise erklärt, wie hohe Seitenspannungen in Sedimentschichten durch einfache, nach oben gerichtete Beanspruchungen erzeugt werden können.

Das Gebirge wird als inhomogenes und anisotropes Material beschrieben, als ein Erzeugnis seiner eigenen Geschichte. Es wird dabei offensichtlich, dass jegliche Voraussagen über die Gegenwirkung des Gebirges auf Laständerungen hin unter Berücksichtigung des gegenwärtigen, vorübergehenden Formänderungszustand des Gebirges gemacht werden müssen.

STRAIN ENERGY IN ROCKS

Charles L. Emery*

TO THE WRITER, an engineer and therefore an applier of science rather than a scientist, the problem of rock mechanics resolves itself into three parts:

1. How do granular heterogeneous aeolotropic media, such as rocks, accept loads and adjust internally and externally to new conditions of equilibrium (or to failure)?

2. How can the intergranular and intragranular adjustments which must be involved be measured or at least observed?

3. How can the measurements and/or observations be used for design and control of rock structures?

No pretense to a solution of the problem will be made in this paper, but certain hypotheses made by the writer concerning part one have led to a new system of observation and measurement in partial answer to part two, and this has resulted in a good measure of preliminary success in part three.

This paper will deal with some hypotheses, some arguments, and some observed and measured data which seem to be pertinent to the arguments. The paper follows others of both theoretical and practical importance which have been published during the past five years. Reference is made to them to avoid repetition.

Much work has been done both in the field and in the laboratory on the measurement of the so-called engineering properties of rocks by various competent and careful investigators.[1-9] Considerable data have been compiled, but the apparent variability of "crushing strength" and

---

*Professor, Mining Engineering Department, Queens University, Kingston, Ontario.

of the "elastic constants" has deterred the engineer from the use of such data in design of rock structures. In the laboratory, the results seem to vary with the orientation of the sample, its size and shape, the time since the sample was taken, the degree of restraint, the method of load application, the load increment interval, and other factors.

The nature of the variations indicates that the cause is a function of the structure and condition of the rock and not of the test. Therefore, the test results will have little meaning unless the initial structure and condition of the rock are considered.

Measurement work done with bore hole transducers and strain gauges of various sorts[10-12] has shown that the distribution of forces in rocks *in situ* is not as predicted from superincumbent load theory.

Tests based on dynamic measurements also have wide variability, and no real correlation has been obtained with similar tests carried out on rock in place.[13] Here also it seems to be the structure and material properties of the rock which cause the variability.

Measurements of convergence of roof and floor or of footwall and hanging wall have shown that convergence does occur, but it is obvious from the observed phenomena that more is involved than simple beam or slab action under superincumbent load.

Rock bursts of considerable violence occur in such varied conditions as granite wallrock at depths of 125 ft in Newfoundland and in greenstone in a drift at the 4000-ft level one mile from the nearest mine workings at Sudbury, Ontario. It is obvious in both cases that substantial strain energies exist in these rocks without relation to their present depth of overburden.

In 1957 the writer developed some ideas about the nature and sources of strains in mine rocks and evolved the concept of a prestressed structure.[14] The mathematics of the concept called for conserved elastic strain energies. Models similar to those used in rheology were shown to explain many hitherto unexplained phenomena. These predicted the importance of intergranular and intragranular strains, and since that time the writer has sought ways to measure, observe, and to work with such strains.[15-20]

Hypotheses

Helmholtz says that the most general motion of a sufficiently small element of a deformable body can be represented as the sum of a translation, a rotation, and an extension (or contraction) in three mutually orthogonal directions. An excellent mathematical development of this is given by Sommerfeld[21] based on the concept of an isotropic homogeneous substance.

This is thought to hold generally when particle size is confined to the dimensions of a crystal spatial lattice. The elementary units of the lattice are the atoms of the substance arranged in lattice planes and, within the planes, in lattice lines. The lattice planes can be arranged in sets of parallel planes, and these sets are not necessarily orthogonal to each other.

At small loads the spatial lattice suffers a continuous strain without essential changes in shape, and the strain disappears as the load is removed. At greater loads a changed shape accompanies strain, and deformation occurs. The lattice planes which contain the most densely occupied lattice lines slide in the direction of these lines and the resultant movement is partially irrecoverable. The strain and the deformation described are controlled by the spatial lattice and are therefore aeolotropic.

Reynolds[22] has shown that a change of shape will ordinarily be accompanied by a change in volume. The Weissenberg effect[23] indicates that simple shearing must have an associated tension or compression. Such second-order phenomena can be very important in large deformations.

The regular arrangement of atoms in a crystal is disturbed by defects. One type of defect is a dislocation which may exist naturally or be induced by loading. Such dislocations tend to migrate and to propagate when force is applied. This movement along a crystallographic plane may cause the crystal to shear. In metals the movement of minute dislocations is restricted by conventional alloying and heat treating techniques. It is worth noting that such metals are no longer thermodynamically stable. This introduces potential dimensional instability.

Most technical solids are more complex than single crystals and are at least polycrystals. In such polycrystals, crystal boundaries constitute interruptions in crystal spatial lattices. The effect of a directed force on a polycrystal will therefore vary from crystal to crystal. The observed over-all behavior will be the result of averaging the reactions of a large number of crystals.

If groups of polycrystals are cemented together, the effect of a directed force will be a function of the primary crystal properties, the polycrystalline properties, the packing pattern of the groups of polycrystals, the distribution of the kinds of polycrystals, the properties of the cementing media, and the transient equilibrium condition of the aggregate prior to the application of the directed face.

Most rocks consist of bonded groups of varied polycrystalline substances and in addition contain particles of water or air or whatever fills the intergranular openings. Terms commonly used to describe the properties of materials must therefore be reviewed if they are to be used to describe rocks.

Isotropy

If a body is isotropic, its elastic, plastic, and fluid properties are the same in all directions. Single crystals do not have this perfect condition, and it would not be likely in complex mixtures. There remains the possibility, however, of statistical isotropy.

Consider a body (Fig. 1) which consists of grains of one material

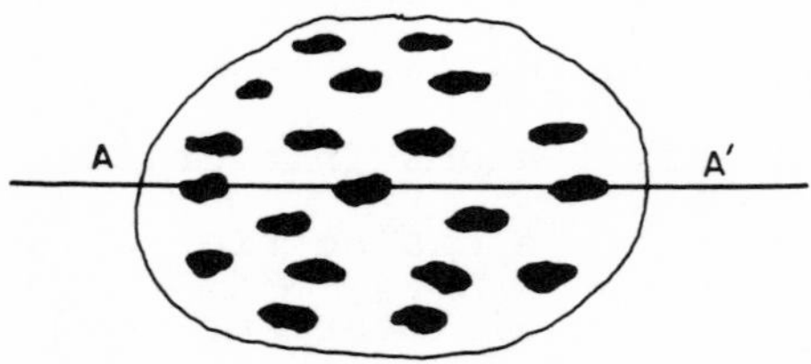

Fig. 1

within and surrounded by grains of another material. If a linear element AA' of length L is taken inside the material and if L is large with respect to the grain size, it will then cut N grains. The body will be statistically isotropic if the ratio N/L is independent of the orientation of AA', provided the grains have completely randomly oriented ores.

This definition rules out bodies composed of flattened particles oriented in planes and bodies of ellipsoidal, spherical, or angular grains arranged in any "pattern" of packing. It rules out all rocks which have a preferred crystal orientation. It is most difficult to imagine a statistically isotropic rock, and the writer would not expect to find one in nature.

Homogeneity

If a body is homogeneous, its matter must be continuously distributed over its volume so that the smallest element cut from its body possesses the same specific physical properties as the body.

This definition excludes bodies such as that shown in Fig. 1. However, we should consider statistical homogeneity. Consider V a volume of the body large in comparison to the volume of a grain. Let V' be the total volume of the grains in V. We could say that the body is statistically homogeneous if the ratio V'/V is a constant independent of the choice of the volume V.

Most rocks are not statistically homogeneous, but there may be local statistical homogeneity in some rocks such as fine-grained, relatively pure limestones or natural glasses. In these cases the grains would have to be oriented completely randomly and the conserved strain energy would have to be isotropically distributed.

Elasticity

A perfectly elastic body is characterized by a fully and instantly recoverable strain resulting from any load being applied, then removed.

Most rocks approximate this condition only when the time duration of the load is measurable in microseconds and when the strains are small. It is often supposed that a rock is elastic because it is possible to obtain a reasonably straight line on a stress-strain diagram. This need not be so,[16] and it can generally be shown not to be so if slow cyclical loading is used.

A better approach is to consider that rocks, and other technical substances, combine the properties of elasticity, liquidity, and plasticity. This introduces variations in rock behavior with time and with magnitudes of loads.

Instead of calculating rock reactions based only on $\sigma = E\epsilon$ (a Hooke solid), we include an allowance for a rate of strain such that $\sigma = \eta(d\epsilon/dt)$ (a Newtonian liquid), and an opportunity to flow plastically at any point where the yield stress, $\sigma = \phi$ (a St. Venant solid), is reached.

Here σ is unit stress,

ϵ is unit strain,

E is Young's modulus,

η is the coefficient of viscosity, and

ϕ is the yield stress.

The total stress short of yielding would then be

$$\sigma = E\epsilon + \eta \frac{d\epsilon}{dt} \quad \text{(a Kelvin solid)},$$

and the rate of strain would be

$$\frac{d\epsilon}{dt} = \frac{\sigma}{\eta} + \frac{d\sigma}{dt} \cdot \frac{1}{E} \quad \text{(a Maxwell liquid)}.$$

The Nature of Rock

From the hypotheses we must think of rock as a granular aeolotropic heterogeneous technical substance which occurs naturally and which is composed of grain of varied polycrystalline materials which are cemented together either by a glue or by a mechanical bond, but ultimately by atomic, ionic, or molecular bonds within the polycrystals, the glue, and at every interface of bonding.

Conserved Strain Energy

It would be expected that any change in load on a rock substance would require the grains, the glue, and their bonds to react to accommodate the load change. Either a new condition of transient equilibrium would be attained or else deformation would continue to failure. Any deformation in a rock can only be a deformation of its grains and glue bonds.

Below the yield stress

$$\sigma = E\epsilon + \eta \frac{d\epsilon}{dt} ,$$

and a solution for this equation is

$$\epsilon = e^{-E/\eta t}\left[\epsilon_o + \frac{1}{\eta}\int_0^t \sigma e^{E/\eta t}\, dt\right] .$$

If σ is constant,

$$\epsilon = \frac{\sigma}{E} + e^{-E/\eta t}\left[\epsilon_o - \frac{\sigma}{E}\right] ,$$

and if

$$\epsilon_o = 0 ,$$

$$\epsilon = \frac{\sigma}{E} [1 - e^{-E/\eta t}] .$$

Therefore the initial elastic strain upon the application of a load continues to increase with time. The total elastic strain will only be attained in infinite time, and if the load is removed, infinite time will be required for complete recovery of the elastic strain.

The deformations which occur under load are in the grains and polycrystals of the rock (as noted in the hypotheses) and are also in the matrix or glue. Different substances react differently, and under load some will deform much more than others. Also, various particles deform at differing rates. It is easy to see how elastic strain energy can be locked in a system of such particles.

In a crystalline rock composed of two substances such as quartz and feldspar, let a typical interface be considered as in Fig. 2. For a small rectangular part at M with sides parallel to and normal to the tangent to M, for equilibrium

$$\sigma_n^a = \sigma_n^b \quad \text{and} \quad \tau_{nt}^a = \tau_{nt}^b .$$

For continuity between the two media,

$$\epsilon_t^a = \epsilon_t^b ,$$

but because $E^a \neq E^b$ and $\mu^a \neq \mu^b$,

$$\epsilon_n^a \neq \epsilon_n^b , \qquad \sigma_t^a \neq \sigma_t^b , \qquad \text{and} \qquad \gamma_{nt}^a \neq \gamma_{nt}^b .$$

An example of this condition is given in the test results in Appendix A.

It is easily seen that under restraint or under a stress condition other than simple normal stress, as assumed in Fig. 2, there are several possible variations of the stress-strain relationships at the interface.

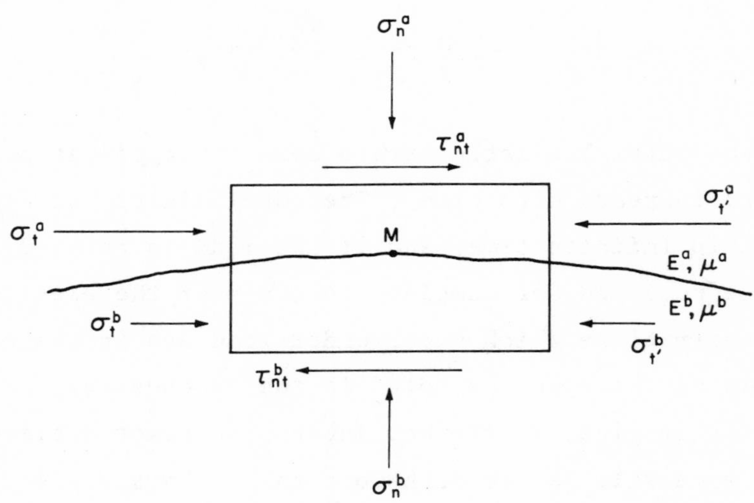

Fig. 2—Strain-stress relations across an interface of two particles such as quartz and feldspar.

Furthermore, the effects of the geometry of the crystals forming the interface can be very important. Figure 3 shows the distribution of strain in two adjoining crystals of aluminum. The effects of the interface and its orientation can be clearly seen.

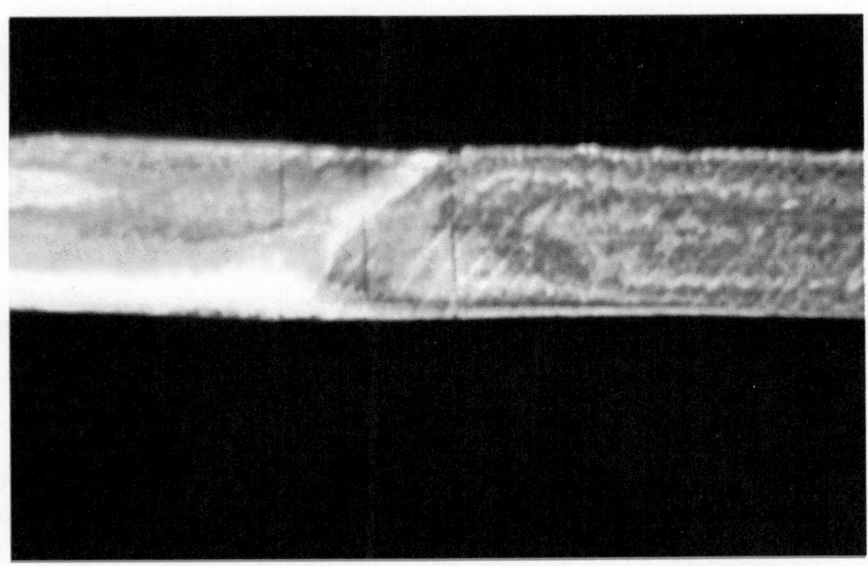

Fig. 3—Strain distribution in aluminum crystals under uniaxial tension (Hounsfield Tensometer).

If a mass of granular material is compressed and if the material contains grains of high yield point and grains of low yield point, or if the grains are oriented or prestressed so that some tend to yield before others, it is readily apparent that elastic strain energy can be stored in the system and that, upon any measure of relief, some of it will be recovered at once; some will be recovered over a period of time; but a very considerable amount will not be recovered as long as the volume of grains involved forms a coherent, inter-reacting, bonded mass.

A simple example could be cited. Suppose two grains of silica are deposited on the sea bottom. They have different properties along their respective sets of axes and are unlikely to fall with their axes respectively parallel. The hydrostatic loading will compress them axially by different amount along each axis and according to their lengths along the axes. As more load is added, the strains will increase. If at some time the grains are bonded by some substance, the two grains and their bonds form a contained strain energy system which will persist in some extent even though all external load is removed. Complete recovery will be possible only if the bond is totally destroyed or annulled.

Such conserved energy can be best demonstrated by compressing two springs of different lengths, different spring constants, and different axial orientations and then embedding them in a polymerizing plastic mass. A condition of transient stability will be reached between the springs and their bonding plastic. It would be possible to have both springs in compression and the plastic in shear. At the other extreme, one spring could end up in compression and one in tension with the appropriate distribution of compression shear and tension in the plastic. Very high energies can be so conserved.

Now if a cut is made into this system, a new state of transient equilibrium will be reached. Continued cutting would cause continued change in the energy system, but full recovery of all elastic strain energy would not be achieved without complete decoupling of the bonding resin from the springs. Alternatively, increases in loading would be additive to the system already present. This should be kept in mind when measuring elastic relaxation by the relief method.

It can be safely said that any rock, because it is a rock, must have in it more or less conserved elastic strain energy and that its present condition is a transient one and represents the sum of all that has happened to the rock. A rock is a product of its own history, and its reaction to new conditions will depend on its history.

Some Effects of Granularity

As noted in the hypotheses, imperfections occur in the arrangement of atoms in a crystal lattice and are governed by the packing pattern of the crystal lattice. The macroscopic behavior of a crystal is the result of a large number of dislocations and the interactions among them. The two main classes of dislocations are (1) edge dislocations where, under stress, part of the crystal is squeezed one atomic spacing over the adjacent part and (2) screw dislocations, in which two parts of the crystal are slipped parallel to the line of dislocation rather than perpendicular to it. Most dislocations contain both a screw and an edge component.

In both kinds of dislocation a certain amount of energy must be supplied to move the dislocation one atomic spacing. During the energy build-up there is shear strain development along the interface where displacement will occur. There is also compression and/or tension and/or torsion applied to the atomic "granules" involved in the propagation of the dislocation. Once a complete atomic spacing has been jumped, residual strain is involved in the system. The dislocation is more easily propagated thereafter because lesser amounts of strain energy must be added to the conserved energy to initiate a second jump. If at any time the region is unloaded, relaxation of some of the conserved energy can occur in the direction opposite to that of initiation.

In polycrystals the interfaces tend to act as barriers to the propagation of dislocations because of the different orientation of the crystal arrangement across the interface. Nevertheless, a crack can propagate across an interface, choosing the path of least resistance according to the strain components generated by the direction of applied force and the geometry of the crystal. A resultant fracture will have at least as many surfaces of crystal displacement as there are crystals affected in the polycrystal, and the surfaces will not all be parallel to the general fracture surface. Such fractures may split into two or more and may come together again or join with other fractures.

Intergranular and Intragranular Strains

In rocks the intragranular strains are explainable in a large part by the dislocation theory applied to polycrystals. If the rock structure is such that it consists of large polycrystals of one kind only, then much of its reaction to load can be predicted based on dislocation theory.

Most rocks contain more than one kind of polycrystal and, as previously noted, there will be differential reaction between polycrystals because of their different physical properties across interfaces.

Our scale of concern should now be shifted from "granular" structures on an atom lattice scale to the somewhat analogous granular structure where polycrystals and their interfaces replace atoms and their planes. In making such a change a much weaker bond system is inferred. Instead of atomic bonds we will now be dealing with molecular or ionic bonds. Now, applied force will tend to cause displacement along interfaces. Grains will tend to slide past each other and to rotate and compress or extend according to their individual natures and according to their packing pattern in the mass. Thus there will be local displacements, local shear strains, local dilations, and a system of conserved elastic strain energy of a transient nature very sensitive to further changes of force, either in direction or magnitude.

We now have a system of polycrystalline grains in which the polycrystals have conserved internal strain energy as well as conserved intergranular strain energy. Figure 4 shows such a strain pattern in a plane parallel to the bedding in a pebble conglomerate cemented with quartzitic material. Examination of the isochromatics shows the variation in shear strain magnitude from point to point. It will be noted that in this specimen there tends to be a lineation of high-order shear strains. There are two sets of such lines and they are not at 90 degrees to each other. These lines are traces of curvilinear surfaces along which high shear strain is already established and along which relatively little more force will increase the shear strain to the point of cracking. The writer has called these "surfaces of preferred shear."

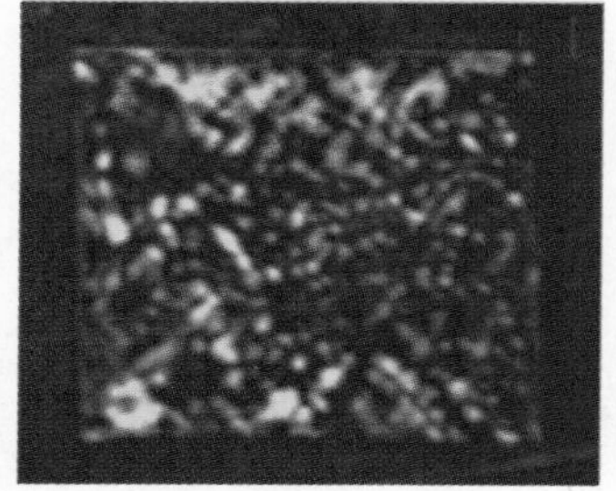

0° isoclinics

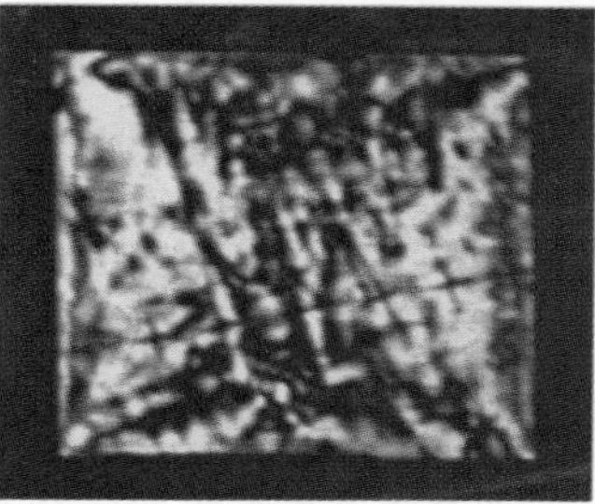

40° isoclinics

Isochromatics

Fig. 4—Strain pattern in pebble conglomerate in plane parallel to bedding.

Therefore, depending on which direction the force is applied, this rock will vary in "strength." Examination of the isoclinics at zero degree and at other angles shows the variability of directions of the principal strains from point to point. If subjected to further loading the isoclinics will change. This indicates local changes of direction of strain as one would expect in a granular mass if particles tend to rotate, to translate and to expand, or to contract from point to point.

At this scale of observation it seems that properly the granular body should be considered to be a vector space with several sets of vectors varying in direction and magnitude according to the packing pattern, the material properties of the particles concerned, and the direction of applied load. Now, if a vector valued field function is applied to the space, one would expect changes in the gradient, divergence, and curl of the field from point to point, and it would likely be a nonconservative field. That the flow lines do change is easily shown by the changes in the isoclinic patterns such as in Fig. 4 when the load is changed. Appendix B gives an example of the strain condition inside and just outside a quartz grain in a conglomerate.

The Effects of Primitive Strains and of Packing Pattern

During a Change of Load. Two disks, each 2 in. in diameter by 0.5 in. thick, were cut from a subpermian coal measure sandstone taken from the interior of a 6-in. diameter drill core approximately one year after it was cored in Lincolnshire. A disk of perspex 2 in. in diameter and 0.5 in. thick was cut at the same time.

Disks of photoelastic plastic 0.12 in. thick were cemented to one of the faces of each rock disk and to the perspex disk.

The rock specimens and the perspex specimen were then each loaded diametrally in compression, as in the Brazil test, with a load well below that necessary to produce rupture. They were first loaded normal to the bedding, then rotated 90 degrees and loaded parallel to the bedding.

It was assumed that the perspex disk would react approximately according to the laws of photoelasticity and would therefore be a suitable control.

A typical set of results is shown in Fig. 5. All the sandstone disks showed evidence of elastic relaxation. Upon loading, the perspex disk showed the expected isoclinic pattern (Fig. 5) both at zero degree and at 15 degrees.

The sandstone disks loaded normal to the bedding did not develop a pattern similar to that in the perspex at any angle. Furthermore, such disks loaded to failure did not have the theoretical distribution of shears, or strains, nor of strain directions at any time under any isoclinic angle. This appeared to indicate that the packing pattern in the direction of the applied load distributed the strains in a way different to those in the perspex disk.

Sandstone disks loaded along the bedding generally developed a recognized isoclinic pattern but not at the same angle as in the perspex control. In Fig. 5 there is no recognizable pattern for the zero isoclinic, but a typical pattern which should have occurred at zero degree did occur at 15 degrees. This was attributed to prestrain in the sandstone, which had the effect of adding to the new strains to give a component 15 degrees off that expected from the new strain.

It is apparent that rock distributes strains differently than does an isotropic homogeneous medium. It seems reasonable from observation of strains in many samples of many kinds of rocks that the packing pattern of the grains and the stored primitive strains in the rock must account in part for the reaction of the rock to new loading conditions.

Wedging and Volume Changing. All rocks are bonded granular aggregates. Their packing patterns will vary depending on the method of their histories thereafter. A simple situation is that shown in Fig. 6 for a hypothetical sandstone. The basis for this is developed in a previous article.(16) It can be seen that the only vertical load will be a resultant one and that there is a resultant horizontal load as well as a hydrostatic load if the pores are filled with water. In nature it is doubtful if the layers of grains are closely packed horizontally. In this case the angle is likely to be smaller and the horizontal component higher than with close spherical packing.

Perspex 0° isoclinics

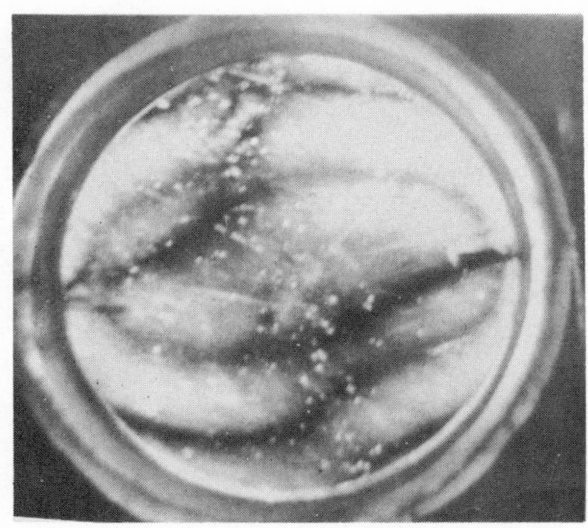

Perspex 15° isoclinics

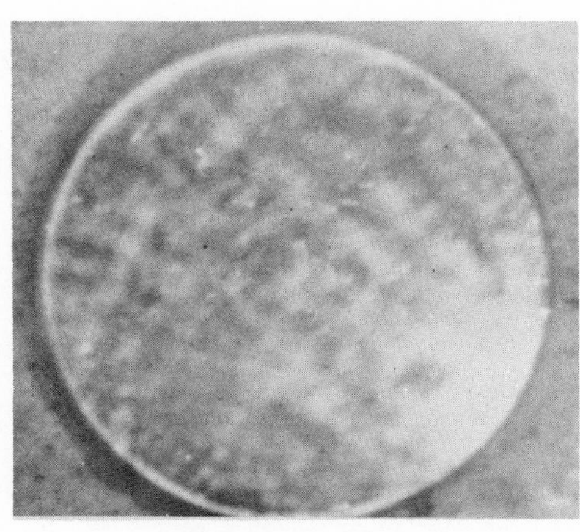

Sandstone isochromatics
load normal to bedding

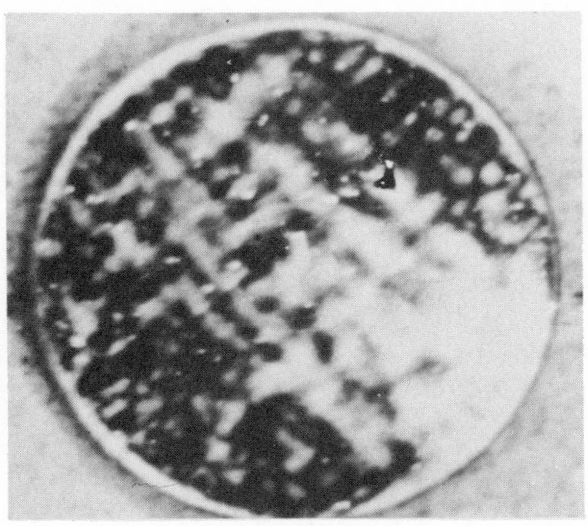

Sandstone 15° isoclinics
load normal to bedding

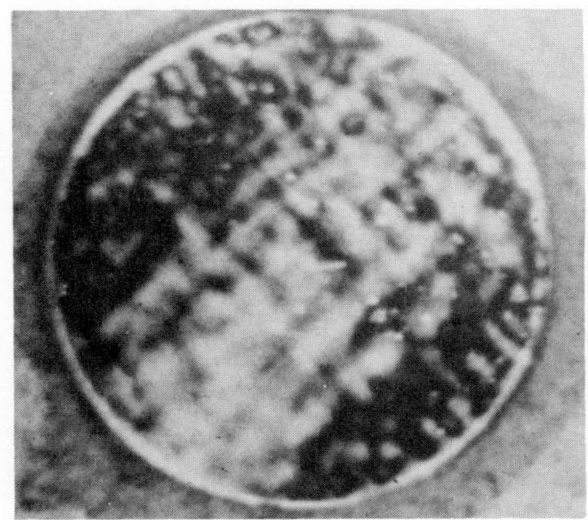

Sandstone 0° isoclinics
load parallel to bedding

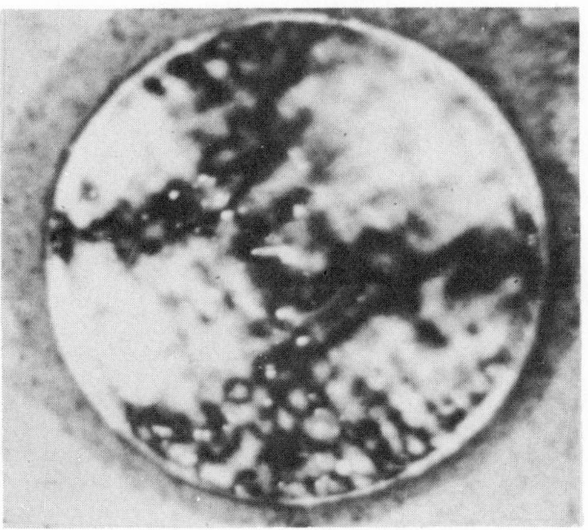

Sandstone 15° isoclinics
load parallel to bedding

Fig. 5—Disks under load.

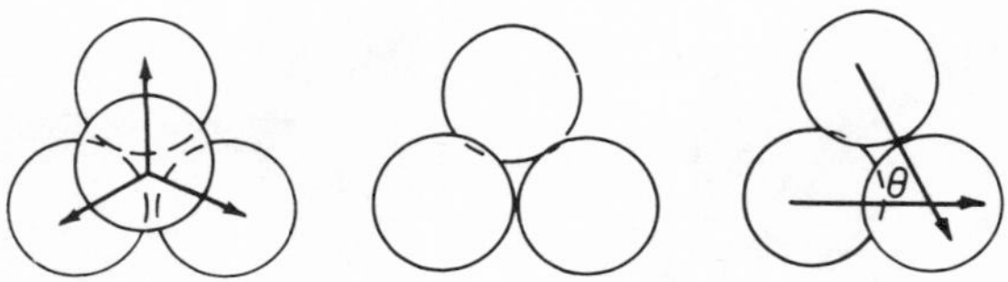

Fig. 6—Close spherical packing.

As loading increases, the points of contact become surfaces of contact and a system of wedging results, with slip occurring on the wedge surfaces. A brief examination of this shows the effect of packing pattern on dimensional change and on stress distribution. Figure 7 shows a corner developed by wedging. It can be seen that up to a point the volume of the rock will decrease under continued loading. However, if the pore water cannot escape, the rock strain will be much less per unit load than it would be if pore water could escape. Under these conditions it seems likely that as the pore water pressure increases, at some stage, hydraulic fracturing of the mass will occur and the internal load will eject pore water as from a sponge into crack systems and into interfaces between beds, etc. This may help

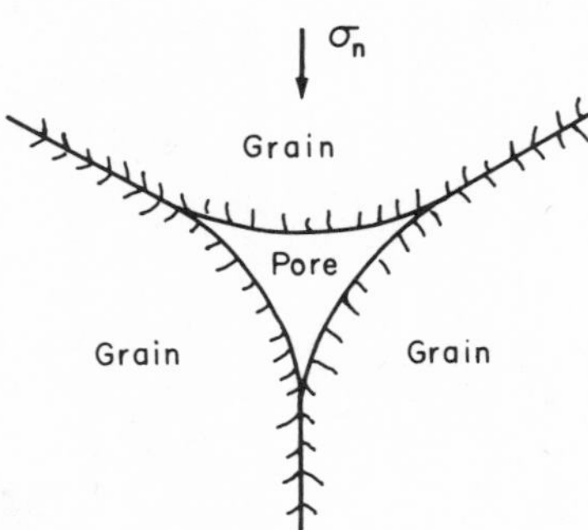

Fig. 7—Grain wedging.

explain some forms of faulting.[24] It would also explain how minerals could be dissolved from a bed in which they are disseminated to be carried into cracks or fractures and re-precipitated in the same bed or at a contact or in another bed. Such a condition, because of the heat and chemical activity of the water, would give the impression that the deposit so formed had been subject to hydrothermal activity normally associated with igneous activity, whereas no igneous activity would be involved and the minerals of the deposit would have originated in a simple source bed.

When such hydraulic fracturing occurred, the rock mass from which the water was expelled would reduce in volume.

After some further loading the wedge effect would require the mass to expand because grains would be pushed apart in some direction. In Fig. 7, for example, continued wedging downward would require the two lower grains to move sideways, and a larger pore space would tend to form. If no water or other material is available to fill the pores, then a change in the strain pattern of the grains would have to take place. The material would have to flow into its own pores. However, if gas or water were available, then the rock mass would tend to increase in volume to reach a new transient equilibrium.

Under these conditions the so-called Poisson's ratio would depend on the angle of contact of the wedging surfaces and upon the original shape of the grains. A Poisson's ratio of 1 could occur when the grains met at 45 degrees. A negative Poisson's ratio could result if rock flow occurred into its own pores upon release of hydrostatic pressure. Since Poisson's ratio, as loosely used for rocks, depends on an assumption as to which direction is longitudinal and which is lateral, the rock in place, undergoing the volumetric strain, would have different ratios if different directions were considered. The engineering use of such a ratio is only valid where the ratio is known--and then only according to the law governing the particular ratio selected under the conditions of its use.

Similarly, engineers should beware of the use of Young's modulus, E. As can be seen from the physics of granular media, a given load will produce quite a different deformation depending on the packing pattern and restraint at the time. It should be further noted that technical substances appear to approach a limiting relative volume when compressed. At the final volume, if it exists, E would be infinitely large. In any event, E cannot be a constant. The notion that it might be seems to have arisen from tension tests on metals. It should be noted that compression tests give different stress-strain curves than do tension tests. An example of the variation of E with load and with a loading cycle is given in Appendixes A and C. Present work at Queen's University indicates that Young's modulus and Poisson's ratio are often quite different in different directions in the same rock under the same load conditions. That is, values parallel to the bedding can be quite different depending on the azimuth of the sample axis, and these can be quite different than the values obtained normal to the bedding.

Stratification

The effects of granularity have been discussed on the scale at which the real movements occur under any load. However, such movements can be grouped and distributed on a larger scale. For example, a sedimentary bed will react differently at its top than it will at its bottom if the grain size, or the kinds of grains, or the glue, or the ratio of one to the other, varies from bottom to top of the bed (as is usually the case).

Figure 8 shows the effects of variation of layers in a bed. The rock is a foliated quartzite and the pictures show, by the effects of compensation, that alternate layers are storing strain energy at different levels. Equilibrium is maintained by certain layers loading other layers. Upon the formation of a new free face, the relaxation pattern is such that some layers have increased strain in compression and some decreased strain in compression. The more pure quartz layers decreased in compression. Under microscopic examination the dislocation patterns in the quartz agreed with the observed strain movements. Furthermore, in the relaxation of the sample the quartzose bands bent somewhat from their original straightness. Stereonet plotting of the directions of principal strains and the surfaces of preferred shears agreed well with the structural analysis of the condition in the field.(25)

The effects of stratigraphy would seem to be that the strain energy stored in a bed would vary from bottom to top and from layer to layer.

In beds of about the same age and history this would give rise to certain pronounced tendencies if further load were to be applied. Some beds would tend to fold, whereas others would tend to shear. Depending on the material in the beds, some would flow while others would absorb increasing elastic strain energy.

Foliated quartzite isochromatics with 10° counterclockwise compensation. (Note that some bands increased and some decreased in apparent shear strain.)

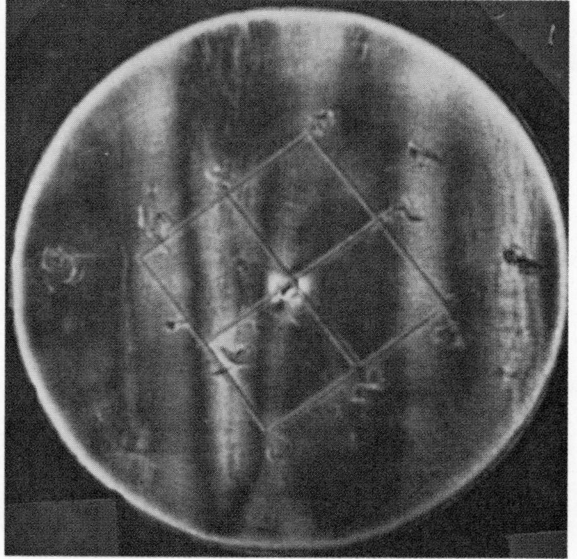

As above. Isochromatics with no compensation.

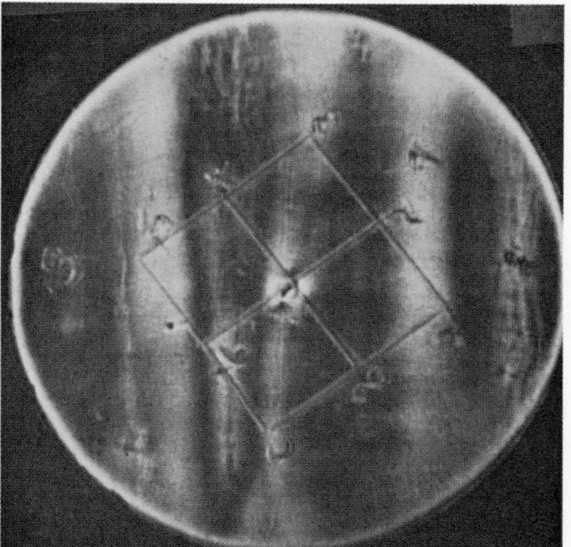

As above. Isochromatics with 10° clockwise compensation. (Note that some bands increased and some decreased in apparent shear.)

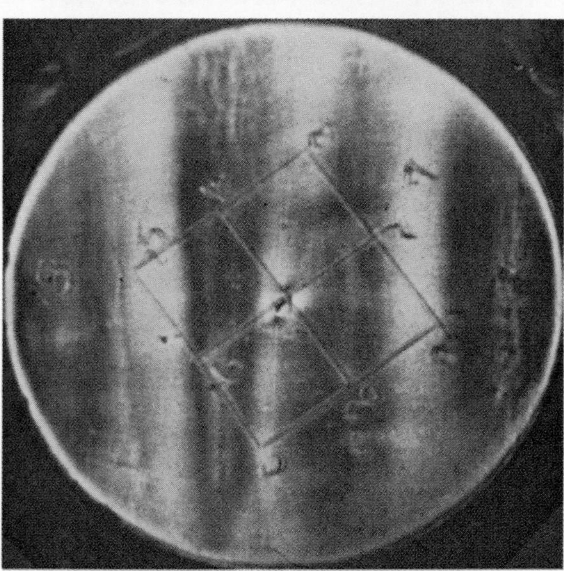

Fig. 8—Isochromatics of strained quartzite.

At horizons which represent defined changes in history, it would be likely that directions of principal strains above and below the horizon would be different because of the different histories of the layers concerned. That strains do differ from point to point in a formation is obvious from the transient configuration of any given bed, or series of beds, and from the changes in configuration and of metamorphism in areas where folding and faulting can be observed.

Following the same reasoning, it is apparent that changes in configuration and of metamorphism are as likely to occur during an uplift as during a down sinking. An area of sedimentation, after long periods of burial such that a broad synclinal configuration had been formed, would, upon being pushed back to a more level position, be subject to ever increasing lateral pressure. Figure 9 shows, in diagram, that a down-bowed region if subject to uplift would have to compress laterally or to override the neighboring rocks. The lower layers would be subject to the greatest restraint and would store the most strain energy. All layers would show the effects of the movement. Although the actual movement might not be great vertically, local relief might produce folding, faulting, and flow structures.

An example of stresses not expected in the bed can be drawn from examination of the Prairie Evaporites in Saskatchewan. In a region where the beds are at a depth of about 3000 ft and practically flat-lying but with gentle local rolls, relaxation strains were measured from drill core which indicated loads in excess of that given by direct superincumbent loading. The loads were estimated at not less than an equivalent stress of 5000 psi. Upon removal of 200 ft of large diameter drill core, the core was measured and observed to increase

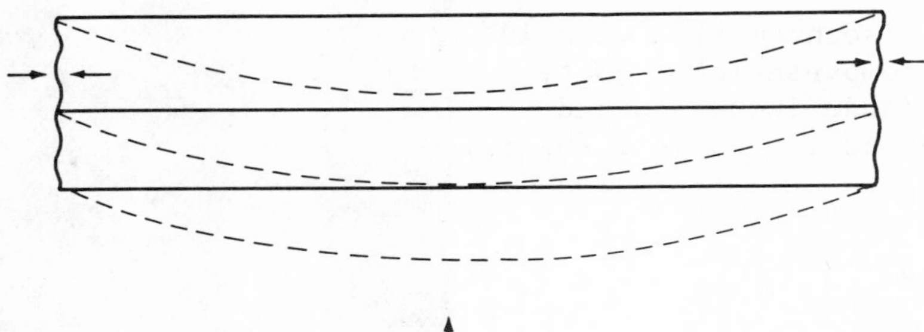

Fig. 9—Effect of uplift on a broad sedimentary basin.

in length by 2.8 ft during two days, although it had been sealed with "Core-Binder" as it was removed from the core barrel.

In the local Black River limestones of the Paleozoic in the Kingston area, it is common to expect upward buckling of certain layers when cover is removed. Local quarries observe this condition at depths as shallow as 25 ft. A local iron mine, during stripping, developed an anticlinal fold overnight with an axis over 300 ft long and an elevation increase at the crest of nearly 15 ft.

Conclusions

If it is accepted that any strain or deformation in a rock mass must involve the particles and their bonds, then it would help resolve design problems if the changes in condition of the particles and their bonds could be observed.

Such changes can now be observed by various methods, including photoelastic transducer applications, X-ray diffraction techniques, simple microscopy, interferometry, and others. It is possible to observe movements underground by the use of many different instruments. All such measurements disclose the time dependency of the strains with load changes.

All rocks observed to date by the writer by the above methods can be shown to contain conserved elastic strain energy, generally in substantial amounts. Most rocks have pronounced measurable aelotropy. By measurement it can be shown that the so-called constants--such as μ, E, η, and others--are really variables and are dependent on the transient state of strain. This does not detract from their use but does require a different approach toward the mechanics of rocks.

Any predictions on the reaction of a rock mass to a load change will have to be made in the light of the present transient state of the rock mass. A considerable amount of such information is now obtainable by good methods. That such data can be used in design is evident from results obtained in many mines in recent years. In any event, it is certain that any design which does not consider the strain energy in the rocks will be more or less unsatisfactory.

Appendix A

AXIAL GAUGES ON QUARTZ AND QUARTZITE
IN THE SAME SPECIMEN[a]

Mean Stress (psi)	Total Microstrains			
	Quartz		Quartzite	
	Load	Unload	Load	Unload
3,200	140	180	230	--
4,100	270	300	395	590
5,000	340	430	630	795
5,600	450	550	760	960
6,900	550	640	880	1,090
8,100	680	760	1,010	1,240
9,200	750	880	1,170	1,350
10,100	890	950	1,280	1,400
11,100	960	1,040	1,455	1,570
11,900	1,050	1,130	1,560	1,710
12,900	1,130	1,210	1,660	1,820
14,000	1,200	1,300	1,780	1,960
14,800	1,300	1,325	1,940	2,000
15,900	1,360	--	2,045	--

[a]Fig. 10 is a graph of these results.

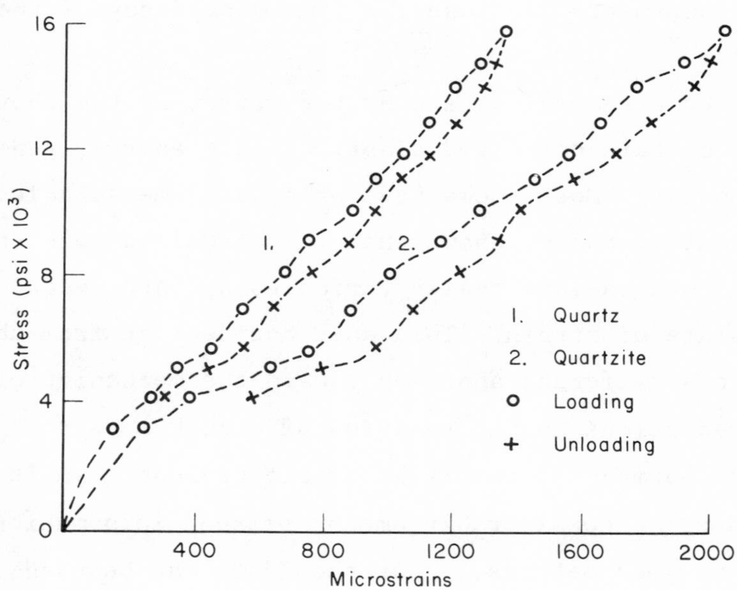

Fig. 10—Graph of test results on quartz and quartzite.

Appendix B

A number of strain measurements were made to determine the nature of the strains involved in early relaxation of a granite specimen. The results of two such measurements are shown here.

Sample. Granite containing about 25 per cent quartz in the form of quartz polycrystals.

Technique. Photoelastic coating.

Instrument Sensitivity (with the plastic used).

± 1 division

$\frac{945}{59} = 16\ \mu''/\text{in.}$

Measurement (1). At a point immediately outside a polycrystal:

$n_n = 23$ divisions, $n_{02} = -48$ divisions,

$n_{01} = +6$ divisions,

whence $\epsilon_1 = -264.6\ \mu''/\text{in.}$ (both compressive)

$\epsilon_2 = -661.5\ \mu''/\text{in.}$

and $(\epsilon_1 - \epsilon_2) = 396.9\ \mu''/\text{in.}$

also $(\epsilon_1 - \epsilon_2)$ calculated from $n_n = 365.7\ \mu''/\text{in.}$

which agrees well considering the instrument sensitivity.

Measurement (2). At a point inside the edge of the same polycrystal near measurement (1):

$n_n = 13$ divisions, $n_{02} = +2$ divisions,

$n_{01} = +25$ divisions,

whence $\epsilon_1 = +382.2\ \mu''/\text{in.}$ (both compressive)

$\epsilon_2 = +213.2\ \mu''/\text{in.}$

$\epsilon_1 - \epsilon_2 = 169.0\ \mu''/\text{in.}$

and $(\epsilon_1 - \epsilon_2)$ calculated from $n_n = 206.7\ \mu''/\text{in.}$

again a reasonable agreement.

It is apparent that the material of the polycrystal at the point examined was expanding in the two principal directions, whereas the material adjacent to it was contracting in the two principal directions.

Samples of this particular granite often break in fragments over a period of time if left on the shelf, although the granite appears ordinary otherwise.

A petrofabrics study made independently confirmed the principal stress directions for this material to be the same as those found by the writer.

The nature of the strains and of the forces that cause the strains shows that

(a) Elastic recoverable strain energy of a high order can exist in rocks.

(b) Rock particles are not strained isotropically.

(c) Strains vary in magnitude from point to point in the mass.

(d) A granular aggregate can develop high shear strains which are functions of the packing pattern and of the strain history of the rock.

(e) Poisson's ratio seems inapplicable if used in the usual sense.

Appendix C

Cameron gives the stress-strain diagram on a specimen of Springhill sandstone shown in Fig. 11. The sample was compressed to 17,000 psi, unloaded to 2000 psi, and then reloaded in the same increments.

Values for E and μ have been calculated from the graph and are tabulated below. Cameron's graph has been used to eliminate test bias.

Mean Stress (psi)	First Loading		Second Loading	
	$E \times 10^6$ psi	μ	$E \times 10^6$ psi	μ
2,500	4.62	0.081	2.40	0.25
5,000	4.62	0.100	3.45	0.33
7,500	4.70	0.129	3.95	
10,000	4.81	0.172	4.45	0.32
12,500	5.03	0.200	4.80	
15,000	5.20	0.241	5.00	0.30
17,000	5.27	0.269	5.27	0.269
20,000	5.09	0.326	5.09	0.326

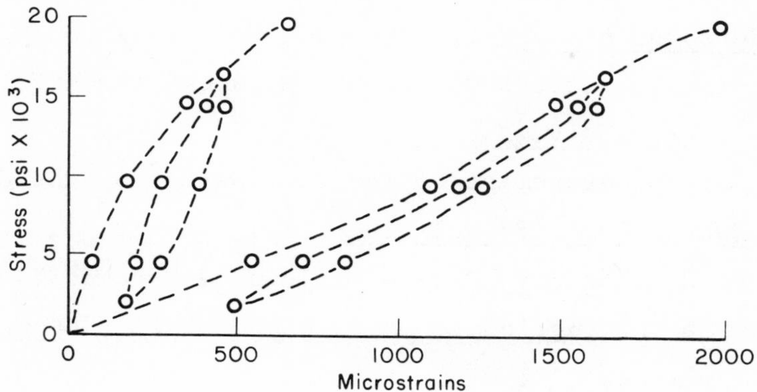

Fig. 11—Graph of test results on Springhill sandstone.

REFERENCES

1. Obert, L., S. L. Windes, and W. I. Duvall, "Standardized Tests for Determining the Physical Properties of Mine Rock," U.S. Bur. Mines Rept. Invest., 3891, 1946.
2. Bridgman, P. W., The Physics of High Pressure, The Macmillan Company, New York, 1952.
3. Griggs, D. T., "Deformation of Rocks under High Confining Pressures," J. Geol., 1936, pp. 44-54.
4. Griggs, D. T., F. J. Turner, I. Borge, and J. Saska, "Deformation of Yule Marble, IV, Effects at 150°," Bull. Geol. Soc. Am., Vol. 62, 1951, pp. 1385-1485.
5. Robertson, E. C., "Experimental Study of the Strength of Rocks," Bull. Geol. Soc. Am., Vol. 66, 1955, pp. 1271-1314.
6. Cameron, E. L., "Physical Properties of Some Canadian Mine Rocks," Trans. Can. Inst. Mining Met., Vol. 57, 1954, pp. 506-509.
7. Wuerker, R. G., "Influence of Stress Rate and Other Factors on Strength and Elastic Properties of Rocks," in 3d Rock Mechanics Symposium, The Colorado School of Mines, Golden, April, 1959.
8. Handin, J., "Experimental Deformation of Rocks and Minerals," Quart. Colo. School Mines, Vol. 52, No. 3, July, 1956, pp. 75-98.
9. Robinson, L. H., "The Effect of Pore and Confining Pressure on the Failure Process in Sedimentary Rock," in 3d Rock Mechanics Symposium, The Colorado School of Mines, Golden, April, 1959.

10. Hast, Nils, "The Measurement of Rock Pressure in Mines," Sveriges Geol. Undersokn. Arsbok, Ser. C, 560, No. 52, 1958, p. 3.

11. Potts, E. L. J., "Stress Distribution, Rock Pressures, and Support Loads," J. Leeds Univ. Min. Soc., Vol. 30, 1954.

12. Olsen, O. J., "Measurement of Residual Stress by the Strain Relief Method," Quart. Colo. School Mines, Vol. 52, No. 3, July, 1957.

13. Nichols, H. R., "In Situ Determinations of Dynamic Elastic Rock Constant," in International Symposium on Mining Research, Rolla, Missouri, February, 1961.

14. Corlett, A. V., and C. L. Emery, "Prestress and Stress Redistribution in Rocks around a Mine Opening," Trans. Can. Inst. Mining Met., Vol. 62, 1959, pp. 186-198.

15. Emery, C. L., "The Strain in Rocks in Relation to Mine Openings," Mining Eng., No. 1, October, 1960.

16. Emery, C. L., "Testing Rock in Compression," Mine Quarry Eng., April and May, 1960.

17. Emery, C. L., "Photoelastic Coatings on Granular Materials," Institute of Physics, 1960.

18. Emery, C. L., "The Measurement of Strains in Mine Rocks," in International Symposium on Mining Research, Rolla, Missouri, February, 1961.

19. Emery, C. L., "The Photoelastic Technique for Studying Rock Strains," Trans. Can. Inst. Mining Met., April, 1962.

20. Roberts, A., C. L. Emery, P. K. Chakravarty, I. Hawkes, and F. T. Williams, "Photoelastic Coating Technique Applied to Research in Rock Mechanics," Trans. Inst. Mining Met., Vol. 71, Pt. 10, 1961-1962.

21. Sommerfeld, Arnold, Mechanics of Deformable Bodies, Academic Press, Inc., New York, 1950.

22. Reynolds, O., Phil. Mag., Vol. 20, 1885.

23. Weissenberg, K., "Continuum Theory of Rheological Phenomena," Nature, Vol. 159, 1947, pp. 310-311.

24. Hubbert, M. K., and W. W. Rubey, "Role of Fluid Pressure in Mechanics of Overthrust Faulting," Bull. Geol. Soc. Am., Vol. 70, No. 2, February, 1959, pp. 165-206.

25. Heidecker, Eric, unpublished Ph.D. thesis (geology), Queen's University, Kingston, Ontario, Canada.

DISCUSSION

G. KIERSCH (USA):

Regarding tectonic stresses in a rock mass: (1) Have any observations or tests been made on the rate of decay or de-stressing of tectonic stresses? (2) Have any engineering techniques been devised to accelerate this decay or de-stressing of rock mass? (3) If such techniques have been devised, what are they?

C. L. EMERY in reply:

I know of no series of measurements recording the decay of conserved tectonic stresses in a rock mass. However, in mining work we have been able to see changes at points in a mine over several years of experience, but these, of course, are not just de-stressing in a natural way. They are functions of the growth of the mines' geometry, and that sort of thing.

De-stressing techniques in mines are used and are valuable. The first instances I know of where blasting was used for de-stressing were in South Africa, and this is a standard procedure with some of the mines. Several of our mines in Canada also make use of blasting techniques for de-stressing. This doesn't necessarily break the rock up, but shakes it and apparently causes some kind of reorientation or change in the pattern which may be equivalent to local stress relief. The stress may build up again over a period of time, in which case one may have to de-stress again, but I don't know of any measurements that have been made on the decay of tectonic stresses in general.

K. F. DALLMUS (South America):*

There has been considerable discussion about the validity of the theory of elasticity with respect to rocks. I want to talk about a cross section which appears in Professor Emery's paper which shows the

---

*This paper was read before the American Association of Petroleum Geologists at New York City, March 31, 1955. Figures 12 through 16 appeared in "Mechanics of Basin Evolution and Its Relation to the Habitat of Oil in the Basin," by K. F. Dallmus (reprinted for private circulation from Habitat of Oil, The American Association of Petroleum Geologists, Tulsa, Okla., June, 1958).

conventional cross section of a basin. There is nothing personal about this, Professor Emery, but that thing just doesn't exist.

Table 1 shows the dimensions of segments of a great circle of the earth using the equatorial radius. It gives the shortening between the arc and the chord and the median height in meters of the center of the arc above the chord.

You will see that for a basin 3 degrees in diameter the difference between the length of the arc and the chord is about 38 meters. The median height is 2185 meters (about 7500 ft). As the length of the arc increases, the median height increases correspondingly.

In Fig. 12 the two cross sections are across the Williston basin. A comparison is made with the conventional type of cross section which shows it as a concave basin. When you draw that to natural scale it appears as the section in the middle, the thin section. The vertical dimension is so small that you cannot see any detail. If you exaggerate the vertical scale of the curve of the earth over that distance by the same amount as you do the vertical dimension of the sediments, you get the picture at the top. You see that the floor of the basin is a chord in the middle, and that it increases in curvature as it approaches the curvature of the earth.

Now, all the basins on this earth have that shape. There are no concave primary basins. The sediments are laid down on a flattening of the curve of the earth, so the floor of the basin must be put into compression.

Figure 13 shows the same thing for the Michigan basin, and again

Table 1

DIMENSIONS OF SEGMENTS OF THE EARTH'S GREAT CIRCLE

Degrees of Arc	Length of Arc (km)	Length of Subtended Chord (km)	Difference in Length (m)	Median Height (m)
1°	111.32387	111.32251	1.355	242.88901
2°	222.64774	222.63648	11.258	971.42849
3°	333.97161	333.93348	38.127	2,185.74600
4°	445.29548	445.20510	90.382	3,885.52261
5°	556.61935	556.44278	176.570	6,070.82213
6°	667.94322	667.63811	305.110	8,741.38940

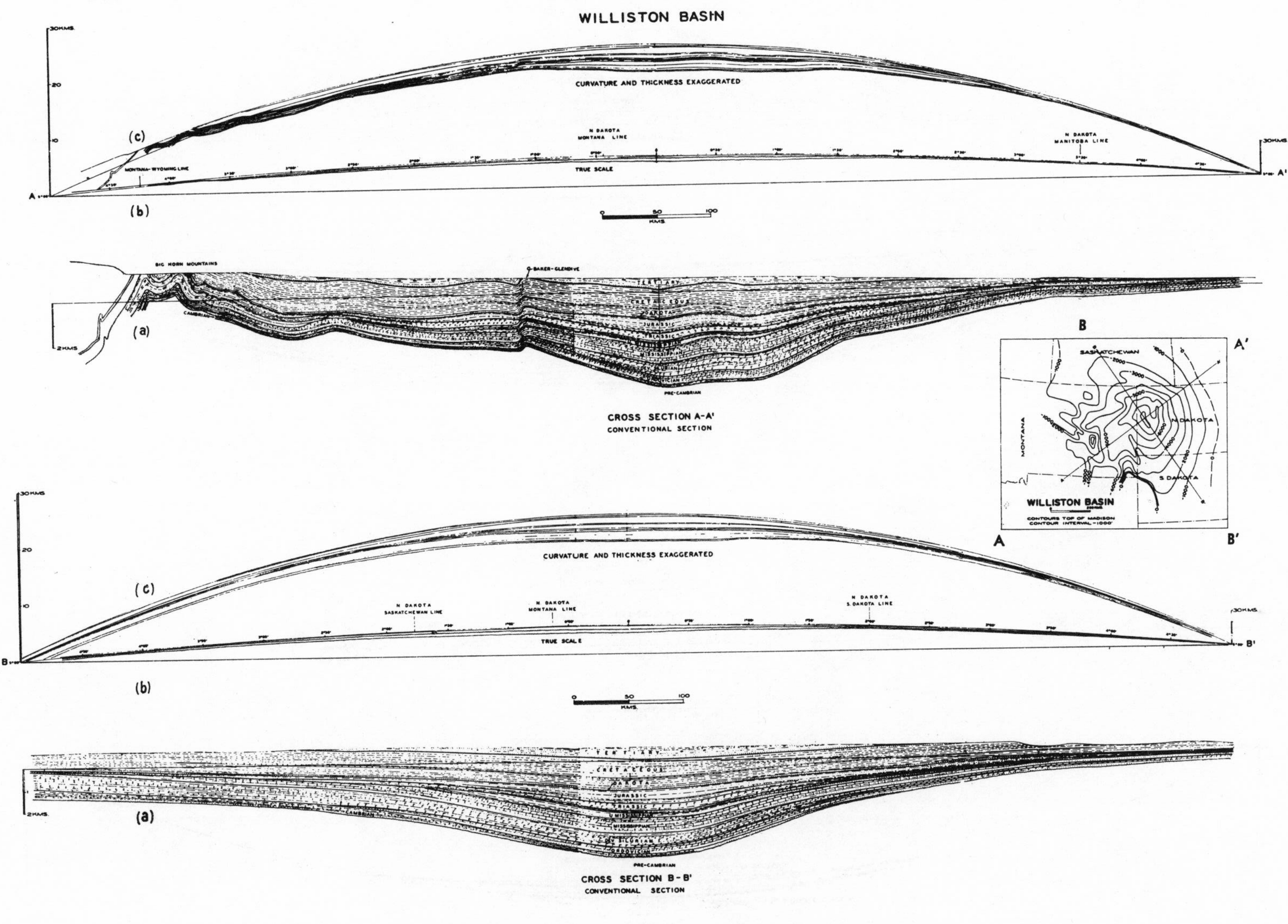

Fig. 12—Cross sections of Williston basin.

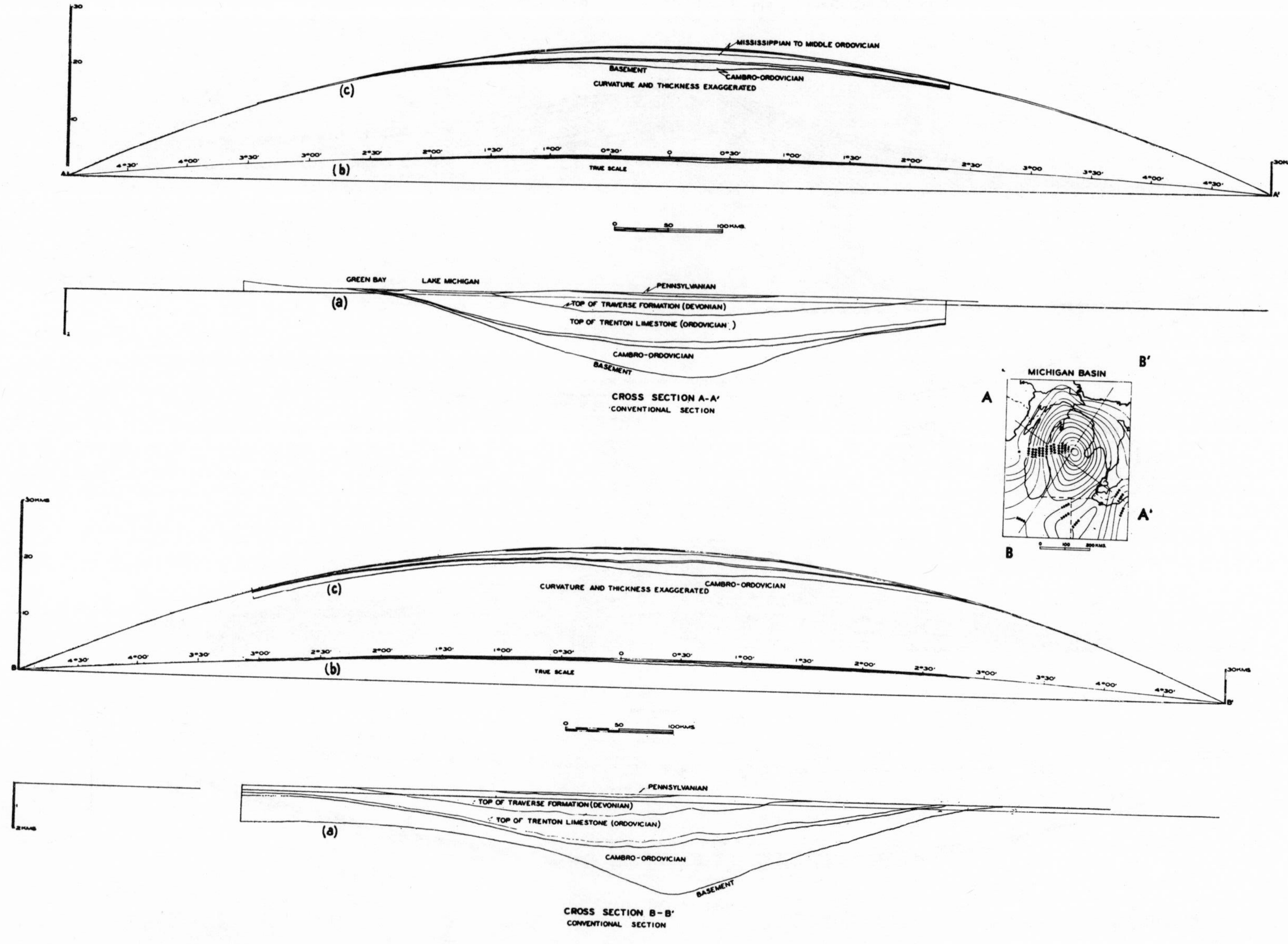

Fig. 13—Cross sections of Michigan basin.

all you have is a flattening of the curvature of the earth upon which the sediments rest.

Figure 14 shows that when you draw a cross section of the Atlantic Ocean and exaggerate the vertical scale you find that there are no concave surfaces any place. On the Newfoundland side there are about 22,000 ft of sediments above the basement according to seismic data, yet the curve is still convex, not concave.

Figure 15 shows a different type of basin. The concentric downbends I have called "primary basins," but this type is a "secondary basin" because growth is on top of the uplifts due to the breakdown of the uplift by extension of the spherical surface upon uplift above its former position.

Consider a granitic surface at sea level and let it drop from sea level to the chord. We can calculate the horizontal compressional stress that is set up in the center of the basin (Table 2). For a basin of 5 degrees the compression is about 20,000 psi. You need not add a foot of sediment to get that pressure. It is generated by the vertical translation of the arc down to the chord.

Figure 16 shows another view of the Michigan basin at a level above the basement at the top of the Trenton formation. It shows nine anticlines across the basin, the average height of which above the synclines is about 300 ft. The distance between the synclines is about 15 mi. Calculations of the shortening show that about 5.6 ft are needed for each anticline. The whole shortening across

Table 2

VALUES AT CENTER FOR ARCS OF 1 DEGREE TO 5 DEGREES

Diameter of Basin (degrees of arc) (km)	Depth to Chord at Center (sea level datum) (m)	Lateral Compression	
		kg/cm^2	psi
1° = 111.3	242.9	59	826
2° = 222.6	970.8	232	3,302
3° = 333.9	2,185.7	519	7,387
4° = 445.2	3,885	935	13,168
5° = 556.6	6,070	1,412	20,097

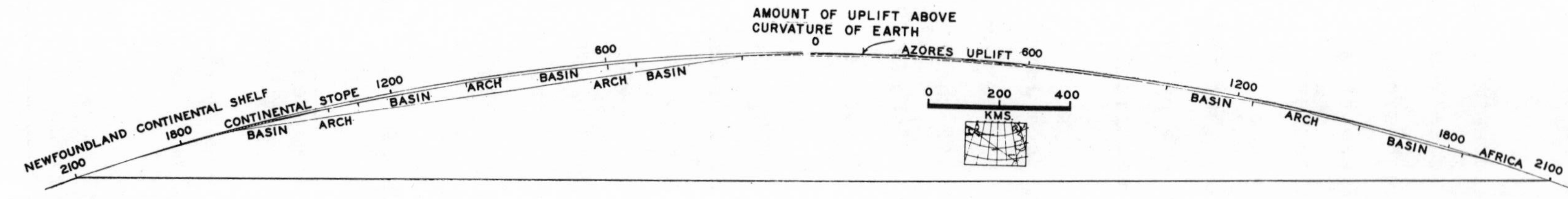

Fig. 14—Cross section, Atlantic Ocean, Newfoundland to Africa.

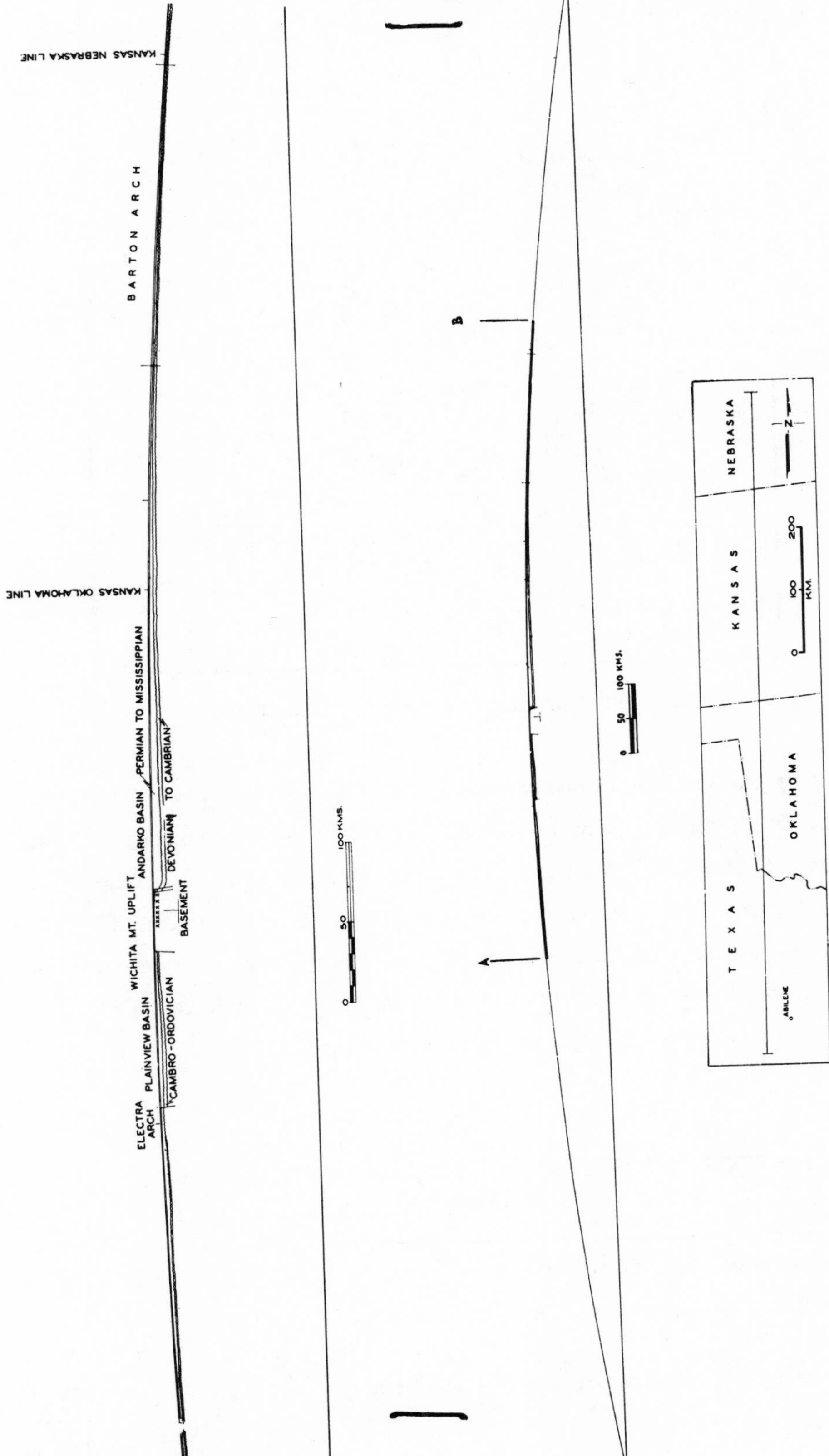

Fig. 15—Cross section, Nebraska to Texas, through Anadarko basin.

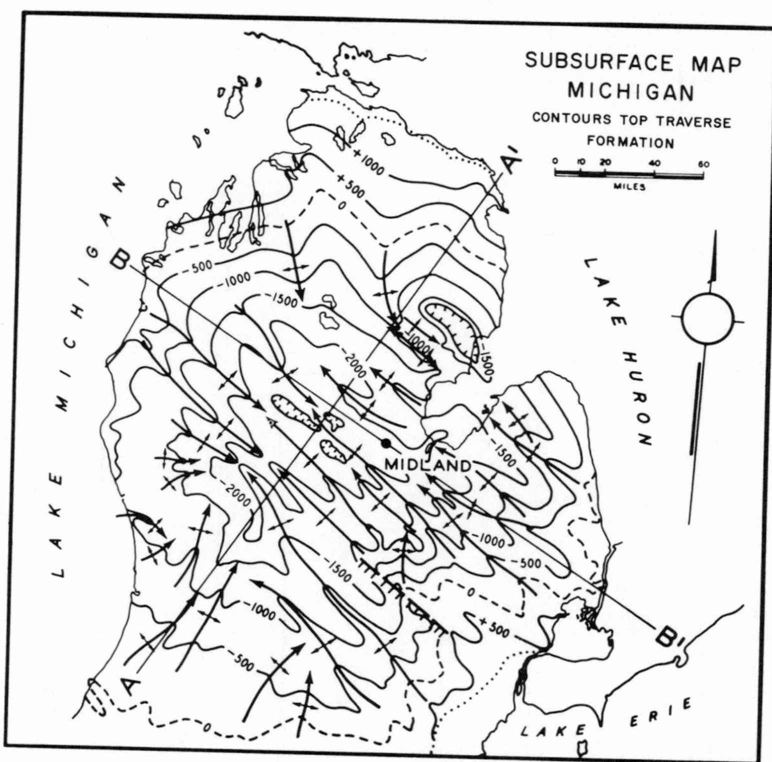

Fig. 16—Subsurface map, Michigan.

the basin is therefore about 50 ft. Table 3 shows the shortening that took place when that basin subsided to its present position. The shortening of the arc to the chord is about 38 meters. The shortening over the distance of 350 km for the points of inflection at the ends of the line is about 82 meters, and the shortening due to change of form is about 48 meters. A total shortening of 168 meters must have taken place when that basin assumed its shape.

Table 3

SHORTENING CAUSED BY SUBSIDENCE OF MICHIGAN BASIN
(meters)

Flattening of arc to chord	38.1
Shortening by subsidence of 1500 m over distance of 350 km	82.3
Shortening by change of form (minimum)	48.0
Total	168.4

Only about 15 meters can be accounted for by folding of the anticlines, so you can see how much shortening has been taken up by some other process, possibly closing of fractures.

C. L. EMERY in reply:

I think that what has been pointed out is quite true, and most of us would have no argument. What I was trying to point out, I believe, was on a much smaller local scale. I merely attempted to show that in some areas of relatively undeformed sediments strain still might exist.

H. J. PINCUS (USA):*

Some of the results obtained from using photoelastic coatings in experiments on rock deformation conducted at the Applied Physics Research Laboratory, U. S. Bureau of Mines, College Park, Maryland, and the Department of Geology, The Ohio State University, Columbus, Ohio, are relevant to Dr. Emery's paper and to other papers presented here. The work has been in progress since June, 1962, and has recently been expanded to include observations in the field in areas of rock burst. The purpose of this program is to investigate relations among rock fabric, properties of rock in deformation and after relief, tectonic framework, local structure, and the like.

Early Experiments with Rock Disks

In one set of experiments, photoelastic coatings, 1 1/2 in. square and 0.071 in. thick, were cemented to 2 1/8 in. (NX) diameter rock disks, 0.5 in. thick. A 5/32 in. diameter relief hole, approximately 0.4 in. deep, was drilled into the center of each disk, from the reverse side, after the photoelastic coating had been applied and the cement had hardened. Precautions were taken to prevent generation of birefringence resulting from excessive heating. The purposes of these experiments were

*This discussion was submitted after the conference as an independent paper entitled "Some Experiments Utilizing Photoelastic Coatings for the Study of Strain in Rocks."

(1) to determine the capabilities of photoelastic coatings in making detailed studies of strain resulting from relief and (2) to analyze such photoelastic relief patterns as might be generated.

This procedure was applied, in this set of experiments, to about twenty specimens. Figures 1 through 4 show examples of isoclinics obtained from four of the specimens.

The Mt. Airy granite (Fig. 1), actually a quartz monzonite, shows a small, sharp pattern. Some first-order yellow and orange isochromatics (not shown here) are present.

The soft Catoctin greenstone (Fig. 2), on the other hand, shows virtually no isoclinic pattern. The birefringence is of very low order. The faces of the disk cut the foliation at an intermediate angle.

Disks of sandstone from the vicinity of Scranton, Pennsylvania, yield considerably different figures, depending on the orientation of the disk with respect to the bedding (Figs. 3 and 4). The disk with faces parallel to the bedding (Fig. 3) gives a clear figure, somewhat more diffuse than that of the Mt. Airy granite; the birefringence of the former is lower than that of the latter. The disk with faces normal to the bedding (Fig. 4) shows virtually no pattern; there is much resemblance here to the behavior of the Catoctin greenstone (Fig. 2). Unfortunately, an exact geological name cannot be assigned to this sandstone, because the slab was donated by a building stone supplier who was uncertain of the specific location of the source. The specimen is probably from the Pottsville or possibly from the post-Pottsville of the Pennsylvanian.

Tentative Results

From the work done on these twenty specimens, several tentative results are summarized below. The supporting data will be reported in detail when additional experiments have been completed; these data will include calculations of principal strain magnitudes from oblique incidence measurements.

Patterns have been generated by most relief holes. Relief patterns in planes of bedding and foliation are significantly more distinctive than patterns in planes perpendicular to these directions. Small holes, 1 mm or less in diameter and depth, generate patterns no sooner than

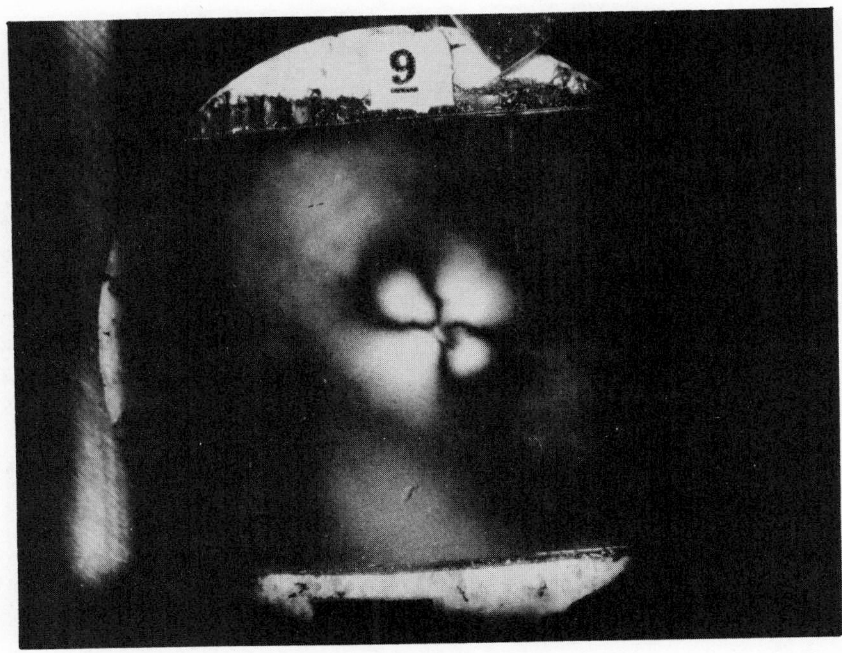

Fig. 1—Disk of Mt. Airy granite. Optics: white light, normal incidence, isoclinics, planes of polarization horizontal and vertical. Scale: disk diameter = 2 1/8 in.

Fig. 2—Disk of Catoctin greenstone. Optics: white light, normal incidence, isoclinics, planes of polarization horizontal and vertical. Scale: disk diameter = 2 1/8 in.

Fig. 3—Disk of sandstone from vicinity of Scranton, Pennsylvania; disk cut parallel to bedding. Optics: white light, normal incidence, isoclinics, planes of polarization horizontal and vertical. Scale: disk diameter = 2 1/8 in.

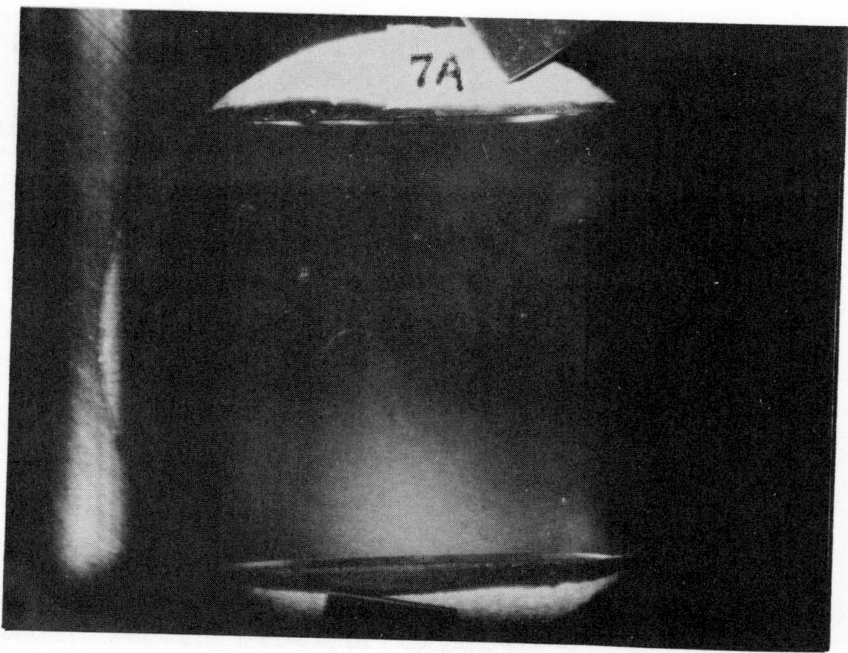

Fig. 4—Disk of sandstone from vicinity of Scranton, Pennsylvania; disk cut normal to bedding. Optics: white light, normal incidence, isoclinics, planes of polarization horizontal and vertical. Scale: disk diameter = 2 1/8 in.

about a day following the drilling of the holes; the standard relief holes (5/32 in. diameter) show patterns immediately after drilling, or not at all. Relief patterns in plutonic rocks are more concentrated and show higher birefringence than is seen in corresponding patterns in sedimentary rocks. Relief patterns in specimens from active rock burst areas show higher birefringence than has been observed in the other specimens. The highest birefringence which can be attributed, with confidence, to relief is in the first order; second-order isochromatics have been observed, but part of this birefringence may be caused by a slight bowing outward of the coating caused by too much pressure on the relief hole drill bit. Some rocks show fine-textured isoclinic patterns clearly similar to rock texture, superimposed upon more generalized patterns which are presumably integrated relief effects; the fine-textured isoclinics can be associated with individual crystals in some specimens and with domains of crystals in others. Thermal effects induced in the laboratory vary from specimen to specimen; thermal effects modify or override almost completely relief effects, and are not completely reversible.

Experiments with Cores

Photoelastic coatings attached to the end-face of rock cores are particularly effective in revealing anistropy when the cores are deformed along the major part of their length by uniform biaxial (hydrostatic) pressure. Taken in conjunction with data obtained from resistance strain gauge rosettes mounted symmetrically on opposite end-faces, much valuable information can be obtained. The pressure cell used here has been developed at the Applied Physics Research Laboratory, U.S. Bureau of Mines.*

An example of this procedure is shown in Fig. 5, displaying an end view of a 2 1/8 in. diameter (NX) core, 8 in. long, of Tennessee sandstone (probably Crossville, or "Crab Orchard"), also from an inexactly known source. The axis of the core is parallel to the bedding. The core is

*J. Fitzpatrick, "Biaxial Device for Determining the Modulus of Elasticity of Stress Relief Cores," U.S. Bur. Mines Rept. Invest. 6128, 1962.

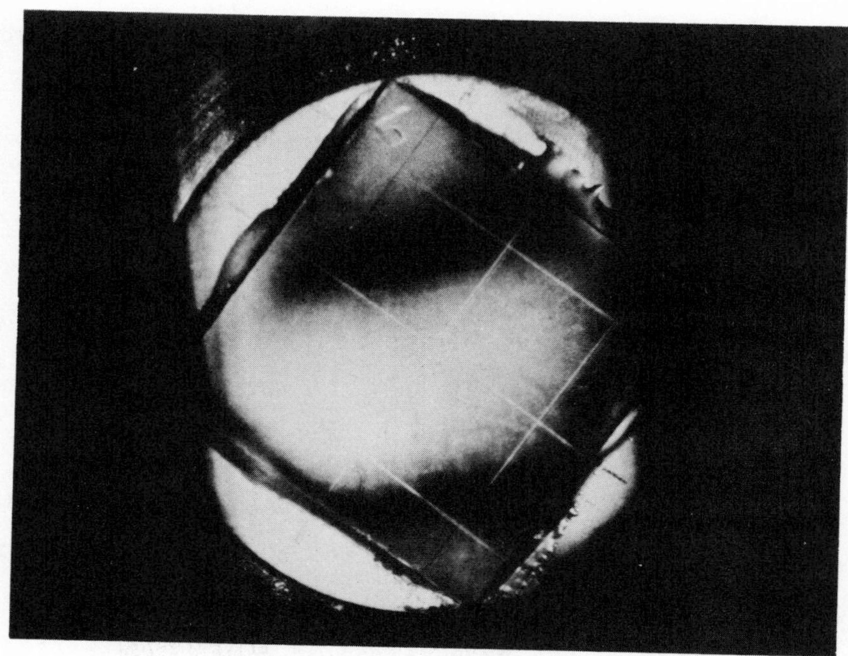

Fig. 5—Cylinder of Crossville (?) sandstone; axis of cylinder parallel to bedding; core is under hydrostatic pressure of 5000 psi along central 6 in. of 8 in. length. Optics: white light, normal incidence, isoclinics, planes of polarization inclined 45 degrees. Scale: cylinder diameter = 2 1/8 in.

under hydrostatic pressure of 5000 psi along the central 6 in. of its length. The photoelastic coating has the same dimensions as those shown in Figs. 1 through 4.

The isoclinic pattern coincides with the orientation of the bedding, and in fact the two dark bands coincide approximately with two identifiable thin beds. The isochromatics, not shown here, consist of first-order yellow and orange, and some gray bands, all parallel to the bedding. The planes of polarization are inclined 45 degrees; in this orientation, the isoclinics are far more distinct than in any other orientation. Strain gauge readings also indicate that principal strain directions are inclined at approximately 45 degrees.

Release of all the pressure in the cell results in the almost complete removal of the photoelastic pattern (Fig. 6). The faint vertical stripes are surface reflections from the projection lens.

Fig. 6—Same specimen as shown in Fig. 5, under no pressure. Scale: cylinder diameter = 2 1/8 in.

Later Experiments with Rock Disks

In another series of experiments with the 2 1/8 in. (NX) diameter disks, 0.5 in. thick, photoelastic coatings covering the entire face of the disks have been used. These coatings are also 0.071 in. thick. The circular coatings provide a much larger usable area of photoelastic readings than are possible with the square patches, for which edge effects are considerable.

Also, steps were taken to develop a simple, standard procedure for drilling small relief holes. In these experiments, holes were drilled through the photoelastic coatings and the specimens. In the earlier set of experiments, illustrated in Figs. 1 through 4, the relief holes were drilled from the reverse side, rather than through the photoelastic coating, in order to avoid destruction of the coating in the critical area surrounding the axis of relief. However, obtaining uniform depths and identical hole bottom shapes in different types of rock is not simple.

In addition, drilling from the reverse side involves a loss of sensitivity because of the bridging effect of the 0.1 in. of rock between the inner end of the hole and the photoelastic coating and because of the dispersal of stress effects arising from the separation of the bottom of the hole from the coating. Finally, stress calculations are far simpler for the case in which the hole passes completely through the specimen.

Both types of relief holes drilled in disks of Massillon sandstone cut both parallel and perpendicular to the bedding show no evidence for strain relief in the rock. Figures 7 and 8 show both sides of the disk with faces cut perpendicular to the bedding. In Fig. 8, the disk as shown in Fig. 7 has been rotated left-right about the vertical axis. Figure 7 shows isoclinics observed in the photoelastic coating, several months after the specimen had been prepared, coated, and drilled. Initially, a 5/32 in. hole was drilled 0.4 in. deep from the reverse side (center, Fig. 8), as in Figs. 1 through 4; this produced no detectable response. Then, 0.25 in. holes were drilled from the coating side; one hole was drilled through the coating only (lower hole, Fig. 7), as a check for parasitic birefringence, and the other hole (upper hole, Fig. 7; larger hole, Fig. 8) was drilled entirely through the coating and the specimen. The small, isoclinic crosses (Fig. 7) around both holes are almost identical, and are presumably from parasitic effects generated in drilling. A similar cross was generated by drilling an identical hole in a coating not cemented to a specimen. Thus there is no evidence for relief in the rock around the 0.25 in. hole which pierces the disk.

Procedures for drilling holes have been improved since the time the holes shown here were made, and it is now rare that this much parasitic birefringence is developed.

Of considerable interest is the correspondence between the banding of the isoclinics (Fig. 7) and the bedding in the specimen. In Figs. 7 and 8, examples of corresponding features are marked A, B, and C. The banded isoclinics, most sharply defined when polarization is inclined at 55 degrees, were not detected in the specimen immediately after it had been prepared. Only after several months, when the specimen was

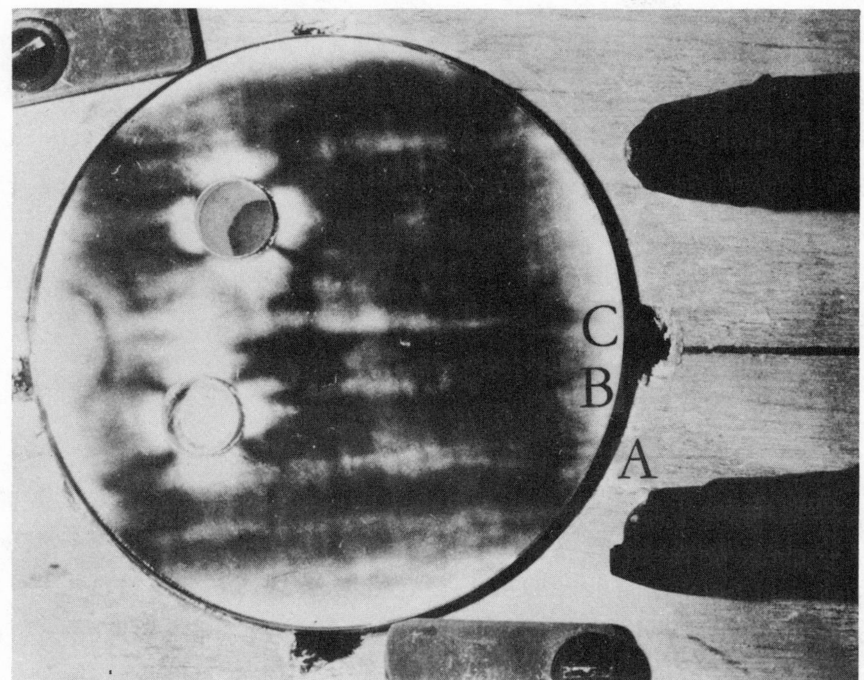

Fig. 7—Disk of Massillon sandstone; disk cut normal to bedding. Optics: white light, normal incidence, isoclinics, planes of polarization inclined 55 degrees counterclockwise. Scale: disk diameter = 2 1/8 in.

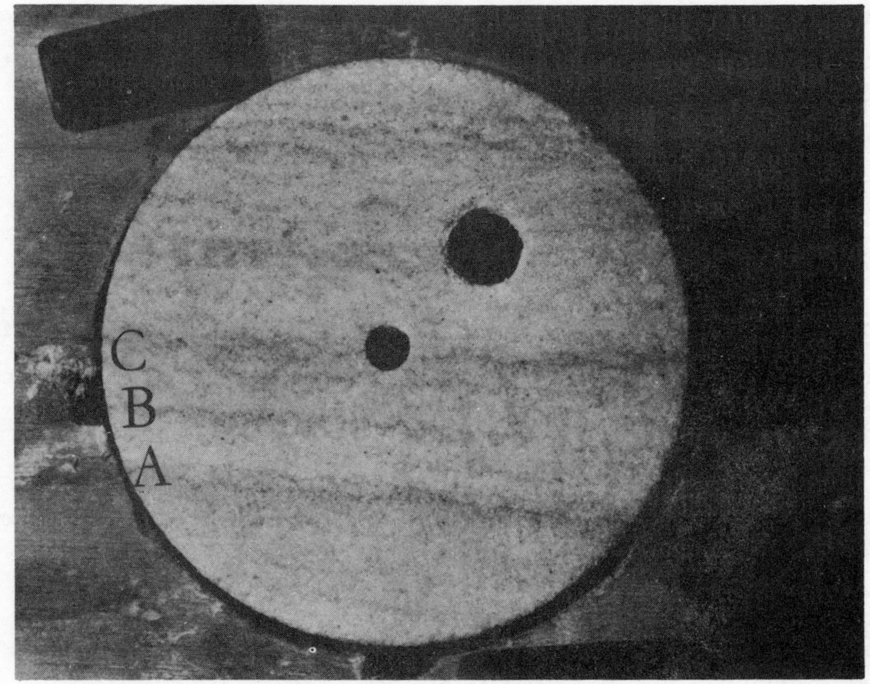

Fig. 8—Reverse side of specimen shown in Fig. 7; specimen has been rotated left-right about vertical axis. Scale: disk diameter = 2 1/8 in.

re-examined, was the pattern detected. (Compare this with the change from Fig. 5 to Fig. 6.)

The semicircle at the left end of the horizontal axis in Fig. 7 is merely the surface reflection of the spotlight lens, and therefore should be disregarded.

Current and Future Work

This work, best described as a feasibility study, is being followed up with a comprehensive program, under sponsorship and direction of the Applied Physics Research Laboratory of the U.S. Bureau of Mines, and directed chiefly toward studies of rock burst problems and materials. The elements of the program include (a) detailed underground studies utilizing photoelastic coatings in bore holes, microseismic instrumentation[*] and borehole strain gauges,[**] and the collection of oriented specimens; (b) laboratory studies of disks and cores, such as those described here, using photoelastic coatings and other strain gauges; (c) petrofabric and petrographic analyses; (d) systematic laboratory investigations of the effects of heat; and (e) detailed structural mapping and regional tectonic syntheses.

Concluding Remarks

From the work described here, in Dr. Emery's paper, and in parts of several other papers presented at this conference, it obviously is not yet possible to describe the mechanics of relief phenomena in simple, consistent terms. Furthermore, it is likely that what has often been identified as strain associated with relief is in fact not always entirely

---

[*] L. Obert and W. Duvall, "The Microseismic Method of Predicting Rock Failure in Underground Mining, Parts 1 and 2," U.S. Bur. Mines Rept. Invest. 3797 and 3803, 1934.

[**] L. Obert, R. H. Merrill, and T. A. Morgan, "Borehole Deformation Gauge for Determining the Stress in Mine Rock," U.S. Bur. Mines Rept. Invest. 5978, 1962.

the result of relief. For example, there should be a systematic investigation of the possibility that relatively small changes in temperature could result in differential thermal expansion of individual crystals, domains of crystals, and beds or laminae, thereby producing strains detectable in photoelastic coatings. We also need rigorous analysis of the as yet unexplained differences in the rates at which strain attributed to relief has developed in the relatively few rocks that have been studied.

The analysis of data from photoelastic coatings, as applied to rocks, is obviously in its infancy. It is, therefore, encouraging to note the increasing activity in this type of investigation and in related techniques.

ABSTRACT

Planar anisotropy can have a pronounced effect on both the strength and the angle of faulting in rocks. For slate experimentally deformed under confining pressures to 2000 bars, curves of differential stress at failure versus inclination of anisotropy are concave upward and parabolic in form. Faults tend to develop parallel to the anisotropy (cleavage) for inclinations of anisotropy from about 15 degrees up to 45-60 degrees to the direction of maximum compression, although faulting for certain orientations tends to occur without loss of cohesion at higher confining pressures. Thus, the mode of deformation changes from shear fracture to ductile faulting with increasing pressure. The type of faulting is, however, a function of orientation as well as of confining pressure.

RÉSUMÉ

L'anisotropie plane peut avoir un effet prononcé à la fois sur la résistance et sur l'angle de faille des roches. Pour une ardoise expérimentalement deformée sous des pressions confinantes atteignant 2000 bars, les courbes de contraintes différentielles à la rupture en fonction de l'inclinaison de l'anisotropie sont concaves vers le haut et de forme parabolique. Les failles tendent à se développer parallèlement à l'anisotropie (clivage) pour des inclinaisons de l'anisotropie comprises entre 15 degrés environ et 45 à 60 degrés avec la direction de compression maximum, bien que les failles pour certaines orientations tendent à se produire sans perte de cohésion à des pressions confinantes plus élevées. Le mode de déformation passe donc de la fracture par cisaillement à la faille ductile lorsque la pression croît. Le type de faille est fonction cependant aussi bien de l'orientation que de la pression confinante.

AUSZUG

Flächenanisotropie kann einen ausgesprochenen Einfluss sowohl auf die Festigkeit als auch auf die Richtung einer Bruchfläche im Gebirge haben. Für Schiefer, der bei Versuchen unter bis zu 2000 kg/cm^2 verformt wurde, ergab sich für das Verhältnis von Spannungsunterschied beim Bruch zur Neigung der Anisotropieflächen eine nach oben konkave, parabolische Kurve. Die Bruchflächen neigen dazu sich parallel zur Anisotropie (Schieferung) zu entwickeln, wenn die Neigung der Anisotropie von 15 bis zu 45 bis 60 Grad zur Richtung der grössten Druckspannung beträgt. Dabei tritt der Scherbruch für bestimmte Richtungen und grössere Seitendrücke häufig ohne Verlust von Kohesion auf. Die Verformungsart ändert sich daher mit zunehmendem Druck vom Scherbruch zur plastischen Scherverformung. Die Art des Bruches ist jedoch sowohl von der Richtung der Anisotropie als auch vom Umschliessungsdruck abhängig.

STRENGTH VARIATION AND DEFORMATIONAL BEHAVIOR IN ANISOTROPIC ROCK

Fred A. Donath*

Introduction

THE CONDITIONS of homogeneity and isotropy generally assumed in theoretical stress analysis are seldom realized in materials of the earth's crust. Most rocks in the upper crust are characterized by some type of foliation and thus are distinctly anisotropic. Rock layers that show obvious differences in composition, grain size or shape, porosity, cementation, and other characteristics can be expected to have distinctive strength and deformational properties. But even where compositional and other characteristics are sufficiently uniform for a rock sequence to be considered reasonably homogeneous, the presence of bedding surfaces or other foliation (for example, cleavage or schistosity) can have pronounced effects on these properties. Theoretical analyses that do not take into account the possible effects of anisotropy may be in error. Similarly, experimentally determined values of strength and observed deformational behavior of anisotropic rocks may not be representative if the effects of anisotropy are not evaluated systematically.

An experimental study is currently in progress at Columbia University to determine the strength variation and deformational behavior in anisotropic rocks. Cylindrical specimens of several rocks with different types of planar anisotropy (bedding, cleavage, and schistosity) are being tested to evaluate these effects at

*Department of Geology, Columbia University.

successively higher confining pressures. Bulk material for the study is collected parallel to the anisotropy so that grain size, cohesion, and other properties are as constant as possible in all specimens of a given rock. The bulk material is sawed into rectangular blocks and cored at angles of 0°, 15°, 30°, 45°, 60°, 75°, and 90° to the plane of anisotropy (Fig. 1); the cores are then ground at both ends to form perfect right cylinders. One-inch diameter by 2½-in. length specimens were used for tests at 35, 105, and 350 bars confining pressure, and ½-in. diameter by 1-in. length specimens were used for tests at 500, 1000, and 2000 bars. Specimen size does not appear to affect the results. (The reader is referred to Figs. 4, 5, and 12 for examples.)

Test specimens are placed between a piston and anvil, jacketed, and inserted in a pressure vessel. Axial load is produced by placing the pressure vessel between the platens of a hydraulic press; confining pressure is produced by pumping kerosene into the vessel. The pressure vessel and specimen assembly are shown in Fig. 2 and the entire test apparatus, in place ready for testing, is shown in Fig. 3.

Axial force and specimen shortening are recorded on a strip chart recorder continuously during a test. The raw data thus obtained are corrected for changes in cross-sectional area of the specimen and elastic distortion of the apparatus to give true stress and specimen strain for any time n during the test. Each orientation is tested at least twice at a given confining pressure to check reproducibility of data. Results for Martinsburg slate (Ordovician; Bangor, Pennsylvania) are discussed here, as this rock illustrates nicely the effects of well-developed anisotropy (cleavage). All data are from compression tests. The differential stress at failure represents the strength of the rock under the imposed conditions of the test and is equal to the axial stress minus the confining pressure.

Fig. 1—Block of slate mounted for coring at 60 degrees to plane of anisotropy; cleavage is parallel to top face of block.

Fig. 2—Pressure vessel and specimen assembly; specimen is mounted between piston and anvil and jacketed with thin-wall copper tubing.

Effect of Anisotropy on Rock Strength

Preliminary results at confining pressures of 35, 105, and 350 bars* indicated that planar anisotropy could have a marked effect on the breaking strength of rocks when failure was by shear fracture.(1) The effect was found to be also very pronounced in subsequent tests at 500, 1000, and 2000 bars confining pressure, although the mode of deformation for certain orientations changed from shear fracture to faulting without loss of cohesion. Curves of differential stress at failure versus inclination of anisotropy for Martinsburg slate are concave upward and roughly parabolic in form for all confining pressures to 2000 bars (Fig. 4). Specimens compressed perpendicular to cleavage sustain the greatest differential stress; those compressed

*1 bar equals 14.5 psi or 1.02 kg/cm^2.

Fig. 3—Test apparatus with pressure vessel in place and ready for testing.

at an angle of 30 degrees to cleavage show the lowest strength. Inasmuch as failure in isotropic substances commonly occurs along surfaces inclined at about 30 degrees to the direction of maximum compression, it might be expected that the 30-degree orientation would

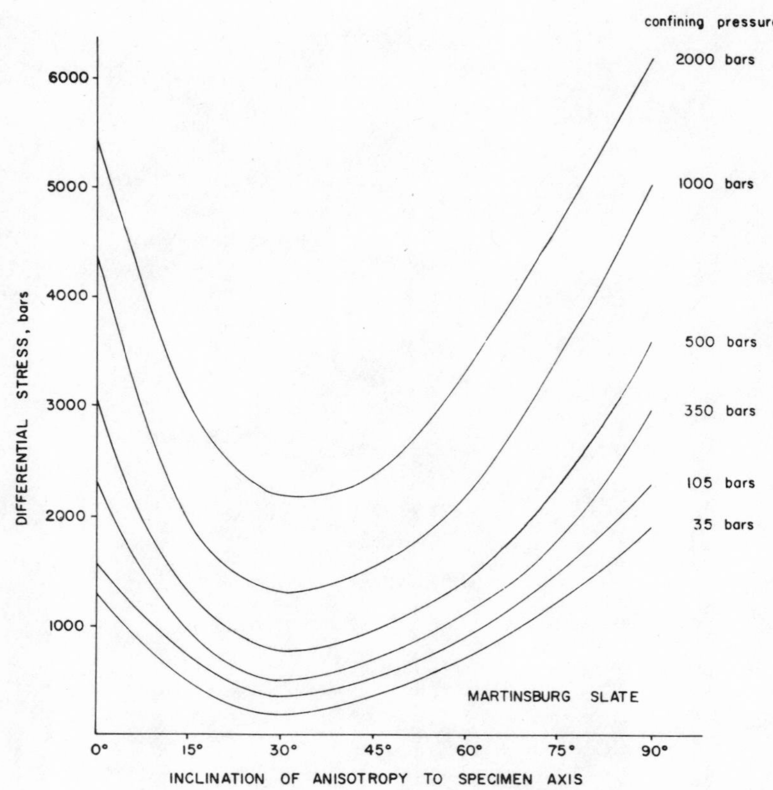

Fig. 4—Ultimate strength versus inclination of cleavage, Martinsburg slate.

be weakest. One might also expect the 90-degree orientation in slate to be strongest. More surprising is the high strength parallel to foliation which, at the higher confining pressures, is nearly equal to that of the 90-degree orientation. There is, in fact, no reason why the strength of certain rocks could not be greater parallel to foliation than perpendicular to it.

The curves of differential stress versus anisotropy are shifted upward with increased confining pressure; the amount of shift is proportional to the increase in pressure. Handin and Hager[(2)] have previously shown that the ultimate strength of several common rock types increases linearly with increasing confining pressure. As seen in Fig. 5, this relationship holds for Martinsburg slate at confining pressures up to 1000 bars, but there is a noticeable departure from linearity at higher confining pressures. The slopes of the curves

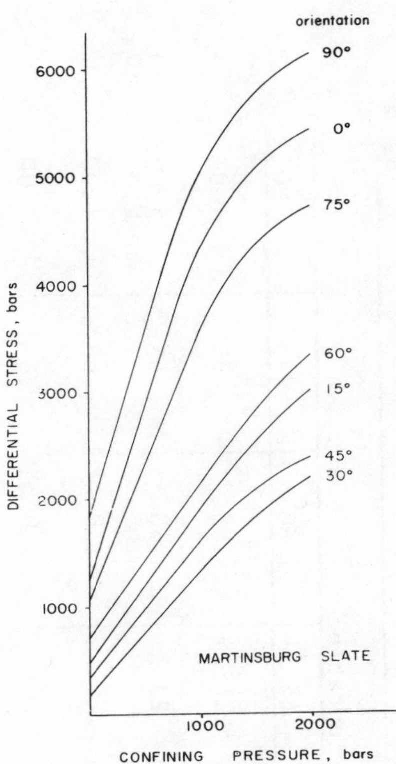

Fig. 5—Ultimate strength versus confining pressure, Martinsburg slate.

for the 15-, 30-, 45-, and 60-degree orientations are very nearly the same and lower than those for the 0-, 75-, and 90-degree orientations. As will be described later, failure tends to be controlled by anisotropy in the first group of orientations, but not in the second.

Anisotropy strength coefficients for Martinsburg slate are shown in Table 1 for several representative confining pressures. Each coefficient is the ratio of strength for a given orientation to that for compression perpendicular to anisotropy, other conditions being held constant. The anisotropy strength coefficients for slate increase slightly with increasing confining pressure. The ratio of strength for the 90-degree orientation to that of the 30-degree orientation ranges from 10:1 at 35 bars confining pressure to approximately 3:1 at 2000 bars. Interestingly, there is little strength difference between the 0- and 90-degree orientations at higher confining pressures even though notable differences exist for the other orientations.

Table 1

ANISOTROPY STRENGTH COEFFICIENTS FOR MARTINSBURG SLATE

Confining Pressure (bars)	Orientation						
	0°	15°	30°	45°	60°	75°	90°
35	.65	.25	.10	.21	.38	.65	1.00
105	.66	.31	.15	.23	.40	.65	1.00
350	.79	.33	.17	.24	.38	.60	1.00
500	.85	.34	.22	.28	.40	.65	1.00
1000	.86	.37	.26	.32	.43	.71	1.00
2000	.89	.49	.36	.39	.54	.78	1.00

Effect of Anisotropy on the Mode of Deformation

Failure in Martinsburg slate was almost invariably by faulting; that is, one part of the specimen was displaced relative to another part along an inclined surface. At confining pressures of 500 bars or lower and for certain orientations at higher pressures, faulting occurred with complete loss of cohesion, release of elastic strain energy, and separation of the specimen into two or more parts--failure was by shear fracture (brittle faulting). At confining pressures of 1000 and 2000 bars, faulting commonly occurred without total loss of cohesion and without release of elastic strain energy; the specimen sustained considerable differential stress while faulting. The phenomenon of faulting without loss of cohesion is called ductile faulting.[3] In some tests, movement occurred along a fault when a minimum differential stress was exceeded; movement ceased when the differential stress fell below this value. In other tests, each increment of displacement on the fault required an additional increment of differential stress (work hardening). Stress-strain curves for ductile faulting thus commonly resemble those for uniform flow with or without work hardening, and the mode of deformation has to be determined from examination of the deformed specimen.

Still another type of stress-strain curve was observed for the 15-degree orientation (Fig. 6). In tests at 200, 400, and 800 bars confining pressure, failure in the 15-degree orientation of slate was marked by a loss of resistance to stress difference until the differential stress was just sufficient to allow movement along a single or multiple cleavage surfaces. At 1200 bars confining pressure and higher, the initial loss of resistance was followed by increased resistance to stress difference. This effect coincided with the development of a kink band at high angles to cleavage (Fig. 7). Thus, with increasing confining pressure the mode of deformation for the 15-degree orientation of slate changes from slip along a single cleavage surface to slip along multiple surfaces to initial multiple slip followed by the development of a kink band. As seen in Fig. 7, the width of the kink band decreases with increasing confining pressure from very wide at 800 bars to quite

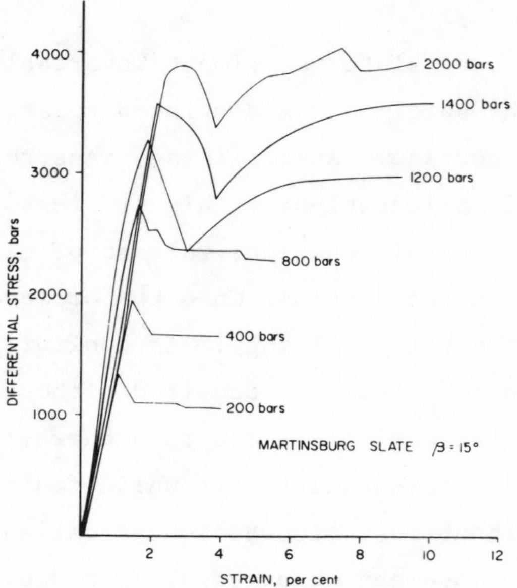

Fig. 6—Differential stress versus strain at several confining pressures, 15-degree orientation of Martinsburg slate.

narrow at 2000 bars. Cleavage is rotated within the kink band until slip is no longer possible along cleavage; further deformation causes faulting within and parallel to the kink band boundaries. The rotation of cleavage within the kink bands is shown in Figs. 8 and 9. Faulting conjugate to cleavage was observed in the 30- and 60-degree orientations as well as in the 15-degree orientation. (See, for example, Fig. 11.)

The type of faulting in slate is clearly a function of the orientation of anisotropy as well as of confining pressure. Specimens with the anisotropy (cleavage) oriented at 15°, 30°, 45°, 60°, and 75° to the direction of maximum compression commonly faulted without total loss of cohesion at 2000 bars confining pressure, whereas faulting with loss of cohesion characterized the 0- and 90-degree orientations.

In a homogeneous, isotropic material, faults would normally develop at approximately 30° to the direction of maximum compression. The angle of faulting in strongly anisotropic rocks varies considerably;

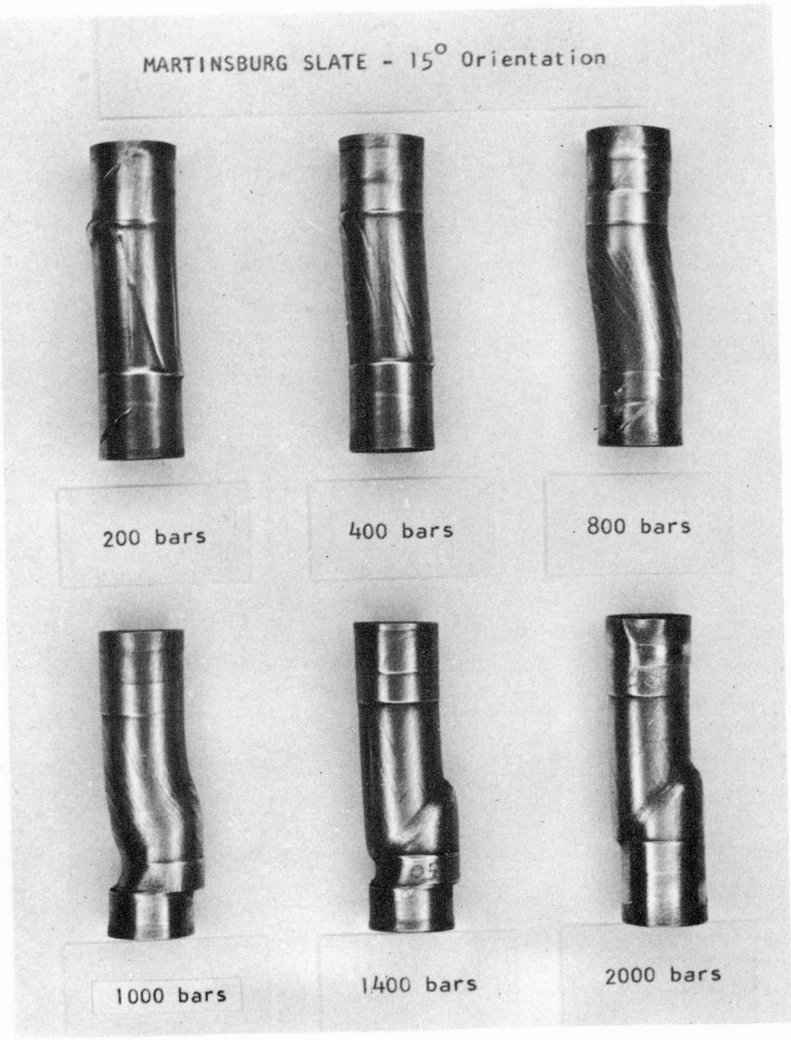

Fig. 7—Effect of confining pressure on mode of deformation, 15-degree orientation of Martinsburg slate. Slip along cleavage is reflected in the copper jackets. Cleavage descends to the right in each specimen.

it is dependent on the inclination of the anisotropy with respect to the principal stress directions. Figure 10 illustrates the effect of cleavage on the angle of faulting in slate specimens deformed at 500 bars confining pressure. Faults have developed parallel to cleavage in the 15-, 30-, and 45-degree orientations; the angle of faulting is strongly affected in the 60- and 75-degree orientations as well. Multiple slip along cleavage in the 15-degree orientation has resulted in external rotation of the specimen.

Fig. 8—Rotation of cleavage in kink bands, 15-degree orientation of Martinsburg slate.

Very similar effects are observed at higher confining pressures (Fig. 11). However, multiple slip along cleavage is more common and kink banding and faulting conjugate to cleavage tend to develop in several orientations at the higher confining pressures. In all tests to 2000 bars pressure the strike of the failure plane is parallel to that of the cleavage; failure in the 0- and 90-degree orientations is invariably by shear fracture.

Fault angles were measured and plotted along with the inclination of cleavage (Fig. 12). In Martinsburg slate deformed at low confining pressures (350 bars or less), failure for all orientations was by shear fracture. Shear fractures were parallel to cleavage for maximum compression inclined up to 45 degrees to cleavage; they were strongly affected even at 60 and 75 degrees. Shear fractures were unaffected by cleavage only for maximum compression perpendicular to cleavage. As stated earlier and indicated in Fig. 12, the effect is essentially the same at confining pressures of 500, 1000, and 2000 bars, although faulting in several orientations occurred without loss of cohesion. The effect of anisotropy oriented at 60 degrees does, however, appear to be somewhat more pronounced than at the lower

Fig. 9—Thin section of 15-degree orientation of Martinsburg slate deformed at 1600 bars confining pressure. (Note rotation of cleavage in kink band.)

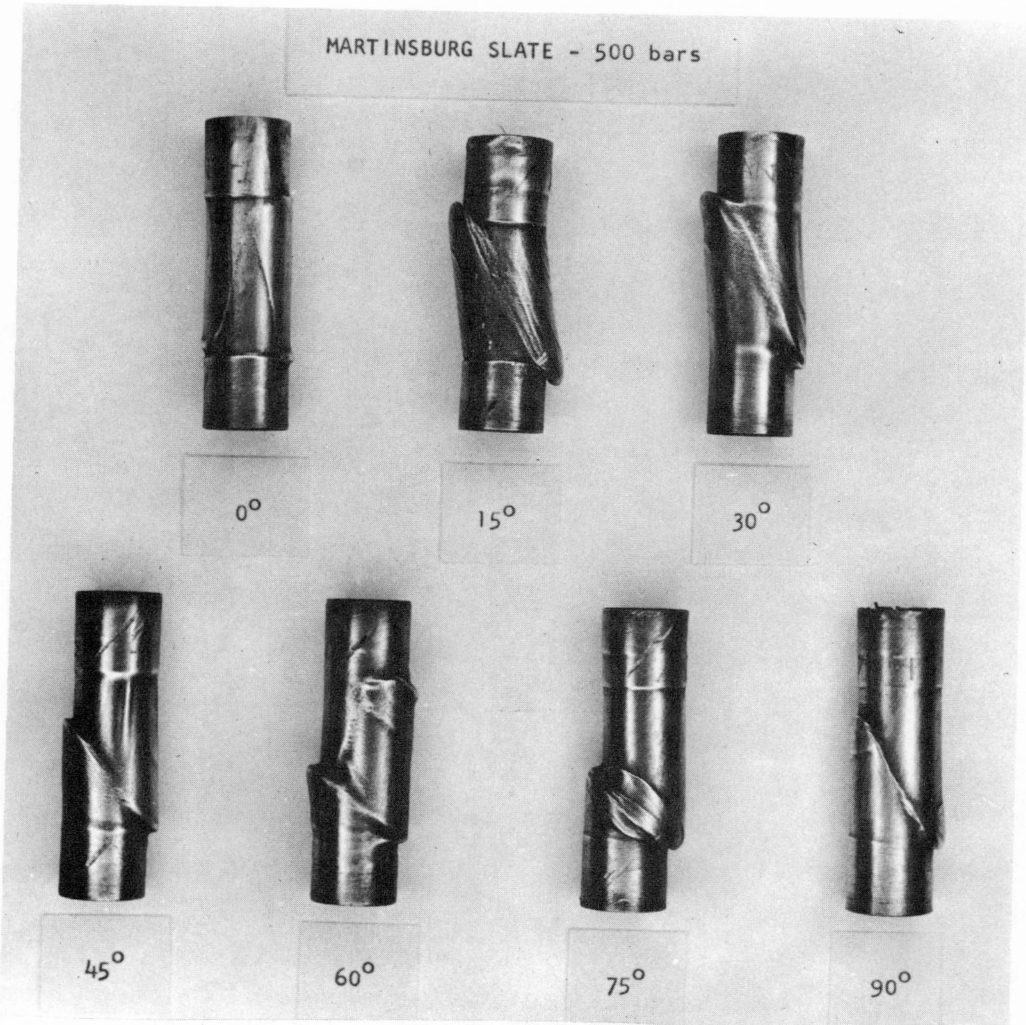

Fig. 10—Effect of anisotropy on the mode of deformation in Martinsburg slate deformed at 500 bars confining pressure.

confining pressures. Again, the angle of faulting is unaffected by anisotropy only in the 90-degree orientation. Geologic implications of these results have been discussed elsewhere.(4)

Conclusions

The differential stress required to produce failure in anisotropic rocks is a function of the orientation of the anisotropy with

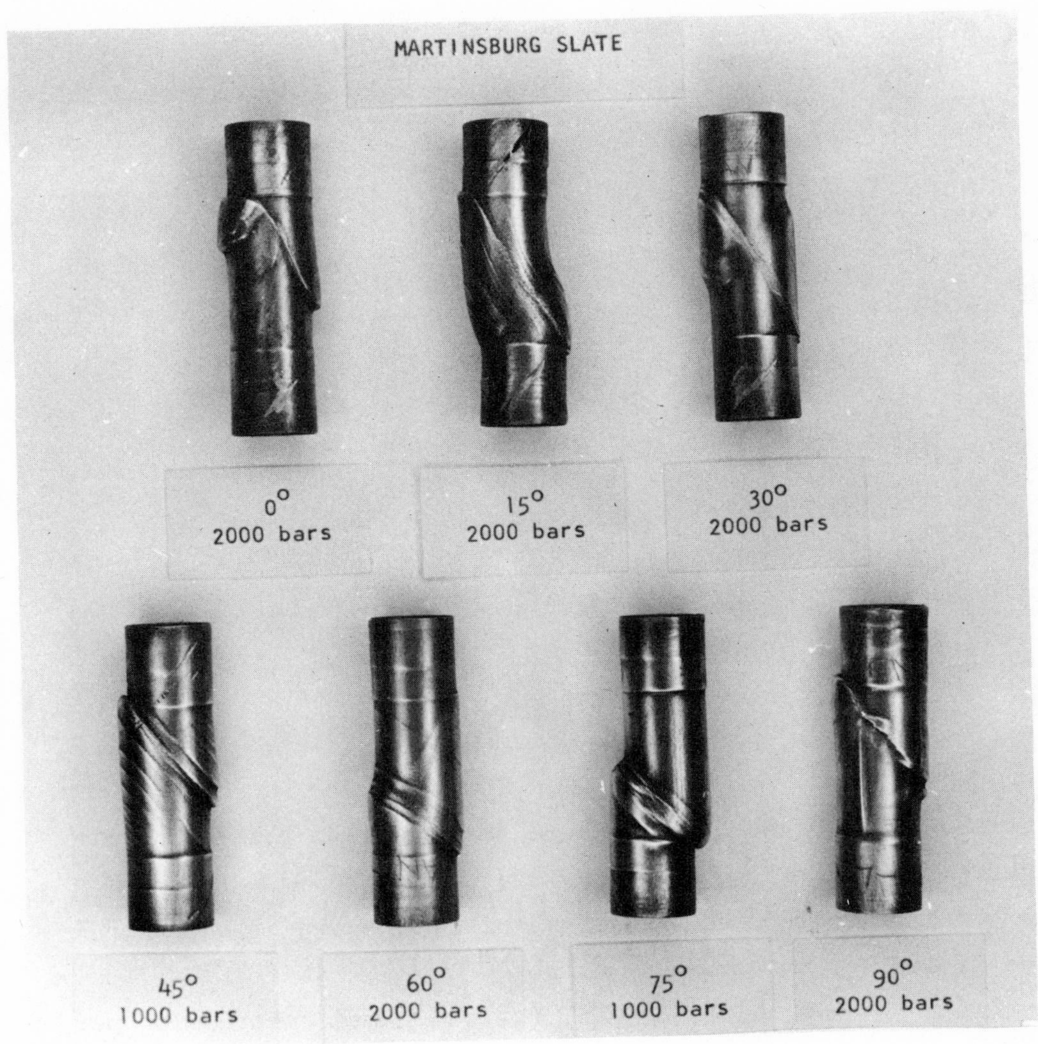

Fig. 11—Effect of anisotropy on the mode of deformation in Martinsburg slate deformed at 1000 and 2000 bars confining pressure.

respect to the principal stress directions; present knowledge suggests that it is low for inclinations between 15 and 60 degrees and at a minimum at 30 degrees to the direction of maximum compression. Moreover, failure can occur parallel to anisotropy for inclinations from about 15 degrees up to 45 to 60 degrees to the direction of maximum compression. Thus, failure parallel to anisotropy is not only possible for these inclinations, but it is likely to occur much more readily than failure across the anisotropy because of the appreciably lower stress differences required.

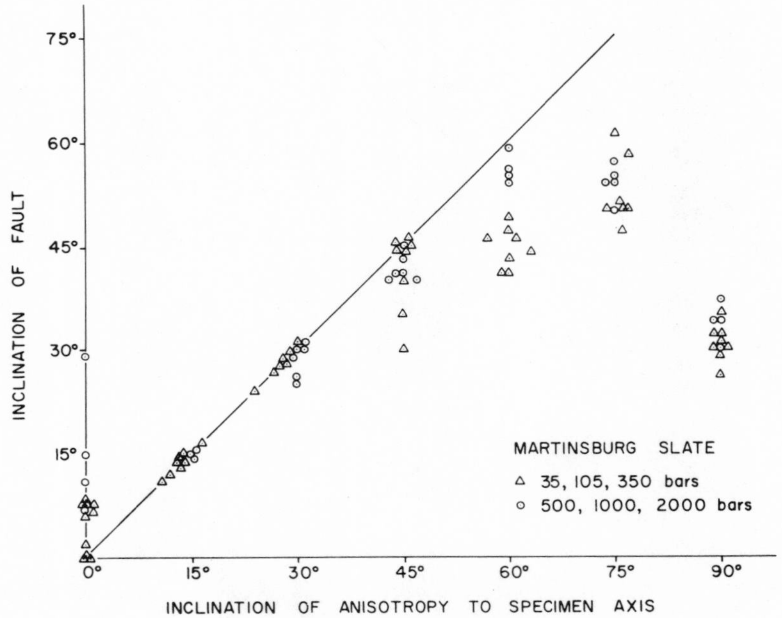

Fig. 12—Effect of cleavage on the angle of faulting, Martinsburg slate.

The effect of anisotropy in slate appears to diminish only slightly with increased confining pressure to 2000 bars. The anisotropy strength coefficient for the 30-degree orientation in slate is still very low (0.36) at 2000 bars; that is, the rock is approximately three times as strong for maximum compression perpendicular to cleavage as for maximum compression directed at 30 degrees to cleavage. The inclination and type of faulting are strongly affected by anisotropy for most orientations even at high confining pressures. Faults are unaffected only for maximum compression perpendicular to cleavage.

It would therefore seem important, if not essential, that the effects of anisotropy be evaluated before attempting either a theoretical analysis of stress in rocks of the earth's crust or a prediction of deformation characteristics.

Acknowledgments

Acknowledgment is made to the donors of the Petroleum Research Fund, administered by the American Chemical Society, for support of this research.

REFERENCES

1. Donath, Fred A., "Experimental Study of Shear Failure in Anisotropic Rocks," Bull. Geol. Soc. Am., Vol. 72, 1961, pp. 985-990.
2. Handin, J. W., and R. V. Hager, Jr., "Experimental Deformation of Sedimentary Rocks under Confining Pressure: Tests at Room Temperature on Dry Samples," Bull. Am. Assoc. Petrol. Geol., Vol. 41, 1957, pp. 1-50.
3. Donath, Fred A., and Rodger T. Faill, "Ductile Faulting in Experimentally Deformed Rocks," Trans. Geophys. Union, Vol. 44, 1963, pp. 102-103 (abstract).
4. Donath, Fred A., "Role of Layering in Geologic Deformation," Trans. N. Y. Acad. Sci., Vol. 24, 1962, pp. 236-249.

DISCUSSION

J. C. JAEGER (Australia):

I would like to make two small comments on Professor Donath's paper. First, I have made similar experiments at confining pressures up to 10,000 psi and found the same sort of behavior in various gneisses and slates. It appears that the behavior described by Professor Donath probably is characteristic of most anisotropic rocks, and, in particular, that they become quite strong in the direction of foliation at relatively low confining pressures. I have met cases in which the rock is a little stronger in the direction of foliation than perpendicular to it at pressures of a few thousand psi. This implies that it is not possible to describe the behavior of an anisotropic rock simply in terms of the variation of the cohesion in a plane with the angle which this plane makes with the plane of foliation. The coefficient of internal friction must vary also, being greatest in the plane of foliation.

Second, in all my experiments, and I assume in all of Professor Donath's, the plane of failure passes through the intersection of the plane of foliation and the plane perpendicular to the greatest principal stress σ_1. I have found that this happens also with unequal principal stresses for the case in which σ_2 and σ_3 are not large. That is, in this case, the plane of failure does not pass through the direction of σ_2 but is determined by the plane of foliation. It remains to be seen whether this is true for all values of the principal stresses and to develop a theory of failure for this case. This point is also made in Dr. Serafim's paper.

ABSTRACT

Stress concentrations about an underground opening and strength of rock per se do not constitute the complete criteria which define initiation and/or extent of failure of openings in idealized homogeneous rocks. The Griffith theory of failure and its application to the failure of materials in tension and shear are the basis of several theoretical and laboratory investigations. There appear to be analogous criteria for crack propagation and failure for materials in compressive uniaxial, biaxial, and triaxial stress fields, although no investigations appear to have been reported concerning crack propagation and velocity in compressive fields.

It is proposed that existing rock compressive (shear?) strength data for uniaxial fields, unconfined, may be utilized to determine the conditions for initiation of cracks at the periphery of an underground opening. Crack propagation phenomena then become the determining factors, in combination with rapidly changing local stress field parameters, in limiting the extent and degree of failure. Such parameters include crack velocity, changing confinement effects, energy flow, and duration and magnitude of stress field, as well as rock properties themselves. Whether a strength exists corresponding to a point on the Mohr envelope for some confining pressure depends on the relative crack velocity of release of confinement ahead of crack tips.

RÉSUMÉ

La concentration d'efforts autour d'une ouverture souterraine et la résistance de la roche en soi ne constituent pas les critères complets qui définissent l'origine et/ou l'étendue de l'affaissement des ouvertures dans des roches homogènes idéalisées. La théorie d'affaissement de Griffith et son application à la rupture des matériaux en tension et en cisaillement servent de base à plusieurs études théoriques et de laboratoire. Il semble qu'il y ait des critères analogues pour la propagation des fissures et la rupture des matériaux dans des champs de force de compression à un, deux ou trois axes, quoiqu'on n'ait pas vu de comptes-rendus d'études concernant la propagation et la vitesse de fissures dans des champs compressifs.

On suggère que les données actuelles sur la résistance en compression (cisaillement?) de la roche pour les champs libres à un axe puissent être utilisées pour déterminer les conditions à l'origine des fissures à la périphérie d'une ouverture souterraine. Les phénomènes de propagation des fissures deviennent alors les facteurs déterminants, en combinaison avec les paramètres à changement rapide du champ local des forces, qui limitent l'étendue et le degré de l'affaissement. De tels paramètres incluent la vitesse de la fissure, les effets changeants de la contrainte, le flux d'énergie, et la durée et la grandeur du champ de forces, ainsi que les propriétés mêmes de la roche. Qu'il existe ou non une résistance correspondant à un point sur l'enveloppe de Mohr pour une pression quelconque de rétention, cela dépend de la vitesse relative du relâchement de rétention pour les fissures devant les pointes des fissures.

AUSZUG

Die Spannungskonzentration um unterirdische Hohlräume und die Gesteinsfestigkeit als solche stellen kein vollständiges Kriterium für den Beginn und/oder den Ausmass des Zusammenbruches von Hohlräumen im idealisierten, homogenen Gebirge dar. Die Bruchtheorie von Griffith und ihre Anwendung auf den Bruch von Material infolge Zug- und Scherbeanspruchungen dienen als Grundlage für verschiedene theoretische und auch Laboratoriumsuntersuchungen. Anscheinend bestehen ähnliche Kriterien für die Fortpflanzung von Rissen und den Bruch von Material in ein-, zwei- und dreiachsigen Druckspannungsfeldern, obgleich anscheinend bisher über keine Untersuchungen über die Fortpflanzung von Rissen in Druckfeldern und deren Geschwindigkeit berichtet wurde.

Es wird vorgeschlagen, die verschiedenen Druckfestigkeitsdaten (Scherfestigkeiten?) für einachsige Felder, ohne Umschliessungsdruck, für die Bestimmung der Bedingungen für das Entstehen von Rissen am Umfang eines unterirdischen Hohlraumes zu verwenden. Die Fortpflanzungsvorgänge an Rissen werden dann zu Bestimmungswerten, die, zusammen mit den sich rasch ändernden Parametern des örtlichen Spannungsfeldes, das Ausmass und Grad des Bruches abgrenzen. Diese Parameter schliessen die Fortpflanzungsgeschwindigkeit von Rissen, sich ändernde Umschliessungsbedingungen, Energiefluss, Zeitdauer und Grössenordnung des Spannungsfeldes und ebenfalls die Eigenschaften des Gesteines ein. Es hängt von der Fortpflanzungsgeschwindigkeit der Haarrisse ab, relativ zum Abfall des Umschliessungsdruckes vor ihren Enden betrachtet, ob für einen bestimmten Umschliessungsdruck eine, einem Punkte der Mohr'schen Hüllkurve entsprechende Festigkeit vorhanden ist.

FAILURE OF HOMOGENEOUS ROCK UNDER DYNAMIC COMPRESSIVE LOADING*

George B. Clark** and Rodney D. Caudle***

Introduction

THE OBJECTIVES of this study were twofold. The primary objective was to establish additional bases for the prediction of failure of cavities in homogeneous rock subjected to impulsive loading generated by a nuclear burst at the ground surface. The second objective was to investigate further by analysis of existing theory and data the failure mechanisms of rock materials *in situ* around underground cavities, and to propose more meaningful failure criteria. The ideal elastic stress distributions around various shapes of cavities in given static stress fields are well known, and while they predict only stress concentration and potential failure, they do provide the best currently available basis for study.

Theoretical analyses and experimentation to determine stress concentration factors in dynamic fields have been limited largely to circular openings. The results of these investigations can be applied in a limited way to other shapes of openings. In this report consideration will be limited to openings of circular cross-sectional shapes. The stress distribution and concentration per se caused by the presence of openings in various stress fields, however, do not offer

---

*From a technical report for Plesset Associates (Contract No. DA-49-146-XZ-073), presented at the conference by G. B. Clark.

**Director, Research Center, University of Missouri School of Mines and Metallurgy, Rolla, Missouri.

***Research Assistant, Research Center, University of Missouri School of Mines and Metallurgy, Rolla, Missouri.

sufficient information to explain or predict the mode of failure of a given geometrical opening in a specified type of rock.

Spalling has been postulated by some investigators as a significant mechanism of failure in rock cavities subject to impact loading. However, the shape and length of the pulse determine whether tensile slabbing may occur although stress concentrations also may cause spalling, particularly where there are incipient fractures in the rock and it is of a type subject to "bursting."

There is a large amount of data available on the physical properties of rock, a large portion of which is reported in Ref. 1. These data can be of significant value if properly interpreted and applied. However, where actual test conditions are not known, the data may be of questionable value.

In conjunction with physical property data the following environmental parameters must be considered: (1) strain, (2) rate of strain, (3) stress, (4) time history of load, (5) particle velocity, (6) acceleration, and (7) wave energy; in addition, the effects of (8) the existing static stress field must be taken into account.

The primary effort of this study was directed toward a correlation of the above parameters with physical property data to define more precisely their relative importance in the mechanism of failure of underground cavities. This led directly to a study of the Griffith theory of crack formation and additional recent theory and data relative to crack propagation.

Basic Assumptions

The basic assumptions are as follows: First, it is assumed that the rock being considered is effectively homogeneous, isotropic, and reasonably elastic in character under nominal loading. Second, the behavior of stress concentration factors under dynamic load is quasi-static in character. As a third assumption, which involves some approximations, dynamic stress fields are taken to be of sufficiently linear characteristics that they can be added to static stress fields by superposition. Fourth, the stress conditions considered are those that would be generated at some distance from a nuclear detonation so that failure would involve little plastic or viscoelastic behavior, but fracture would occur in the elastic range.

Rock Failure

General

Aside from crushing of (ideal) rock due to intense overpressures very near a detonation, two types of rock failure may occur. The first is tensile failure, in which most rocks, being of brittle materials and weak in tension, fail along a plane of tensile stress. This process is different from that in ductile materials, such as some metals, which fail along planes of shear, even though subject to pure tensile loads. However, with few exceptions, when unconfined rock is loaded in compression it actually fails along a shear plane whose direction is defined by the failure angle from Mohr's circle. Hence, the following development is concerned almost entirely with tensile and shear failure.

Tensile strength of unconfined rock is usually evaluated in terms of the modulus of rupture, indirect tensile tests, or by direct axial tensile strength measurements. The values of tensile strength by axial tests range from 1.2 to 3.0 times the values obtained by the modulus of rupture method. The compressive strength varies from 3 to 10 times the axial tensile strength, and from 5 to 15 times the modulus of rupture values.

The stresses induced around a circular opening when a long duration pulse with a not too abrupt rise time impinges upon it are considered to be quasistatic. Resulting stress fields become approximately the sum of the existing static stress field and the superimposed pulse field. At depths of interest the static stress field is also assumed to have both vertical and horizontal components, as will the pulse stress field. Horizontal components of total value of one-third the vertical component will be considered herein because it is believed that this may approach some actual field conditions.

As a basic criterion for failure it can be said that as failure occurs the opening will tend to assume a shape which will relieve, but not necessarily minimize, the critical stresses. In the "doming theory," which has been used to explain the static failure behavior of some types of rocks, it has been postulated that an opening will assume a shape such that the critical stresses for a given stress field will be minimized. For long time stresses in terms of months or years, creep and viscoelastic failure could well result in redistribution of stresses and ultimate formation of domes which could well create an opening of such a shape that critical stresses would be minimized. However, for failure mechanisms under impact loading this does not appear to be true.

The factors which may govern the initiation and the extent of damage to homogeneous rock around underground openings are

1. The magnitude and duration of the total stress field, i.e., the sum of static plus dynamic fields.
2. The character of the stress field, i.e., the relationship between the two perpendicular (principal) stresses in the free field.
3. The resulting stress concentrations created, their relation to the strengths of rock and the distribution of strain energy.
4. The level of stress and strain energy required to initiate crack propagation, the velocity of crack propagation, and the level of energy required to maintain crack propagation and the relationship of stress, strain, and energy which are critical for cessation of crack propagation.
5. The role of confinement and release of confinement and crack propagation.
6. The rate of flow of energy from the stress field about an opening to the area of the opening as it fractures.
7. The tendency of fractured rock to "key" and provide partial confinement, support, and to otherwise inhibit fracturing.

Theories of Strength

The principal classical theories of strength which have found some degree of substantiation are as follows:[2]

1. Maximum stress theory.
2. Maximum elastic strain theory (St. Venant's "equivalent" stress theory).
3. Theory of constant elastic energy of deformation.
4. Theory of constant elastic strain energy of distortion or of constant octahedral shearing stress.
5. Maximum shearing stress theory.
6. Mohr's theory of strength.
7. Griffith's theory of fracture.
8. Theory of slip of loose granular material.
9. Theory of octahedral stress as a function of mean normal stress.

Of these, Mohr's theory and Griffith's theory appear to have found the greatest application.[3-7]

Fracture Processes

There are a large number of papers on fracturing,[8] including the original work in this area by Inglis[5] and Griffith.[4] Much of the important theory and data relative to the kinetics of fracturing has been summarized by Berry,[9] McClintock and Sukhatme,[10] and Craggs.[11]

The Griffith theory postulates that the strength of materials as calculated from atomic bonding is much greater than that exhibited by various substances, and that the very small relative strength shown by actual material is due to small cracks. Inglis[5] considered these as ellipses and Griffith[4] utilized the work of Inglis as a basis for his well-known theory of crack initiation.

Most of the extensions of the Griffith theory, particularly those dealing with tension cracks, assume a stress condition in a plate which is held by "fixed grips." Thus, the total energy of the system is also fixed, and stress and strain conditions, both about a crack and at its tip, are governed by this boundary condition, both at initiation and during the process of propagation.

However, Yoffe[12] assumed that a constant tensile stress field exists during initiation and propagation. Craggs,[11] on the other hand, studied the mathematics of crack propagation in a biaxial stress field whose principal stresses are P and S and in an infinite elastic body in a state of plane (tensile) strain. His analysis further assumes that the velocity of crack propagation is associated with the velocities of longitudinal and shear waves in the material, and is less than that of a Rayleigh surface wave.

That is,

$$u_x = \frac{\partial \phi}{\partial x} + \frac{\partial \psi}{\partial y} , \qquad u_y = \frac{\partial \phi}{\partial y} - \frac{\partial \psi}{\partial x} \tag{1}$$

and

$$\frac{\partial^2 \phi}{\partial t^2} = c_1^2 \nabla^2 \phi = c_1^2 \left(\frac{\partial^2 \phi}{\partial x^2} + \frac{\partial^2 \phi}{\partial y^2} \right) , \tag{2}$$

$$\frac{\partial^2 \psi}{\partial t^2} = c_2^2 \nabla^2 \psi , \tag{3}$$

where

$$\rho c_1^2 = (\lambda + 2\mu) , \qquad \rho c_2^2 = \mu , \tag{4}$$

u_x, u_y are displacements

$$c_1 = \left(\frac{\lambda + 2\mu}{\rho} \right)^{\frac{1}{2}} \text{ or longitudinal wave velocity,}$$

$$c_2 = \left(\frac{\mu}{\rho} \right)^{\frac{1}{2}} \text{ or shear wave velocity,}$$

and ϕ and ψ are potentials defined by Eq. (1).

Solutions of Eqs. (2) and (3) were assumed to be of the form

$$\phi = \phi(x - Vt, y) , \qquad \psi = \psi(x - Vt, y)$$

$$X = x - Vt , \tag{5}$$

where V is the (constant) crack velocity.

A solution is obtained in complex numbers for the stress components around the crack, and energy balance considerations indicate that energy fed into the material by the normal forces P and S must escape through the singularities at the head of the advancing crack. Also, if the rate at which external forces do work is equated to the rate of creation of new surface energy, 2TV, its value is defined by the equation

$$\frac{4aV^2}{c_2^2} \{(1 - V^2c_1^{-2})^{\frac{1}{2}}\rho^2 + (1 - V^2c_2^{-2})^{\frac{1}{2}}S^2\} = 2\pi\mu\ DT$$

$$= 2\pi\mu T\{4(1 - V^2c_2^{-2})^{\frac{1}{2}}(1 - V^2c_1^{-2})^{\frac{1}{2}} - (2 - V^2c_2^{-2})^2\} . \tag{6}$$

This development assumes that V is constant and can never be larger than C', the velocity of a surface Rayleigh wave, which is the same concept as that adopted by Yoffe.[12]

Table 1 shows the variation of P with V for S = 0 and S with V for P = 0. The second line represents the direct stress required to maintain the crack in absence of shear, and the third line the tangential stress in the absence of normal stress. It is noted that as V increases, the force required to maintain the crack decreases and the crack may continue even when the applied stress is decreased. Catastrophic cracking may occur, then, under decreasing load. Also as $V \to C'$ the rate of work of a finite force tends to infinity and the cracking process would appear to be self-maintaining. However, Craggs maintains that crack splitting or division will occur before this velocity is reached, at some critical velocity.

Table 1

($\nu = 0.25$)

V/c_2	0.0	0.1	0.2	0.3	0.4	0.5
$a^{\frac{1}{2}}P/(\pi\mu T)^{\frac{1}{2}}$	0.8165	0.8137	0.8002	0.7878	0.7629	0.7284
$a^{\frac{1}{2}}S/(\pi\mu T)^{\frac{1}{2}}$	0.8165	0.8150	8.8098	0.8004	0.7860	0.7675
V/c_2	0.6	0.7	0.8	0.9	C'/c_2	
$a^{\frac{1}{2}}P/(\pi\mu T)^{\frac{1}{2}}$	0.6799	0.6092	0.4957	0.2305	0.0000	
$a^{\frac{1}{2}}S/(\pi\mu T)^{\frac{1}{2}}$	0.7363	0.6893	0.6026	0.3228	0.0000	

NOTE: $C' = 0.9194c_2$ is the velocity of a Rayleigh wave on a free surface.

The critical velocity may be associated with maximum stress (V_1), maximum strain (V_2), hydrostatic stress (V_3), or shearing stress (V_4). Possible values are given in Table 2.

Table 2

ν	V_1/c_2	V_2/c_2	V_3/c_2	V_4/c_2
1/4	0.629	0.611	0.565	0.772
1/3	0.667	0.621	0.571	0.810
1/4*	0.612	0.598	0.562	0.762
1/3*	0.629	0.597	0.565	0.772
1/2*	0.667	0.571	0.571	0.810

NOTE: * denotes plane stress.

Experiments were performed by Wells and Post[13] with CR-39 to observe stress distributions surrounding a running crack photoelastically using Schardin's spark system. The mechanical system used for testing was a plate in tension under "fixed grip" conditions with the crack starting from a notch in one edge of the plate.

Crack velocity measurements were also made and the results compared with the theoretical velocity based on work by Roberts and Wells:[14]

$$V = 0.38\left(\frac{E}{\rho}\right)^{\frac{1}{2}}\left(1 - \frac{c_o}{c}\right) \quad . \tag{7}$$

There was reasonably close agreement except in the accelerating stage.

Wells and Post also concluded that for a brittle transparent material the stress disturbance grows in all directions with the increase in length of the crack. Also, the dynamic stress distribution approximates that for the static case where the crack is extended by a fixed displacement. In this experiment the crack velocities did not exceed 0.38 of the longitudinal velocity.

Irwin,[(15)] however, points out that the velocities measured by Wells and Post are subcritical, that is, below branching velocity. Branching velocity can be attained if the extension force is increased sufficiently.

Berry[(9)] has analyzed some of the aspects of Griffith fracture in tension and cleavage. He assumes that a crack has (1) a ground potential energy, (2) a total strain energy, and (3) a kinetic energy while the crack propagates. Expressions for each are developed and final equations for crack velocity and acceleration derived without actual reference to the Griffith criterion.

When these expressions are related to the Griffith criterion for fracture, the velocity equation becomes

$$V_c = V_m\left(1 - \frac{c_o}{c}\right) , \tag{8}$$

where V_c = crack velocity,
V_m = maximum velocity,
c_o = initial crack length, and
c = crack length corresponding to V_c.

Energy partition is plotted diagrammatically for a state of tensile stress and cleavage, into surface energy, strain energy, and kinetic energy. The above equation corresponds in form to equations derived by others in which $V_m = 0.38(E/\rho)^{\frac{1}{2}}$.

Fracture in a Compressive Stress Field

An intensive literature search has revealed no crack velocity investigations in a compressive field, either on a theoretical basis or in laboratory investigations.

Inasmuch as most fractures of brittle materials under compression are through shear, it is proposed to investigate the phenomena involved in compressive shear fracture. Mohr's envelopes have been determined for a large number of rocks under biaxial conditions, but no effort has been made to evaluate crack phenomena in compressive biaxial or triaxial stress fields.

Some hypotheses have been put forward by Clark and Caudle[16] concerning the relation of crack velocities to the extent of failure in underground openings subjected to compressive impact loads, but these hypotheses were limited to time effects rather than to basic crack phenomena. In any event, in fracture stress fields just above critical the extent of failure will probably be determined by the length of dynamic pulse and the crack velocity.

Fracture under Compression and Confinement

Virtually all of the fracture investigations reported in the literature are concerned either with tensile (reflection) slabbing, which is not under direct consideration here, or with fracture in tension or shear under unconfined conditions. That is, boundary conditions for other investigations described above are for the relatively simple case of uniaxial tension, a type of "tearing shear" or for splitting cleavage. While none of these are directly applicable to the present problem, the processes of fracture probably have many analogous parameters which are similar in character.

Consider, for example, a circular opening located in homogeneous rock at a depth of 3500 ft. The lithostatic pressure would be approximately 3500 psi and would create a biaxial stress field with a horizontal component $\sigma_x = [\nu/(1 - \nu)]\sigma_z$. This would produce a zero tangential stress concentration at the roof and floor of the opening. For purposes of analysis the imposed load from an explosive source is assumed to have the same ratio between vertical and horizontal components, and the duration of the load sufficiently long that it may be analyzed as a static condition. Thus, a strong enough dynamic load will serve to multiply the existing stress until fracture ensues.

Inasmuch as the surface of the underground opening is unconfined on one side, fracture should theoretically begin at the surface and progress outward. The ribs have the highest stress concentration, the stresses are compressive, and hence the initial failure stress should be defined by the Mohr shear stress value at or near zero confinement. Thus, if Mohr's envelope is assumed to be approximately a straight line (Fig. 1) the initial fracture stress will be determined by a

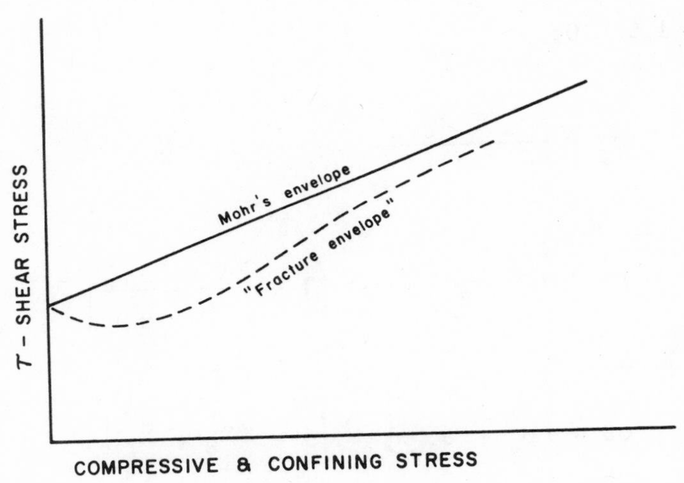

Fig. 1—Hypothetical "fracture envelope" analogous to Mohr's circle for running fractures.

point on this curve near the vertical axis. It has been found in the kinetics of both tensile and cleavage crack formation that the stresses necessary to maintain crack propagation are less than those required to initiate the crack. Hence, a "fracture envelope" is proposed to define qualitatively the resistance of a rock to the propagation of a moving crack, the resistance being lower than the shear strength of the rock defined by Mohr's envelope. Considering a cross section of a tunnel with an element at some distance in the rock along a horizontal plane (Fig. 2), the stresses acting on the element at any time during

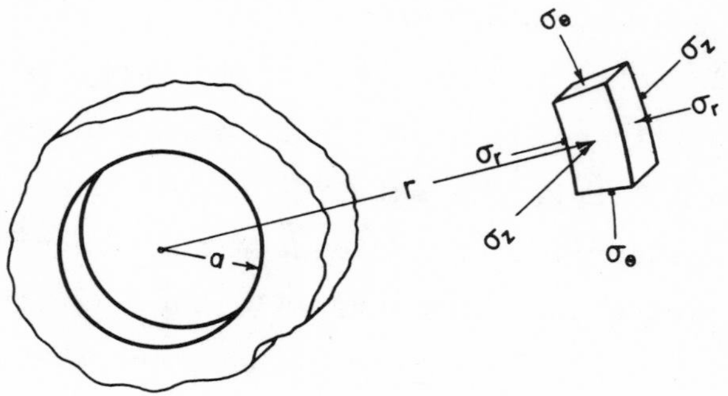

Fig. 2—Sketch showing the confining and active pressure on an element in the solid near an underground opening.

the application of the pulse and prior to fracture will be, for Poisson's ratio $\nu = \frac{1}{4}$ and $\theta = 0$:

$$\sigma_z = \frac{(\sigma_s + \sigma_{dyn})}{3} \; , \tag{9}$$

$$\sigma_r = \frac{(\sigma_s + \sigma_{dyn})}{3} = \left[1 + \frac{2a^2}{r^2} - \frac{3a^4}{r^4}\right] , \tag{10}$$

$$\sigma\theta = (\sigma_s + \sigma_{dyn}) = \left[1 + \frac{2a^2}{3r^2} + \frac{a^4}{r^4}\right] . \tag{11}$$

The confining pressure parallel to the axis of the opening is constant while the tangential and radial stresses are functions of r and θ, σ_θ having a maximum value at $r = \sqrt{6}a$ and $\theta = 0°$, and zero for θ equal to $\pi/2$ or $3\pi/2$.

Thus, as a crack is initiated and progresses outward from an opening, it will experience a constant confining pressure with respect to the z direction and a variable pressure in the r and θ directions. The relative values of these stresses on horizontal, 45- and 90-degree planes are indicated in Fig. 3 for an assumed value of Poisson's ratio $\nu = \frac{1}{4}$. For practical purposes the radial (σ_r) and axial (σ_z) pressures on the horizontal diameter are equal except inside a distance of $\frac{1}{4}$a from the periphery of the opening into the rock wall. The plot

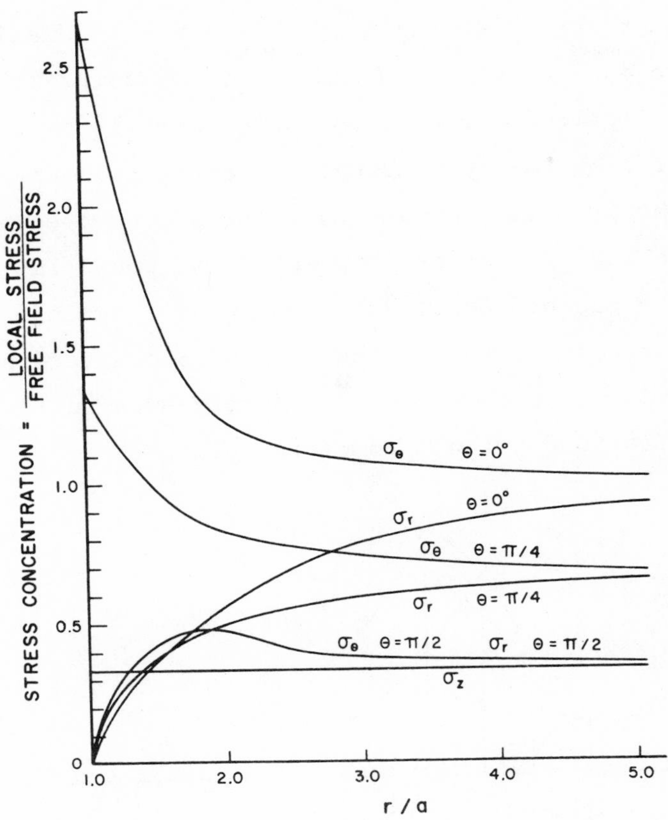

Fig. 3—Tangential and radial stress concentrations for horizontal field component equal to one-third of vertical (i. e., $\nu = \frac{1}{4}$).

of stresses along planes inclined at 45 degrees and vertically shows that the tangential stresses go to zero at the roof, but increase into the rock at this point. The radial stress concentration never exceeds a factor of one.

In relation to Mohr's circle the point of the greatest difference in principal stresses together with maximum stress is critical. The greatest stress difference as well as the maximum stress concentration occurs at the center of the rib (horizontal diameter), both at the surface of the opening and back into the wall rock. A large stress difference is also found in the rock at the boundary of the opening where $\sigma_\theta - \sigma_r$ is largest to a distance of $r/a \cong 1.25$ and $\sigma_\theta - \sigma_z$ is greater beyond this point, but not significantly so. Hence, a homogeneous rock would be expected to begin to fail by shear when the stress at the center of the rib (2.67 times the free field stress) exceeded the "compressive" strength of the rock. If the free field stress (ffs) were only slightly in excess of the compressive strength divided by 2.67, little fragmentation would normally be expected. However, as the ffs became increasingly greater, fracture and fragmentation would occur. As soon as the shape of the opening is changed by failure of the rib, the stress concentration would increase (Fig. 4), and fracture would proceed until "keying in" of fragments and blocks gave sufficient support to reduce the stress concentration, or until the ffs was reduced. The above assumes an ffs only sufficiently strong to cause severe fracture but not complete collapse.

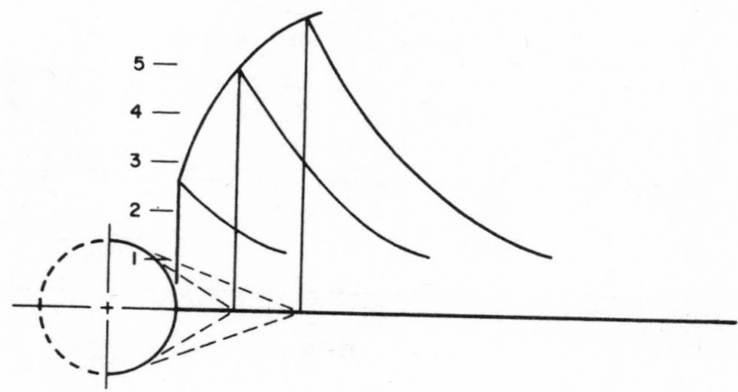

Fig. 4—Hypothetical increase of stress concentration with axes in shape of circular opening as fracture occurs.

In a recent study, Logcher[17] has extended the mathematical solution for stresses around a circular opening originally proposed by Baron <u>et al.</u>[18,19] The stress distribution was determined by computer for a unit step function, and a sawtooth wave with a decay time of five transit times. Logcher noted that there was a tensile reflection from the top of the cavity, being greater for the sawtooth wave. He concludes that "this early time tensile wave probably accounts for most spalling failures." The maximum stress concentration factor was found to be 2.96 for Poisson's ratio equal to 1/4.

However, for "waves with no sharp fronts, ... failure (is) likely only due to oblique fracture." Hence, the basic assumption that the effect of a long pulse with a slow rise time from a large weapon would appear to have the same effect as a quasistatic field, except that the stress concentration is increased from 2.67 to 2.96.

Table 3 gives some pertinent physical property data on representative rock types. No property is characteristic of a particular rock

Table 3

COMPRESSIVE STRENGTH

Rock	E (psi)	ρ (gr/cm)	Compressive Strength (psi)	Shear Strength (unconf.)	$(E/\rho)^{\frac{1}{2}}$ (ft/sec)
Granite	5.2×10^6	2.63	21.6×10^3	10.8×10^3	12,100
Gneiss	15.0×10^6	3.36	31.7	15.8	18,200
Sandstone	3.9	2.28	7.47	3.73	11,300
Granite	3.9	2.63	10.46	5.23	10,500
Granite	10.6	2.63	23.3	11.7	17,400
Sandstone	1.1	2.14	14.2	7.1	6,200
Limestone	9.1	2.82	14.7	7.3	16,200

type. However, bar velocities vary from 6200 to 18,200 ft/sec. This means that for every millisecond of application of supercritical stress fields above a certain level the cracks (of bar velocity) may progress into the wall rock at distances varying from about 2 ft to 7 ft per millisecond. Thus, the pulse duration above critical levels, as well as the magnitude, can be related to the rock properties and resultant failure in at least a semiquantitative manner. Also, in homogeneous rock, cracks would be initiated at the surface of the opening, and the unconfined compressive strength of the rock is believed to be a reasonably accurate measure of the strength of the surface rock <u>in situ</u>.

Gilvarry and Bergstrom[(20)] have made a statistical study of Griffith flaws, their distribution and effect upon brittle fracture. For a single fracture a distribution function for fragment size was obtained using the Griffith theory and based on the assumptions that (1) fracture proceeds by the activation of flaws in the volume, in fracture surfaces and in fracture edges and (2) the flaws are distributed independently (Poisson's law) of each other when activated. This leads to a unique expression for the probability of the formation of a particle of a given size. Further simplifying assumptions lead to well-known expressions for particle size distribution. Experimental results indicate that the degrees of activation of the different types of flaws do not differ greatly.

The question arises as to the relative speed with which confinement is released. Due to the fact that most of the failure indicated will be by shear (the cracks will proceed with a velocity less than $(E/\rho)^{\frac{1}{2}}$) the release in confinement probably will not precede the crack tips by an appreciable distance where broken fragments are immediately removed, and less so when fragments are held in place even momentarily. In general, it could be assumed that confinement is released with the invasion velocity of the fracture front into the rock or at a lower velocity. If this is the case, then the strength of the rock is defined by a proposed "fracture stress" curve analogous to Mohr's envelope (Fig. 5). The confining pressure is provided by σ_r and σ_z, which are approximately 0.3 of the vertical component of the free field stress along a horizontal plane, which are almost equal in value, and which are characteristic values only for the type of stress field indicated.

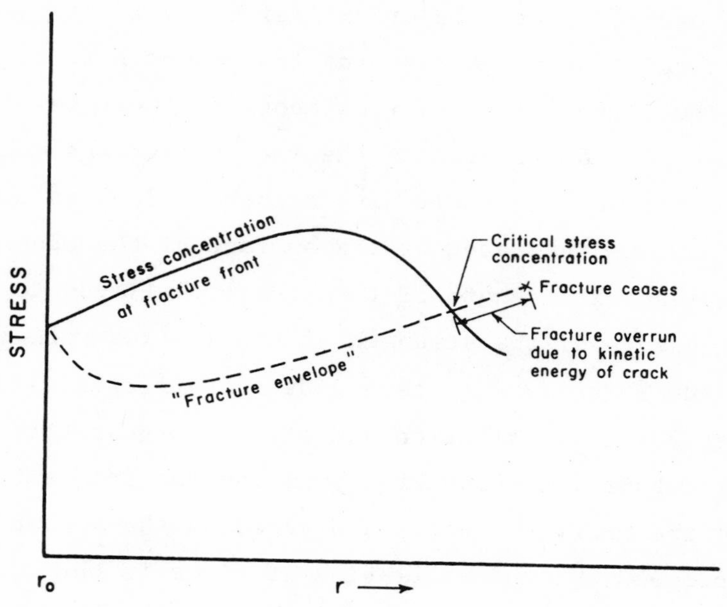

Fig. 5—Schematic of fracture in rock under confining pressure near a circular opening, with particular reference to beginning and termination of fracture process.

From data similar to that shown in Fig. 3, it is possible to draw a curve in the σ_x and σ_y plane with lines representing the possible ratios between σ_x and σ_y (Fig. 6). In this case, if the rock curve is below a given stress ratio line, the rock will fracture.

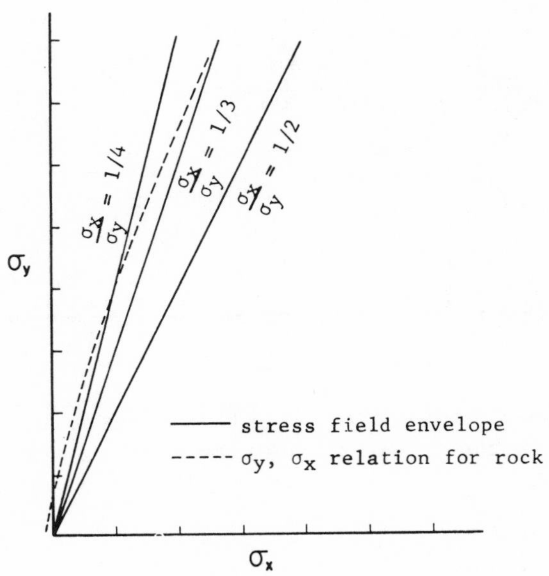

Fig. 6—Example rock failure envelope and stress field envelopes for various ratios of active to confining stresses. For given value of stress field ratio, rock will fail if its curve is to right of this value.

Figure 7 is a schematic for fracture history of rock subject to long pulses. In general, the time t_c', probably shorter because of confinement, is critical because it determines the amount of failure that can actually take place. Although collapse of openings will probably never proceed as far as propagated fractures, their depth of propagation into the wall rock will probably also determine the extent of collapse, the latter being an appreciable fraction of the former, particularly if the peak stress exceeds the initiating stress by a critical amount.

During the period in which quasistatic conditions are assumed, the stress concentration under nonfailure conditions may also be employed as a measure of the strain energy distribution. As fracture takes place in the rock near the opening, however, energy is consumed

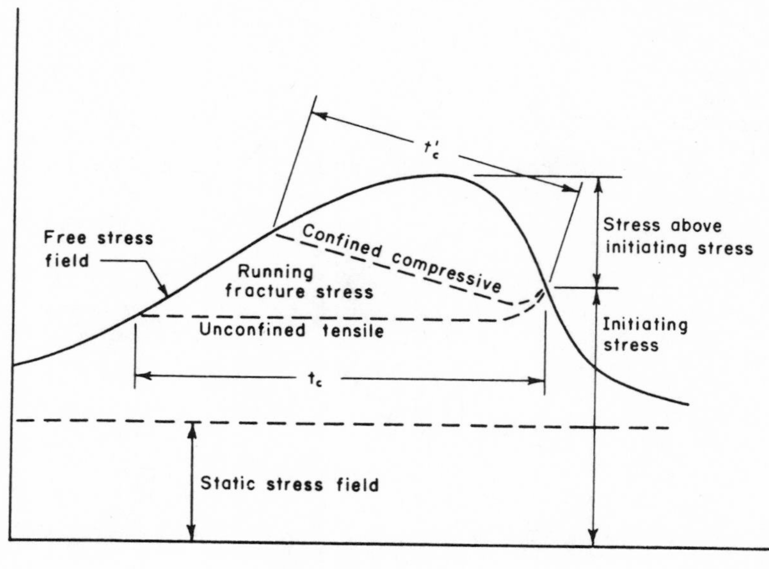

Fig. 7—Schematic of fracture histories in a long period stress field superimposed on a static stress field.

by fracturing and permanent deformation. Also, while the "long" quasi-static pulse is active, the volume of rock strained by the pulse is very large compared to the size of the openings. Hence, the strain energy may flow toward the opening probably with a velocity equal to the dilatation velocity. In other words, the opening acts as an energy "sink" during failure. Hence, the rate of flow of strain wave energy may also be one of the important rock failure parameters.

Conclusions

Based upon the existing theories of fracture of solid materials, it appears that the existing reliable data on rock strength as determined by a standard unconfined compressive test may be utilized in conjunction with stress concentration data for openings in homogeneous rock to define the stress at which fracture will be initiated.

Also, existing Mohr's envelope data may be employed as a guide to define the strength of rock under confined conditions. The actual rock strength for running cracks is probably less than that for initiation.

The critical portion of the rock around a circular tunnel is that extending about one tunnel radius into the wall. However, if failure begins, the stress concentration may increase due to a change in the shape of the opening. It is not known whether the release of confinement is coincident with the movement of the fracture front into the rock. However, the failure and collapse of openings appear to be restricted by the fact that fragments may fracture but remain "keyed" in place due to irregularities in the fracture pattern. This behavior would hold only where pressures in the free stress field were not of catastrophic magnitudes.

Crack propagation velocities appear to be important parameters in determining the distance of failure into the rock from the opening, their distance of travel being determined by the time of application of stress fields above a critical level.

REFERENCES

1. Clark, G. B., and R. D. Caudle, Geologic Structure Stability and Deep Protective Construction, AFSWC TDR-61-93, November, 1961.
2. Nadai, A., Theory of Flow and Fracture of Solids, McGraw-Hill Book Company, Inc., New York, 1950.
3. Balmer, G., "A General Analytic Solution for Mohr's Envelope," Am. Soc. Testing Mater. Proc., Vol. 52, 1952, pp. 1260-1271.
4. Griffith, A. A., "The Phenomena of Rupture and Flow in Solids," Phil. Trans. Roy. Soc. London, Ser. A, Vol. 221, 1921, p. 163.
5. Inglis, C. E., "Stresses in a Plate due to the Presence of Cracks and Sharp Corners," Trans. Inst. Naval Architects, London, Vol. 55, Pt. I, 1913, p. 219.
6. Griffith, A. A., "Theory of Rupture," First Internatl. Congr. Appl. Mech., Delft, Vol. 55, 1924.
7. Clausing, D. P., "Comparison of Griffith's Theory with Mohr's Failure Criteria," Quart. Colo. School Mines, Vol. 54, July, 1959.
8. Averbach, B. L., et al. (eds.), National Academy of Sciences Conference on Fracture, John Wiley & Sons, Inc., New York, 1959.

9. Berry, J. A., "Some Kinetic Considerations of the Griffith Criterion for Fracture, I and II," J. Mech. Phys. Solids, Vol. 8, No. 3, 1960, pp. 194-216.

10. McClintock, F. A., and S. P. Sukhatme, "Traveling Cracks in Elastic Materials under Longitudinal Shear," J. Mech. Phys. Solids, Vol. 8, No. 3, 1960, pp. 187-193.

11. Craggs, J. W., "On the Propagation of a Crack in an Elastic-Brittle Material," J. Mech. Phys. Solids, Vol. 8, No. 1, 1960, pp. 66-75.

12. Yoffe, E. H., "The Moving Griffith Crack," Phil. Mag., Vol. 42, 1951, pp. 739-750.

13. Wells, A. A., and D. Post, "The Dynamic Stress Distribution Surrounding a Running Crack," Proc. Soc. Exptl. Stress Anal., Vol. 16, 1958, pp. 69-92.

14. Roberts, D. K., and A. A. Wells, "Velocity of Brittle Fracture," Engineering, Vol. 178, 1954, pp. 820-821.

15. Irwin, G. R., "The Dynamic Stress Distribution Surrounding a Running Crack--Discussion," Proc. Soc. Exptl. Stress Anal., Vol. 16, 1958, pp. 93-96.

16. Clark, G. B., and R. D. Caudle, Failure of Cavities in Homogeneous Rock under Dynamic Loading, Nuclear Explosions, Report for Plesset Associates (Contract No. DA-49-146-XZ-073).

17. Logcher, Robert D., "A Method for the Study of Failure Mechanisms in Cylindrical Rock Cavities due to the Diffraction of a Pressure Wave," Mass. Inst. Tech. Rept. T 62-5, July, 1962.

18. Baron, M. L., et al., "Diffraction of a Pressure Wave by a Cylindrical Cavity in an Elastic Medium," J. Appl. Mech., Vol. 28, No. 3, 1961, pp. 347-354.

19. O'Sullivan, J. J. (ed.), Protective Construction in a Nuclear Age: Proceedings of the Second Protective Construction Symposium, Vols. 1 and 2, The Macmillan Company, New York, 1961.

20. Gilvarry, J. J., and B. H. Bergstrom, "Fracture of Brittle Solids, I and II," J. Appl. Phys., Vol. 32, No. 3, 1961, pp. 391-410.

DISCUSSION

D. U. DEERE (USA):

Because the title of this conference is "The State of Stress in the Earth's Crust," and because Clark and Caudle make reference to this state of stress, I wish to quote two statements from their paper and then to state my objections to their method of evaluating the initial stress state. I realize that these quotations will be out of context, and I would ask Professor Clark for his comments.

> Horizontal components of total value of one-third of the vertical component will be considered herein because it is believed that this may approach some actual field conditions.
>
> Consider, for example, a circular opening located in homogeneous rock at a depth of 3500 ft. The lithostatic pressure would be approximately 3500 psi and would create a biaxial stress field with a horizontal component $\sigma_x = [\nu/(1 - \nu)]\sigma_z$.

(In the above, σ_x is the horizontal stress, σ_z is the vertical stress taken equal to the overburden pressure of 1 psi per foot of depth, and ν is Poisson's ratio.)

I question these statements and, in particular, the validity of the use of an equation from the theory of elasticity involving Poisson's ratio for determining the initial state of stress in a natural rock mass.

First, a rock mass is a jointed, discontinuous medium which does not fully satisfy the requirements of an elastic, homogeneous, isotropic medium. It would appear that the above equation cannot be expected to apply under these conditions.

Second, if the equation were to apply, I see great difficulty in evaluating Poisson's ratio. The value obtained in a laboratory test on an intact sample of the rock, whether by static or dynamic means, would not necessarily be applicable to the jointed rock mass in the field. The value obtained in the field by seismic methods might likewise be questioned because of the small amount of energy involved and the small displacements induced in the rock mass by the seismic test with respect to those likely to be experienced by the rock under design loads. The spacing, orientation, and type of filling in the joints of the rock mass, as well as the stress level, would appear to me to be of overwhelming importance in the evaluation of an equivalent Poisson's ratio.

Third, the stress history of a rock mass would also appear to be a controlling factor in determining the initial state of stress. The stress history of a given rock mass may have been a complicated one involving tectonic forces with associated folding, faulting, jointing and recrystallization, or it may have been a rather simple one involving only erosion of superincumbent materials (or melting of an overlying glacier). The removal or reduction of the overlying load would certainly result in the reduction of the vertical stress in accordance with the laws of statics. The magnitude of the horizontal stress would also decrease, but very possibly at a slower rate, so that at a given moment at a point in the rock mass the horizontal stress might conceivably be equal to or even greater than the vertical stress.

The ratio σ_x/σ_z is essentially the same as K_o, the coefficient of earth pressure at rest, as used in soil mechanics. The factor $\nu/(1 - \nu)$ used by Clark and Caudle is a theoretical expression of the factor K_o. A value of Poisson's ratio of 0.25, as Clark and Caudle take in their example, would give K_o equal to 0.33. However, if the stress history concept is valid, K_o might be equal to and even considerably greater than 1.

A research program has been under way for the past two years at the University of Illinois by Hendron and others investigating the effect of the stress history on the value of K_o for sands and clays loaded slowly to high pressures (up to 2500 psi) and then unloaded slowly. The test equipment has been specially designed so that lateral strain of the sample is prevented, a necessary feature for the evaluation of K_o. During the loading cycles the value of K_o is found to be quite constant (about 0.35 to 0.45 for sands and 0.4 to 0.7 for clays, depending upon their physical properties). However, in the unloading cycle, the resulting horizontal stress is found to exceed the vertical stress, and the factor K_o quickly approaches and exceeds unity and even reaches values of 2.0 to 2.5 or greater as the vertical stress approaches zero.

On the basis of the experimental results obtained to date, we believe that horizontal stresses can be built-in to sediments (and very likely into some sedimentary rocks) by the simple process of loading by continuing sedimentation, followed by erosion. It is also highly probable that rocks of all types--regardless of the source of their original high stress state--may exhibit a high ratio of horizontal stress to vertical stress following erosion simply because the horizontal stresses do not relieve themselves to the same extent as the vertical stresses.

In conclusion, when one considers the geological character of rock masses in light of the requirements for elastic theory, the effect of the stress history of the rock mass as may be deduced from the results of laboratory tests on sand and clay, and the experimental evidence of actual field measurements of stress in rock masses (as pointed out by Mr. Judd in his opening paper), it would appear that the use of the factor $\nu/(1 - \nu)$ from the theory of elasticity--without qualification--for evaluating the magnitude of the horizontal stress in terms of the overburden weight may be invalid in rock mechanics.

J. W. HANDIN (USA) in reply:

Let us clearly understand the issue here. The relative magnitudes of the vertical and horizontal principal stresses have been calculated for the one-dimensional compression of a perfectly elastic, homogeneous, isotropic material with a definite, constant Poisson's ratio. As applied to rocks in their natural state, this procedure involves the untested assumption that these rocks behave as the idealized material under severely limited boundary conditions. I feel sure that Professor Clark, along with most of us, doubts that this assumption is generally valid or geologically realistic. However, if theory is to advance, some reasonable supposition about the static stress field must be made.

Professor Deere has mentioned the wide range of values for ratios of vertical to horizontal stresses measured in tests of unconsolidated materials. Values for similarly tested consolidated rocks would almost certainly be more restricted, but in any event a better criterion, in my opinion, must be empirical and can be discovered only by in situ measurements. For example, the stress ratio will depend among other variables on the strength of the rock, i.e., the ability of the rock to sustain the required stress difference for a long time. To account for this upper limitation we should have to know all about the rheological properties of rocks, and Dr. Robertson's discussion has indicated how little we yet know.

I believe that the assumption of complete lateral restraint is untenable on the geological scale. Small soil masses may sometimes be nearly rigidly bounded, but in the large basins described by Mr. Dallmus, the horizontal strains clearly can not vanish.

The answer depends on more and better field measurements. In the meantime, Professor Clark's prestress field is <u>one</u>, perhaps representing the probable lower limit of the stress ratio, that can be used as a first approximation. Rock mechanics is not as simple as perhaps some once naively thought. Much work remains to be done before we can assess the geological realism of much of current theory and before we can decide which more sophisticated theory is worth developing.

D. U. DEERE (USA):

Professor Serata has a few comments about the ratio of lateral stress to vertical stress in the earth's crust as he has arrived at it by some laboratory experiments.

S. SERATA (USA):

Dr. Deere has pointed out very clearly that estimating horizontal stress from vertical stress and Poisson's ratio is very erroneous, and he has also reported that the ratio of horizontal stress to vertical stress could be greater than one. He has found the ratio to be even more than two. I am supporting his evidence with a theory that the ratio could be any value ranging from nearly minus infinity to plus infinity in underground formations. This may sound a little strange; however, if you assume there is some residual stress left in the horizontal direction while you remove vertical stress, then this is when the ratio becomes nearly infinity.

From this evidence, I am trying to point out that expressing the state of stress by using the ratio is not logical. Therefore we need something better than this. It seems to me, octahedral shearing strength indicates more precisely the relationship between the vertical and lateral stresses in the earth's crust. Octahedral shearing strength is, to put it simply, shearing strength of materials in a three-dimensional scale. By using this concept, lateral stress should be restricted in the range of axial stress plus or minus 2.1 times octahedral shearing strength.

This theoretical estimation has been checked by using rock salt, shale, dolomite, marble, and granite. It seems to be providing supporting evidence, although the harder the rock, the more difficult it is to control the experiment.

ABSTRACT

Rock mechanics research in South Africa began with an interest in studying and controlling the rock bursts that occur in deep-level gold mines. This study is also fundamental to problems of strata control and underground water seepage in hard- and soft-rock mines.

The principal research activities in the field of rock mechanics in South Africa include theoretical studies of the behavior of the rock surrounding mining excavations; model studies designed to extend the scope of theoretical investigations; laboratory studies of the properties of rock samples; underground observations and measurement of rock stress in particular areas; measurements of strata movement and surface subsidence, designed to study the over-all disturbances caused by mining; and statistical analyses of data obtained from underground observations and records of rock burst sites.

RÉSUMÉ

Les recherches en mécanique des roches en Afrique du Sud ont commencé par un intérêt dans l'étude et le contrôle des éclatements de roche qui ont lieu dans les mines d'or aux niveaux profonds. Cette étude est aussi fondamentale pour les problèmes de contrôle des couches et d'infiltration d'eau souterraine dans les mines en roche molle et dure.

Les activités principales de recherche dans la mécanique des roches en Afrique du Sud comprennent des études théoriques du comportement de la roche qui entoure les excavations des mines; des études sur modèles destinées à étendre la portée des études théoriques; des études de laboratoire des propriétés des échantillons de roche; des observations et des mesures souterraines des forces dans la roche de certaines regions; des mesures de mouvement de couches et d'affaissement de surfaces, destinées à l'étude des perturbations générales causées par l'exploitation des mines; et l'analyse statistique des données obtenues des observations souterraines et des documents des sites des éclatement de roche.

AUSZUG

Gebirgsmechanische Forschung begann in Südafrika mit dem Erwachen des Interesses an der Erforschung und Beherrschung von Gebirgsschlägen, die dort in tiefen Goldgruben auftreten. Diese Untersuchung ist für Probleme des Gebirgsdruckes und des Grundwasserflusses in Bergwerken in festen und weichen Gebirgen ebenfalls grundlegend.

Die grundsätzliche Forschungsarbeit auf dem Gebiete der Gebirgsmechanik in Südafrika schliesst folgende Teilgebiete ein: theoretische Untersuchungen über das Verhalten des bergbauliche Öffnungen umgebenden Gebirges; Modellstudien zum Zwecke der Ausweitung des Bereiches der theoretischen Untersuchungen; Bestimmungen der Gesteinseigenschaften im Prüfraum; Beobachtung und Messung von Gebirgsdrücken in bestimmten Bereichen untertage; Messung von Gebirgsbewegungen und Bergsenkungen, mit dem Zwecke die durch bergbauliche Massnahmen verursachten Gesamtstörungen zu erkunden; ferner statistische Auswertungen von Beobachtungen untertage und Aufzeichnungen über Gebirgsschläge.

ROCK MECHANICS RESEARCH IN SOUTH AFRICA

Evert Hoek*

Introduction

ROCK MECHANICS RESEARCH in South Africa began some 10 years ago when it was decided to initiate a systematic study of the causes and possible means of controlling rock bursts that occur in deep level gold mines. It was soon realized that a vast amount of fundamental information would have to be accumulated before the problem could be solved, but that significant contributions to the science of rock mechanics could be made during the course of the investigations.

Rock mechanics is not only fundamental to the study of rock bursts but also to problems of strata control and underground water seepage in both hard and soft rock mines. During recent years the scope of the South African rock mechanics research activities has been widened to include such investigations as surface subsidence studies in coal mines and more general studies in hard rock mines.

In the beginning, the research was undertaken by the South African Council for Scientific and Industrial Research. Later, the efforts were joined by the Bernard Price Institute for Geophysical Research and by the Mining Department of the University of the Witwatersrand. More recently, some of the individual mines have set up small research groups to solve *ad hoc* problems which occur during the routine operation of the mine.

---

*National Mechanical Engineering Research Institute, Council for Scientific and Industrial Research, Pretoria, South Africa.

The principal research activities of the Rock Mechanics Research Team of the South African Council for Scientific and Industrial Research are as follows:

Theoretical studies of the behavior of the rock surrounding mining excavations. To date, all these studies have been based upon the theory of elasticity since the behavior of the hard quartzitic rocks in which the rock bursts occur is predominantly elastic. The application of the theories of plasticity and rheology, however, is also indicated particularly in the case of softer strata.

Model studies, designed to extend the scope of the theoretical investigations and to study the influence of chosen parameters under controlled conditions. Two- and three-dimensional photoelastic studies have played an important part in this work, and the birefringent layer technique has proved particularly useful in the study of fracture phenomena.

Models made from brittle materials and loaded to destruction in a large centrifuge are used to study the fracture of the rock surrounding excavations.

Laboratory studies to determine the properties of rock samples. These studies include the determination of elastic constants, uniaxial and triaxial strength determination, and the study of creep and time dependent failure. These studies are designed to provide the basic information upon which the theoretical and model investigations can be based.

Underground observations and measurements of rock stress in areas of particular interest. These measurements are designed to determine the state of stress in the virgin rock and the changes in stress induced by mining.

Mention must also be made of the work done by the Bernard Price Institute of Geophysical Research on the seismic location of rock burst foci.

Strata movement and surface subsidence measurements, designed to study the over-all disturbances induced by mining. These studies are particularly important in shallow mines in soft strata such as coal mines.

Statistical analysis of data obtained from underground observations and records of rock burst sites. These analyses are designed to establish the effects of geological factors and of mining methods upon the incidence of rock bursts.

A detailed bibliography of South African publications on rock mechanics and related subjects has been prepared by Dr. H. G. Denkhaus, Director of the National Mechanical Engineering Research Institute of the South African Council for Scientific and Industrial Research, and is available upon request.

DISCUSSION

J. A. TALOBRE (France):

According to a very common opinion, the Mohr's envelope of concrete and rocks should be a parabola. Griffith considered this opinion as granted, although it is far from being confirmed by facts.

Handin and Hager* and many authors have shown that the relation between principal stresses S_1 and S_3 was practically linear for many rocks tested under moderate confinement.

I have myself plotted several hundreds of results obtained from numerous confined compression tests performed on the same rock, and found that the relation existing between shear strength and mean compression was also linear. When using the results of the very carefully performed tests of Bredthauer,** I obtained two different linear relations, the first one being relevant to brittle fracture, the other one to ductile failure.

In Fig. 1 the results of the tests of Heard*** on limestone were used. These tests were performed at diverse temperatures. The Mohr's

---

*J. Handin and R. V. Hager, Jr., "Experimental Deformation of Sedimentary Rocks under Confining Pressure," Bull. Am. Assoc. Petrol. Geol., Vol. 41, January, 1957, pp. 1-50.

**R. O. Bredthauer, "Strength Characteristics of Rock Samples under Hydrostatic Pressure," Trans. ASME, Vol. 79, May, 1957, pp. 695-708.

***H. C. Heard, "Transition from Brittle Fracture to Ductile Flow in Solenhofen Limestone as a Function of Temperature, Confining Pressure, and Interstitial Fluid Pressure," Geol. Soc. Am. Mem. 79, 1960, pp.193-226.

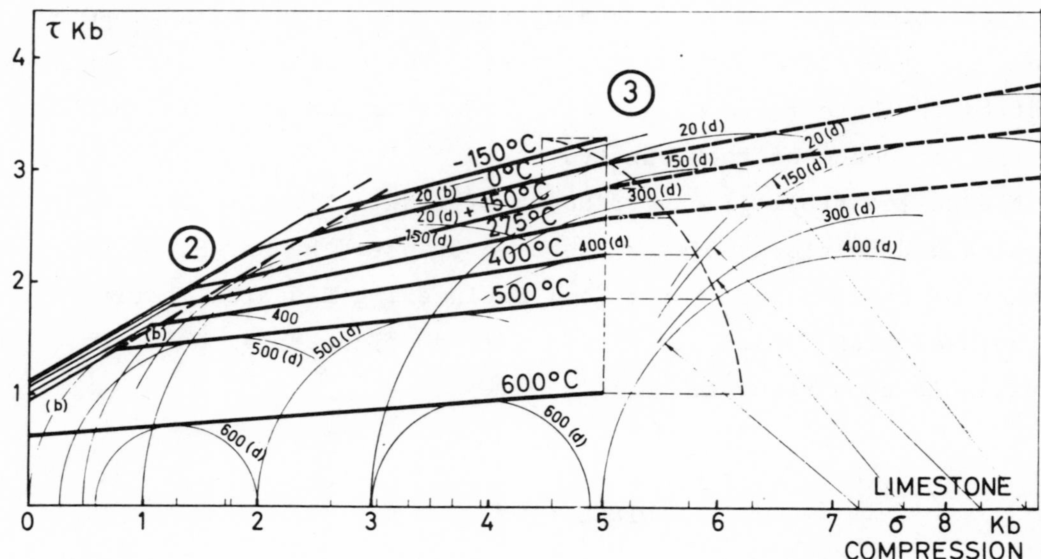

Fig. 1—Transition from brittle failure (2) to ductile failure (3) on Mohr's envelope according to Talobre theory.

envelope is formed, at each temperature, by two straight lines. This does not comply with the Griffith theory.

But let us admit that this theory applies especially to brittle failure. Consequently, we may assume that the theory of elasticity is adapted to the cracked body. The influence of the length of the cracks on the peak stresses, in this case, appears rather doubtful. The gap between the theoretical strength of perfect crystals and of the real solid is so large that the explanation given by the Griffith theory for this gap cannot be considered as satisfactory. So, we come back to the necessity of some other explanation. This explanation is given by modern physics. The commonplace according to which the molecular bonds at any time and in any case are supposed to be extremely high is quite unrealistic. In a work to come, I intend to show to what extent it is unworthy of belief.

E. HOEK in reply:

Without knowing more about the rock type and the test conditions under which the Mohr circles illustrated by Dr. Talobre were obtained, it is very difficult to comment on these particular results.

However, to comment generally, I regard the criterion proposed by Griffith and modified by McClintock and Walsh as a guide to the fracture behavior that may be expected of a brittle rock. The actual relationship between the principal stresses at fracture must be determined from tests on specimens of the rock itself.

My experience with hard rocks, such as the quartzites encountered in South Africa, leads me to believe that the modified Griffith theory does indeed provide an excellent prediction of their fracture behavior. However, I am prepared to accept that there are factors which cause some rock types to deviate appreciably from this predicted behavior.

GENERAL DISCUSSION TO PART II

W. I. DUVALL (USA):

It seems to me that we have had enough material presented that we should now consider a major question. The papers that have been presented have been concerned in one way or another with the physical properties of the rock, that is, their elastic, viscoelastic, plastic, and strength properties, and we have had a discussion of various modes of failure. Now one of the problems that faces a mining engineer is the design of his structure so that it will stand up under the load imposed. Therefore, he has to estimate the ultimate strength, say of pillars, or the ultimate strength of roof slabs overlying an open stope in a bedded formation. My question is simply this: How can we make use of the basic theories that have been discussed here today to predict the ultimate strength of rock pillars or roof slabs? I think it would be interesting to have Drs. Brace, Emery, and Robertson attempt to give us a brief explanation of how they would propose to use their material.

C. L. EMERY in reply:

Most mining jobs, of course, are not short-time things in the sense of short-time testing or in the laboratory, and while one may sometimes get elastic conditions, one would generally get some flow. In the work that I do, where actual applications have been made of our notions, I use the hand samples that you have seen and core samples to try to define what kind of energy pattern exists in the rock. In other words, where are the shear surfaces and what are the directions of the principal stresses? These I can do something about. And, if I take a sufficient number of samples and plot them, I can at least put down stress trajectories for the area where I want to work, and this will help me orient the openings in a favorable direction. The matter of the preferred shear planes is useful if we are working with room-and-pillar design. I make use of these to set up rock bolting patterns if they are required, and also use the various measuring devices to see the effect of what I have done. I can usually check myself out to some extent. In predicting the size of a pillar, one still has to fly to some extent by the seat of one's pants. Until very recently, mining has been an art, and for all practical purposes, we are just now beginning to introduce a little science into it. As in all new things, we have a long way to go. I don't think that anybody can use the kind of results that we get in compression tests to predict the strength of the pillar without a lot of good intuitive mining knowledge to go along with it. However, one can do something about directions, and learn by so doing. I also use the preferred shear planes for developing drilling and blasting patterns, and these are effective very often. In some rocks where the amount of strain energy is not great, we find the differences are not great. In others where there is a substantial amount of strain energy, then one might make use of theory. In most ore bodies, there is a substantial amount of strain energy because ore bodies very often are formed where there has been some kind of major disturbance.

E. C. **ROBERTSON** in reply:

As to the application of this viscoelastic parameter that I have looked into, the fact is that in mines the rocks are near room temperature, and therefore, high stresses are required to attain even slow rates of strain. You recall the diagrams in which for Yule marble a stress difference of the order of 3 kilobars is required even for low strain rates of 10^{-8} per second. What this means, it seems to me, is that you can't apply creep properties of these rocks. You are more or less constrained to accept the elastic analysis as a first-order approximation, and from there on, you must measure the properties of the rock *in situ* to find out what the effects really are. In such cases you may find a creep behavior which you can analyze in somewhat the same fashion that I have done here, and perhaps observe that at low stresses you will have creep and at high stresses you will have fracture and failure. Perhaps it would be worth while, then, to make time dependent creep studies of the mine rock to find this out. Any measurement of the mechanical properties of rock certainly can be useful.

W. F. BRACE in reply:

We have been discussing primarily properties of small rock samples. Obviously, a large mass of natural rock differs in important respects from these. The question is, how does it differ, and can we take this difference into account?

A pillar or roof slab, or nearly any mass of natural rock over a few feet in dimension, usually appears "broken up," that is, to contain numerous cracks and small fissures. We might expect that this characteristic would play an important part in determining strength of the pillar or roof slab, and that, in fact, the strength would be quite different from that of a more or less perfect laboratory sample. Now, how can we take this important characteristic into account, and obtain a reasonable estimate of strength of our pillar or roof slab?

If the rock in question is brittle and fairly homogeneous, one possibility is to estimate strength using the Griffith theory. To do this one needs to know how big the cracks and fissures are in the rock in question. Using relations such as those given in the section of my paper entitled "Discussion," one could either extrapolate strength from laboratory measurements, or estimate it directly by using reasonable values of specific surface energy and elastic modulus. Griffith crack length would be set equal to the length of the biggest cracks or fissures in the natural material.

PART III--MEASUREMENTS

Charles L. Emery, Chairman

Evaluations of (1) approaches for determining stress *in situ* using relief techniques and the assumptions that must be considered; (2) methods used in exploration geophysics for petroleum and civil engineering work; (3) petrofabric techniques for mapping principal stress directions in naturally deformed rocks.

MESURES

Evaluation (1) des méthodes qui permettent de déterminer les contraintes *in situ* au moyen des techniques du relief et des hypothèses qui doivent être envisagées; (2) des méthodes utilisées en géophysique d'exploration pour des travaux pétroliers ou de génie civil; (3) des techniques dites de "petrofabric" utilisées pour tracer les directions des contraintes principales dans les roches naturellement déformées.

MESSUNGEN

Auswertungen von (1) Möglichkeiten zur Bestimmung von *in situ* Spannungen mittels Entlastungsverfahren und der hierzu erforderlichen Annahmen, (2) Geophysikalischen Erkundungsverfahren bei Erdöl- und Tiefbauvorhaben und (3) Verfahren zur Bestimmung der Hauptspannungsrichtungen imnatürlich verformten Fels, auf Grund von Gefügedaten.

CONTENTS

INTRODUCTION

Charles L. Emery*

IF IT HAS DONE nothing else, Part II, on the fundamentals of rock mechanics, must have impressed the reader with the amount of cerebration currently involved in rock mechanics studies. He may also be disconcerted by the apparent conflict of opinions.

Because of the scope of rock mechanics as a study, investigators find themselves working along many different paths, each according to the line of thought that leads to his particular path. Each is noting phenomena of some sort or is developing a hypothesis, together with its attendant formulas; and each may be describing some aspect from a different point of view from the others. In the process, each is required to evaluate his work as he proceeds, and each must rely on some form of measurement.

In Part III, on measurements, we will try to cover the broad spectrum of measurement techniques. This will be done by examining several quite different techniques and by discussing as many phases of measurement as space permits.

Much has been accomplished by known measurement techniques. Most of us do not know exactly what we are measuring, except that observation shows we are measuring something that appears to be a characteristic phenomenon of a rock. One aim of measurement will be, then, to define roughly what we are measuring and to attempt to standardize the techniques involved. It is not necessary, however, to identify fully the phenomenon -- only to know that it is characteristic and that it is a parameter of some sort. People using gravity

*Assistant Professor, Mining Engineering Department, Queen's University, Kingston, Ontario, Canada.

as a parameter, for example, do not know what it is; and the history of the development of electricity is not based on a specific knowledge of what an ampere is, but rather of what it does. In the same way, the most important use of measurements at present is to measure parameters of some sort so that we can check our hypotheses and judge their applicability to rock mechanics.

ABSTRACT

Relief techniques have been used by several investigators to determine the stresses at various depths from the edge of underground openings. These techniques and the associated instrumentation, assumptions, and considerations that must be accepted are reviewed and evaluated. An experiment in which two relief techniques were used concurrently is briefly described and the results compared for agreement. These techniques, namely, borehole deformation and flatjack methods, were also used to test some of the assumed versus the *in situ* conditions. Results of these experiments indicate that the flatjack and borehole deformation methods can give satisfactory results, even though some of the assumptions may not be correct.

RÉSUMÉ

Les techniques du relief ont été utilisées par plusieurs chercheurs dans le but de déterminer les contraintes à diverses profondeurs a partir des bords d'ouvertures souterraines. Ces techniques et l'appareillage associé, ainsi que les hypothèses et les considérations qui doivent être faites, sont ici passés en revue et évalués. Un essai est brièvement décrit, dans lequel deux techniques du relief ont été parallèlement utilisées, les résultats étant comparés pour accord. Ces techniques, à savoir celle de la déformation d'un trou de forage et celle de "flatjack" ont aussi été utilisées pour vérifier l'exactitude de certaines hypothèses vis-à-vis des conditions *in situ*. Les résultats de ces essais indiquent que les méthodes de vérin plat Freyssinet et de la déformation du trou de forage peuvent donner des résultats satisfaisants bien que quelques-unes des hypothèses puissent être incorrectes.

AUSZUG

Entlastungsverfahren sind von verschiedenen Forschern zur Bestimmung der Spannungen in beliebigen Abständen von den Wänden unterirdischer Hohlräume angewendet worden. Diese Methoden und die dabei verwendeten Instrumente, Annahmen und Überlegungen, die dabei notwendig werden, werden hier zusammengefasst und ausgewertet. Ein Versuch, in dem zwei verschiedene Entlastungsverfahren durchgeführt wurden, wird kurz beschrieben und die Versuchsergebnisse miteinander verglichen. Die beiden Verfahren, die Bohrlochverformungsmethode und das Druckkissenverfahren wurden auch dazu benutzt, um die Annahmen mit den tatsächlichen *in situ* Bedingungen abzugleichen. Die Ergebnisse dieser Untersuchungen zeigen, dass beide grundsätzliche Verfahren zufriedenstellende Ergebnisse liefern, obwohl einige ihrer Voraussetzungen nicht der Wirklichkeit entsprechen mögen.

IN SITU DETERMINATION OF STRESS BY RELIEF TECHNIQUES

Robert H. Merrill*

Introduction

DURING THE LAST THIRTY YEARS a number of methods have been developed to determine *in situ* stress in rock by relief techniques. Relief techniques are defined as methods or procedures of wholly or partially isolating a specimen of rock from the stress field in the surrounding rock. Examples of *in situ* stress-relief techniques used in past investigations include

1. isolation of a cylindrical or rectangular specimen of rock by drilling an annular ring of overlapping drill holes,
2. cutting the top of a pillar free from the roof,
3. removing rock from a slot (in which a flatjack is later cemented), and
4. coring the rock by adaptations of diamond core drilling techniques.

This paper briefly reviews some of the general procedures and instruments that have been used in relief techniques. The assumptions and considerations associated with each method are discussed collectively in a separate section. A comparison of the stresses determined by the use of two different relief methods at the same site is presented together with the results of studies performed to compare the assumed and *in situ* conditions.

*Denver Mining Research Center, U.S. Bureau of Mines, Denver, Colorado.

Various Relief Techniques[*]

The first determination of *in situ* stress in rock by relief techniques was performed by Lieurance in 1932.[1,2] Lieurance placed reference pins in the wall of a tunnel (Fig. 1) and measured the distance between the pins with a commercial extensometer calibrated in units of strain. The stress in the rock around the pins was relieved by cutting a slot with overlapping drill holes. The drill holes were 30 in. deep around all four sides of the section of rock containing the pins. This section of rock (within the overlapping holes) was about 4 ft square. The change in distance between the reference pins (located on the circumference of a 20-in.-diameter circle) was measured as the slot was cut. To obtain rock for laboratory evaluation of the modulus of elasticity, 3-in.-diameter cores were drilled from the relieved section. Conventional strain-rosette formulas were used to compute the stress in the rock. The results of Lieurance's computations are shown as stress ellipses on Fig. 2.

Interesting features of this study are the corrections made for stress concentrations around the tunnel and the determination of the ratio between the vertical and horizontal stress. The ratio between the computed vertical (or nearly vertical) stress and the adjusted vertical stress is three, which indicates that the investigator assumed a stress concentration of three at the edge of the circular tunnel. This suggests that the stress field was assumed to be uniaxial in the plane perpendicular to the major axis of the tunnel. No adjustments were made regarding the magnitude of the horizontal stress parallel to the axis. The correction for stress concentrations in the vertical direction resulted in a ratio between the horizontal and vertical stress in the wall of from two to four. The average vertical stress was 1200 to 1500 psi.

---

[*]No attempt is made to recognize all persons or methods associated with relief techniques. New techniques, or changes in techniques, are recognized by name or reference.

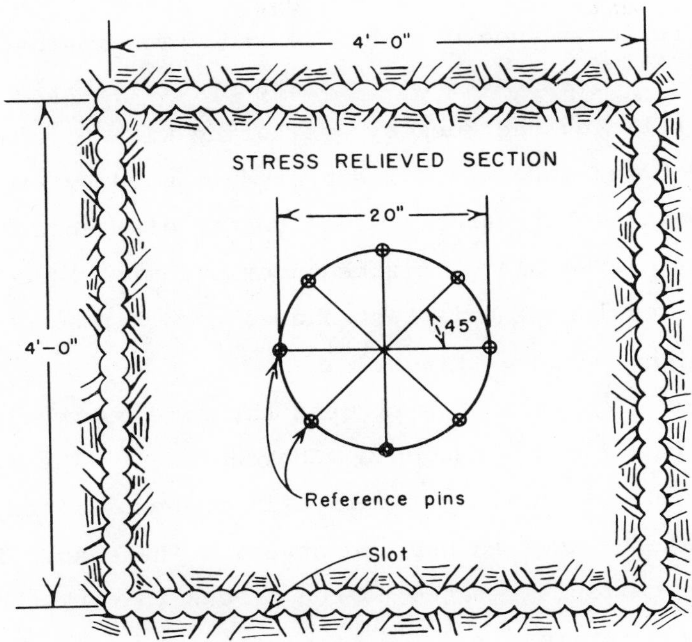

Fig. 1—Relieved section of rock and instrumented points used by Lieurance.

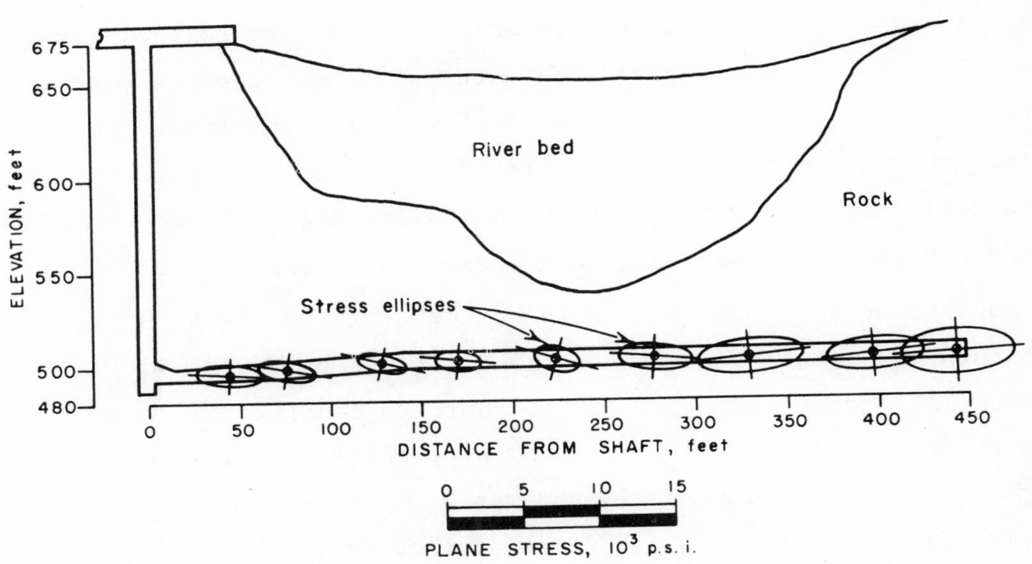

Fig. 2—Approximate profile of topography and openings and stress ellipses near Boulder (Hoover) Dam.

The stress-relief method used by Lieurance was improved by Olsen in 1949.[3,4] Olsen attached resistance-wire strain gauges to the rock and relieved the rock by overcoring with a relatively large diameter diamond core bit. The procedure used by Lieurance was used to compute the stresses. Olsen tested his techniques by stress relieving a large block of rock under uniaxial load in the laboratory. The results of this test showed good agreement between the computed and the applied stress.[4]

In the period 1954 to 1958 Hast used the same overcoring technique to relieve the rock around a gauge in a borehole.[5] The transducer in the gauge is a nickel cell that varies in magnetostrictive properties with differences in applied stress. The change in these properties in the cells is related to the stress in rock through formulas, laboratory-determined modulus of elasticity, and various calibration techniques.

In 1958 and 1959 the Federal Bureau of Mines developed a similar technique.[6,7] The change in deformation of the borehole, measured as the rock is overcored, is related to the stress in the rock by deformation-rosette formulas and the laboratory-determined modulus of elasticity. A photograph of the gauge and borehole is shown in Fig. 3. The primary difference between the Hast and Bureau methods is that the Hast gauge has relatively low compliance (that is, the gauge is comparatively "stiff") and the Bureau gauge has a relatively high compliance.[8] Examples of stress computations using the high compliance gauge are presented in this paper.

Another approach to the problem of determining stress by relief techniques was first used extensively in France by Mayer, Habib, Marchand,[9] and Tincelin.[10] This approach consists of measuring the displacements in the rock created by cutting a nearby slot. A flatjack (a thin, approximately square, hydraulic-pressure cell) is cemented into the slot and pressures applied to the hydraulic fluid in the cell until the displacements created by cutting the slot are "cancelled" by the pressure in the flatjack. An approximate plan and section of the instruments, slot, and flatjack are shown in Fig. 4. The pressure in the cell required to cancel the deformation in the rock is presumed to be equal to the stress in the rock before the slot was cut. This method is commonly called the flatjack method of stress determination.

More recently the flatjack method has been used in the United States and Australia.[11-14] Various instruments have been used to measure the change in strain or deformation in the rock as the slot

Fig. 3—Borehole gauge being inserted in borehole.

is cut and the pressure at which the changes are cancelled. Panek used an embedded resistance-wire strain gauge, and Tincelin, Mayer, and others used reference pins and extensometers. However, in both cases the basic method is the same. To obtain the stresses in a biaxial field, some investigators use two flatjacks oriented at right angles. The pressures in the flatjacks are substituted in

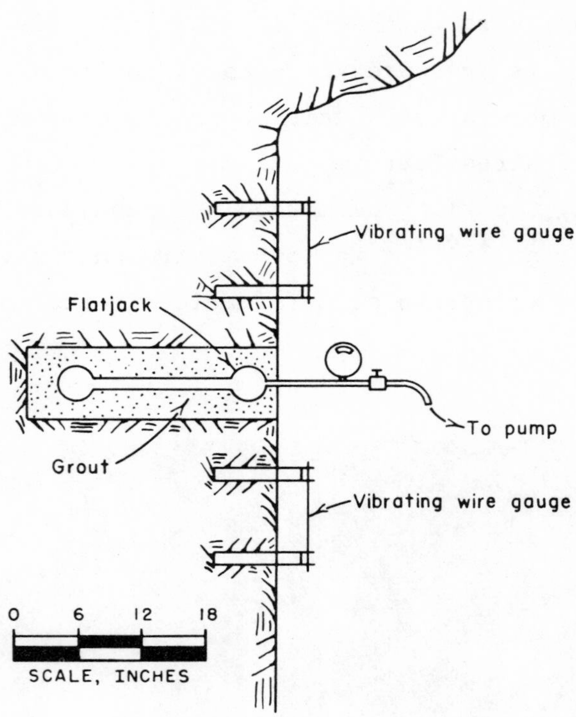

Fig. 4—Idealized section of flatjack and instrumentation.

the following relationships to compute the orthogonal stresses in the rock:[10-12]

$$S = aP_h + bQ, \tag{1}$$

$$Q = aP_v + bS, \tag{2}$$

where S, Q = orthogonal stresses in the rock,

P_h = cancellation pressure of horizontal flatjack,

P_v = cancellation pressure of vertical flatjack, and

a, b = constants depending on the gauge geometry.

Examples of stresses obtained by direct determination and through use of the above relationships are presented in a later section of this paper.

Assumptions and Considerations Concerning Relief Techniques

Some of the assumptions and considerations associated with the foregoing stress-relief techniques are as follows:

1. The change in stress "released" from the rock when isolated by overcoring or other relief techniques is equal (and opposite in direction) to the stress in the rock *in situ*.
2. The values of the modulus of elasticity and Poisson's ratio determined by laboratory methods are identical to the *in situ* values.
3. The relationships between stress, strain, and deformation for rock are identically equal to the relationships for perfectly elastic, homogeneous, isotropic media subject, in most cases, to plane stress or plane strain. Further impositions are necessary when specific geometric requirements (as thin plates with specific sizes) or boundary conditions are used to obtain the stress-strain, or other, relationships.
4. The instruments accurately measure the intended quantities; strain-measuring devices accurately measure only the strain, and borehole-deformation gauges accurately measure only deformation.

Some of the assumptions and considerations associated with the flatjack method are as follows:

1. The cutting of the slot causes a change in stress distribution in the rock, which, in turn, produces a corresponding strain or displacement at the instrumented points.
2. The strain or displacement produced by pressure in the flatjack is equal, but opposite, to that listed under item 1.
3. The pressure in the cell, at cancellation, is equal to the stress in the rock (before the slot was cut) normal to the plane of the flatjack.

4. As in the case for stress techniques, the instruments accurately measure the intended quantities.

Each method has specific limitations that must be considered. In both the stress-relief method employing strain gauges and in the flatjack method, measurements are usually taken at or near the surface of the opening. Although there are instances when a surface determination is desirable, the determination of stresses at depth by these methods would undoubtedly be very costly and time consuming. In two cases where stress-relief procedures were used at depth (with resistance-wire strain gauges),[15,16] the investigators found that the gauge placement became very difficult at depths greater than 1 ft. In this method the gauge (at depth) is at the bottom of the hole created by overcoring, and stress concentrations at the bottom of the hole are added to the field stresses to be determined. As a consequence, the stresses determined from the relief are subject to error. Hence, measurements are usually made at or near the surface, and these measurements are used with relationships from the theory of elasticity to obtain the stress outside the zone of influence of the opening. The necessity for these relationships has led to considerable laboratory work to determine the stress concentration at the surface of an opening.[12,13]

The flatjack method has the advantage that the relationship between stress and strain or stress and deformation in the rock need not be linear; however, strain or deformation in the rock caused by cutting the slot must follow the same relationship as the strain or deformation caused by the pressure applied to the rock by the flatjack, whether this relationship is linear or not. The pressure applied by the flatjack is assumed to be more or less uniformly transmitted to the rock; however, the rock may be subject to a stress gradient between the edge of the opening and the back of the flatjack. Hence, the displacement and rotation of the instrumented points would be different when the slot is cut and when the pressure is applied to the rock by the flatjack.

If the stress determined by the cancellation technique is not parallel to the orthogonal stresses in the rock, the procedure for computing the stresses will involve the theory of elasticity and the corresponding assumptions. Also, a minimum of three flatjacks, placed along different orientations, are necessary to compute the magnitude and direction of the orthogonal stress. Hence, the advantage that the method does not require a "perfect" rock or a knowledge of the physical properties is lost by the necessity to use equations that require a knowledge of the environment.

The borehole stress-relief methods have the advantage that stress determination can be made with relative ease to depths of 20 ft or more from the opening. Both the magnitude and direction of the stress field can be computed at specific distances in the rock. The determinations are made relatively fast when compared with other methods. In regard to this advantage, the average time required to perform a complete flatjack cycle in an amphibolite is about 6 weeks; using the Bureau of Mines gauge and technique, borehole stress reliefs to depths of 10 ft can be completed in 24 hours or less. Using the overcoring techniques with resistance-wire strain gauges, the time required for a complete relief cycle is 3 to 4 days; however, a part of this time is consumed by "inactive" periods during which cements are curing, temperatures are stabilizing, or other factors that require specific periods are involved.

Although the overcoring stress-relief techniques are fast and efficient, the rock must core well and be reasonably elastic. Further, the area of rock under investigation by overcoring techniques is very small compared with areas covered by flatjack methods. Therefore, in rocks containing many defects (unbonded joints or faults) or large differences in properties from point to point in the rock, the overcoring method (whether with strain or borehole gauges) is very difficult, if not impractical, to use.

Results of Stress Determination Using Different Relief Techniques in the Same Rock

In a cooperative investigation by the Federal Bureau of Mines and the State of California, stress determinations were made at five sites using both the flatjack and the borehole-deformation stress-relief methods. The determinations were made in the walls of openings, about 6 ft wide and 7 ft high, in a fine-grained amphibolite. The measurements were made at the ends of cutouts or at intersections, as these locations provided more space for the necessary equipment. The approximate plan and profile of the opening and topography over the opening are shown in Figs. 5 and 6.

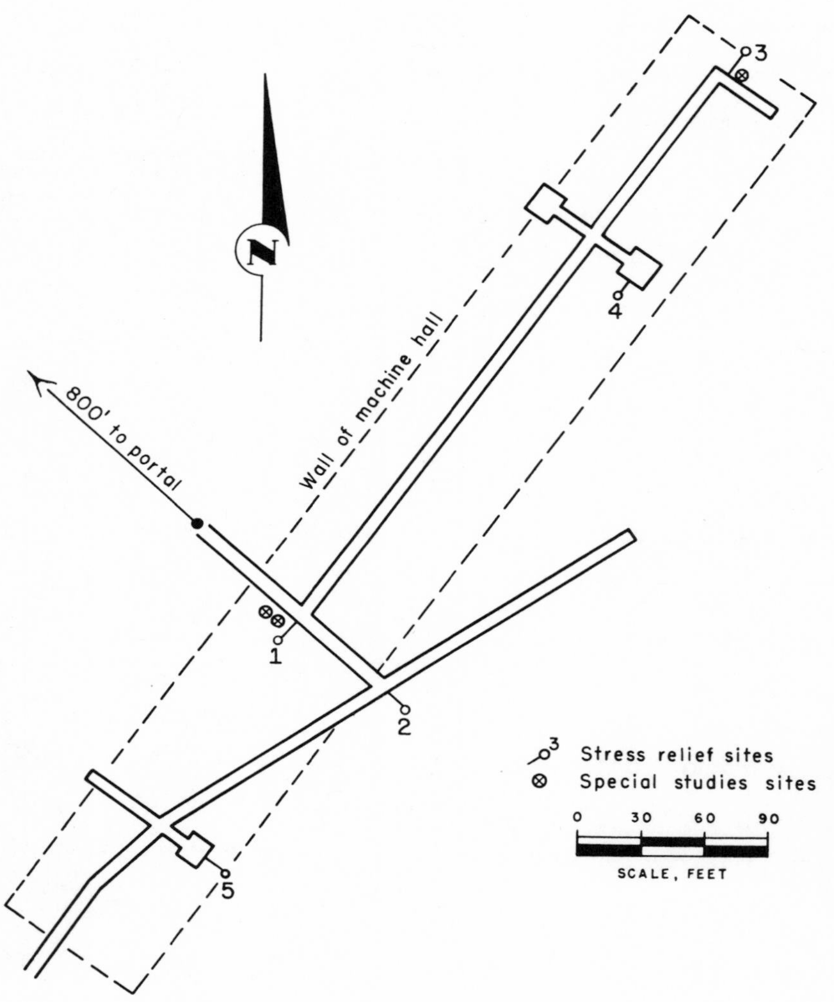

Fig. 5—Sites of flatjack and borehole measurements.

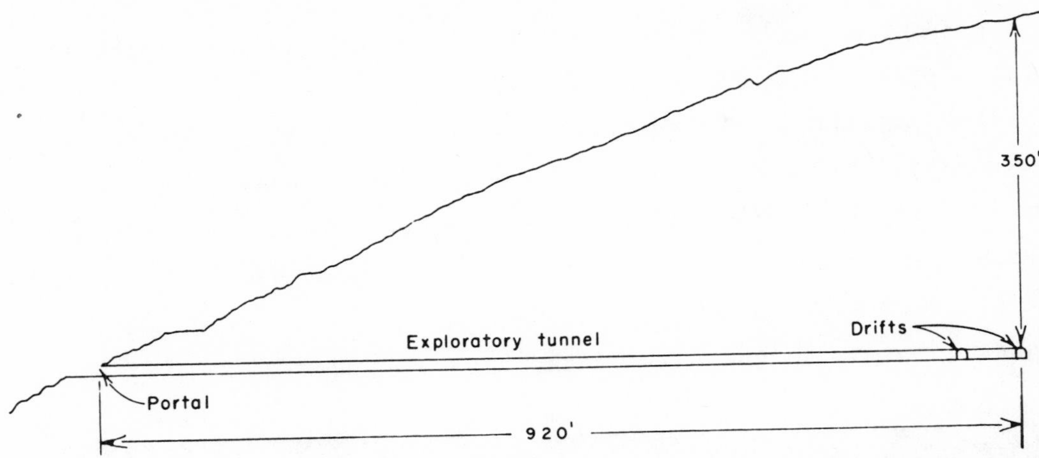

Fig. 6—Approximate profile of topography and exploratory openings.

The procedure for borehole stress relief has been described briefly in other reports(6,7) and is summarized as follows:

1. An EX-size diamond drill hole is drilled to a depth (for this investigation) of 10 ft.
2. The measuring point of the borehole gauge is placed 6 in. deep in the hole and oriented 30 deg counterclockwise from horizontal.
3. The rock around the gauge is overcored coaxially with a 6-in.-diameter coring bit and the resulting borehole deformations recorded concurrently. The overcoring is stopped when the gauge readings indicate that the stress in the rock around the gauge is relieved (usually not more than 2 in. beyond the points of measurement).
4. The gauge is placed 4 to 6 in. beyond the first location, oriented at 90 deg (vertical), and step 3 repeated.
5. Step 4 is repeated with the gauge oriented at 150 deg.
6. The cycle is repeated until the deformations are nearly equal from point to point in the hole, along any of the three orientations of the gauge; that is, U_{90} (deformation at orientation of 90 deg), at perhaps 5 ft, is about the same as for 6.3 and 7.6 ft. When overcoring produces about the same deformations from point to point, it is presumed that

the borehole gauge is outside the zone influenced by the opening from which the tests are being made.

A photograph of the equipment is shown in Fig. 7. The data obtained by the above procedure are used, with the modulus of elasticity of the rock, in deformation-rosette formulas. The stresses, as calculated from these formulas, are shown graphically in Figs. 8 to 12. These stresses are the secondary principal stresses* normal to the axis of the borehole.

Fig. 7—Stress-relief drill and associated equipment.

The procedure used in the flatjack tests is also described in detail elsewhere,[12,13] but is briefly summarized as follows:

1. The distance between each set of pins (A, B, C, D, E, and F; Fig. 13) is measured with the extensometer. To increase the accuracy and reliability of the measurements, the

*See Max M. Frocht, Photoelasticity, Vol. II, John Wiley & Sons, Inc., New York, 1948, Chap. 10.

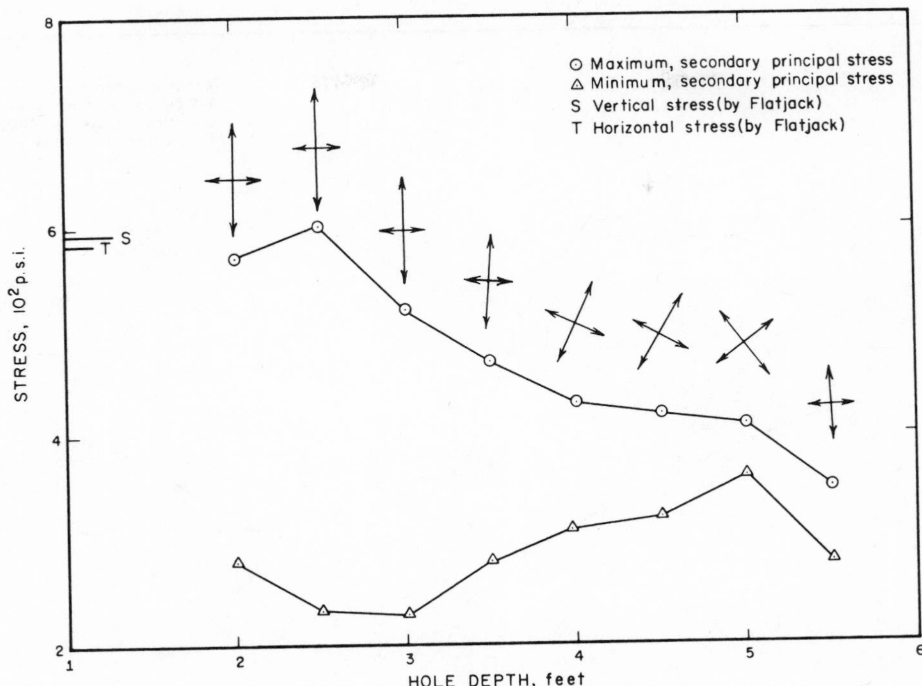

Fig. 8—Magnitude and direction of stress versus depth, site 1.

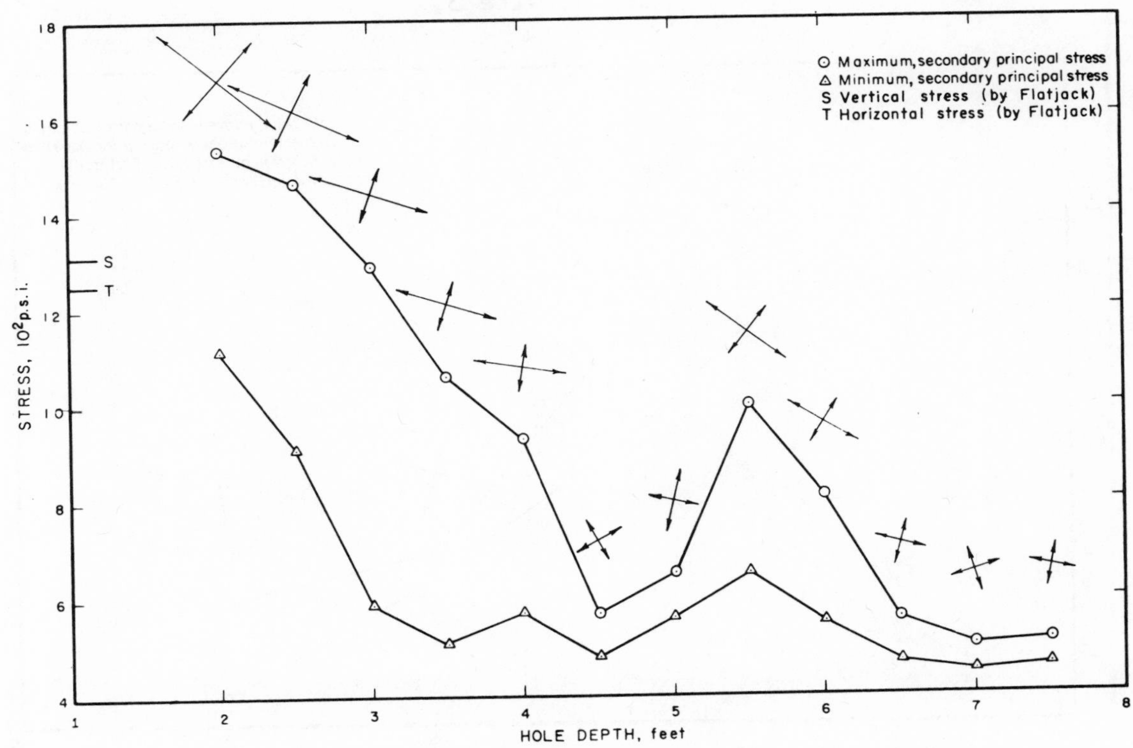

Fig. 9—Magnitude and direction of stress versus depth, site 2.

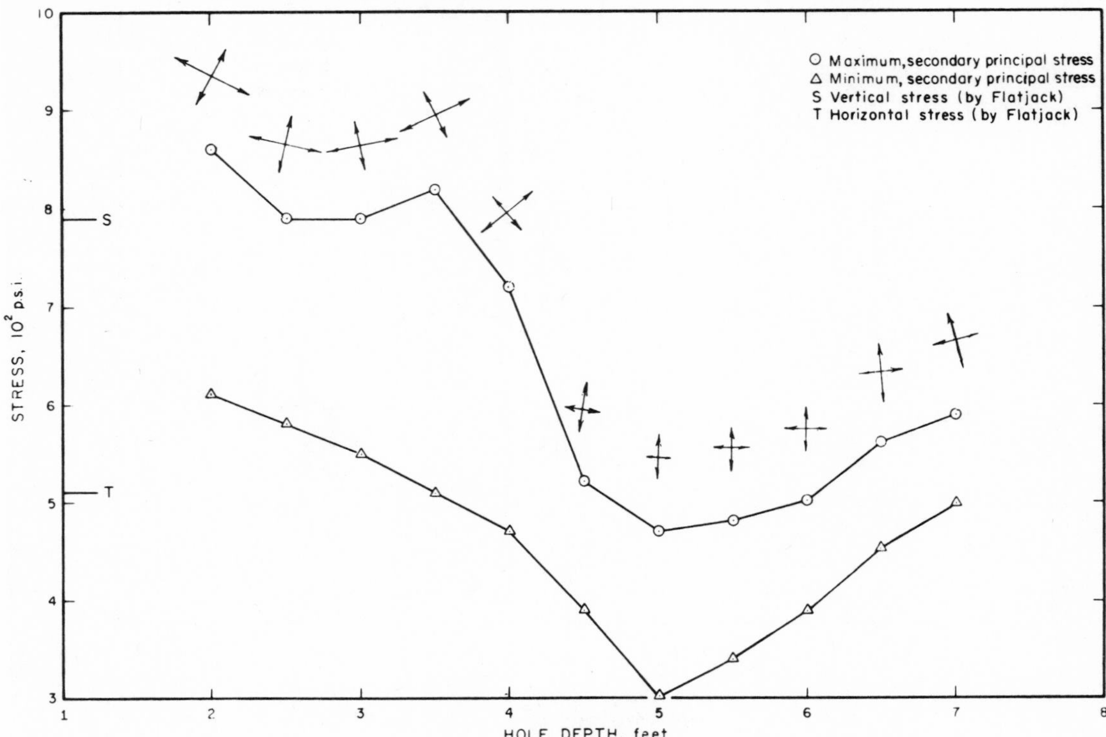

Fig. 10—Magnitude and direction of stress versus depth, site 3.

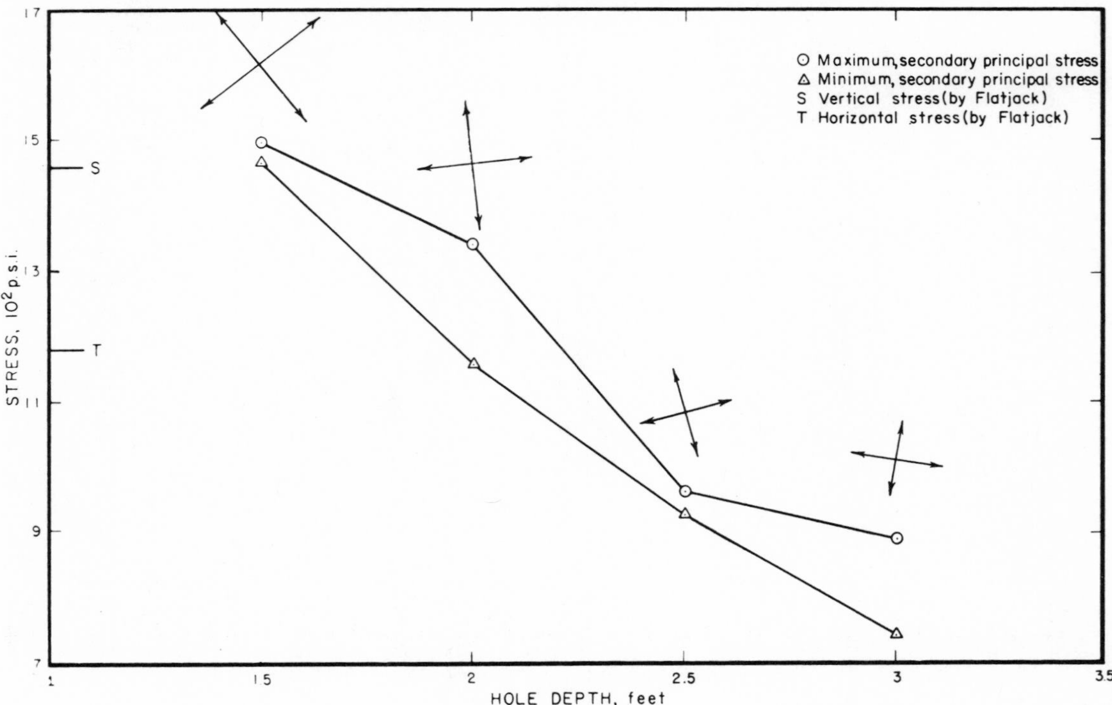

Fig. 11—Magnitude and direction of stress versus depth, site 4.

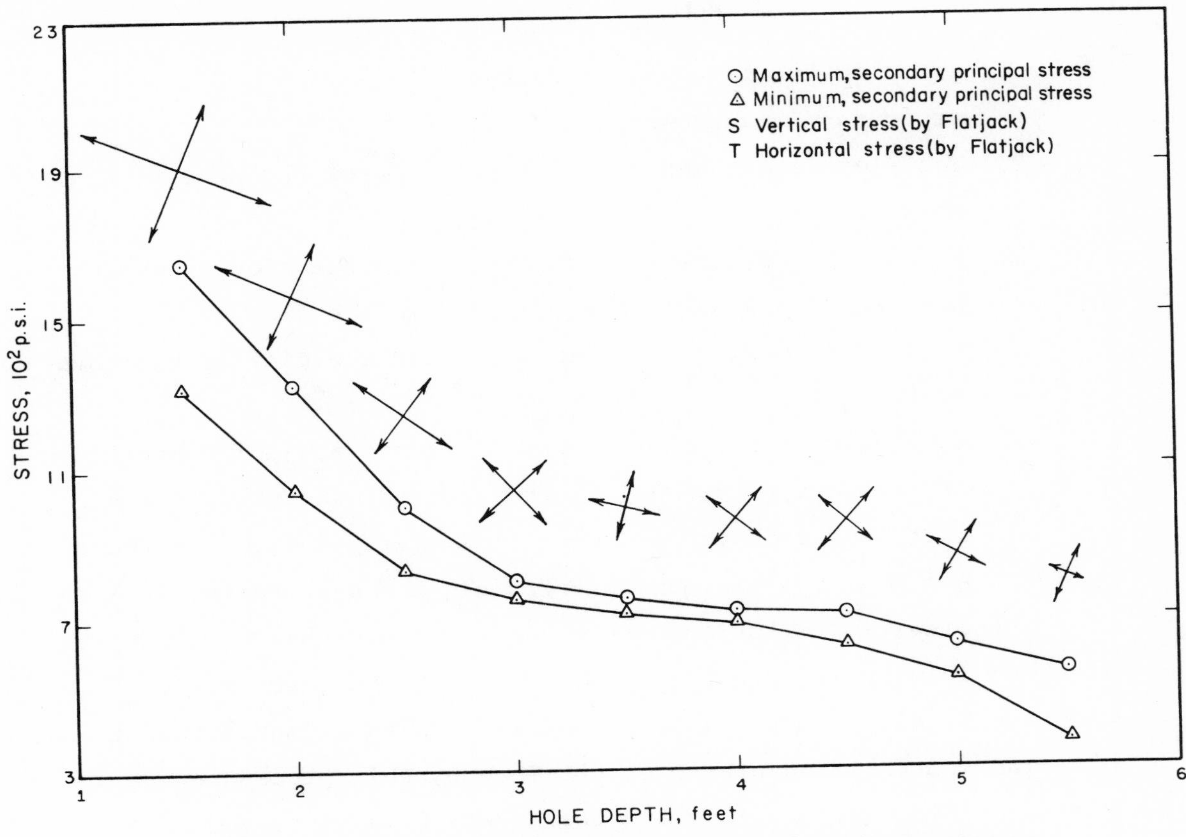

Fig. 12—Magnitude and direction of stress versus depth, site 5.

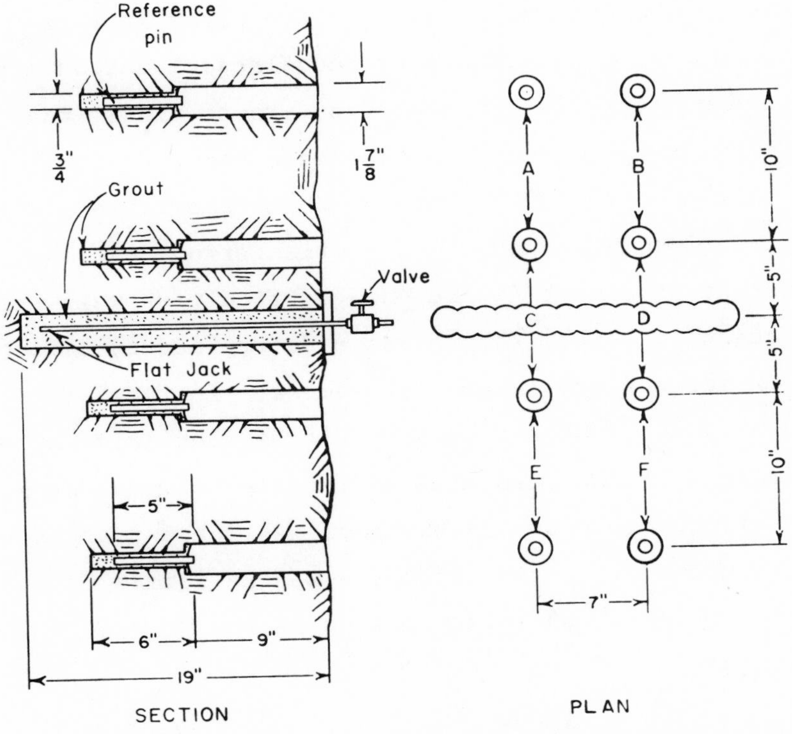

Fig. 13—Plan and section of flatjack and reference pins.

extensometer is checked against a "standard" before and after each measurement.

2. The slot is cut and the distance between the pins is again measured.
3. A flatjack (Fig. 14) is cemented into the slot, and after the cement has cured the pressure in the flatjack is increased until the distance between the pins is the same as in step 1.
4. Generally, the different distances are not all cancelled at the same pressure. In these cases, the pressures are increased or decreased until the average of the distances A + B/2, C + D/2, and E + F/2 are the same for steps 1 and 3.

A typical installation of the reference pins and the flatjack is shown in plan and section in Fig. 13. The extensometer used to measure the distance between the reference pins is shown in Fig. 15.

The cancellation pressures obtained by this procedure are inserted in Eqs. (1) and (2), and the stress S is computed. The cancellation pressures and computed stresses from the flatjack method, and the vertical and horizontal stresses as determined by borehole deformation measurements, are given in Table 1.

The stresses determined by the flatjack method are assumed to be at a depth of 9 in. (depth of reference pins); the stresses determined by the borehole method were computed for a depth of 18 in., which is the approximate depth at which the third stress relief in the cycle (30-, 90-, and 150-deg orientation) was made.

Except at side 5 (where some difficulties were experienced with the flatjack method), the stresses determined by borehole and by flatjack methods are in good agreement. In fact, for engineering design purposes, the agreement could be classified as excellent.

Because of the suspected difference between the stress distributions in the rock before the slot was cut and after the flatjack was pressurized, a series of tests was performed to obtain an estimate of the stress distribution in the rock above and below the flatjack.

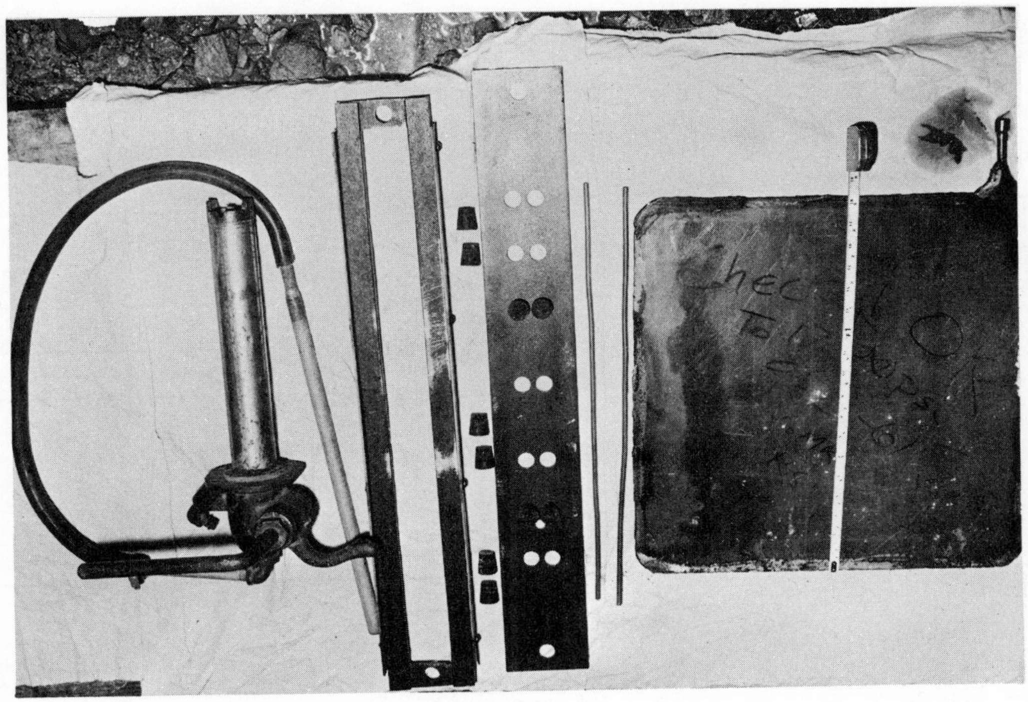

Fig. 14—Flatjack, grout pump, and other accessories (State of California, Dept. of Water Resources, photo).

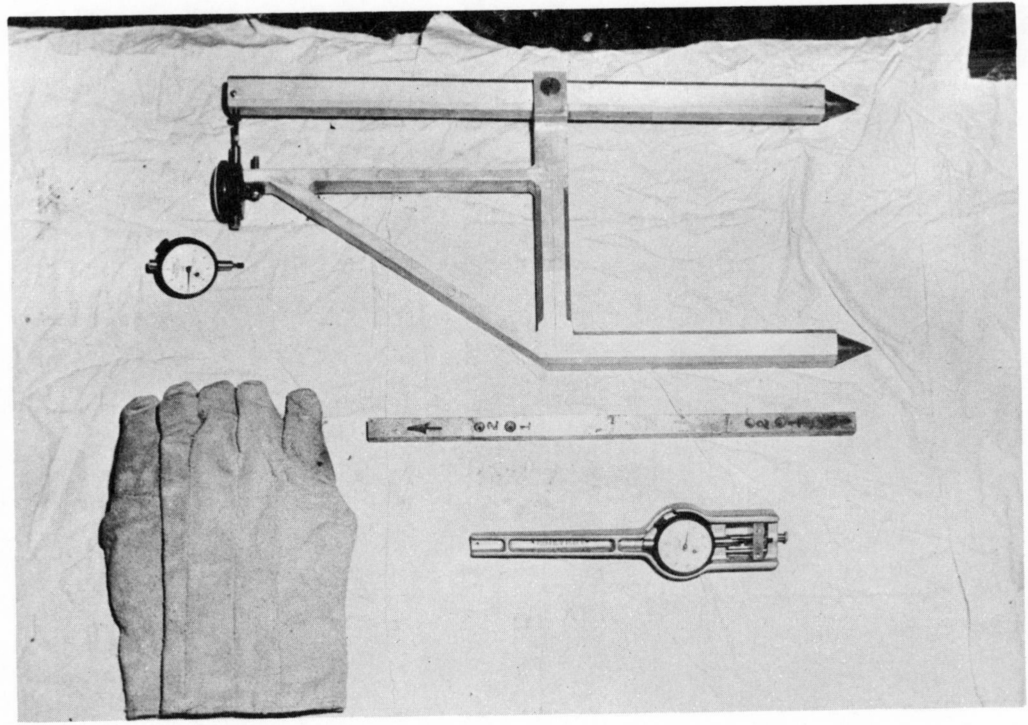

Fig. 15—Extensometer used with flatjack method (State of California, Dept. of Water Resources, photo).

Table 1

STRESSES DETERMINED BY BOREHOLE-DEFORMATION AND FLATJACK METHODS (psi)

Site No.	Borehole-deformation Method[a]		Flatjack Method[b]			
	Vertical Stress	Horizontal Stress	Cancellation Pressure		Computed Stress[c]	
			Vertical	Horizontal	Vertical	Horizontal
1	570	280	690	450	595	585
2	1530	1110	1840	1290	1310	1250
3	860	610	660	490	790	510
4[d]	1490	1460	1460	1130	1410	1180
5[e]	1660	1310	3800	2350	3560	2180

[a]At 18-in. depth.
[b]At 9-in. depth.
[c]Computed from Eqs. (1) and (2).
[d]Results from flatjacks in roof and wall.
[e]Only three of eight pin-displacement combinations cancelled.

The following procedure was used to obtain this estimate:

1. EX-size boreholes were drilled on a pattern as shown in Fig. 16.
2. Borehole-deformation gauges were inserted in the holes at a depth of 2 in. behind the front edge of the flatjack, oriented normal to the major axis of the flatjack.
3. Pressure was applied to the flatjack and held until borehole deformations were recorded; the pressure was then released and deformation again noted. This step was repeated three or more times for each location of the borehole gauges.
4. Step 3 was repeated with the gauges at depths of 9 in. (center of flatjack) and 16 in. (2 in. from the back edge of the flatjack).

This procedure was used at three flatjack sites. The average deformations* are presented numerically at points above and below the flatjack (Figs. 17, 18, and 19). An inspection of these figures shows

*In 90 per cent of the cases, the changes in the deformations measured before and after a pressure increase or decrease were the same. In the other 10 per cent of the cases the differences were less than 15 per cent. For simplicity, these data are averaged.

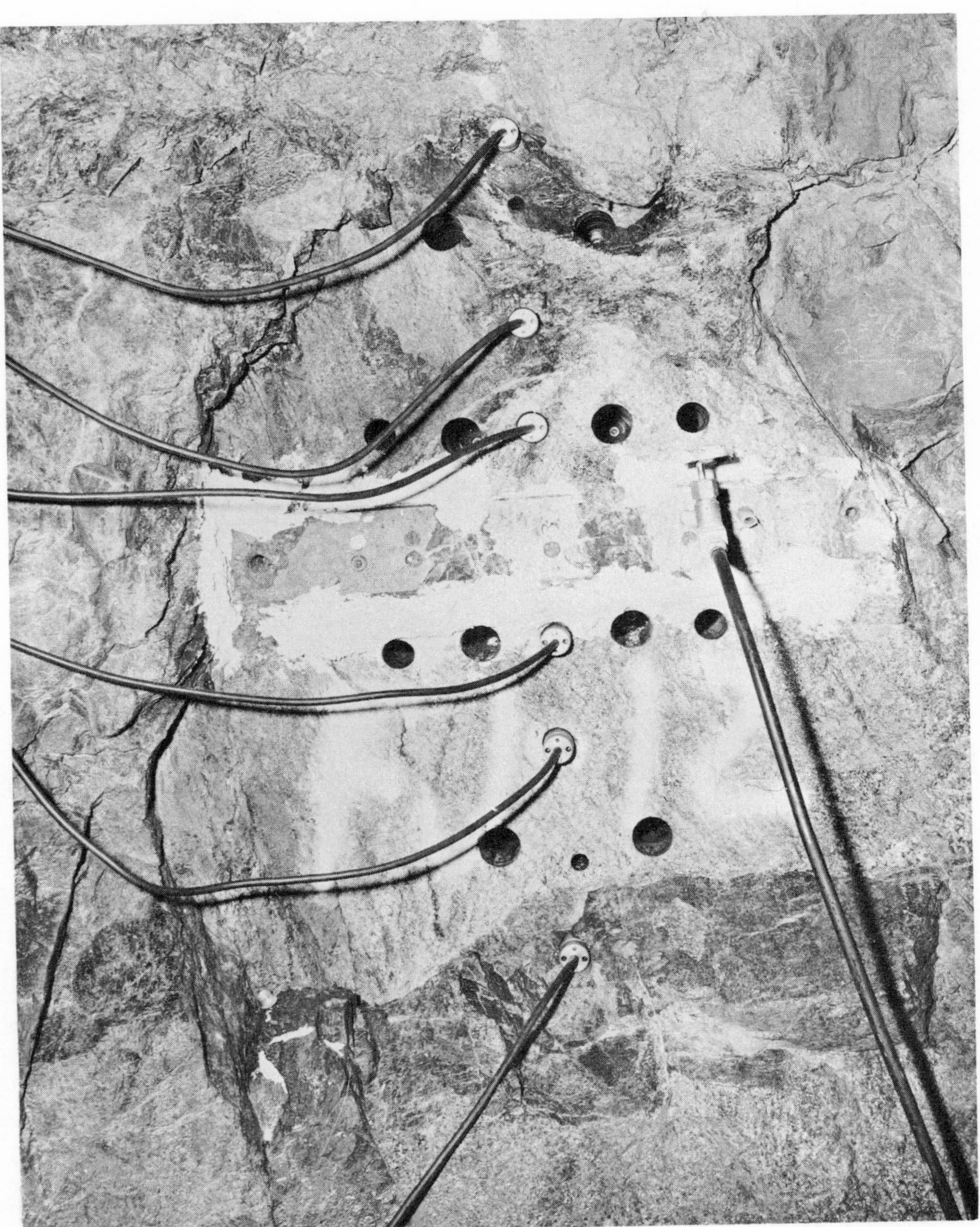

Fig. 16—Boreholes above and below flatjack (State of California, Dept. of Water Resources, photo).

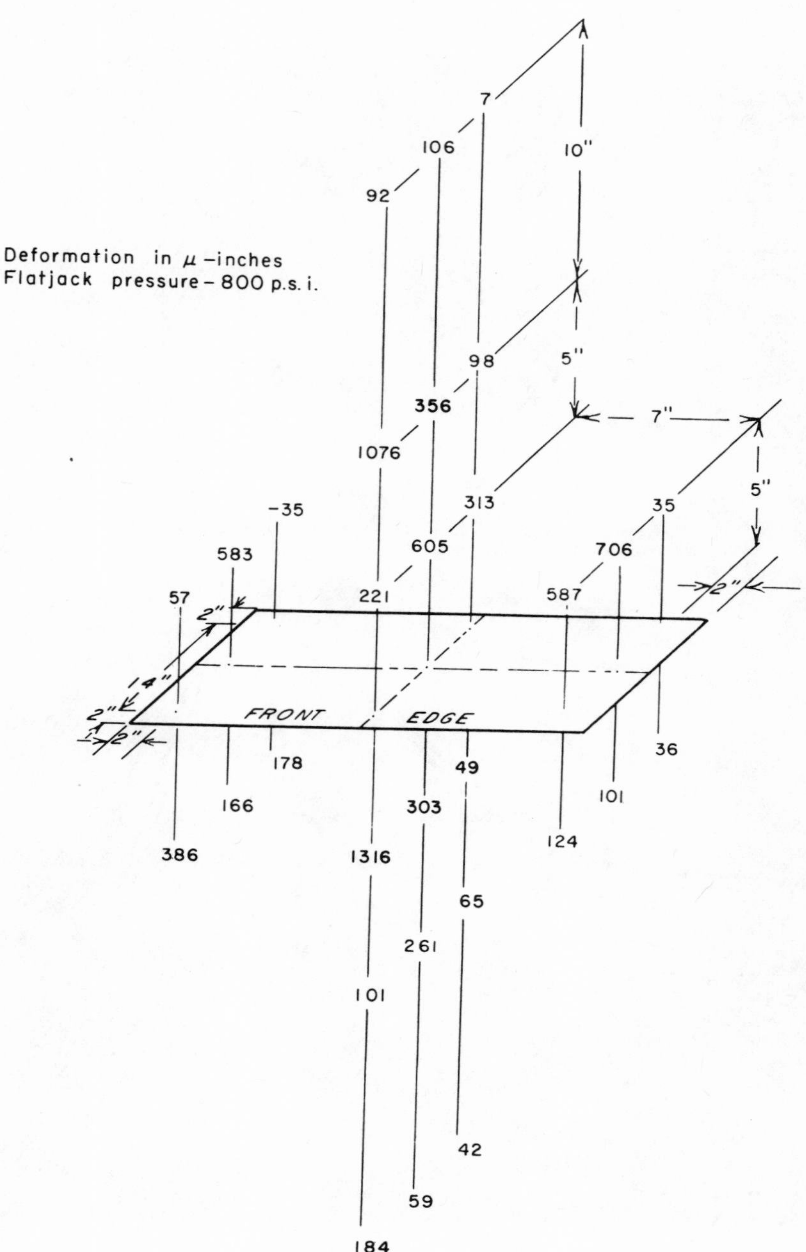

Fig. 17—Deformations above and below horizontal flatjack, site 3.

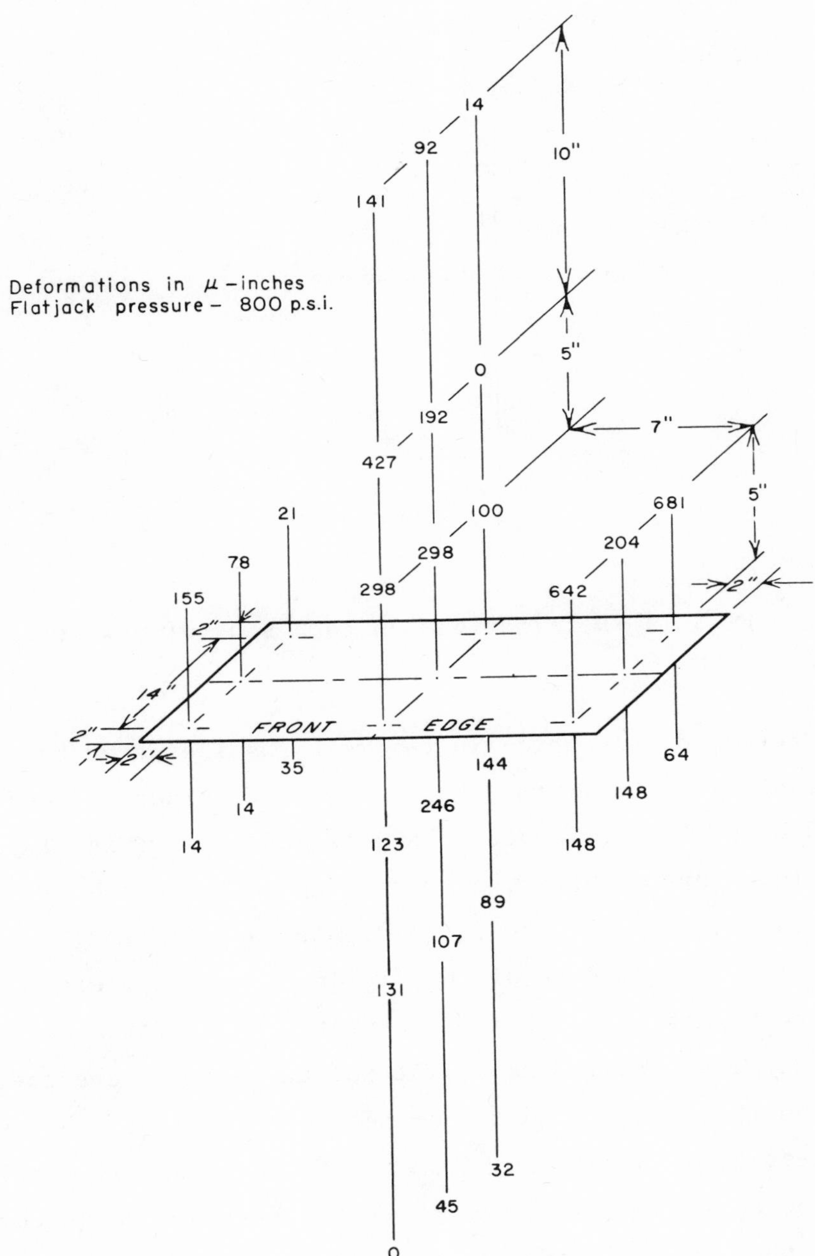

Fig. 18—Deformations above and below horizontal flatjack, site 1.

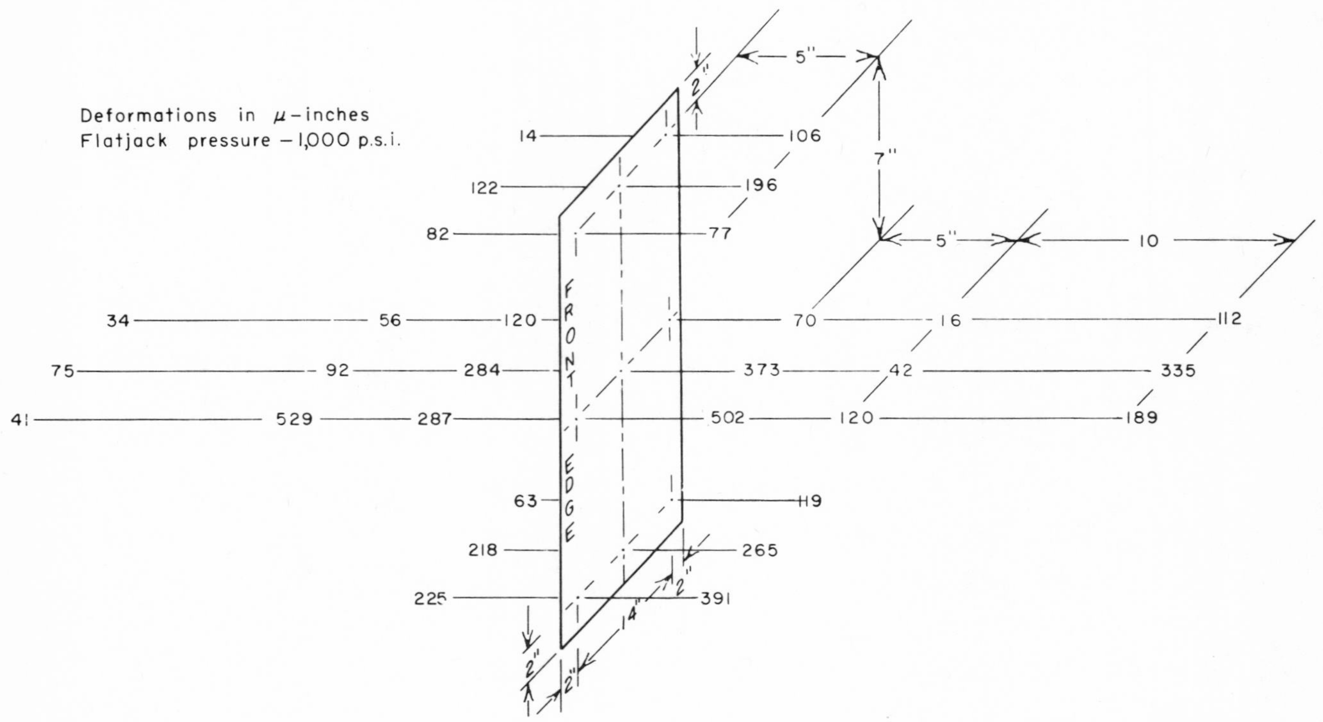

Fig. 19—Deformations on each side of vertical flatjack, site 1.

that the stress distribution in the rock (as indicated by borehole deformation) is not uniform. For example, consider the deformations above and below the horizontal flatjack at site 3 (Fig. 17). At points 5 in. above and 2 in. from the right edge of the flatjack, the deformations are 587, 706, and 35 μin. at points 2, 9, and 16 in. from the front edge of the flatjack, respectively. Directly below these points (5 in. below the flatjack), deformations of 124, 101, and 36 μin. were recorded. At points 10 in. above the center line of the flatjack, the deformations were 1076, 356, and 98 μin. at distances 2, 9, and 16 in. from the front of the flatjack. At 5 in. below these points (5 in. above the flatjack), the deformations were 221, 605, and 313 μin. At a point 5 in. above, 2 in. from the left edge, and 16 in. from the front of the flatjack, a deformation of -35 μin. was reproducibly measured over the three cycles. Hence, at that point, vertical stress in the rock must have been tensile or subject to a decrease in compression. Inspection of other points around the horizontal flatjack at site 1 and around the other areas tested suggests similar nonsymmetrical stress distribution.

These results are not in agreement with the assumption that the stress distribution in the rock (created by the flatjack) is symmetrical.

Summary and Conclusions

The relief techniques treated in this paper are subject to assumptions and considerations that are not exactly in agreement with *in situ* conditions.

The relief technique involving the isolation of a specimen of rock from its environment must satisfy the requirement that the stress in the specimen be removed upon isolation, and the stress so removed be identically equal to the stress in the environment from which the specimen was taken. A comprehensive study of the change in strain in rock, measured over relatively long periods *after* the rock has been removed from its environment, has shown this assumption to be false.[17] However, the ratio of the change in strain in rock *during relief* to the change in strain after relief has not, to the author's knowledge, been documented. The overcoring types of relief techniques require a determination of the modulus of elasticity and, in some cases, Poisson's ratio; investigations have shown that the *in situ* properties of rock may or may not be the same as the laboratory values.[18] The computation of the stresses involves relationships that were derived using assumptions and boundary conditions that may not be true for the rock under study.[19]

The flatjack method requires that the stress relieved by the slot be identical to the stress created by pressure in the flatjack, a condition that laboratory tests show to be false for some materials.[20] The *in situ* stress in the rock may also change over the distances covered by the flatjack, whereas the pressure in the flatjack is presumed to be evenly distributed over a relatively large area. The estimated distribution of stress in rock above and below the flatjack briefly described in a preceding section lends some doubts concerning this assumption.

The experiments by Lieurance indicated that the overburden pressures were about 1300 psi for an overburden depth of only 175 ft -- a stress much higher than expected.[19] This computed stress would have been much lower had Lieurance used a higher value for the stress concentration at the edge of the opening. Whether or not these computed stresses are reasonably accurate is unknown; however, the fact that Lieurance was forced to assume a stress concentration certainly contributed to any possible inaccuracies. Lieurance's work also disclosed a horizontal stress that was much greater than the vertical stress, a result that has also been observed by other investigations.[5,16] Olsen, in his tests in the laboratory, found that the stresses determined by relief techniques were about equal to the stress applied to the specimen by a press. A similar study to determine borehole deformation versus applied stress also resulted in reasonably good agreement.[7] The studies in which borehole-deformation and flatjack methods were used at the same location showed that the stresses determined by the two methods were, for engineering purposes, in good agreement. Because this agreement was good for distances close to the edge of the opening, the measurements at depth (by borehole methods) are probably reasonable estimates of the stress in the rock outside the zone of influence of the opening. For the most part, this stress field is essentially hydrostatic with stress values of 300 to 500 psi. This stress field is consistent with computed stress fields for depths of 350 to 400 ft and overburden weight densities of about 185 lb/ft^3.[19] These results by both flatjack and borehole methods indicate that the stress determinations were reasonable regardless of all the assumptions.*

All of the foregoing methods must be used within the limitations of the assumptions, or the investigator must be willing to accept the

---

*The results obtained in these comparison studies represent only one series of tests in one rock type, and tests in other rocks at other sites may or may not show similar agreement.

errors that exist when the methods are used outside the limitations. In addition, the different methods require specific instruments, techniques, materials, and equipment, all of which are items that contribute to the cost of stress investigations. Some of the methods require personnel specially trained in the techniques involved and personnel qualified to reduce the measurements to stresses. All of these factors contribute to the scope of the investigation, the required accuracy, and the costs that must be evaluated by the design engineer who needs the information. In some cases the engineer may have to weigh the specific measurement errors against the accuracy and cost requirements. The method he wishes to use is obviously his choice, a choice that will be influenced by the characteristics of the rock involved (for example, will it core in lengths required for overcoring techniques?), the degree of elasticity of the rock, the type of stresses to be determined (surface or field stresses), and cost and accuracy limitations.

REFERENCES*

1. U.S. Bureau of Reclamation, Boulder Canyon Project Final Report, Part V (Technical Investigation), Bull. 4, 1939, pp. 265-268.
2. Lieurance, R. S., "Stresses in Foundation at Boulder Dam," Tech. Memo. 346, Bur. Reclamation, Denver, Colo., 1933, 12 pp.
3. Olsen, O. J., "Residual Stresses in Rock as Determined from Strain Relief Measurements on Tunnel Walls," thesis, University of Colorado, Department of Civil Engineering, 1949, 61 pp.
4. Olsen, O. J., "Measurement of Residual Stress by the Strain Relief Method," Quart. Colo. School Mines, Vol. 52, July, 1957, pp. 183-204.
5. Hast, Nils ("The Measurement of Rock Pressure in Mines"), Sveriges Geol. Undersokn., Arsbok 52, 3, Stockholm, 1958, 183 pp.
6. Merrill, R. H., "Static Stress Determinations in Salt," U.S. Bur. Mines Rept. APRL 38-3.1, July, 1960, 30 pp.

*Titles enclosed in parentheses are translations from the language in which the item was originally published.

7. Merrill, R. H., and J. R. Peterson, "Deformation of a Borehole in Rock," U.S. Bur. Mines Rept. Invest. 5881, 1961, 32 pp.

8. Obert, L., R. H. Merrill, and T. A. Morgan, "Borehole Deformation Gauge for Determining the Stress in Mine Rock," U.S. Bur. Mines Rept. Invest. 5978, 1962, 11 pp.

9. Mayer, A., P. Habib, and R. Marchand ("Underground Rock Pressure Testing"), Conférence internationale sur les pressions de terrains et le soutènement dans les chantiers d'exploitation, Liège, du 24 au 28 avril, 1951, pp. 217-221.

10. Tincelin, E. ("Research on Rock Pressure in the Iron Mines of Lorraine"), Conférence internationale sur les pressions de terrains et le soutènement dans les chantiers d'exploitation, Liège, du 24 au 28 avril, 1951, pp. 158-175.

11. Alexander, L. G., "Field and Laboratory Tests in Rock Mechanics," Third Australia-New Zealand Conf. on Soil Mech. and Foundation Eng., 1960, pp. 161-168.

12. Kruse, G. H., Stress Determinations in Pilot Tunnel, California Department of Water Resources, Sacramento, Calif. (in print).

13. Moye, D. G., "Rock Mechanics in the Investigation and Construction of T.1 Underground Power Station, Snowy Mountains, Australia," Geol. Soc. Am., Eng. Geol. Case Histories 3, 1958, pp. 13-44.

14. Panek, L. A., "Measurement of Rock Pressure with a Hydraulic Cell," Trans. AIME, Vol. 220, 1961, pp. 287-290.

15. Merrill, R. H., "Design of Underground Openings, Oil Shale Mine, Rifle, Colo.," U.S. Bur. Mines Rept. Invest. 5089, 1954, 56 pp.

16. Utter, S., "Stress Determination around an Underground Mine Opening," Intern. Symp. Min. Res., Vol. 2, Pergamon Press, New York, 1962, pp. 569-582.

17. Roberts, A., C. L. Emery, P. K. Chakravarty, and F. T. Williams, "Photoelastic Coating Techniques Applied to Research in Rock Mechanics," Trans. Inst. Mining Met., Vol. 71, Pt. 10, 1961-1962, pp. 581-617.

18. Merrill, R. H., and T. A. Morgan, "Method of Determining the Strength of a Mine Roof," U.S. Bur. Mines Rept. Invest. 5406, 1958, 22 pp.

19. Obert, L., W. I. Duvall, and R. H. Merrill, "Design of Underground Openings in Competent Rock," U.S. Bur. Mines Bull. 587, 1960, 36 pp.

20. Jaeger, J.C., "Technical Discussion No. 9: Rock Mechanics," Third Australia-New Zealand Conf. on Soil Mech. and Foundation Eng., 1960.

DISCUSSION

J. A. TALOBRE (France):

May I add some remarks to the very good paper of Mr. Merrill? He was right in making no attempt to recognize all persons utilizing relief techniques. However, since I have this opportunity, I will recall that the methods of Lieurance were introduced in Europe by Professor Oberti many years before 1939. After the second World War, I tried to raise a new interest in the stress measurements and initiated them in France in 1949. With the collaboration of Professor Berthier, our company put in operation for the first time, this same year, 1949, a gauge using the overcoring technique in deep boreholes. (See Fig. 20.) It was used in a 6000-ft-deep tunnel. I improved this technique in the following years, and the uncertainties about the elasticity modulus were no more a problem when Mr. Tincelin, in the year 1951, proposed to use flatjacks. Much experience has been gained in overcoring techniques during these last fifteen years. It is now sufficient to affirm that the results of the Lieurance measurements near Boulder Dam were largely erroneous. He certainly used a value of modulus of elasticity many times too high, since the rock, at the walls of the adits in which he made his experiments, was obviously stress relieved, and since its behavior was no more elastical.

The stress concentration at the walls of the tunnel was not three, as assumed, but rather one, or even less. The stress field was not uniaxial, and the decreasing dimensions of the stress ellipses, when going from the gallery face to the access shaft, was only a consequence of the time elapsed since the completion of the excavation work.

Mr. Merrill made mention of the process of Professor Hast. It has some drawbacks that our devices, although older, did not present.

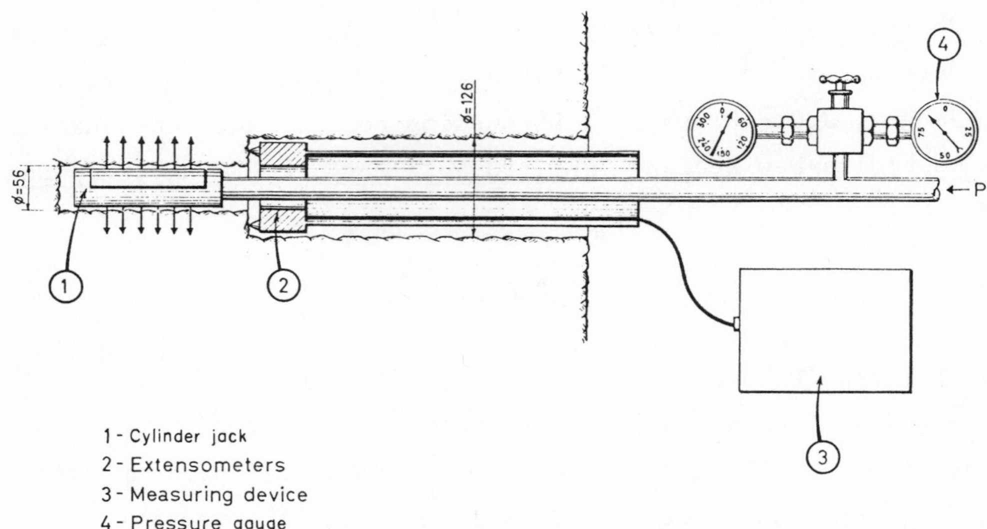

Fig. 20—Stress gauge using overcoring technique ("Les Conduites forcées souterraines dans le monde," Monde Souterrain, Feb.-Apr., 1954, pp. 664-668).

R. H. MERRILL in reply:

Dr. Talobre states that in regard to where we have been with relief techniques, there have been a number of people in countries other than the United States that should be recognized, and I agree with him that they should.

He makes a note of the fact that the methods first used by Lieurance in the United States were also used by Professor Oberti before the date of 1939. We have been able to establish Lieurance's efforts through the obscure literature on this subject. Incidentally, the literature can be expected to be obscure for an infant science, but we should make every effort to keep it from becoming obscure in the future. However, the earliest record we have of a stress-relief technique being used was by Lieurance in 1932, and this record is noted by a technical memorandum in the U.S. Bureau of Reclamation's Library in Denver, Colorado.

Is there any other point that you would like to make, Dr. Talobre?

J. A. TALOBRE (France):

Relief techniques are not the only existing methods for the measurement of stresses in rocks. I intend to use very soon a new method that allows the direct measurement of stresses. It will be faster and simpler to use, and will eliminate the large risk of error resulting in the translation of deflection readings to stresses.

R. H. MERRILL in reply:

I wish to thank Dr. Talobre for his compliments concerning the engineering approach used in the reported investigation and the confirming laboratory study made to check the accuracy of the borehole deformations and stress.* These investigations, that is, the work we did to use concurrently the flatjack and the borehole method, and the fundamental laboratory investigation to prove the borehole theory versus response in rock, demonstrate general approaches I attempt to follow in my Bureau of Mines research. First, one should use the design theory that he is capable of treating, that provides a good engineering estimate, and is economic. For example, the State of California, Department of Water Resources, has an engineering problem to solve. To obtain their solution they need to know, among other things, the stress in the rock and the stress concentrations created by the openings for the power generating equipment, transformers, penstocks, etc. We recognize that a theory may be developed in the future that will uniquely predict the response of the stress field in the rock to the opening. Meanwhile, however, this opening must be designed, and the design engineers must accept a theory that we have today. The question remains, what theory other than the theory of elasticity can these engineers use to design their opening? I am sure that most design engineers value this approach; but the approach does require one to make allowances for unknowns as yet not included within the truths of our science. Therefore, his results must, by necessity, be tempered by judgment. While we all recognize that the design will not be a unique solution, we also recognize that the design should represent a better estimate than one could achieve without this approach and, most certainly, is better than a guess.

Second, one should attempt to make his theory fit the rock rather

*Merrill, R. H., and Peterson, J. R., "Deformation of a Borehole in Rock," U.S. Bur. Mines Rept. Invest. 5406, 1958, 22 pp.

than make the rock fit his theory. In the tests of the deformation of a borehole versus applied stress, reasonable evidence was assembled to establish that the rock was not ideally elastic. However, the differences between the applied stress and the stress computed from the borehole deformations were, for engineering purposes, in good agreement. Because the theory for the deformation of a borehole in a thin, infinite, elastic plate was in good agreement with the results from tests in rock, this theory was accepted as the mathematical model to relate deformation versus stress for the borehole method of stress relief. In our Bureau of Mines programs we have assembled a number of stress determinations from various locations, in different rock types at different depths, etc. To date, we have found that the stresses are never inconsistent with observation; that is, we have never observed (1) a stress (in a location where the rock was visibly capable of withstanding the stress) that was greater than the rock could withstand, or (2) a vertical stress that was not closely equal to the stress presumed owing to the gravity load on the rock above the point of measurement. Consequently, we are satisfied that this particular theory fits the rock; that is, it will provide a reasonable engineering estimate provided the rock conditions are not greatly different from the assumed conditions.

K. BARRON (Canada):

Mr. R. H. Merrill's discussion of _in situ_ stress measurement by relief techniques compares admirably the relative advantages and disadvantages of using borehole deformation meters and flatjacks. Instruments based on the rigid inclusion principle* may also be used

---

*The reader is referred to the following references: A. N. May, "Instruments To Measure the Stresses in Rocks Surrounding Underground Openings," _Intern. Conf. Strata Control_, Paris, May, 1960. A. N. May, "The Application of Measurement Instrumentation to the Determination of Stresses Encountered in Rocks Surrounding Underground Openings," _I.S.A. Trans_., Vol. 1, No. 2, April, 1962. E. L. J. Potts and N. Tomlin, "Investigations into the Measurement of Rock Pressure in the Mine and the Laboratory," _Intern. Conf. Strata Control_, Paris, May, 1960. A. H. Wilson, "A Laboratory Investigation of a High Modulus Borehole Plug for Measurement of Rock Stress," _Proc. 4th Symp. Rock Mech_., Pennsylvania State University, 1961.

for these measurements, and a comparison of their virtues, or otherwise, with the above instruments is of interest. The rigid inclusion meter is compared with the borehole deformation meter since its relative merits and demerits in comparison to the flatjacks are similar to those of the deformation meter.

The rigid inclusion meter has two definite advantages over deformation meters:

1. In the case where the elastic modulus of the meter is greater than twice that of the host rock, the determined stresses are practically independent of the modulus of the rock. In such cases the possible errors, due to rock inhomogeneity and error in modulus determination, are drastically reduced compared to the deformation meter. Where the ratio of moduli is less than two then an accurate rock modulus must be determined and, in this case, the rigid inclusion offers no advantage.
2. By virtue of the calibration method used, that of calibrating instrument response directly against applied stress, the instrument can be said to determine accurately the stress change directly.

On the other hand there are major disadvantages of rigid inclusion meters compared to deformation meters:

1. No less than three instruments are required for stress determination at a point in a plane, and further these occupy a finite length of borehole.
2. The rigid inclusion meters must be fixed in one position of the borehole, whereas, as Mr. Merrill points out, the deformation meter can be used to take measurements along the whole length of the borehole.
3. Rigid inclusion meters require prestressing in the hole. Thus, to prevent boundary displacement due to prestress, a larger diameter core must be relieved than is the case with the deformation meter. A ratio of instrument diameter to core diameter of 6:1 is adequate to prevent boundary displacement (for example, the Mines Branch Stressmeter has

a diameter of 1 1/2 in. and a 12-in.-diameter core is extracted). In consequence of the above, one pass overcoring becomes difficult and ring drilling is probably better. Thus, the drilling costs and the time taken are both greater than for the borehole deformation meter (for example, a 12-ft-deep, 12-in.-diameter core in iron ore can be completed in 3 days drilling on 2 shifts per day).

Digressing slightly, to consider the measurement of stress changes with time, as mining proceeds the major advantages of the rigid inclusion meters become apparent. Not only can they be made extremely stable with time but, since they are firmly fixed in position, they are not as subject to the effects of blasting in their vicinity as are the deformation meters. In fact, in certain conditions they appear to be the only practical measurement method (for example, Leeman* reported the use of a deformation meter for measuring stress changes as a stope approached the instrument). As the stress level increased, the hole started to spall on the inside preventing any further measurements. Such would not have been the case had a rigid inclusion meter been used.

It is evident from both Mr. Merrill's paper and from the above comments that no one method of stress determination is ideal for all circumstances. For each specific project, geological conditions, cost, etc., must be considered and the best available method selected for the measurements. It would, therefore, seem that the well-equipped rock mechanics laboratory should have all the instruments and techniques at its disposal and should not just rely on one method.

R. H. MERRILL in reply:

I have only one statement to make in regard to your presentation, Mr. Barron. Professor Potts in his discussion paper makes a definition that I consider very adequate. I shall quote it for him

---

*E. R. Leeman, "The Measurement of the Stress in the Ground Surrounding Mining Excavations," Assoc. Mine Managers South Africa, Papers and Discussions, 1958/1959, pp. 331-356.

if I may. "The hydraulic stress meter, which is a borehole-deformation meter of the type which measures the resistance to deformation of a borehole"--I shall end the quote there. In other words you must deform the "stress meter" to get a representative change that you translate in one way or another. I think that we should all recognize that the stress meter is a deformation meter calibrated in terms of stress.

J. FIDLER (Australia):

The author has prepared an interesting survey of experience in the measurement of stress in solid rock by relief techniques.

The Hydro-Electric Commission of Tasmania has recently completed excavation of a large underground power station (Poatina), for which a large number of flatjack measurements were taken. It was found that the flatjack readings were influenced by changes in the condition of rock with changes in the temperature and humidity of the circulating air.* The rock is a mudstone and contains about 1.5 per cent moisture that can be evaporated by oven drying. Air drying of the rock is accompanied by shrinkage, the shrinkage from the *in situ* condition, in the direction parallel to the bedding, being 0.045 per cent after 12 days at 75°F and 45 per cent relative humidity. Under conditions of constant strain, shrinkage is equivalent to a change in stress. At Poatina a lineal shrinkage of the above amount was estimated to be equivalent to a reduction in compressive stress of about 2000 psi.

It would appear that in certain rocks even a small decrease in relative humidity could lead to significant errors in measurement.

The author's comments on this would be of interest.

Photoelastic studies of the cross sections of a circular drive, where stress measurements were taken, showed that the effect of irregularities of the rock surface, in this case, was relatively small at a depth equal to the half depth of the flatjack. Thus the normal

---

*L. A. Endersbee and E. O. Hofto, "Civil Engineering Design and Studies in Rock Mechanics for Poatina Underground Power Station, Tasmania," *J. Inst. Engrs. Australia*, Vol. 35, No. 9, September, 1963, pp. 187-209.

elastic stress distribution for a smooth opening could be used for the analysis of the measurements.

The author has commented on the increasing number of observations of high horizontal ground stresses. This was also observed at Poatina underground power station, the two horizontal stresses at a depth of 500 ft being about 2400 and 1800 psi.

R. H. MERRILL in reply:

The only comment requested was in regard to the flatjack technique, and in the investigation that I reported engineers from the State of California were responsible for the use of the flatjack method. Therefore, any questions regarding the jacks should be answered by one of these investigators.

However, to conserve time, I shall make the following comment and ask these engineers to nod approval or disapproval. The walls around the drift shown in plan in Fig. 5 of my paper are relatively wet. However, the amount of moisture in and around the drift does not change significantly. Also, the temperature measurements made in the drift do not indicate any large differences in temperatures. Consequently, if there were errors owing to temperature or humidity, these errors did not appear within the resolution of the measurements.

I have noted a nod of approval from Mr. Kruse of the State of California, so apparently my response to your question is correct.

H. W. BURKE (USA):

There is one point that came to mind with this mention of the effect of a change in water content. We had an interesting rock in a dam site in Formosa. It was a dirty graywacke. It had a modulus of elasticity of about 1,000,000 psi, a compressive strength of something like 7000 psi. The most interesting part about it, however, was that on a change of water content from 0 to about 15 per cent, the rock expanded about 2 per cent in volume, unrestrained. If restrained it exerted an expansive stress of up to 750 psi. Such a rock, if any of the stress-relief methods were used, might give rather strange results.

C. L. EMERY (Canada) in reply:

Thank you, sir. Your point is well taken. It is probable that the water droplets in a normal rock must be considered as grains of the rock. If you remove them you get a different kind of condition.

You can load them. You can, if you increase the quantity, push the other grains apart and hydraulically fracture the rock.

G. EVERLING (Germany)*:

The results of my studies on problems of primary stresses in the tectonically deformed Carboniferous sedimentary formations of central Europe, particularly on the order of magnitude of their horizontal components, which were completed in 1960, are in exceptional agreement with Albert Heim's propositions.

In rock mechanics the horizontal pressure p_h is frequently determined from the vertical surcharge pressure $p_v = \gamma \cdot H$ and the Poisson number m** by means of the equation $p_h = p_v/(m - 1)$. This relationship is not applicable for untouched but tectonically strained rock masses. It would only be valid if the rock mass in its present condition would have been formed in an unstressed state and would then have been subjected to compression due to its own weight only. Even if the rock mass were perfectly elastic at the present time, the state of stress at the time of sedimentation could not have at once been converted to the state represented by the above equation. Also to be considered are the effects of the temporary action of high tectonic horizontal pressures, which have been verified beyond doubt in the central European Carboniferous formations and at other locations. During the diminishing of the tectonic pressures, which were superimposed at a later time, the horizontal compression cannot have decreased to less than $p_h = p_v$, regardless of its original magnitude. This "hydrostatic" state of stress, with equal stressing in all directions, is the only ultimate state without shear stresses, which, therefore, does not give rise to further stress adjustments.

---

*In absentia. Contribution translated from the German by K. W. John.

**The "Poisson number" is the reciprocal of Poisson's ratio.

The commonly used assumption that the horizontal pressure is less than the vertical pressure does not, therefore, generally hold true. Of all assumptions as to the horizontal pressure, that of equal pressure in any direction is the most probable. This applies also to loose soils in which the hydrostatic state of stress lies, using the principles of soil mechanics, between the limiting states of active and passive earth pressures.*

Adapting of the equation indicated above to these results by assuming Poisson numbers for rock masses of m = 4 to 5 is not believed feasible. This conclusion is reached on the basis that in determining the stresses in the vicinity of underground openings, higher values of m should be used throughout, as shown, for example, in the summary of test data by Link.** Principally, the Poisson number is only defined for elastic bodies; it is rendered meaningless in masses subject to stress relaxation and plastic deformations.***

---

*The following authors are quoted for verification of the thoughts as presented here: R. D. Mindlin, "Stress Distribution around a Tunnel," Proc. Am. Soc. Civil Engrs., Vol. 65, 1939, pp. 619-642. S. G. Awerschin, "Erfahrungen aus der Gebirgsdruckforschung," Internationale Gebirgsdrucktagung Leipzig, 1958, Deutsche Akademie der Wissenschaften zu Berlin, Akademie-Verlag, Berlin, 1958, pp. 43-52. S. G. Awerschin, "Neue Methoden und Ergebnisse von Versuchen über die beim Abbau von Kohleflözen auftretenden Druckerscheinungen.," Vortrag Internationaler Kongress für Gebirgsdruckforschung, Paris, 1960, Société de l'Industrie Minérale, Saint-Etienne/Loire. M. A. Slobodov, "Erfahrungen bei der Anwendung der Entlastungsmethode bei der Erforschung der Spannungen in der Tiefe des Gebirgsmassivs," Ugol, No. 7, 1958, pp. 30-35. N. Hast, "The Measurement of Rock Pressure in Mines," Sveriges Geol. Undersokn., Arsbok, Ser. C, No. 560, 1958. J. A. Talobre, "Dix ans de mesures de compression interne des roches: progrès et résultats pratiques," Geol. Bauw. (Vienna), Vol. 25, 1960, pp. 148-165. C. L. Emery, "Testing Rock in Compression," Mining Quarry Eng., April, 1960. O. Jacobi, "Der Spannungszustand im unverritzten Gebirge," Glückauf, Vol. 96, 1960.

**H. Link, "Über die Querdehnungszahl des Gebirges," Geol. Bauw., Vol. 26, No. 4, July, 1961, pp. 246-257; "Zur Querdehnungszahl von Gestein und Gebirge," Geol. Bauw., Vol. 27, No. 2, November, 1961, pp. 89-100.

***For detailed presentation, reference is made to Glückauf, Vol. 96, 1960, pp. 1199-1202.

ABSTRACT

A method of determining the stress in rock by means of measurements in diamond drill holes is described. It uses flatjacks bent into the form of a quadrant of a circle so that they can be inserted into the ring cut by a diamond drill bit. Jacks in a central hole are used as the measuring element. A hole of larger diameter is drilled around the central hole and the fall in pressure in the central jacks is observed. Finally, jacks are inserted in the outer ring and pumped until the pressure in the central jacks is restored to its original value.

The principal directions of stress can be determined by pumping jacks in a drill hole to pressures sufficient to cause failure of the rock; such failures will occur in the direction of the maximum principal stress and can be seen in the core extracted when the hole is drilled over in a larger diameter.

A complete theory of the systems used is given, assuming elastic behavior. Some preliminary measurements made in short holes under mining conditions at a depth of 8000 ft are described.

RÉSUMÉ

Une méthode de détermination de la contrainte dans la roche est décrite, qui se fonde sur des mesures dans des trous forés par une perceuse à diamant. Cette méthode utilise des vérins plats courbés en quart de cercle de façon à pouvoir être insérés dans l'anneau découpé par une perceuse au diamant. Les vérins dans un trou central servent d'élément de mesure. On perce un trou de plus grand diamètre et concentrique au premier, et l'on observe la baisse de pression dans les vérins du centre. Enfin on introduit des vérins dans l'anneau extérieur et on en élève la pression jusqu'à rétablir celle des vérins centraux.

On peut déterminer les directions des contraintes principales en élevant la pression de vérins dans un trou de perçage jusqu'à produire la fracture de la roche. Ces fractures ont lieu dans la direction de la contrainte maximum et elles peuvent être observées dans les roches extraites lorsqu'on agrandit le diamètre du trou.

On donne une théorie complète des systèmes employés, en supposant un comportement élastique. Quelques mesures préliminaires sont décrites, faites dans de courts trous, dans des conditions minières, à une profondeur de 2500 mètres.

AUSZUG

Ein Verfahren zum Messen von Spannungen in Diamantbohrungen mit Hilfe von gekrümmten Druckkissen wird hier beschrieben. Druckkissen werden in die Form von Kreisquadranten gebogen, sodass sie in den Ringschnitt einer Diamantkrone eingeführt werden können. Druckzellen in einer zentralen Bohrung werden als Messelemente benützt. Eine Bohrung von grösserem Durchmesser wird um die Zentralbohrung niedergebracht und der Druckabfall in den zentralen Druckzellen gemessen. Dann werden die gekrümmten Druckkissen in den äusseren Ring eingebaut und abgepresst, bis der Druck in den zentralen Druckzellen wieder erreicht ist.

Man kann die Hauptspannungsrichtungen bestimmen, indem man die Druckzellen in einem Bohrloch bis zu einem Druck füllt, der den Bruch des Gebirges verursacht. Solche Brüche werden in der Richtung der grössten Hauptspannung auftreten. Sie können an den Bohrkernen beobachtet werden, nachdem das Loch mit einem grösseren Durchmesser überbohrt wurde.

Eine vollständige Theorie der benutzten Systeme unter Annahme elastischen Verhaltens wird gegeben. Einige vorläufige Messungen in kurzen Bohrlöchern, unter Bedingungen eines Bergbaues in einer Tiefe von 2500 Metern, werden beschrieben.

THEORY AND APPLICATION OF CURVED JACKS FOR MEASUREMENT OF STRESSES

J. C. Jaeger* and N. G. W. Cook**

Introduction

THE WELL-KNOWN FLAT JACK METHOD of measuring stress has two serious disadvantages: first, it can only be applied in the wall of an excavation and therefore in a region which may be irregularly stressed or partially destressed; second, because of the complicated geometry of the system it is virtually impossible to study its behavior theoretically. On the other hand, it has the great advantage that it is an absolute method, although an approximate one, in the sense that it measures a substitute system of stresses which gives the same displacements as the original set. It can be verified by laboratory experiments that this is so.[(1)] All other methods depend to a much greater degree on calibration and measurement of moduli in the laboratory.

At a depth of several thousand feet it is necessary to drill at least 10 ft from a drive to reach sound rock. This implies that the minimum requirement for any method of measuring stress is that it should be operable in a diamond drill hole, say, 20 ft deep and preferably not more than 4 in. in diameter.

The present work is essentially an attempt to adapt flat jack techniques for use in diamond drill holes by means of thin elongated flat jacks bent into an arc of a circle so that, for example, they can be made to fit snugly into the groove cut by a diamond drill bit.

---

*This paper was presented at the conference by J. C. Jaeger, Department of Geophysics, Australian National University, Canberra, Australia.

**Bernard Price Institute of Geophysical Research, University of the Witwatersrand, Johannesburg, South Africa.

Clearly, when such a jack is pumped up to pressure p, the boundary condition at the surface of the rock with which it is in contact is fairly accurately that of normal pressure p; and, since the surfaces involved are coaxial cylinders, a complete theory of the elastic behavior of the system can be written out.

The simplest situation is the application of normal pressure p over portions of the internal surface of a drill hole of radius a. Here AB and CD are a pair of thin curved jacks J,J, symmetrically placed and each subtending an angle 2β at the axis of the hole, as shown in Fig. 1(a). They are supported on a central core and pumped with oil to the same pressure p. The radial stress σ_r at the surface of the hole is then

$$\left.\begin{aligned} \sigma_r &= p, \ -\beta < \theta < \beta, \ \pi - \beta < \theta < \pi + \beta, \\ \sigma_r &= 0, \ \beta < \theta < \pi - \beta, \ \pi + \beta < \theta < 2\pi - \beta. \end{aligned}\right\} \quad (1)$$

Usually 2β is taken to be 90 degrees, so that each jack forms a quadrant of a circle, and in this case they will be described as "quadrantal" jacks. For the present, however, the general angle β will be retained. If β is not too large it is probably adequate to replace the system of Fig. 1(a) by the slightly simpler one of Fig. 1(b), in which a flat jack EF pushes two metal cheeks AFEB and CEFD against the surface of the hole; however, in this case the boundary condition is approximate rather than accurate.

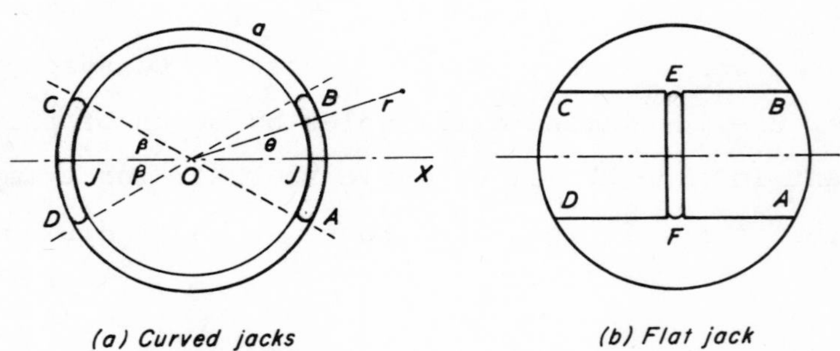

Fig. 1—Placement of jacks in bore holes.

Some useful preliminary measurements can be made by using the pressure elements of Fig. 1 in a drill hole; these will be described in the following three sections. However, the most elaborate procedure of the present type, which is the extension of the flat jack technique referred to above, consists of drilling a hole of radius a and inserting a pair of quadrantal jacks J,J which are used as a sensing element; an annulus of internal radius $b > a$ is then overdrilled to relieve the stresses; and finally two pairs of quadrantal jacks are inserted in this annulus and pumped to restore the stresses in J,J. A complete elastic theory of the system and its possible uses is given under "Displacements for Coaxial Systems of Quadrantal Jacks," "Pressure Changes in Systems of Jacks," and "Experimental Methods." Details of the apparatus used and results of some preliminary experiments are given in the final two sections of the paper.

Theory for a Pair of Curved Jacks

As before, suppose the hole is of radius a and that the jacks subtend 2β at its axis. The stresses and displacements are most readily found by using the complex variable (see Refs. 2 and 3). They follow immediately by substituting the Fourier series for Eq. (1) in Section 51 (14) of Ref. 3. Now, if we write

$$\rho = \frac{a}{r} , \tag{2}$$

the following is obtained:

$$\frac{\pi\sigma_r}{p} = 2\beta\rho^2 + 2 \sum_{n=1}^{\infty} \frac{1}{n} \rho^{2n}(n + 1 - n\rho^2) \cos 2n\theta \sin 2n\beta , \tag{3}$$

$$\frac{\pi\sigma_\theta}{p} = -2\beta\rho^2 + 2 \sum_{n=1}^{\infty} \frac{1}{n} \rho^{2n}(n\rho^2 - n + 1) \cos 2n\theta \sin 2n\beta , \tag{4}$$

$$\frac{\pi\tau_{r,\theta}}{p} = 2(1 - \rho^2) \sum_{n=1}^{\infty} \rho^{2n} \sin 2n\theta \sin 2n\beta \ . \tag{5}$$

When $\rho = 1$, the series can be summed, and it is found that the tangential stress in the surface is

$$\begin{aligned} [\sigma_\theta]_{r=a} &= p\left(1 - \frac{4\beta}{\pi}\right), \quad \text{in the arcs AB and CD,} \\ &= \frac{-4p\beta}{\pi}, \quad \text{in the arcs BC and AD.} \end{aligned} \tag{6}$$

The radial displacement u_r at the surface $r = a$ is

$$\frac{2G\pi u_r}{ap} = -2\beta - \sum_{n=1}^{\infty} \frac{1}{n}\left(\frac{\varkappa}{2n - 1} + \frac{1}{2n + 1}\right) \cos 2n\theta \sin 2n\beta \ , \tag{7}$$

where $\varkappa = 3 - 4\nu$ for plane strain and $\varkappa = (3 - \nu)/(1 + \nu)$ for plane stress, and G and ν are the modulus of rigidity and Poisson's ratio, respectively. In all cases stresses are positive when compressive, and displacements are positive when toward the origin.

For a concentrated load $P = 2\beta pa$ with β small, the displacement at $\theta = \pi/2$ is

$$[u_r]_{\theta=\pi/2} = (\varkappa - 1)\frac{P}{8G} \ . \tag{8}$$

If pressure is applied over two quadrants with $\beta = \pi/4$, Eq. (7) becomes

$$\frac{2G\pi u_r}{ap} = -\frac{1}{2}\pi - \sum_{m=0}^{\infty} \frac{(-1)^m}{(2m + 1)}\left(\frac{\varkappa}{4m + 1} + \frac{1}{4m + 3}\right) \cos 2(2m + 1)\theta \ , \tag{9}$$

and in this case the displacement when $\theta = \pi/2$ is

$$[u_r]_{\theta=\pi/2} = (0.151\kappa - 0.203)\,\frac{ap}{G}\,. \tag{10}$$

Modulus Measurements In Situ

The measurement of G from the radial displacement $u_r = -ap/2G$ of a circular hole with internal hydrostatic pressure p is well known,[4] but its instrumentation is not easy. In the situation of Figs. 1(a) and 1(b), measurement of displacement in the direction $\theta = \pi/2$ by means of a displacement transducer running through the core is a relatively easy matter. However, the expressions (8) and (10) for the displacement involve κ, and hence both ν and assumptions about the axial stress. If the displacement is measured in the direction $\theta = \pi/4$, then the term in κ vanishes.

Determination of Principal Directions by Fracture

We consider only the components of stress in a plane perpendicular to the drill hole and suppose that σ_1 and σ_2 are the principal stresses at a considerable distance from the hole, σ_1 being at an angle φ to the direction OX in Fig. 1(a). The least tangential compressive stress in the surface of the hole will be $3\sigma_2 - \sigma_1$ in the direction $\theta = \varphi$. If the jacks in the hole are now pumped up, a tensile component $-4p\beta/\pi$ has to be added to the tangential stress in the arcs BC, AD. Thus if the direction of σ_1 lies in either of these arcs, tensile failure will occur in this direction if p is increased sufficiently. Such tensile failures can be observed in the core if the original hole is overdrilled to a larger diameter. The principal directions can readily be found in this way. This procedure is of course a generalization of the hydraulic fracturing method of Hubbert and Willis[5] to include directional effects.

It can be verified by stressing blocks in a testing machine that failure does occur in the direction of σ_1, independent of the angle the jacks make with it, provided it lies in the arc AB.

It might be hoped that the magnitudes of the principal stresses could be found in this way, but, judging by laboratory experiments, cracking may not occur until the calculated surface tensile stress is considerably greater than the uniaxial tensile strength. However, some inferences can frequently be drawn. For example in Rand quartzite at a depth of 8000 ft, cracking was observed (in overdrilled core) in a vertical plane after horizontally oriented jacks had been stressed to 3000 psi; but no cracking occurred after vertically oriented jacks had been stressed to 15,000 psi. This implies that the principal directions are horizontal and vertical. Also by Eq. (6) with $\beta = \pi/4$ for quadrantal jacks, if cracking takes place at a tensile stress of T_0,

$$-\sigma_1 + 3\sigma_2 - 3000 < -T_0 \ ,$$

$$3\sigma_1 - \sigma_2 - 15{,}000 > -T_0 \ .$$

The first of these implies that

$$\frac{\sigma_2}{\sigma_1} < \frac{1}{3} - \frac{T_0 - 3000}{3\sigma_1} \ ,$$

and since σ_1 may be expected to be of the order of the gravitational load, 10,000 psi, this suggests that $\sigma_2/\sigma_1 < 1/3$.

Even under laboratory conditions it is difficult to observe the occurrence of cracking without either the use of crack-indicating substances or the use of overdrilling. However, it is found in practice that whenever two jacks are used in a system the existence of a crack can usually be detected because of a change in their interaction.

Displacements for Coaxial Systems of Quadrantal Jacks

In the present work the essential measuring element is a component (usually a pair of quadrantal jacks in the region $r < a$) which at some stage is sealed off containing a constant mass of fluid. Subsequent changes, Δp of pressure and Δv of volume, in it are related by

$$\Delta v = -Cv\Delta p \ , \tag{11}$$

where v is its volume and C an over-all compressibility.

In order to calculate the change in volume, the integral of the radial displacement over the periphery of the jack is needed. A number of such quantities can be calculated according to the positions in which the pressure is applied and the displacement is measured.

First, for a pair of quadrantal jacks with unit pressure in a hole of radius a in an infinite medium, as in Fig. 1(a), we define

$$U_a = 4a \int_0^{\pi/4} [u_r]_{r=a} \, d\theta \ , \tag{12}$$

where u_r is given by Eq. (9). It follows that

$$U_a = -(0.33\varkappa + 0.90) \frac{a^2}{G} \ . \tag{13}$$

The integrated displacement over the opposite pair of quadrants may also be needed. This will be denoted by V_a, where

$$V_a = 4a \int_{\pi/4}^{\pi/2} [u_r]_{r=a} \, d\theta = \frac{-\pi a^2}{2G} - U_a \ . \tag{14}$$

Similar coefficients can be defined for the region $0 \leq r < a$ if the behavior of the core supporting the jacks is in question.

The most complicated system which will be considered here is that of Fig. 2(a) in which there is a pair of quadrantal jacks J,J inside the cylinder r = a, a second pair K,K in the same orientation outside the cylinder r = b, and a third pair L,L in the perpendicular orientation outside the cylinder r = b. For the case of the hollow cylinder $a < r < b$, eight coefficients will appear and a double suffix notation will be used, the first suffix being the radius at which pressure is

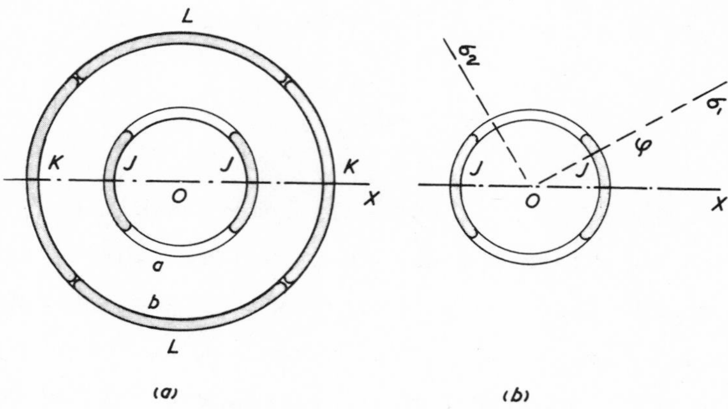

Fig. 2—Application of three pairs of quadrantal jacks.

applied and the second that at which the displacement is taken. Thus for unit pressure in jacks such as J,J at r = a, $U_{a,a}$ and $U_{a,b}$ are the integrated displacements over the jack quadrants for r = a and r = b, respectively; and $V_{a,a}$ and $V_{a,b}$ are those for the opposite pair of quadrants. Similarly $U_{b,a}$, $U_{b,b}$, $V_{b,a}$, and $V_{b,b}$ refer to unit pressure in a jack at r = b. The theory for the hollow cylinder is developed by Muskhelishvili,[2] and series for all eight coefficients can be written down.

In our experimental situation, a/b = 0.375, and with this value the coefficients needed here are

$$\frac{G}{(\varkappa + 1)a^2} U_{b,a} = 1.575 \ ,$$

$$\frac{G}{(\varkappa + 1)a^2} V_{b,a} = -0.66 \ , \qquad (15)$$

$$\frac{G}{(\varkappa + 1)a^2} (U_{a,a} - U_a) = -0.35 \ .$$

Finally, formulas for the displacement of the cylinder r = a due to principal stresses σ_1, σ_2 at infinity are needed for the case in which σ_1 is inclined at φ to OX, as shown in Fig. 2(b). The radial displacement u at r = a is given by

$$\frac{8Gu}{(\varkappa + 1)a} = (\sigma_1 + \sigma_2) + 2(\sigma_1 - \sigma_2) \cos 2(\theta - \varphi) .$$

The integrated displacement U over the quadrants $\pi/4 > \theta > - \pi/4$ and $5\pi/4 > \theta > 3\pi/4$ is

$$U = 2a \int_{-\pi/4}^{\pi/4} u \, d\theta = \frac{a^2(\varkappa + 1)}{8G} [\pi(\sigma_1 + \sigma_2) + 4(\sigma_1 - \sigma_2) \cos 2\varphi] . \quad (16)$$

Pressure Changes in Systems of Jacks

The changes in pressure in a number of important situations can now be written down in terms of the coefficients U and V.

The Relative Effect of Perpendicular Jacks

Suppose that the jacks J,J in Fig. 2(a) are pumped to pressure P so that their volume becomes $v - PU_{a,a}$ and that they are then sealed off. If pressure p_1 is now applied at the jacks K,K and p_2 at the jacks L,L, the pressure in J,J will change to some unknown value P^* and the volume in them to $v - P^* U_{a,a} - p_1 U_{b,a} - p_2 V_{b,a}$. Therefore by Eq. (11),

$$(Cv - U_{a,a})(P^* - P) = p_1 U_{b,a} + p_2 V_{b,a} . \quad (17)$$

This is a linear relation between the values of p_1 and p_2 necessary to produce a specified rise of pressure in J,J.

It follows that the changes in pressure in J,J caused by unit changes in K,K and L,L are in the ratio $U_{b,a}/V_{b,a}$, and by Eq. (15) this ratio is independent of the material in the region $a < r < b$. Equation (17) implies that the quantities $U_{b,a}/(Cv_1 - U_{a,a})$ and $V_{b,a}/(Cv_1 - U_{a,a})$ are measurable experimentally.

Complete Destressing for the Region r > a

Suppose that the jacks J,J in Fig. 2(b) are pumped to pressure P and the values of the principal stresses σ_1, σ_2 at infinity are applied. The volume in the jack will be $v - PU_a - U$, where U is given by Eq. (16). If the jacks J,J are now sealed off and the stresses at infinity are reduced to zero, the pressure in the jacks will fall to some value P^* and the volume in them will be $v - P^* U_a$. Hence by Eq. (11),

$$(Cv - U_a)(P - P^*) = \left[\frac{a^2(\varkappa + 1)}{8G}\right] \cdot [\pi(\sigma_1 + \sigma_2) + 4(\sigma_1 - \sigma_2) \cos 2\varphi] \ . \tag{18}$$

The changes in pressure in the principal directions are in the ratio

$$\frac{[\pi(\sigma_1 + \sigma_2) + 4(\sigma_1 - \sigma_2)]}{[\pi(\sigma_1 + \sigma_2) - 4(\sigma_1 - \sigma_2)]} \ . \tag{19}$$

Destressing by Overdrilling at Radius b

The only change from the previous case is that Eq. (18) is replaced by

$$(Cv - U_{a,a})(P - P^*) = P(U_a - U_{a,a}) + \left[\frac{a^2(\varkappa + 1)}{8G}\right] \cdot [\pi(\sigma_1 + \sigma_2) + 4(\sigma_1 - \sigma_2) \cos 2\varphi] \ . \tag{20}$$

Restoration of Pressure After Destressing

Suppose that after destressing as in the example immediately preceding, jacks K,K and L,L are pumped to pressures p_1, p_2, respectively, until the pressure in J,J is restored to its original value P. Combining Eqs. (17) and (20) gives

$$p_1 U_{b,a} + p_2 V_{b,a} = P(U_a - U_{a,a}) + \left[\frac{a^2(\varkappa + 1)}{8G}\right] \cdot [\pi(\sigma_1 + \sigma_2) + 4(\sigma_1 - \sigma_2) \cos 2\varphi] \ . \tag{21}$$

Symmetrical Systems of Jacks

A simple and important case occurs if there are jacks in all quadrants at r = a connected together (or, what is the same thing, a cylindrical pressure cell). Suppose this system is pumped to pressure P and the stress is then relieved by overdrilling at radius b and that, finally, jacks K,K and L,L at r = b are pumped to that pressure p_1 which restores the internal pressure to its initial value P; then

$$p_1 = \frac{a^2}{b^2} P + \frac{1}{2} (\sigma_1 + \sigma_2)(1 - \frac{a^2}{b^2}) . \qquad (22)$$

Experimental Methods

Possible experimental methods based on the solutions of the previous section will now be discussed. Use of the coefficients calculated on elastic theory shows the order of magnitudes of the errors in this case.

Measurement of $\sigma_1 + \sigma_2$ by Stress Restoration

The relevant equation is Eq. (22). It appears that p_1 is only equal to the mean stress $\frac{1}{2} (\sigma_1 + \sigma_2)$ if b is very large. For the ratio a/b = 0.375, which we have used, the value of p_1 is 14 per cent too low if the initial pressure P is small, and the error decreases as P is increased. Even if elastic behavior is not assumed, it may be expected that p_1 will be rather less than $\frac{1}{2} (\sigma_1 + \sigma_2)$, and it seems reasonable to correct it by using Eq. (22).

Determination of Principal Directions by Destressing

Using the values of Eqs. (15) as the coefficients, Eq. (20) becomes

$$\frac{2G(P - P^*)(Cv - U_{a,a})}{a^2(\varkappa + 1)} = 0.7P + 0.79(\sigma_1 + \sigma_2) + (\sigma_1 - \sigma_2) \cos 2\varphi . \qquad (23)$$

In principle, measurement of $\Delta P = P - P^*$ with the jacks oriented in three different directions gives φ. This would be a cumbersome process, and is greatly inferior to the simple method given in the section entitled "Determination of Principal Directions by Fracture."

Determination of Principal Stresses by Destressing

Assuming that the principal directions have been found, separate experiments may be made with jacks J,J oriented in each of these directions and the corresponding changes of pressure $\Delta P_1 = P_1 - P_1^*$ and $\Delta P_2 = P_2 - P_2^*$ in the two cases found. Then by Eq. (23)

$$\frac{\Delta P_1}{\Delta P_2} = \frac{0.7P_1 + 0.79(\sigma_1 + \sigma_2) + (\sigma_1 - \sigma_2)}{0.7P_2 + 0.79(\sigma_1 + \sigma_2) - (\sigma_1 - \sigma_2)} \,. \tag{24}$$

If $\sigma_1 + \sigma_2$ has been found by the method given under "The Relative Effect of Perpendicular Jacks," then Eq. (24) gives $\sigma_1 - \sigma_2$. For example, with $\sigma_1 + \sigma_2 = 17{,}000$, ΔP_1 was 1200 with $P_1 = 2000$ in one direction, and ΔP_2 was 300 with $P_2 = 1000$ in the other. It follows from Eq. (24) that $\sigma_1 - \sigma_2 = 8300$. It appears from the elastic theory that the effect of the terms in P_1 and P_2 in Eq. (24) is not important, so Eq. (24) can probably be used in all cases.

The Relative Effect of Pressures in Perpendicular Jacks

By Eq. (17), this effect is $U_{b,a}/V_{b,a}$, which in elastic theory is -2.38 for a/b = 0.375. This is independent of $\varkappa$ and so should be the same for all materials. Laboratory experiments on a mild steel cylinder give -2.6, and for a hole in a marble block, -2.3. These small discrepancies are probably caused by minor deviations from the ideal case: for example, the jacks in practice exert pressure over an angle smaller than 90 degrees.

A quartzite cylinder, however, gave a value of -1.25. On careful examination it was found to be covered with a network of fine fractures, and so the low value measured was probably caused by the rock's not behaving elastically. Thus measurements of this type provide a simple method of deciding whether rock *in situ* is behaving elastically.

Stress Restoration

In elastic theory with the values in Eqs. (15) as the constants, Eq. (21) becomes

$$p_1 - 0.42p_2 = 0.22P + 0.25(\sigma_1 + \sigma_2) + 0.32(\sigma_1 - \sigma_2)\cos 2\varphi \,. \qquad (25)$$

With the jacks oriented in the perpendicular direction, the corresponding result would be

$$p_1' - 0.42p_2' = 0.22P + 0.25(\sigma_1 + \sigma_2) - 0.32(\sigma_1 - \sigma_2)\cos 2\varphi \,. \qquad (26)$$

Equations (25) and (26) provide a method for finding $\sigma_1 + \sigma_2$ and $\sigma_1 - \sigma_2$ that may be used as an alternative to those described under "Measurement of $\sigma_1 + \sigma_2$ by Stress Restoration" and "Determination of Principal Stresses by Destressing."

When this work was begun it was hoped, by analogy with Eq. (22) and the behavior of flat jacks in general, that if two pairs of jacks, K,K and L,L, were oriented so that $\varphi = 0$ and stressed so that $p_1/p_2 = \sigma_1/\sigma_2$, the value of p_1 for restoration might be a reasonable approximation to σ_1. In fact Eq. (25) gives for this case

$$p_1 = 0.38P + 0.86\sigma_1 \,, \qquad \text{for } \sigma_1 = \sigma_2 \,;$$

$$p_1 = 0.25P + 0.63\sigma_1 \,, \qquad \text{for } \sigma_1 = 3\sigma_2 \,.$$

Thus, regarded as absolute measurements, the values obtained are too low.

Apparatus and Experimental Procedure

We have constructed apparatus of this type and, by courtesy of East Rand Proprietary Mines Limited and Cementation (Africa) Limited, have made a number of experiments at a depth of 8000 ft in a drive several thousand feet from other mining excavations.

The jacks were made of 16-gauge copper tube, pressed flat and silver-soldered at the ends. Lengths of either 6 in. or 3 in. were used. Jack pressures up to 15,000 psi were used without trouble. For simplicity, pressure measurements were made directly on pressure gauges, connections and taps being of small diameter, high pressure components. The connecting tube from the jacks J,J to the pressure gauge was run through the drill rods to permit overdrilling. It might be expected that trouble would be experienced with this system due to changes in temperature, but no effects attributable to this cause were observed, except that when the drilling water was turned on, a small change in pressure occurred, but this rapidly settled down to a steady value. The jacks were attached to the end of a long rod for insertion, and no difficulty was experienced with this process in holes 10 to 20 ft deep. Trouble can be experienced in removing them, and this process is assisted by applying vacuum to the jacks.

Ordinarily, three separate measurements would be made: (1) determination of principal directions as in the section entitled "Determination of Principal Directions by Fracture"; (2) insertion of center jacks J,J in the σ_1 direction and pumping to P, overdrilling to observe stress relief, insertion of outer jacks and pumping to restore the pressure in J,J to P; (3) _either_ repetition of step (2) with J,J oriented in the σ_2 direction _or_ determination of $\sigma_1 + \sigma_2$ with a cylindrical center cell as in "Measurement of $\sigma_1 + \sigma_2$ by Stress Restoration." These operations would take an average time of about 2 hr each, so the times involved compare very favorably with the flat jack method.

One difficulty has been experienced which is probably peculiar to the great depth at which we were working, namely that because of the phenomenon of "disking" it was impossible to leave the rock annulus in the hole after overdrilling. This annulus usually broke away at some stage and was found to contain a regular series of transverse fractures at approximately 1-in. intervals. This phenomenon we have discussed elsewhere.[6] From the present point of view it is not necessary in elastic theory to study stress restoration _in situ_, since it can be done equally well in a laboratory experiment; but in view of the possibility of nonelastic behavior of the rock, it is most desirable to make the complete measurement underground.

REFERENCES

1. Jaeger, J. C., Discussion of a paper by Mr. L. G. Alexander, Third Australia-New Zealand Conference on Soil Mechanics and Foundation Engineering, 1960.
2. Muskhelishvili, N. I., <u>Some Basic Problems of the Mathematical Theory of Elasticity</u>, Noordhoff, 1953.
3. Jaeger, J. C., <u>Elasticity, Fracture and Flow</u>, 2d ed., Metheun, 1962.
4. Talobre, J., <u>La Mécanique des Roches</u>, Dunod, 1957.
5. Hubbert, M. King, and D. G. Willis, "Mechanics of Hydraulic Fracturing," <u>Trans. AIME</u>, Vol. 210, 1957, p. 153.
6. Jaeger, J. C., and N. G. W. Cook, "Pinching-off and Disking in Rocks," <u>J. Geophys. Res.</u> (to be published).

ABSTRACT

Complex mining developments make *in situ* stress measurements and subsequent analysis extremely difficult. The problems are intensified by the inability to be certain that the rock "in the solid" has the same characteristics as that surrounding the "opening" and the difficulty of obtaining access to zones of interest.

Mining conditions, however, permit and frequently necessitate the measurement of stress change as mining development proceeds. In this respect, the application of "relief techniques" for *in situ* stress evaluation are, in many cases, either impractical or impossible. The discussions therefore should consider other than "relief techniques."

The scale and size of the mining development allow the application of two fundamentally different techniques, one based on the deformation of a borehole and the other on the measurement of the axial displacement of fixed plugs in boreholes drilled "into the solid." Under these conditions, the approach is less sophisticated than would normally be demanded in science.

The conversion of the results into stress components in either approach requires the acceptance of the laboratory determinations of the physical characteristics of the rock materials.

The geometry of the mining design may assist in the preassessment of principal stress directions, which simplifies the analysis of the results. The technique based on deformation of the borehole may allow free measurement of the deformation or depend on the measurement of resistance to this deformation. The latter technique is the one adopted in our investigations with the "hydraulic stressmeter." The borehole extensometer has been developed to conduct investigations using the second approach based on rock mass movement.

RÉSUMÉ

Les exploitations minières complexes rendent extrêmement difficiles les mesures des contraintes sur les lieux et leur analyse subséquente. Le fait qu'on ne peut être sûr que la roche "dans la masse" a les mêmes caractéristiques que celles entourant "l'ouverture" et la difficulté d'accéder aux zones intéressantes accentuent encore ces problèmes.

Les conditions minières, cependant, permettent et souvent nécessitent la mesure du changement des contraintes au fur et à mesure que l'exploitation minière progresse. De ce point de vue, l'application de "techniques du relief" pour l'estimation des contraintes "sur les lieux" sont, dans bien des cas, ou peu pratiques ou impossibles. Les discussions devraient par conséquent considérer autre chose que des "techniques du relief."

L'échelle et les dimensions de l'exploitation minière permettent l'application de deux techniques fondamentalement différentes, l'une fondée sur la déformation d'un trou de sonde et l'autre sur la mesure du déplacement axial de goupilles fixées dans des trous de sonde forés "dans la masse." Dans ces conditions, ce problème est abordé d'une façon moins évoluée que ne l'exigerait normalement la science.

Pour convertir les résultats en composantes des contraintes dans l'une ou l'autre méthode, on devra accepter les déterminations en laboratoire des caractéristiques physiques des matériaux rocheux.

La géométrie du projet minier peut aider dans la pré-estimation des directions des contraintes principales, ce qui simplifie l'analyse des résultats. La technique fondée sur la déformation d'un trou de sonde peut permettre la libre mesure de la déformation ou bien elle peut se fonder sur la mesure de la résistance à cette déformation. Cette dernière technique est celle qui a été adoptée dans nos recherches avec "l'appareil hydraulique de mesure des contraintes." L'extensomètre du puits de sonde a été mis au point afin de conduire des recherches utilisant la seconde méthode fondée sur le mouvement de la masse rocheuse.

AUSZUG

Komplexe Bergbaubetriebe erschweren sowohl *in-situ* Spannungsmessungen als auch deren spätere Auswertung. Die Schwierigkeiten vermehren sich weiterhin, da man nicht immer sicher sein kann, ob das unverritzte Gebirge die gleichen Eigenschaften als das aufgeschlossene Gebirge besitzt. Weiterhin ist der zu beobachtende Bereich meist nur unter Schwierigkeiten zugänglich.

Die Verhältnisse im Bergbau erlauben aber, ja erfordern es häufig, die Änderung der Spannungsverhältnisse mit dem Fortschreiten des Abbaues zu beobachten. In dieser Hinsicht ist die Anwendung von Entlastungsverfahren zur *in-situ* Spannungsermittlung in vielen Fällen unpraktisch oder gar undurchführbar. Die Diskussionen hier sollten daher auch andere als die Entlastungsmethoden berühren.

Das Ausmass von Bergwerksanlagen erlaubt die Anwendung von zwei grundlegend verschiedenen Methoden: die eine beruht auf der Verformung eines Bohrloches, die andere auf der Messung der achsialen Lageänderung eines, in einem Bohrloch befestigten Stopfens. Unter diesen Verhältnissen ist die Annäherung an das technische Problem nicht so vollkommen als man gewöhnlich in der Wissenschaft verlangt.

Die Umwandlung der Messergebnisse in Spannungskomponenten erfordert bei beiden Methoden, dass Laboratoriumswerte für die physikalischen Eigenschaften des Gebirges angenommen werden.

Die Geometrie der Anordnung des Abbaues kann bei der Abschätzung der Hauptspannungsrichtungen helfen, was die Ermittlung der Endergebnisse vereinfacht. Bei dem, auf der Bohrlochverformung beruhenden, Verfahren kann man entweder die unbehinderte Formänderung oder den ihr entgegengesetzten Widerstand messen. Die letztere Methode wurde für unsere Untersuchungen mit der "hydraulischen Spannungsmessdose" angewendet. Das "Bohrloch-Extensometer" beruht auf dem Messen von achsialen Längenänderungen von Bohrungen; es wurde entwickelt, um Formänderungen in einer Gebirgsmasse zu messen.

THE IN SITU MEASUREMENT OF ROCK STRESS BASED ON DEFORMATION MEASUREMENTS

Edward L. J. Potts*

THE CONFERENCE has directed particular attention in this measurement discussion to approaches for determining stress *in situ* using relief techniques.

To the mining engineer, the evaluation of the stress distribution around productive mine openings is of paramount importance, since failure theories are all based on stresses, and the understanding of the phenomena of rock failure is a prerequisite to the effective design of underground workings.

The Rock Mechanics Laboratory in the Department of Mining Engineering at King's College, Newcastle upon Tyne, England, has been engaged in investigating phenomena around mine openings for several years.

The measurement of stress is only possible through its effects; therefore, the selection of the effect most likely to lead to successful *in situ* measurement is basic to the technique and vitally important. A technique that does not disturb the *in situ* stress condition is attractive. The measurement of sonic velocity change has been attempted, as has change in porosity of the rock material. It is considered, however, that the measurement of these parameters is not likely to be as successful as the classical method of stress measurement based on the deformation of the rock material itself. The *in situ* determinations in our studies are all based on the measurement of rock deformation.

The successful interpretation of deformation measurements in terms of stress components is dependent on the ability to establish a stress-

*Professor, Department of Mining Engineering, University of Newcastle upon Tyne, England.

strain relationship. This relationship in many rocks is very complex and its determination presents great difficulties. However, stress evaluations based on other parameters face similar difficulties, and, at least in the analysis of displacement, we can be assisted by the vast theoretical and experimental data collected in the field of stress and strain analysis.

To the mining investigator, the problems requiring stress measurement are made even more difficult by the complex influence of the mining development itself, the inability to be certain that the rock "in the solid" has the same characteristics as that surrounding the "opening," and the difficulty in getting access to the zone where measurements are to be made.

The technique of measurement requires the evaluation of strain components near the point considered within the mass of rock. These strain components may have either been induced by mining or may have existed prior to mining. In the latter case relaxation or relief techniques would be applied. The majority of investigations in mining engineering require intermittent or continuous evaluations to be made as mining proceeds, and the influence of the extraction of support and load transfer affects the over-all stability of the mining area.

The measurement "in the solid," *in situ*, can only be carried out if access to the point is established. This access is usually provided from one or more boreholes from the underground roadway or development and offers the choice of two different techniques. The borehole will induce a local change in the general stress or strain pattern that can be used with advantage and forms the basis of one technique. In the other method, the approach is based on the measurement of the relative displacement of two points in the borehole, the distance between these points being established using borehole plugs and selected so that the effects of the local disturbance can be neglected.

Both of these techniques suffer from certain general difficulties. The stress or strain at a point can be defined by three principal components and at least two angles correlating their directions with respect to a coordinate system. Thus the number of unknowns at any one point is five, the determination of which would demand the execution of five independent measurements at all points of interest. In mining engineering it should be emphasized that this requirement, at the present stage of our development in rock mechanics, is formidable and generally not adhered to. It is usual to make some reasonably acceptable assumptions, such as in the assessment of the direction of the principal components, so that the number of unknowns is reduced, at the expense of losing some accuracy in the determinations.

The method of approach that utilizes the measurement of borehole deformation as a means of stress evaluation can be successfully applied if the relationship between the principal stress components of the stress field surrounding the borehole, and the measured deformation, can be established.

Since this technique is dependent on the measurement of localized effects of stress change, it is possible to conduct a laboratory calibration of the instrument in the rock in which it is to be inserted. The relationship may be established theoretically, but the direct calibration is the most satisfactory method.

The deformation meter, however, by the very nature of its application to the measurement of localized disturbance--caused mainly by the presence of the hole itself, and not by the general strain field of the rock mass in which the hole is located--can only provide results of limited value if the simulation of the stress field in the laboratory calibration is not in accordance with *in situ* conditions.

The second and simpler technique, which involves the measurement of the change in length between points along the axis of a borehole, eliminates the influence of the hole itself, since the length between points is sufficiently greater than the hole diameter. The magnitudes of the displacements induced by mining are usually small in relation to the scale of the over-all mining operations, which permits the application of the classical principles of the analysis of small displacements.

The borehole measuring network, incorporating a series of points within the rock mass, if sufficiently comprehensive can provide a general displacement field, and from it the strain field can be deduced.

The stress-strain relationship of the rocks can, theoretically, define the stress field from the strain field. The measurement network, however, will only provide average strain values, which under certain conditions in mining may include bed separation, etc.

The laboratory stress-strain determinations can only be defined by small-scale specimens, probably with the addition of core examinations from the holes themselves. The results could be misleading, especially if the base length of the borehole measurement extends through several strata beds of different physical characteristics.

These objections raise considerable doubt as to the accuracy of the interpretation of the measurements in terms of stress, but the evaluation involves a great deal less effort and critical acceptance of "reasonable" assumptions than with the borehole-deformation approach. The technique will provide a good over-all appreciation of the general displacement and strain fields, to which the borehole-deformation technique is certainly less applicable.

It is, therefore, logical in comparing these methods of stress evaluation in the rock, at least in mining and in certain civil engineering investigations, that the design of investigation should include the two techniques to obtain an over-all strain distribution that can be reassessed in stress terms and that can be enhanced by carefully selected measurements using the borehole-deformation technique.

The determination of stress from diametral deformation of a borehole can be made by two methods: (1) measurement of deformation of the borehole, or (2) measurement of the load induced in an inclusion by the deformation.

At King's College we use an instrument called a "stressmeter," which is based on the second method. This method has two advantages over the first: (1) The instrument is less sensitive to disturbing influences, for example, shot-firing, rockbursts, etc., and (2) the interpretation is less dependent on the module of the rock.

The hydraulic stressmeter, which is a borehole-deformation meter of the type that measures the resistance to deformation of a borehole, is illustrated in Figs. 1 and 2. In Fig. 1 the stressmeter blade, C, coupled to a pressure-sensitive head, is provided with a slit that is filled with a soluble oil and water mixture, the slit being coupled

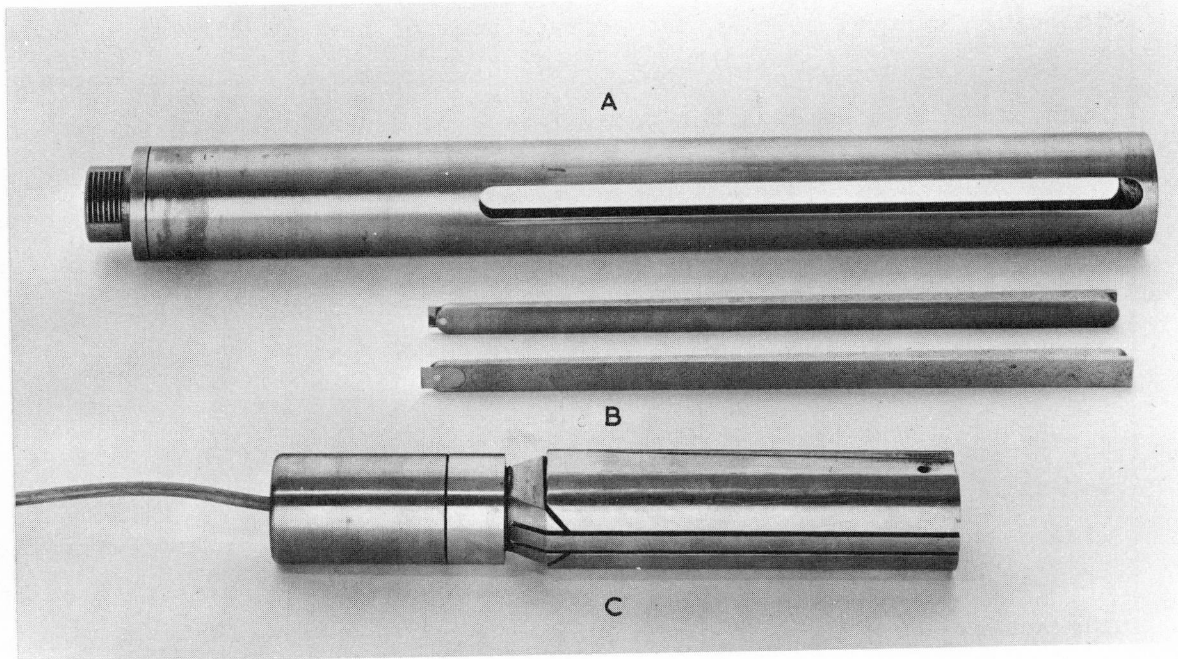

Fig. 1—Hydraulic stressmeter.

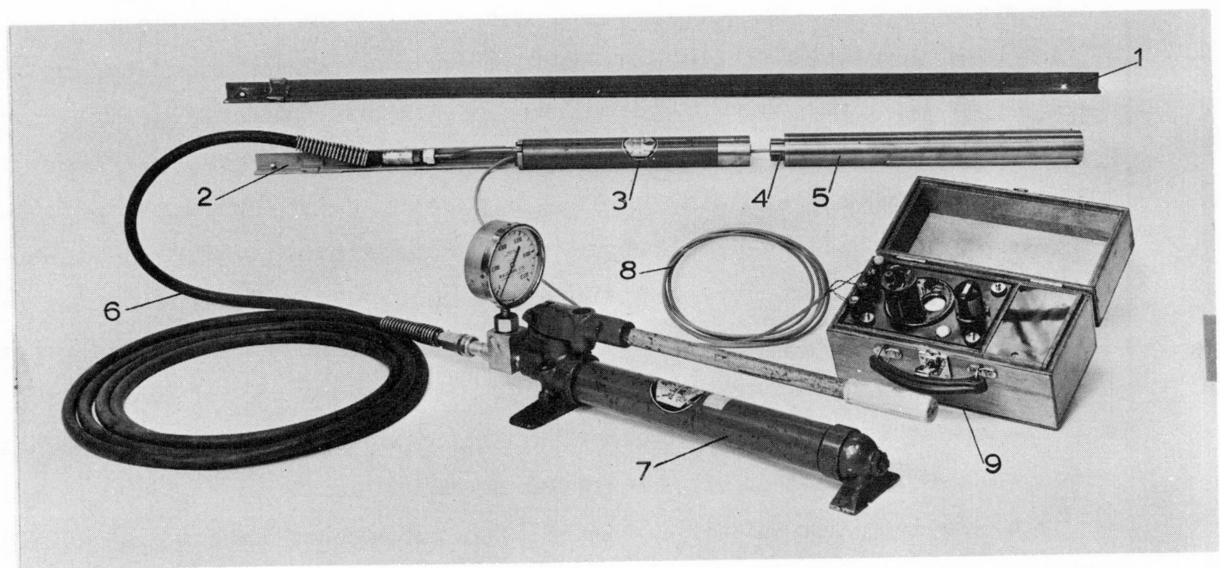

Fig. 2—Hydraulic stressmeter with its installation equipment.

through a central hole in the screwed boss of the blade to the diaphragm of the pressure-sensitive head. This head contains a complete bridge with two active arms on the outer face of the diaphragm and a compensating pair on a disc, under no load, within the head. The strain gauges are of the foil type, with a radial outer compression arm and a spiral central tension arm. The blade illustrated has been slotted on each side of the slit to increase the stressmeter sensitivity. Two tapered keyways are cut on each side of the blade in a plane at right angles to the slit. The blade is inserted in the sleeve, A, loose tapered keys, B, being inserted in the machined slots on the sleeve, A, so as to engage in the tapered keyways of the blade. The end of the sleeve remote from the insertion end is solid, the other end being fitted with a screwed boss on which a coarse quadrant thread has been machined. The stressmeter, 5 (as shown in Fig. 2), is coupled to a hollow hydraulic ram cylinder, 3, with a matching quadrant thread adaptor, 4, by inserting the stressmeter into the ram head and turning it through 90 deg until the opposing quadrant threads engage. The cable from the stressmeter is passed through the hollow ram, which is fitted with an angle iron attachment to take a simple insertion rod assembly as shown at 1 and 2 and which may also be used to orientate the meter. The hydraulic ram hose, 6, lies in the base of the angle rod as each length is added on insertion in a borehole. The hand-operated pump, 7, is used to advance the ram through the end of the sleeve, so prestressing the stressmeter through the tapered keys, which make contact with the periphery of the borehole. The stressmeter may be prestressed to a predetermined value and subsequent observations of stressmeter output made, using the indicating unit, 9. The stressmeter (Fig. 1) may also be coupled to a continuous recorder of the potentiometric type. The bridge arms have a resistance of 30μ and are operated on a constant current input of 65 mA. Experience has shown that this meter can be inserted satisfactorily in horizontal holes up to 60 ft long with this equipment, and the meters have been installed under widely different rock conditions.

In many rocks and in most circumstances it is reasonable to assume that in a small zone around the stressmeter the rock is isotropic and elastic and thus that the stressmeter loading is proportional to the deformation of the empty borehole. The proportionality factor is dependent on the modulus of the rock and of the stressmeter, and it is normally determined by calibration of the stressmeter in the laboratory.

A rectangular block of rock is prepared, a hole is drilled perpendicular to one face, and the loading surfaces are accurately dressed. Calibration is made by inserting the stressmeter into the loaded rock, with its wedges in a vertical plane, prestressing the stressmeter by means of the hydraulic jack, and varying the uniaxial vertical load applied to the block to correspond to the stress changes expected.

In anisotropic rocks and under some circumstances in viscoelastic rock, for example, rock salt and potash, the assumptions previously stated are not valid and the stressmeter can only be used as a qualitative instrument. Nevertheless, valuable information can be obtained from a stressmeter on the location or positions of maximum and minimum stress with respect to time and the location of the extraction zone in the mining development.

The output from the stressmeter in the same rock underground is given by

$$\gamma = \alpha(1 - \mu^2)\left\{[P + Q] + [2(P - Q) \cos 2\phi] - \frac{\mu}{1 - \mu^2} R\right\},$$

where γ is the electrical output of the stressmeter,

α is the constant proportionality factor for the stressmeter in the rock,

P,Q are principal stresses in a plane perpendicular to the axis of the borehole,

R is a principal stress acting along the axis of a borehole,

ϕ is the angle between the P stress direction and the wedge contact of the stressmeter with the hole periphery, and

μ is Poisson's ratio of the rock.

If R is small compared with (P + Q) and μ is relatively small, say, below 0.3, then the R term can be neglected.

The evaluation assumes that prior to installation, a preassessment has been possible from which the planes of the principal stresses are approximately defined. The insertion of the stressmeters is then carried out in holes perpendicular to these planes.

The borehole extensometers shown in Figs. 3 and 4 have been developed along with other auxiliary equipment to measure displacements in boreholes. The two Mark I units in Fig. 3 are a normal and reversed type, those illustrated being used with stainless steel or other metal ungraduated tape 3/8 in. x 0.02 in., which is located in the borehole using a modified expanding, shell type, plug adapted from a roof bolt. The extensometers A and B comprise a capstan headed screw, 5, which is

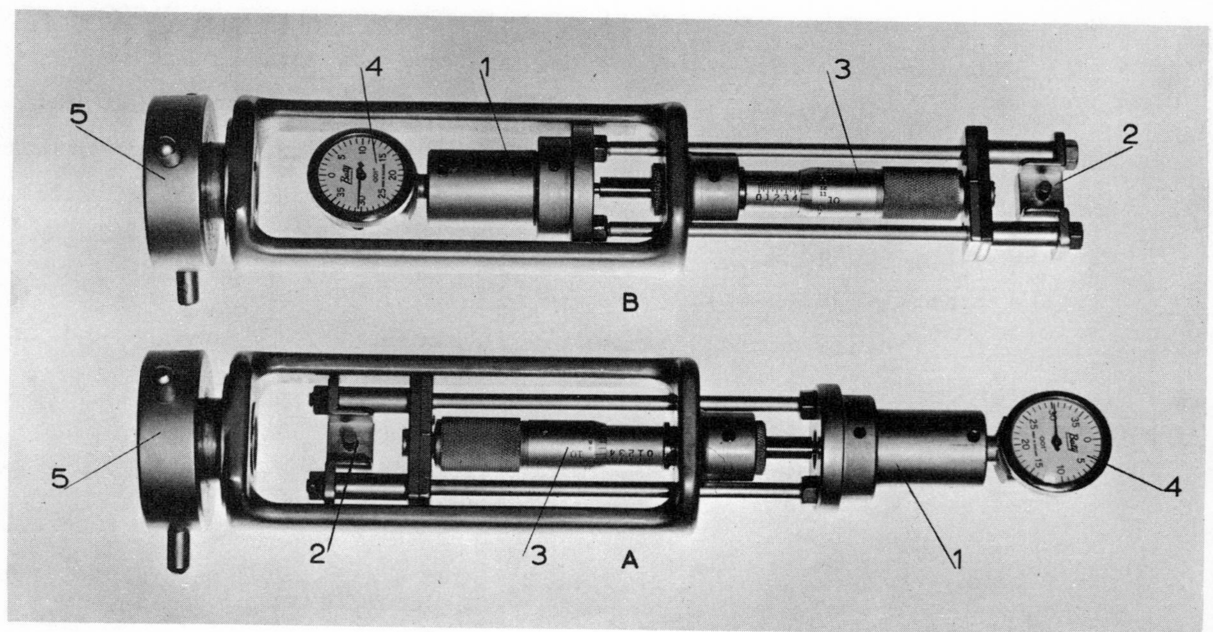

Fig. 3—Borehole extensometers Mark I.

attached to a threaded borehole collar plug, and a frame, into which is built a barrel type micrometer, 3, and an anchor pin, 2, to take the punched tape from the hole that is loaded with a helical spring in the section, 1, to a predetermined tension using the dial gauge, 4. The extensometer may be used with wire, in which case the pin anchor, 2, is replaced by a saddle into which is fitted a metal thimble that is securely attached to the wire with a socket screw. This latter method is used in holes of AX-size when using several anchor plugs, and up to four wires have been installed satisfactorily in holes 35 ft long.

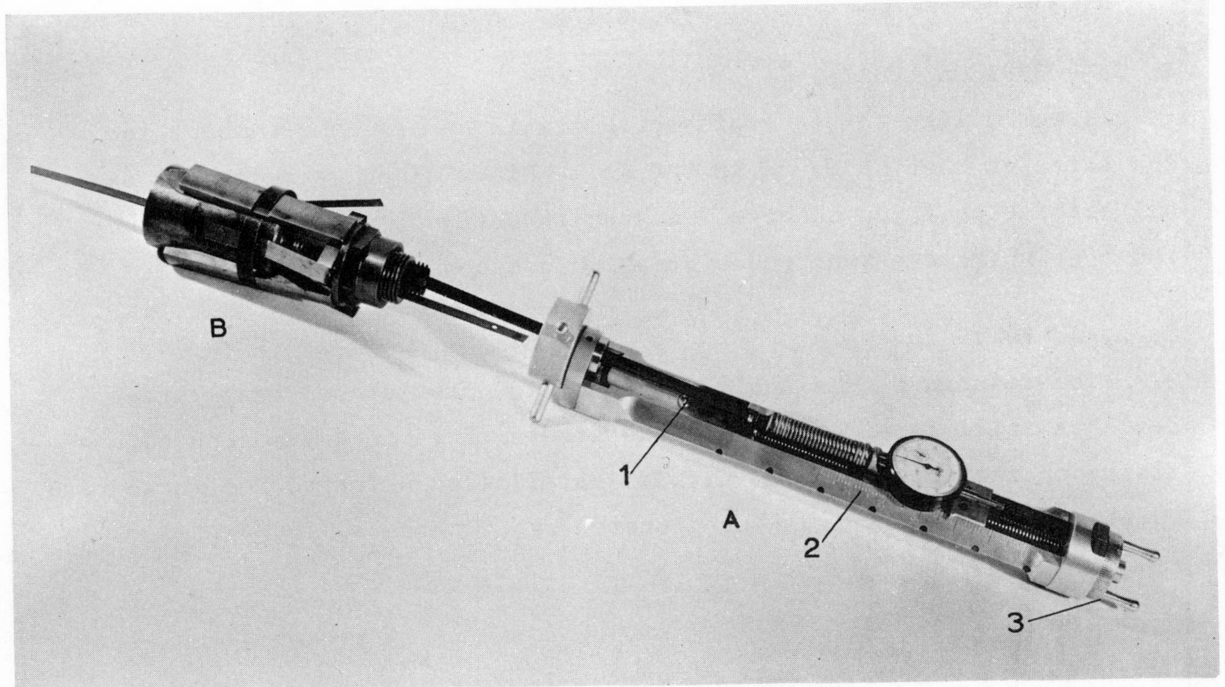

Fig. 4—Borehole extensometer Mark II.

The Mark II extensometer in Fig. 4 is shown with a specially designed expanding plug used in 3-in.-diameter down holes. With this anchor plug the central hole in the bolt is large enough to pass three tapes, 3/8 in. wide x 0.02 in. section, without major interference. The extensometer is basically the same as the Mark I, but has a range of 6 in. instead of 1 in. The body is made in duralumin, the dial gauge carrier phosphor bronze, sliding in key steel guides, and is operated through a stainless steel screw to a graduated head. The body is marked in inches and 1/10ths, the head being graduated to 0.001 in.

This assembly is also provided with auxiliary equipment of special borehole collar plugs to take the capstan headed screw or collar bolt attachments for lateral surveys along a tunnel.

The basic equipment described has been satisfactorily used in a recent interstrata borehole survey in holes up to 150 ft deep using three plugs.

DISCUSSION

E. C. ROBERTSON (USA):

I would like to ask Professor Potts if he has some numbers for the stresses that he found in the Kolar Mine. What were the stress differences that you observed in your investigation? Were they near those at which the rock would break in a laboratory test?

E. L. J. POTTS in reply:

The results of the analysis are so far provisional and a great deal has still to be done. All the stressmeters installed showed a rising characteristic from initial installation prior to the rockbursts. Number 9 meter (see Fig. 5) registered rapid changes.

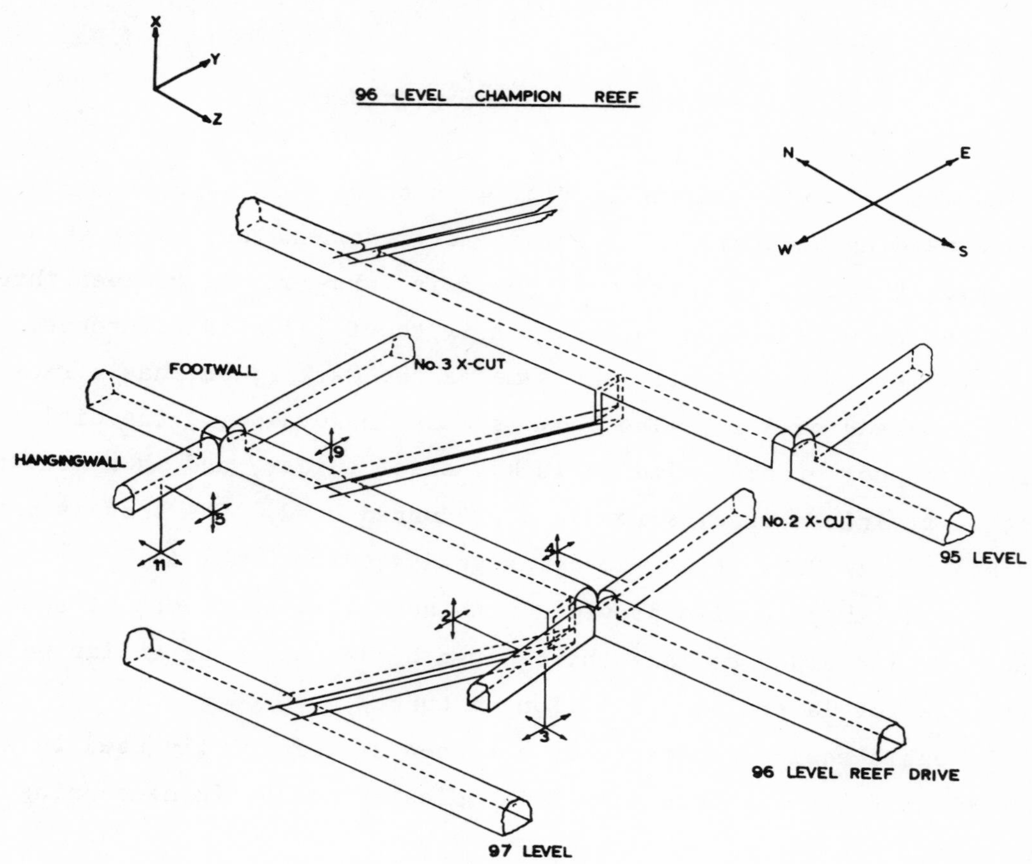

Fig. 5—Stressmeter layout.

A primary burst occurred, and 25 days later a major rockburst damaged drives and crosscuts in both wings of the ore shoot over 3000 ft in depth and with a maximum span on strike of 1000 ft.

Stressmeter No. 9 in the vicinity of the primary burst registered a stress change of 15,790 psi in the X direction (vertical) and 3910 psi in the Y direction (normal to the reef). After this burst, the Y direction increased by a further 12,370 psi, and the meter on the X component went off scale.

None of the remaining stressmeters showed any change at the time of the primary burst but the stress recorded continued to rise until the major burst. After the major burst a general relaxation occurred in the horizontal directions, and in the vertical direction the stress recorded showed an increase, probably due to the re-establishment of the gravity loading. This stress redistribution was substantiated by the analysis of the extensometer borehole measurements, indicating the general strata movements. Clearly the footwall rock had failed and the stress changes recorded were in excess of the values of stress difference required to cause rupture.

ABSTRACT

Equations have been known for many years that relate the elastic properties of rock in place and the speed of travel of artificial seismic waves through that rock. The idea of measuring quantitatively and *in situ* the elastic modulus of a major portion of the abutment of a dam by measuring the velocity of seismic waves in and through the abutment was first put into practice in this country in 1954 by the U.S. Bureau of Reclamation, Department of the Interior. This initial test was made at the Monticello Dam, California. Similar field determinations of the elasticity of rock in place have been made at five other projects.

The elastic modulus and the Poisson's ratio are two physical properties of rock used by design engineers to evaluate the anticipated deformation of a dam foundation. Accurate information on abutment deformation permits more precise analysis of stress distribution in the dam and foundation, with a resultant increase in safety and economy of construction materials.

The major objective of the work in the field of rock mechanics presented in this paper has been to establish controlled comparisons between the values of the elastic properties of rock obtained from seismic field measurements of wave velocity and jack loading tests in exploration tunnels, and laboratory tests of cores.

RÉSUMÉ

On connaît depuis de nombreuses années les équations qui expriment la relation entre les propriétés élastiques de la roche sur place et la vitesse de déplacement des secousses sismiques artificielles à travers cette roche. L'idée de mesurer quantitativement et sur place le coefficient d'élasticité d'une portion importante d'une butée de barrage, en mesurant la vitesse d'ondes sismiques dans et à travers la butée, a été tout d'abord mise en pratique dans ce pays, en 1954, par le Bureau de Défrichement des Etats-Unis, Ministère de l'Intérieur. Le premier essai a été fait au Barrage de Monticello, Californie. Des déterminations semblables, faites sur les lieux, pour évaluer l'élasticité de la roche sur place ont été faites pour d'autres projets sérieux.

Le coefficient d'élasticité et le coefficient de Poisson représentent deux propriétés physiques de roche utilisées par les ingénieurs d'étude, pour évaluer la déformation escomptée des fondation d'un barrage. Des renseignements précis sur la déformation d'une butée permettent une analyse plus précise de la répartition des contraintes sur le barrage et les fondations, résultant en outre en une augmentation de sécurité et d'économie de matériaux de construction.

Le but fondamental, dans le domaine de la Mécanique des Roches, du travail présenté dans cet exposé, a été d'établir des comparaisons contrôlées entre les valeurs des propriétés élastiques de la roche fournies par des mesures, sur les lieux de séisme, de la vitesse de la secousse, par des essais de chargement par vérins effectués dans des tunnels d'exploration, et enfin par des essais de laboratoire sur carotte-témoin.

AUSZUG

Seit vielen Jahren sind die Zusammenhänge zwischen den elastischen Eigenschaften des Gebirges und der Fortpflanzungsschwindigkeit von künstlichen Schockwellen bekannt. Der Gedanke, den Elastizitätsmodul von grösseren Teilen des Gebirgswiderlagers einer Talsperre durch Messen der Geschwindigkeit von Schockwellen zu bestimmen, wurde erstmalig in diesem Lande im Jahre 1954 durch das U.S. Bureau of Reclamation, Department of Interior, in die Tat umgesetzt. Dieser erste Versuch fand an der Monticello Sperre im Staate California statt. Ähnliche Bestimmungen der Elastizität von Gebirgsmassen durch Feldmessungen wurden inzwischen an fünf weiteren Projekten durchgeführt.

Der Elastizitätsmodul und die Querdehnungszahl sind zwei physikalische Eigenschaften des Gebirges, welche von den Entwurfsingenieuren zur Voraussage der Verformung von Sperrengründungen benötigt werden. Genaue Aussagen über die Formänderung der Widerlager erlauben bessere Ermittlung der Spannungsverteilung in der Mauer und ihrer Gründung, woraus sich grössere Sicherheiten für das Bauwerk und eine bessere Wirtschaftlichkeit seiner Ausführung ergeben.

Das Hauptziel der in diesem Aufsatz dargestellten gebirgsmechanischen Arbeiten war die Erstellung eines strengen Vergleiches zwischen den Kennziffern der elastischen Eigenschaften, die sowohl durch Feldmessungen der seismischen Fortpflanzungsgeschwindigkeiten und Belastungsversuche in Erkundungsstollen als auch in Laboratoriumsversuchen an Bohrkernen ermittelt worden waren.

GEOPHYSICAL MEASUREMENTS OF ROCK PROPERTIES IN SITU

Dart Wantland*

Introduction

THE ARTICLE by Messrs. P. D. Brown and J. Robertshaw published in 1953[(1)] was one of the first examples of field measurements of the speed of seismic waves in foundation rock to determine the elasticity of the rock in place. At the site of a proposed arch dam in Ducan Gorge, on the Lesser Zab River in northern Iraq, they established the following conditions in respect to foundations:

1. That the underlying massive-bedded and jointed dolomite of the foundation had an elastic modulus of 7,500,000 psi (530,000 kg/cm^2) in place as measured parallel to the bedding and an elastic modulus of 3,500,000 psi (247,000 kg/cm^2) vertical to the bedding.
2. That the overlying comparatively thin-bedded, closely jointed, and finely crystalline limestones had an elastic modulus of 1,200,000 psi (84,000 kg/cm^2) parallel to the bedding and 1,000,000 psi (71,000 kg/cm^2) vertical thereto.

The dolomite and the limestone beds at the site were essentially flat lying. The seismic tests showed significant differences in the elasticity along the bedding and perpendicular to it in these two types of rock.

Directional differences in the elasticity of rocks and formations have been observed repeatedly and measured during seismic field studies at dam sites. This form of anisotropy is a characteristic of a foundation that is important in the design of a dam, although such data are

*Head, Geophysics Section, U. S. Bureau of Reclamation, Denver, Colorado.

not ordinarily secured during the testing of the foundation rock at dam sites.

Rock Quality and Seismic Wave Velocity

General Considerations

The basis of seismic field measurements of the properties of foundation rock in place is the fact that the better the quality of the rock, the faster (artificial) seismic waves travel through it. For example, seismic wave velocity in a weathered metadiabase at one dam site was about 2000 ft/sec (600 m/sec). In fresh, hard metadiabase at the same site seismic wave velocity was over 14,000 ft/sec (4300 m/sec).

The term "artificial seismic waves" is used to indicate that in applying the seismic technique seismic waves from explosions of dynamite are employed--in contrast to natural or earthquake waves. The character of rock as measured seismically can be expressed in terms of seismic wave velocity or in terms of the modulus of elasticity of the rock; the latter combines the measured velocity and the density of the rock. The seismic technique for measuring the elastic modulus of rock at a dam site is shown in Fig. 1.

Field Tests at the Oroville Dam Site, California

Seismic field investigations by the Bureau of Reclamation in 1950 at the site of the Oroville Dam,[2,3] on the Feather River in northern California, demonstrated the relationship between rock quality and seismic wave velocity. The State of California is now constructing the Oroville Dam at this locality.

Five seismic lines were established to test the basic metadiabase that formed the right abutment of the Oroville site. These lines were some 400 ft (122 m) in length and were located at points on the axis about 400 ft apart so that a distance of about 1500 ft (455 m) on the right abutment was studied. Each seismic line was located along the contour so that the elevation of points on them would be approximately uniform.

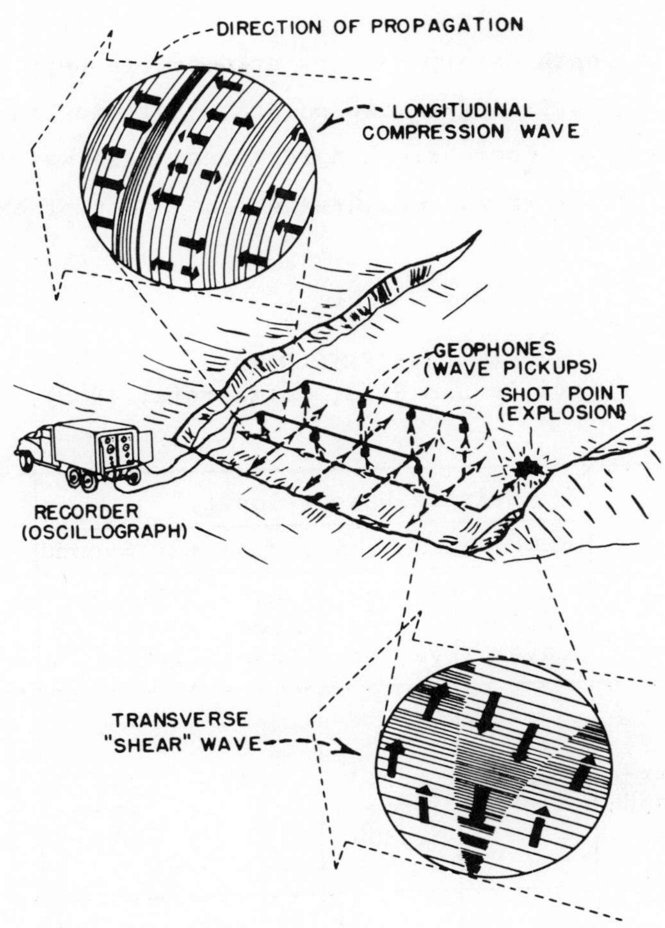

Fig. 1—Seismic modulus determinations.

In carrying out the investigation, dynamite charges were exploded at a total of 18 points on the five seismic lines. Data on the speed of seismic waves in the layers present at and below the surface of the rock of the abutment were secured, as well as information on the thickness of and depth to certain of these layers, as shown in Table 1.

The seismic field measurements at the Oroville site show that the quality of a rock is reflected by the speed of travel of seismic waves in it. The engineering objective of the work there, however, was to test the ability of the seismic method to measure the thickness of weathered and unsound rock at and between drill holes. It was desired to ascertain if the depth to sound rock that would be suitable for the foundation of a dam could be determined with an accuracy useful for planning and design purposes.

The seismic depth determinations at Oroville were correlated directly with core drilling at two points. Seismic Line 1 passed through Drill Hole 4, where fresh hard metadiabase was found at 18 ft. A seismic depth of 19 ft was measured at the point of explosion, or

Table 1

SEISMIC DATA
(Oroville Dam Site, California, 1950)

Layers	Seismic Wave Velocity				Remarks
	Feet per Second		Meters per Second		
	Av for Layer	Range of Values	Av for Layer	Range of Values	
V_1 layer:					
Surface and near-surface soil and weathered rock	2,050	1,720 to 2,300	620	525 to 700	Rock tested was basic metadiabase on the right abutment
V_2 layer:					
Underlying fresh hard "sound rock" with stained joints	14,375	12,200 to 16,500	4,370	3,920 to 5,050	Depths to the V_2 layer averaged 39 ft (11.8 m) and ranged from 15 ft (4.6 m) to 64 ft (19.5 m)

shot point, nearest that hole. Seismic Line 4 traversed Drill Hole I.F.2 where the seismic depth was 64 ft compared with a drill log depth of 58 ft to fresh rock. This was not as close an agreement as in the previous case but other seismic depths at points along Line 4 were 56, 58, and 60 ft.

It was concluded that the usefulness of the seismic method was demonstrated to determine the depth to fresh hard metadiabase and/or the thickness of weathered rock at drill holes and at points between. In preparing the site of a major structure, disintegrated or weathered rock is, of course, removed. Accurate data on the yardage of such material at a given site are important in preparing cost estimates.

Geophysical Depth to Sound Rock

Field Tests at the Folsom Dam Site, California

In February 1950 seismic and electrical resistivity field studies were also made at the site of the Folsom Dam, on the American River near the town of Folsom, California.[4,5] The objective was to establish the accuracy of seismic and resistivity measurements of the depth to sound unweathered rock by geophysical check measurements at and near core drill holes.

The Folsom Dam site is entirely in granite and except along the stream channel is covered by a thin mantle of soil that grades into weathered, decomposed granitic rock. The depth of weathering as shown by core drilling averages about 30 ft. In places, along local jointed zones, weathering is known to depths up to 80 ft.

Geophysical depth determinations for checking purposes were made at points located near core drill holes put down during the detailed exploration of the site by the Corps of Engineers of the U.S. Army. The cores obtained were carefully analyzed by the geologists of that organization to determine specifically the degree of weathering of the rock, and on the detailed geological logs that were made, the sound rock line was marked.

As shown in Table 2, nine direct comparisons were obtained at drill holes. These indicated that the geophysically measured depth to sound rock averaged 42.4 ft in comparison with an average depth from drilling of 39.4 ft, a difference of only 3 ft. The figures indicate that the geophysical measurements placed the firm rock at a slightly greater depth, on the average, than drilling. The close agreement shown permits the conclusion that in addition to the seismic method the resistivity technique is also applicable to determining the amount of weathered rock that would have to be stripped at a dam site.

Table 2

COMPARISON OF DEPTH TO SOUND ROCK BY CORE DRILLING AND GEOPHYSICS, FOLSOM DAM SITE, CALIFORNIA

Core Drill Hole	Depth to Sound Rock, Drilling		Geo-physical Point	Depth to Sound Rock, Geophysical		Difference		Remarks
	ft	m		ft	m	ft	m	
84	56	17.0	R 24	56	17.0	0	0.0	R = Resistivity measurement
48	50	15.2	S 8	58	17.7	+8	2.4	
82	35	10.6	R 21	30	9.2	-5	1.5	S = Seismic measurement
52	25	7.6	S 6	39	11.9	+14	4.3	
52	25	7.6	R 20	31	9.5	+6	1.8	Greater geo-physical depth is taken as +
60	36	11.0	R 19	37	11.3	+1	0.3	
47	51	15.5	R 25	59	18.1	+8	2.4	
51	38	11.6	S Line 6	41	12.5	+3	0.9	Data projected
51	38	11.6	R 20	31	9.5	-7	2.1	Data projected
Average	39.4	12.0		42.4	12.9	3.1	1.7	

Seismic Field Measurements

Seismic Lines

To apply the seismic refraction procedure, a number of seismic wave receivers, or geophones, are placed along a straight line--on the surface--that traverses a section of the foundation or abutment of a dam. The geophones are located at measured distances from a shot point, where a charge of explosives, buried in a shallow hole, is detonated. The explosion, or shot, puts seismic wave energy into the ground (as noted previously), which travels along the (seismic) line to the geophones, where it is picked up and its arrival recorded photographically. These arrangements are shown in Fig. 2.

This figure shows the paths followed by the seismic waves where two layers, designated as V_1 and V_2, are present. From the shot point, the seismic waves (1) move downward through the near surface, or low velocity layer; (2) are refracted along the top of the underlying layer, in which they travel at greater speed; and (3) then return to the surface.

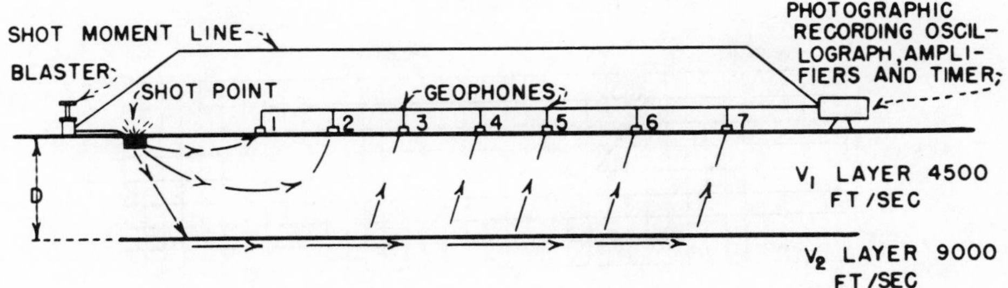

Fig. 2—Seismic investigation line.

Seismic Records

A seismic record, or seismogram, is made for each shot by means of a multichannel, photographic, recording oscillograph, which is usually mounted in a field truck. (See Fig. 3.) The time required for the seismic wave to travel from the shot point to each geophone can be determined from the seismogram. Three necessary items are shown on it, namely:

1. The shot moment mark that indicates when the explosion occurred and/or when the seismic wave started.
2. An indication of the arrival of the seismic wave at each geophone--"refraction event" of Fig. 3.
3. A set of time lines that permit the time to be measured, to 0.001 of a second, between the start of the wave and when it reaches a geophone.

Traces. The seismic wave, or waves, from an explosion is in the form of vibration energy that travels from the shot point by causing motion of particles in the layer through which it is propagated. A familiar example of this is the set of circular ripples, or waves, that expand outward over the surface of a still pond of water when a stone is thrown into it.

When the on-coming seismic wave strikes a geophone, it causes the moving element in it to vibrate, which sets up a varying electrical impulse. This impulse is carried to the recording unit where it is amplified and transferred to one of the galvanometers in the oscillograph. The movement of a small mirror in the galvanometer faithfully reproduces the variations of the electrical current generated by the

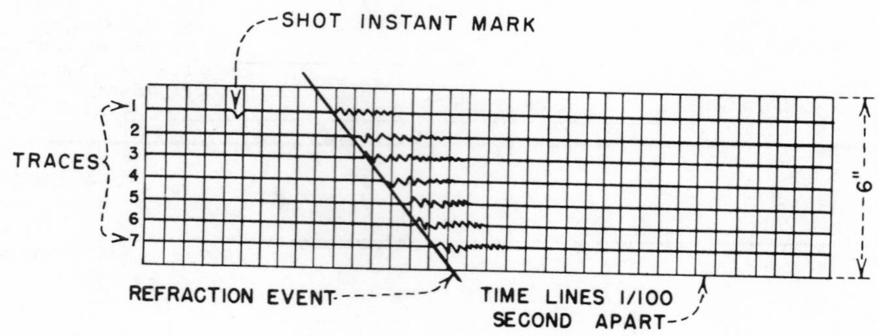

Fig. 3—Seismic record or seismogram.

geophone that is photographed as a trace, or line, which appears on the seismogram. Each geophone on the seismic line is connected to its own amplifier and galvanometer.

The refraction event appears on each of the traces in succession as a sharp downward movement and records the progress of the seismic wave along the seismic line. From the seismic record, the time of travel of the seismic wave from the shot point to each geophone can be measured to 0.001 of a second. This is done by counting the number of 0.01-sec intervals and fractions thereof between the shot moment mark and the refraction event on the different traces.

Travel Time Curves

The data on the time of travel of seismic waves from the seismic record and the measured distances of the geophones from the shot point are assembled on a travel time curve or graph (Fig. 4). From such a graph the seismic wave velocity in the different layers below the surface can be determined.

The depth (D) to the top of the underlying or V_2 layer (in which the waves travel at a relatively high speed) is found by the formula given in Fig. 4. This formula relates the layer velocities and the slope change or break point distance X of the travel time graph.

In many cases a third and yet lower layer is present that transmits seismic waves at a velocity greater than the V_2 layer shown in

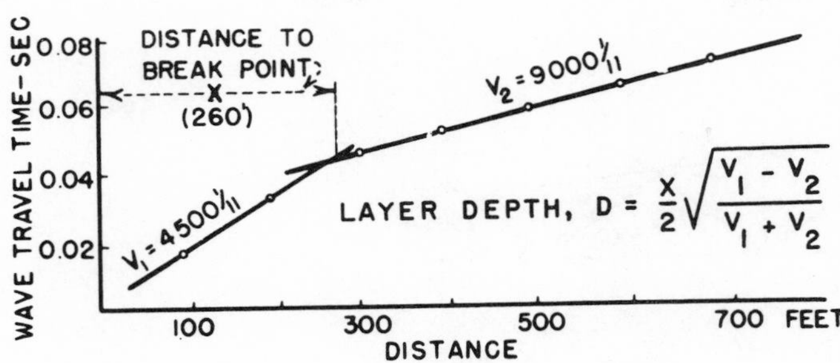

Fig. 4—Travel time curve.

Fig. 4. A formula involving additional terms is required to determine the depth D_2 to the top of the V_3 horizon. This formula involves an X_2 distance to the slope change point, on the travel time graph, between the V_2 and V_3 segments. Details on these matters are available in a number of standard works on exploration geophysics, such as that by the late Dr. C. A. Heiland[4] and by J. J. Jakosky.[5]

Data Secured

From seismic field measurements the speed of seismic waves in the layers below the surface and the thickness and depth to certain of these layers can be measured. The underlying layer in which seismic waves travel at a high speed may represent the sound rock line, or firm unweathered rock of good quality. As previously noted, when combined with determinations of the density of the rock along a seismic line from tests of samples, the measured seismic wave velocity provides a quantitative value of the Young's modulus (E) of the rock in place, as will be discussed subsequently.

Multiple Shot Points

In carrying out field measurements, after a seismic line has been selected and the geophones placed thereon, it is often advantageous to establish shot points at intermediate locations. An example of such multiple shot points is shown in Fig. 5, which covers seismic rock weathering depth studies at the Lower Ashland Dam site, located about two miles below the Folsom Dam site referred to previously. In

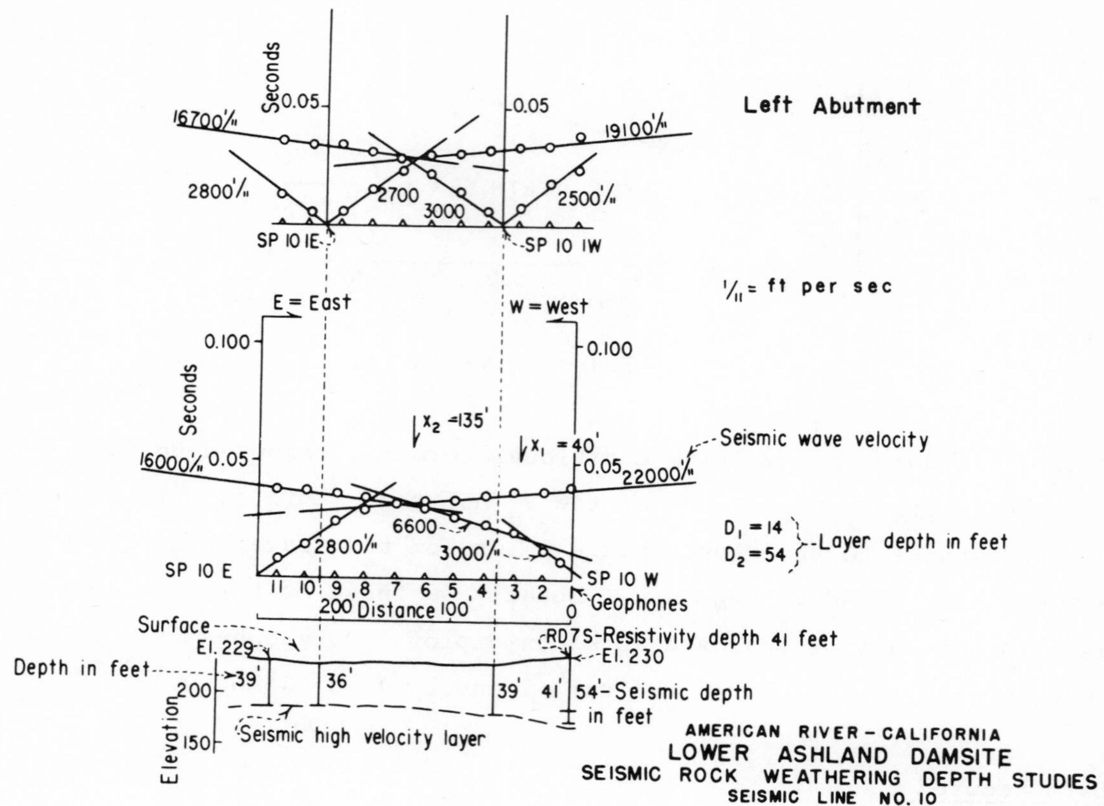

Fig. 5—Travel time curves for multiple shot points on seismic line.

this figure the six travel time curves resulting from four shots are presented--at Shot Points 10 E, 10 IE, 10 IW, and 10 W, on Seismic Line 10. The letters E for East and W for West indicate the direction in which the seismic waves travel from the shot points. The letter I indicates intermediate shot points and is coupled with the direction designation.

Geophone Spacing

In making seismic wave velocity measurements in tunnels it is desirable to place the geophones at a short distance from each other on the seismic line selected. An interval of 10 to 15 ft (3.5 to 4.6 m) is satisfactory. Such an interval permits the depths to layers to be determined even though they may be only a few feet from the surface of the wall of the tunnel. The depths to such shallow layers and the velocities at which seismic waves travel in them are important.

Seismic Measurements in Drill Holes

It is often possible to make seismic measurements in the drill holes put down in the normal course of the exploration of a site to secure samples of core for testing in the laboratory. This can be accomplished in two ways, both of which are equally effective:

1. A geophone is placed on the ground surface near the mouth of the drill hole. Shots are fired at different depths in the drill hole, and the time of seismic wave travel is measured for each shot.
2. A borehole geophone is lowered into the drill hole to different depths. A series of shots is fired from points on the surface, near the mouth of the drill hole, and the time of travel of the seismic wave is measured for each geophone setting.

The results are presented in a travel time curve in which the distance axis is plotted vertically downward and the time axis is horizontal. An example is shown in Fig. 6.

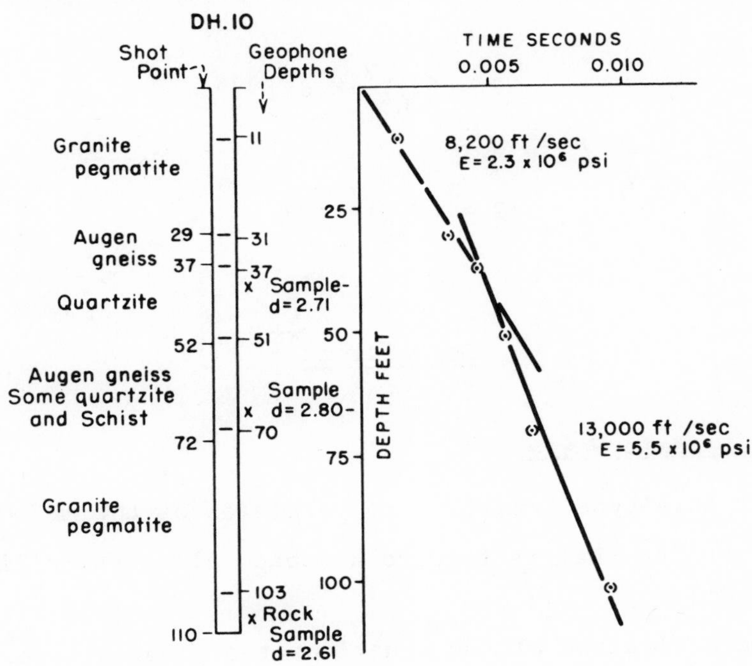

Fig. 6—Drill hole velocity measurements.

Equations*

The elastic properties of rocks and formations control the velocity with which seismic waves travel in them. The relationships involved are expressed in two equations, which, in their simplified form, are

$$V_L = \sqrt{\frac{E}{d}} , \quad (1)$$

$$V_{TR} = \sqrt{\frac{G}{d}} . \quad (2)$$

Equation (1) states that the speed of travel of longitudinal seismic waves (V_L), which are compressional waves, equals the square root of the Young's modulus of elasticity (E) of the rock divided by its density (d). Equation (2) states that the speed of travel of transverse seismic waves (V_{TR}), which are shear waves, equals the modulus of rigidity (G) of the rock divided by its density (d).

Solving these equations for E and G gives

$$E = (V_L)^2 \times d , \quad (3)$$

$$G = (V_{TR})^2 \times d . \quad (4)$$

Definitions**

Longitudinal Seismic Waves

Waves of this type transmit, to particles of the medium in which they travel, motion that is forward and back along their line of

*For the derivation of the equations that follow, see Ref. 4, pp. 442-449.

**For further details, see Ref. 6.

propagation. Longitudinal waves advance by causing alternating momentary compressions and rarefactions in the material through which they are passing. They are alternations of stress--or pressure--and tension, and their velocity (symbol V_L) is measured in feet per second or in meters per second.

Transverse Seismic Waves

Waves of this type transmit, to particles of the medium in which they travel, motion that is to and fro in a direction at right angles to their line of propagation. They are applied (pure shear) stresses, in wave form, which tend to cause slipping along parallel planes of slices, which are of molecular thickness, in the medium involved.

Note. Longitudinal and transverse waves are known as body waves, because they pass through the body of the material in which they are propagated.

Seismic wave energy, from an explosion of dynamite, consists of longitudinal, transverse, and other kinds of waves. The longitudinal waves travel faster than the transverse waves and therefore reach the geophones on a seismic line ahead of the transverse waves. The longitudinal wave energy is recorded first on a seismogram. The wave motion evidenced by a trace or traces (on a seismic record) from the longitudinal waves may obscure the later arriving transverse wave so that the latter cannot be clearly distinguished and its time of reaching a geophone determined. For this reason, it is ordinarily necessary to employ special means of generating transverse seismic waves.

Modulus of Elasticity

The Young's modulus of elasticity of a rock, or other substance (symbol E), is defined as the ratio of stress, or force applied, to the strain, or deformation it produces in the material tested. The relationship that E = stress/strain holds only if the elastic limit of the substance is not exceeded.

The elastic modulus is expressed in pounds per square inch, that is, psi, or kilograms per square centimeter, and in some cases in other units. The stress involved is a pressure--like that applied by a piston. Such a force tends to cause a change in volume of the body under stress.

Note. It is unfortunate that as they are generally used, the words "stress" and "strain" have rather similar connotations. In terms of rock mechanics the elastic modulus E might be best defined as pressure/deformation, rather than stress/strain.

Modulus of Rigidity

The modulus of rigidity of a material is also defined as the ratio of the force applied to the deformation produced (symbol G), and the units of its measurement are, as in the previous case, in terms of psi or kg/cm^2. The force or stress, however, that is involved is a shear stress that tends to cause a change in the shape of the body stressed.

Density

The density of a material is its mass per unit volume. The symbol for density is d, pounds per cubic foot, or grams per cubic centimeter.

Mass. The mass of a material is its weight divided by the acceleration of gravity, symbol M. In the English system the gravity acceleration is 386 in./sec/sec. Using w for weight and a for acceleration, mass $M = w/a$.

Comment

The elasticity of a rock or portion of a foundation is more particularly its ability to resist deformation. For example, an abutment composed of rock of high elasticity such as a massive quartz diorite with an elastic modulus of 12.3 million psi should resist without difficulty the pressure put on it by an arch dam. Such an abutment could support a greater load per unit area than if the rock was a shattered quartz diorite whose elasticity was 1.2 million psi. These values of E, for the two diorites, are factual, as they were measured during field studies in California.

Poisson's Ratio

Although Poisson's ratio is not involved in any of the four equations presented previously, it will be defined in this section. As will be shown subsequently, this ratio is related to the ratios of the velocities of longitudinal and transverse seismic waves.

This ratio--symbol σ (sigma)--as applied to a cylindrical sample, is the change in diameter of the sample per unit diameter to its change in length per unit length when placed in tension. With di = diameter and L = length, and if Δdi and ΔL = change in diameter and length,

$$\sigma = \frac{\Delta di/di}{\Delta L/L} \ . \tag{5}$$

These relationships apply equally to a sample under pressure where its length decreases and its diameter increases. Poisson's ratio for rock ordinarily ranges from 0.1 or less to not more than 0.3.

Example of Calculations

The velocity of longitudinal seismic waves was measured in the quartzite of the left abutment of the Morrow Point Dam on the Gunnison River in western Colorado. The determination was carried out along the floor of an investigation tunnel in that abutment and was found to be 12,800 ft/sec. The density of the quartzite was 2.70 as established by laboratory tests of samples obtained at the site.

The equation used to calculate the modulus of elasticity was

$$E = (12V_L)^2 \times 9.3 \times 10^{-5} \times d \text{ psi} \ . \tag{6}$$

The terms E, V_L, and d have been defined. Multiplying V_L by 12 converts feet per second to inches per second. The factor 9.3×10^{-5} is the mass of 1 in.^{3} of water and is derived as follows: As noted, density is mass per unit volume and the latter, in this case, is 1 in.^{3} Mass is weight divided by the acceleration of gravity, or

386 in./sec^2. The weight of 1 in.^{3} of water is 0.036 lb. The mass, therefore, of 1 in.^{3} of water is 0.036 lb divided by 386 in./sec^2 = 9.3×10^{-5} lb sec^2/in. Data: V_L = 12,800 ft/sec Morrow Point quartzite, d = 2.70 density of quartzite, E = 5.3×10^6 psi,* E = 373×10^3 kg/cm^2.

Formula Involving Poisson's Ratio

For calculating the elastic modulus of foundation rock from seismic wave velocity measurements and density determinations of samples, Eq. (6) provides a value of sufficient accuracy for many purposes, particularly for preliminary estimates of the order of magnitude of the elastic modulus. However, for comparison between the values of the elastic modulus of rock obtained from tests by different procedures, the accuracy of the results is significant and Poisson's ratio should be considered.

The velocity of longitudinal seismic waves on this basis is

$$V_L = \sqrt{\frac{E}{d} \times \frac{1 - \sigma}{(1 + \sigma)(1 - 2\sigma)}}\ , \qquad (7)$$

and in a form suitable for calculations

$$E = (12V_L)^2 \times d \times 9.3 \times 10^{-5} \times \frac{(1 + \sigma)(1 - 2\sigma)}{(1 - \sigma)}\ \text{psi}\ . \qquad (8)$$

To apply Eq. (8) a value of the Poisson's ratio of the rock must be assumed or obtained by other means, such as laboratory tests of cores. Poisson's ratio can also be determined by measuring the velocity of transverse seismic waves along the same seismic lines, or in the same drill holes, in which the longitudinal wave velocity was measured. The relationship of the longitudinal and transverse wave velocities and the Poisson's ratio are

---

*Corrected for a Poisson's ratio of 0.2.

$$\sigma = \frac{\frac{1}{2}\left(\frac{V_L}{V_{TR}}\right)^2 - 1}{\left(\frac{V_L}{V_{TR}}\right)^2 - 1} . \tag{9}$$

Relative Importance of Factors

Equation (8) shows that the elastic modulus of rock measured seismically varies as the square of the wave velocity and (only) as the first power of the density and of a factor involving the Poisson's ratio. This indicates that the seismic wave velocity and its accurate measurement are quite important, in applying the refraction procedure, to determining the elastic modulus of rock in place. It is also indicated that the value of the density of the rock and the value of its Poisson's ratio have relatively less influence on the calculated value of the elastic modulus.

The amount of change in the elastic modulus caused by varying one of these quantities while the other two are kept constant is shown in Table 3 on page 426. The values of velocity and rock density were obtained during field tests at the site of the Bureau of Reclamation's Morrow Point Dam. The value of Poisson's ratio was consistent for rock of the type found at the Morrow Point site.

Evaluation of Tunnel Rock Quality

Objective

Seismic measurements were made in the fall of 1958 at six locations in the Clear Creek Tunnel, a feature of the Bureau of Reclamation's Central Valley Project in California. The objective was to test the ability of the seismic method to evaluate the quality and character of tunnel rock by measuring the velocity of seismic waves in the rock along the floor and walls of a tunnel. It was desired to ascertain if such measurements could establish quantitatively how good or how poor the

Table 3

EFFECT OF CHANGES IN (1) SEISMIC WAVE VELOCITY, (2) ROCK DENSITY, AND (3) POISSON'S RATIO ON CALCULATED VALUE OF ELASTIC MODULUS

(1) Change in Wave Velocity (density constant)			(2) Change in Density (wave velocity constant)			(3) Change in Poisson's Ratio (wave velocity and density constant)		
Wave Velocity, V_L	Density, d	E, 10^{-6} psi	Wave Velocity, V_L	Density, d	E, 10^{-6} psi	Velocity, V_L, 11,000 Density, d, 2.0 — Poisson's Ratio	$\frac{(1 - \sigma)}{(1 + \sigma)(1 + 2\sigma)}$	E, 10^{-6} psi
14,000	2.71	7.1	14,000	2.81	7.3	0.28	0.78	3.4
13,000		6.1		2.71	7.1	0.25	0.83	3.6
12,000		5.2		2.61	6.8	0.20	0.90	4.0
11,000		4.4		2.51	6.6	0.15	0.94	4.1
10,000		3.6	9,000	2.81	3.1	0.12	0.97	4.3
9,000		3.0		2.71	3.0			
				2.61	2.8			
				2.51	2.7			

rock of a given section of tunnel might be as related to the amount of reinforcement that might be required in the tunnel lining. As a corollary, if seismic measurements could show the strength of the rock through which a tunnel had been driven, they might provide a basis for criteria as to what portion of such pressure as might be transmitted to the tunnel lining could be carried by the rock.

In conducting these field tests seven grades of rock were designated and described by the project geologist. These ranged from excellent to very poor and were based on careful study and many months of geological mapping relating to the driving of the tunnel.

Field Operations

The Bureau's seismic field truck was taken into the tunnel, and at certain locations seismic lines were laid out along the tunnel floor and shots were fired at points along these lines. To make measurements at other sites, the seismic equipment was removed from the truck and taken into the tunnel on a small flatcar. Where the location to be studied was some distance from the tunnel portal, equipment was loaded on one of the locomotives used in tunneling. A recording station was set up close to the tunnel wall so that men and equipment would be out of the way of moving trains. In one instance seismic measurements were successfully made within a few hundred feet of the "face" where active tunneling operations, drilling, loading broken rock, etc., were going on.

Results

The results of seismic measurements in the Clear Creek Tunnel are shown in Table 4. It was found that the seismic wave velocity and the modulus of elasticity increased as the quality of the rock increased, and conversely these values decreased where relatively poorer rock was involved. These findings suggest that seismic wave velocity measurements provide a means of evaluating the quality of tunnel rock on a quantitative basis.

A unique advantage of the seismic method for measuring the quality of rock in a tunnel is the fact that sections of rock of considerable

Table 4

CORRELATION OF ELASTIC MODULUS AND CHARACTER OF TUNNEL ROCK, CLEAR CREEK TUNNEL, CALIFORNIA

Seismic Line; Tunnel Stations; Length	Seismic Wave Velocity		Density from Rock Samples	Elastic Modulus		Character of Tunnel Rock and/or Rock Quality
	ft/sec	m/sec		$E \times 10^6$ psi	$E \times 10^3$ kg/cm^2	
C 462+40 to 463+40 100 ft (30.5 m)	18,250	5,550	2.84	12.3	865	Excellent Rock Massive Quartz Diorite Unaltered, coarse grained
K 558+20 to 560+15 195 ft (59.5 m)	18,400 15,450	5,600 4,690	2.68	12.3 8.6	865 605	Good Rock Metarhyolite Porphyry Schistose greenstone also present. Rock is blocky; some fractures and shears
M 555+37 to 556+66 129 ft (39 m)	14,650	4,460	2.70	7.8	550	Good to Fair Rock Metarhyolite Siliceous greenstone in this section. Rock heavily foliated, highly sheared, and fractured
J 562+65 to 563+70 105 ft (32 m)	13,850	4,220	2.69	7.0	490	Fair to Poor Rock Metarhyolite Porphyry Subschistose and jointed in numerous directions; some crushing
B 563+20 to 563+90 70 ft (21 m)	11,000	3,350	2.78	4.5	316	Moderately Poor to Poor Rock Metarhyolite Porphyry Interbedded volcanics and greenstone present. Rock massive and blocky; some crushing and shearing
A 569+05 to 570+00 95 ft (29 m)	9,500	2,890	2.68	3.2	225	Poor Rock Weathered Rhyolite Porphyry Massive, prominently jointed

length are studied. The seismic lines at the Clear Creek Tunnel were from 70 to 195 ft long. Furthermore, a measured seismic wave velocity sums up the good and bad qualities of a section of rock.

Selection of Seismic Layer Wave Velocities

When data from seismic measurements of rock properties are analyzed, the question must be answered as to which wave velocity should be selected--from among several velocities that apply to different subsurface layers--to calculate the elastic modulus. For the seismic work at the Clear Creek Tunnel the highest measured value of velocity on the different seismic lines was used to determine the moduli presented in Table 4.

Using the maximum velocity has been the practice at the Bureau of Reclamation for several reasons; namely (1) it was not certain if the value of elasticity obtained with wave velocities below 2000 to 3000 ft/sec (610 to 915 m/sec) was significant and valid, and (2) it was known from available comparisons between seismically determined values of rock elasticity and values from laboratory tests of cores of the same rock that the agreement for hard, sound rock was relatively close.

The values and ranges of seismic wave velocities and the depths and thicknesses of layers from the measurements in the Clear Creek Tunnel are shown in Table 5. In this table V_1, V_2, V_3, and V_4 designate values of seismic wave velocity in layers 1, 2, 3, and 4, which lie at progressively greater depth beneath each other. It is to be noted that for the six grades of rock tested, three showed only two values of wave velocity, two grades showed three values, and one grade showed four different values.

In the table the symbol D_1 indicates the depth to the top of the V_2 layer (which is also the thickness of the V_1 layer); the symbol D_2, the depth from the surface to the top of the V_3 layer (thickness of V_2 layer equals $D_2 - D_1$); and the symbol D_3, the depth to the top of the V_4 layer. For values of these quantities in the metric system see Table 6.

Table 5

SEISMIC WAVE VELOCITIES AND LAYER DEPTHS, CLEAR CREEK TUNNEL, CALIFORNIA

Seismic Line; Tunnel Stations; Length of Line	Average Values of Seismic Wave Velocities (ft/sec)				Section Showing Average Velocities and Layers
	V_1 Layer	V_2 Layer, D_1 Depth to Top of V_2 Layer	V_3 Layer, D_2 Depth to Top of V_3 Layer	V_4 Layer, D_3 Depth to Top of V_4 Layer	
C 462+40 to 463+40 100 ft	2,900 Range of 4 2,700-3,200	None	18,200 Range of 4 15,000-20,000 D_1 = 3 Range 2-3		Excellent Rock 0-3 2,90 at and > 3 18,25 (> = below)
K 558+20 to 560+15 195 ft	2,050 Range of 7 1,600-3,000	10,750 Range of 4 9,600-12,000 D_1 = 4 Range 4-5	15,450 Range of 3 14,800-16,000 D_2 = 6 Range 4-7	18,400 Range of 3 18,400-19,000 D_3 = 18 Range 12-25	Good Rock 0-4 2,05 4-6 10,75 6-18 15,45 at and > 18 18,40
M 555+37 to 556+66 129 ft	2,825 Range of 6 2,500-3,900	9,850 Range of 3 9,200-10,800 D_1 = 4 Range 3-4	14,650 Range of 4 13,300-16,000 D_2 = 9 Range 5-16		Good to Fair Roc 0-4 2,82 4-9 9,85 at and > 9 14,65
J 562+65 to 563+70 105 ft	None	4,325 Range of 4 3,200-5,100	13,850 Range of 4 12,200-15,200 D_2 = 4 Range 3-6		Fair to Poor Roc 0-4 4,32 at and > 4 13,85
B 563+20 to 563+90 70 ft	2,500 Range of 4 2,500	None	11,000 9,600-14,500 D_2 = 2 Range 1-4		Moderately Poor t Poor Rock 0-2 2,50 at and > 2 11,00
A 569+05 to 570+00 95 ft	1,850 Range of 7 17,000-20,000	5,500 Range of 8 4,900-6,200 D_1 = 2 Range 1-3	9,500 Range of 4 9,000-10,000 D_2 = 15 Range 13-17		Poor Rock 0-2 1,85 2-15 5,50 at and > 15 9,50

Table 6

SEISMIC WAVE VELOCITIES AND LAYER DEPTHS, CLEAR CREEK TUNNEL, CALIFORNIA (metric system)

Seismic Line; Tunnel Stations; Length of Line	Average Values of Seismic Wave Velocities (m/sec)				Section Showing Average Velocities and Layers
	V_1 Layer	V_2 Layer, D_1 Depth to Top of V_2 Layer	V_3 Layer, D_2 Depth to Top of V_3 Layer	V_4 Layer, D_3 Depth to top of V_4 Layer	
C 462+40 to 463+40 30.5 m	885 Range of 4 825-975	None	5,550 Range of 4 4,570-6,090 D_1 = 0.9 0.6-0.9		0-0.9 885 at and > 0.9 5,550 (> = below)
K 558+20 to 560+15 59.5 m	625 Range of 7 488-915	3,270 Range of 4 2,920-3,660 D_1 = 1.2 Range 1.2-1.5	4,720 Range of 3 4,500-4,975 D_2 = 1.8 Range 1.2-1.5	5,600 Range of 3 5,600-5,800 D_3 = 5.5 Range 3.7-7.6	0-1.2 625 1.2-1.8 3,270 1.8-5.5 4,720 at and > 5.5 5,600
M 555+37 to 556+66 39.3 m	860 Range of 6 765-1,190	3,040 Range of 3 2,800-3,300 D_1 = 1.2 Range 0.9-1.2	4,460 Range of 4 4,060-4,880 D_2 = 2.7 Range 1.5-4.9		0-1.2 860 1.2-2.7 3,040 at and > 2.7 4,460
J 562+65 to 563+70 32.0 m	None	1,320 Range of 4 975-1,550	4,225 Range of 4 3,720-4,625 D_2 = 1.2 0.9-1.8		0-1.2 1,320 at and > 1.2 4,225
B 563+20 to 563+90 21.4 m	765 Range of 4 765	None	3,350 Range of 4 2,920-4,420 D_2 = 0.6 Range 0.3-1.2		0-0.6 765 at and > 0.6 3,350
A 569+05 to 570+00 29.0 m	565 Range of 7 518-610	1,670 Range of 8 1,500-1,890 D_1 = 0.6 Range 0.3-0.9	2,900 Range of 4 2,750-3,050 D_2 = 4.6 Range 4.0-5.2		0-0.6 565 0.6-4.6 1,670 at and > 4.6 2,900

Special Case

In the Clear Creek Tunnel a section of very poor rock was tested seismically on Line D, 95 ft long, between tunnel stations 378+55 to 377+60. The rock at this location was a shattered quartz diorite, which was soft due to crushing but showed no alteration. Several steel ribs between the stations noted were warped and moved a few inches into the tunnel by pressure from the crushed section of the tunnel wall. It was considered that the crushed and fragmented condition of the rock represented relief of stress developed as the tunnel was driven. This would explain the fact that the rock, although very soft where it could be observed, was not altered or strained. Seismic data on Line D are shown in Table 7.

The elastic modulus for the average velocity of 5300 ft/sec, using a density of 2.57, was $E = 1.3 \times 10^6$ psi, 92×10^3 kg/cm^2. For 12,800 ft/sec, using a density of 2.80, $E = 6.4 \times 10^6$ psi, 450×10^3 kg/cm^2. In this case the 5300-ft/sec velocity was considered to be representative of the crushed rock zone. This zone, however, was only about 13 ft thick on the average, as high velocity was measured at and below 16 ft. The thickness of 13 ft is obtained by subtracting from the 16 ft depth to the V_3 layer the 3 ft thickness of the V_1 layer, that is, 3 ft.

Table 7

SEISMIC WAVE VELOCITIES AND LAYER DEPTHS, CLEAR CREEK TUNNEL, CALIFORNIA, SEISMIC LINE D

Seismic Line; Tunnel Stations; Length	Average Seismic Wave Velocities						Average Depth to			
	V_1 Layer		V_2 Layer		V_3 Layer		V_2		V_3	
	ft/sec	m/sec	ft/sec	m/sec	ft/sec	m/sec	ft	m	ft	m
D	1,400	440	5,350	1,630	12,800	3,900	3	1	16	4.9
378+55	Range									
377+60	1,300-	400-	3,800-	1,160-	12,200-	3,730-	2-	0.6-	14-	4.3-
95 ft	1,700	510	5,600	1,720	13,400	3,800	3	1	17	5.2

Practice in Yugoslavia

A recent translation of an article by Branislav Kujundzič and Bratislav Colić, Engineers of the Institute of Hydraulic Engineering, Belgrade, Yugoslavia, shows how seismic measurements are employed in detailed investigations in pressure tunnels in that country.[7] The authors point out that, in most cases, in hydraulic pressure tunnels the internal pressure of the water is ordinarily not entirely supported by the lining of the tunnel but in part is transferred to the rock behind the lining. Since the rock works in conjunction with the lining, it is necessary to have a thorough knowledge of the elastic properties of the body of the rock through which the tunnel is driven to properly design the lining.

The geoseismic method is described for determining the elasticity modulus of the rock in a tunnel *in situ* from measurements of the velocities of propagation of longitudinal seismic waves. Special attention is given to the use of travel time hodographs or travel time curves and the solution of the problem of locating the contact between sections along a tunnel of rock with different elasticity characteristics. Another problem discussed is that of measuring the depth of the disaggregated rock, also through the use of these curves.

It appears that the disaggregated zone is formed mainly as a result of relieving the rock mass as a whole from natural tensions as a consequence of driving a tunnel through it. The term "dubine rastresene," which is translated as disaggregated, is taken to mean the strain relieved and/or cracked or shattered zone of rock surrounding a tunnel.

Tests made at several sites have shown that the values of the elastic modulus determined seismically are several times greater than the modulus of the same material determined by static loading of a portion of the tunnel wall and measuring the resulting deformations. The differences are considered to relate to the nature of the two procedures. The seismic method is based on the deformation of small particles of rock material from explosions where the force is applied for a very short time. The *in situ* pressure tests involve pressure lasting for several hours, or even days. Since rock is an elastoplastic substance, the deformation depends on the amount of pressure and also on the length of time it is applied.

It is the opinion of these authors that the elastic modulus determined by seismic measurements represents the upper limit or maximum value for the rock tested. As used in studies of pressure tunnels the modulus of elasticity of the disaggregated zone as measured by the seismic technique is considered as most significant and helpful in solving problems of injection grouting behind the lining of pressure tunnels.

The data secured in the geoseismic examination of three tunnels in connection with hydroelectric developments are presented in tabular form. These are the development at (1) Vrutok, involving two tunnels, 3164 m (10,400 ft) and 3000 m (9900 ft) in length; (2) Raven, 1400 m (4600 ft); and (3) He-Gojak, 9.5 km (31,200 ft).

Comparison of Results by Different Methods

Seismic Determinations versus Static Tests of Cores

An important question in the use of seismic determinations of the elastic modulus of foundation rock, in place, is how closely do these values compare with those determined by other means of testing. Such a comparison for seismic field investigations at three Bureau of Reclamation dams made during the period from 1955 through 1957 is shown in Table 8.

The table shows that for the quartzite of the Flaming Gorge site the seismic measurements and the static tests of cores gave closely similar results. For the sandstone and shale at the Monticello site and the sandstone at the Glen Canyon site, the seismic elasticity was from 1 to 4 times the static determinations.

Explanation

At the Monticello Dam site the modulus of elasticity of the sandstone determined seismically at $E = 2.85 \times 10^6$ psi was an average of $E = 3.47 \times 10^6$ psi, over 219 ft of surface seismic line, normal to the

Table 8

COMPARISON BETWEEN VALUES OF THE MODULUS OF ELASTICITY OF FOUNDATION ROCK IN PLACE DETERMINED BY SEISMIC FIELD MEASUREMENTS AND STATIC LABORATORY TESTS OF CORES

Location of Test	Type of Rock	Modulus of Elasticity, $E \times 10^6$ psi	
		Seismic Av	Static Av
Monticello Dam site, California	Sandstone Shale	2.85 2.35	1.54 1.94
Flaming Gorge Dam site, Utah	Quartzite	2.53	2.37
Glen Canyon Dam site, Arizona	Sandstone	1.97	0.45

bedding of the sandstone, and $E = 2.22 \times 10^6$ psi, over 193 ft of line along the bedding. The value for purposes of comparison, $E = 1.54 \times 10^6$ psi, was the average of static (pressure) tests on 23 cores, 75 per cent saturated, at 0 to 400 psi, for the first cycle of loading. The cores were selected from four drill holes.

The seismically determined average elasticity of the shale at this site was 2.35×10^6 psi. It was an average of $E = 3.18 \times 10^6$ psi, over 184 ft of surface seismic line normal to the bedding of the shale, and $E = 2.52 \times 10^6$ psi, over 45 ft of line along the bedding. The value $E = 2.35 \times 10^6$, for the shale, compares with the value $E = 1.94 \times 10^6$ psi from tests on three samples that were also 75 per cent saturated, at 0 to 400 psi, and for the first cycle of loading.

At the Flaming Gorge Dam site the average seismic elasticity of the quartzite was $E = 2.53 \times 10^6$ psi, which was obtained from measurements along 270 ft of drill hole. For comparison the average value $E = 2.37 \times 10^6$ psi was secured from static tests on eighteen samples, air dry, at 0 to 1000 psi for the first cycle of loading.

Although not shown in Table 8, at this site measurements on 183 ft of surface seismic line normal to the bedding gave a value of $E = 1.99 \times 10^6$ psi, and for measurements on 589 ft of line along the bedding of the quartzite a value of $E = 2.85 \times 10^6$ psi. The average of these surface seismic measurements is $E = 2.42 \times 10^6$ psi.

At the Glen Canyon Dam site the average seismic value of the elasticity of the sandstone was $E = 1.97 \times 10^6$ psi. This value was an average from measurements over a length of 276 ft made in four drill holes in the right abutment which gave a value of $E = 2.09 \times 10^6$ psi, and measurements over a length of 483 ft in eight drill holes in the left abutment which gave a value of $E = 1.76 \times 10^6$ psi.

For comparison the core test value of $E = 0.45 \times 10^6$ psi is based on static tests of ninety-three samples of core, at 75 per cent saturation, and 0 to 400 psi for the first cycle of loading. It is a weighted average value of the above tests considering types or groups of cores by strength and percentage of their occurrence at the dam site.

Seismic versus Core Tests over a Section of an Abutment

The characteristic of a seismic field measurement to sum up the quality of the rock over a portion of an abutment is illustrated by data secured at the site of Morrow Point Dam in Colorado. The Morrow Point Dam is a feature of the Bureau of Reclamation's Colorado River Storage Project.

During field work at that dam in the spring of 1962, Seismic Line F consisted in a measurement of seismic wave velocity through a section of rock on the right abutment some 176 ft in length. As shown in Fig. 7, Line F extended from a shot point at a location about 95 ft from the portal of an exploratory tunnel in the right abutment upward through the abutment rock to the site of Drill Hole 31 on the surface. This drill hole had not been drilled at the time the seismic measurement was carried out. In the course of additional seismic testing, during December 1962, it was planned to measure seismic wave velocity in that drill hole. This was, however, not possible as the drill hole was not open.

As a result of the program of static laboratory tests of drill cores from the Morrow Point site and the petrographic study of some sixty-five specimens of core, five groups or types of rock were established. The average value of the elastic modulus, and its range, for

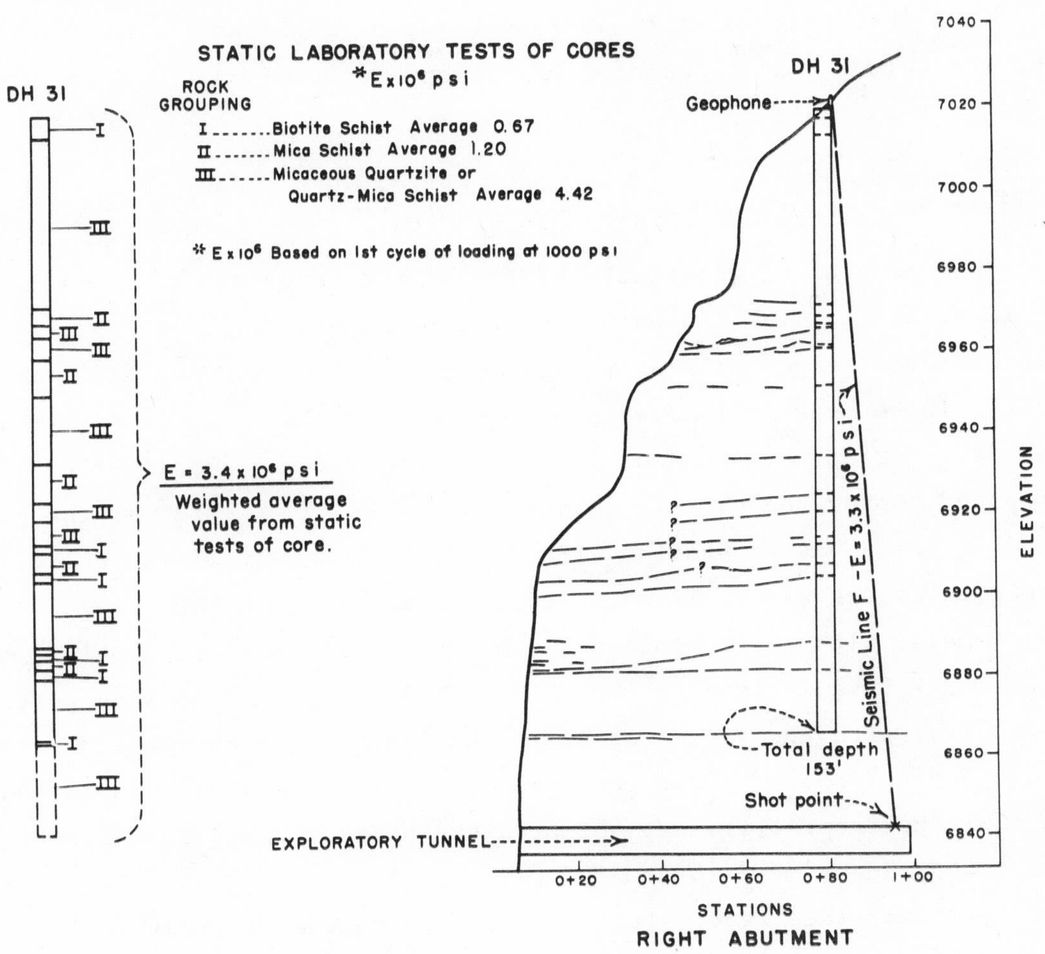

Fig. 7—Seismic measurements of rock elasticity, Morrow Point Dam.

the three of these types present in Drill Hole 31 and the section below the drill hole are shown in Fig. 7.

The seismic wave velocity along Line F was 10,000 ft/sec. Using a value of density of the rock (from tests of samples) of 2.75, the elastic modulus obtained was $E = 3.3 \times 10^6$ psi. To establish a comparison the weighted average value of the elastic modulus of the rock along Line F was calculated. In the calculation the average values of the elastic modulus of the rock from laboratory tests of core for the intervals involving rock of different types in the drill hole and the section of rock below the drill hole were used. These intervals and the types of rock, designated by roman numerals I, II, and III, are shown in Fig. 7.

The resulting weighted average value of the elastic modulus of the rock along Line F based on laboratory tests of sample cores was $E = 3.4 \times 10^6$ psi.

Seismic Determinations versus Jack Loading Tests

The modulus of elasticity of the Navajo sandstone of the left abutment at the Glen Canyon Dam, on the Colorado River at Page, Arizona, was measured seismically at the sites of two jack loading tests in June 1962. These measurements were made in a foundation tunnel in the left abutment, at an elevation of 3480. This tunnel was 235 ft below the proposed top of the dam at elevation 3715 ft. The center of the jack loading tests was at Station 0+87.9. Through a pair of hydraulic jacks, pressure was applied to two circular concrete pads 24 in. (61 cm) in diameter with their centers 36 in. (91.5 cm) apart in 200-lb (91 kg) increments to a maximum of 600 lb (272 kg). The jack loading tests were made first with the jacks vertical, the pressure being applied to the floor and the roof. Tests were then made with the jacks mounted on suitable horizontal frames, the pressure being applied to the walls of the tunnel.

Seismic lines were established along the tunnel floor and on the left wall, some 4 ft above the floor. These lines extended between tunnel stations 0+22 and 1+06.9. Using a 10-ft geophone spacing, eight shot points were used on each line in order that details of the shallow (seismic) layers at and near the surface of the rock and the seismic wave velocities in these layers could be secured. Data on Line I along the tunnel floor and on Line II on the left wall of the tunnel are summarized in Table 9.

In determining the applicable value of elasticity of the rock at the jack loading test sites as determined seismically, it was assumed that the rock below the depth of 15 ft was not affected by the loads used in the tests carried out. The weighted average value of the elasticity was calculated at the vertical jack loading site (using a 2-ft section with a value of $E = 0.07 \times 10^6$ psi, a 7-ft section with $E = 1.0 \times 10^6$ psi, and a 6-ft section with $E = 2.0 \times 10^6$ psi). The

Table 9

SEISMIC WAVE VELOCITIES, LAYER DEPTHS, AND VALUES OF THE MODULUS OF ELASTICITY--LEFT ABUTMENT FOUNDATION TUNNEL, ELEVATION 3480, AT JACK LOADING TEST SITES, GLEN CANYON DAM, PAGE, ARIZONA

Seismic Cross Sections					
LINE I Tunnel Floor	$E \times 10^6$ psi	$E \times 10^3$ kg/cm^{-2}	LINE II Tunnel Wall	$E \times 10^6$ psi	$E \times 10^3$ kg/cm^{-2}
0-2 ft 1650 ft/sec 0-0.6 m 518 m/sec	0.07	5	0-3 ft 1960 ft/sec 0-0.9 m 595 m/sec	0.1	7
2-9 ft 6125 ft/sec 0.6-2.7 m 1865 m/sec	1.0	70	3-8 ft 5520 ft/sec 0.9-2.4 1680 m/sec	0.8	56
At and below 9 ft 8750 ft/sec 2.7 m 2660 m/sec	2.0	140	At and below 8 ft 9000 ft/sec 2.4 m 2660 m/sec	2.0	140

resulting weighted average value of $E = 1.3 \times 10^6$ psi. From the vertical jack loading the value of $E = 1.4 \times 10^6$ psi.

The weighted average value of elasticity from the seismic measurements on the tunnel wall was $E = 1.2 \times 10^6$ psi, which can be compared with the value of $E = 1.1 \times 10^6$ psi obtained from the horizontal jack loading at the same site.

Discussion

In making such comparisons it is, of course, necessary to consider the fundamental characteristics of different procedures and to recognize the fact that merely because two things bear the same name they may not necessarily be equivalent. A case in point is the term Young's modulus of elasticity, the theory of which is based on the assumption that the material involved is uniform and isotropic. That rocks and formations seldom approach this criterion is common knowledge.

However, we can accept the premise that the elastic properties of rock do control the speed at which seismic waves travel in and through it. We can assume that rock, considered in a gross sense, is for practical purposes uniform. On a basis of these assumptions useful field data can be secured for engineering purposes. We can, furthermore, take advantage of the unique features that seismic field measurements test the rock *in situ* and in relatively large sections.

Conclusion

It is our opinion that seismic measurements of the elastic modulus of the rock in foundations and abutments of dam sites can be employed advantageously in the design of the structures to be built and in connection with and supplemental to other types of field tests at sites and laboratory tests of samples of foundation rock. Our program to establish quantitatively the degree of agreement and/or the differences in the values of rock properties by different procedures is not complete. A number of additional seismic field studies are planned for the near future.

Possible Improvements

There are a number of possibilities for improvement in the application of the seismic refraction method to the problem of measuring the elastic properties of foundation and tunnel rock *in situ*. Two such improvements are tools, presently available, for continuous borehole velocity logging and radiometric logging to measure rock density and the amount of moisture in the rock encountered in a drill hole.

Borehole Velocity Logging[8]

The continuous measurement of seismic wave velocity from point to point in a drill hole was developed for use in seismic prospecting for oil structures. The procedure in essence consists of applying a pulse of wave energy from a continuous or a pulse vibrator incorporated into a logging tool that is lowered into a drill hole. This energy is received and recorded at a geophone placed at a fixed distance above (or below) the vibrator and also made a part of the logging tool. The distance of the wave source and vibrator may be 5 or 10 ft (1.5 to 3.0 m). The continuous graphic log that is made shows the changes in velocity of seismic waves through the section of rock traversed. Such a velocity log supplies an amount of detail on the wave velocity character of the rock that cannot be secured by the step-by-step method of shots and recordings on a borehole geophone placed at different depths in the drill hole.

The difficulty in applying the velocity logging technique to rock properties studies at dam sites lay in the fact that such logging tools were not available commercially in a size that could be used in small-diameter drill holes. It is understood that the U. S. Geological Survey now has developed and field-tested a velocity logger for use in a 3-in.-diameter (11.8 cm) drill hole.

Borehole Radiometric Logging[9]

Based on earlier oil industry logging, the quantitative measurement of the density and moisture content of rock by radioactive logging has been brought to a high state of perfection. Many of the later developments in gamma ray and neutron logging have been made (in recent years) by the geophysicists, geologists, and physicists associated with the U. S. Atomic Energy Commission. Field-tested radiometric instruments for borehole logging are readily available commercially. These tools can improve our determinations of rock density for use in elastic modulus of foundation and tunnel rock *in situ*.

A New Type of Seismograph[10]

In the last few years, a new type of seismic field equipment has come on the market that departs markedly from what may be called the conventional type in several respects, namely:

1. The blow of a sledge hammer is substituted for the charge of dynamite as a source of seismic wave energy.

2. An electronic timer is substituted for the photographic recording oscillograph.

3. An inertia switch attached to the sledge hammer starts the timer, and the arrival of the seismic wave at a geophone stops it.

4. Wave travel time, from the striking point on the ground surface to the geophone (placed at a measured distance from the strike point) is recorded by a series of small electric lights that are turned on and that represent milliseconds and fractions of milliseconds.

This type of seismic equipment, which for want of a better name may be called a sledge hammer seismograph, cannot accomplish anything that cannot be accomplished with a 12-trace standard equipment, in making seismic wave velocity measurements. In fact, the single geophone must be placed at twelve different distances from the striking point and twelve travel times measured, one at a time, to equal the data secured from one shot and one seismogram. However, the new instrument requires only two men to operate it and costs only about one-fourth as much as a more sophisticated set of seismic instruments.

It is pertinent also that the new instrument is light in weight and very simple to operate and with certain improvements should be quite useful in rock property studies at dam sites and in tunnels. The timing accuracy of the scaler unit is, in fact, in terms of 1/2 to 1/4 of a millisecond. That is an advantage in measuring short seismic lines. Although as far as is known to the writer the instrument has not been used in borehole velocity measurements, it has a good potential in that application and also in the measurement of the velocity of transverse seismic waves.

REFERENCES

1. Brown, P. D., and J. Robertshaw, "The _In-situ_ Measurements of Young's Modulus for Rock by a Dynamic Method," _Géotechnique_, Vol. 3, No. 7, 1953, pp. 283-286.
2. Wantland, D., "Seismic Investigations of the Depth of Weathered Rock at the Oroville Dam Site--Feather River--Central Valley Basin, California," _Geology Report No. G-110_, October, 1950, U. S. Bureau of Reclamation, Denver, Colorado.
3. Wantland, D., "Geophysical Measurements of the Depth of Weathered Mantle Rock," _Am. Soc. Testing Mater._, _Spec. Tech. Publ_. No. 122, 1951.

4. Heiland, C. A., Geophysical Exploration, Prentice-Hall, Inc., New York, 1940.
5. Jakosky, J. J., Exploration Geophysics, Trija Publishing Company, Los Angeles, California, 1950.
6. Leet, L. Don, Earth Waves, Harvard Monographs in Applied Science, No. 2, Harvard University Press, Cambridge, Massachusetts, 1950.
7. Kujundzič, B., and B. Colić, "Determination of the Elastic Modulus of Rocks and of the Depth of the Disaggregated Zone in Hydraulic Pressure Tunnels by Means of the Seismic Refraction Method," Trans. Inst. Hydraulic Eng., No. 8, Belgrade, Yugoslavia, 1959.
8. Kokesh, F. P., "The Long Interval Method of Measuring Seismic Velocity," Geophysics, Vol. 21, No. 3, July, 1956.
9. Scott, J. H., P. H. Dodd, R. F. Droullard, and P. J. Murda, "Quantitative Interpretation of Gamma-Ray Logs," Geophysics, Vol. 26, No. 2, April, 1961.
10. Fahnestock, C. R., "Use of Seismic Techniques in Analyzing Subsurface Materials," Preprint No. 61L45, Meeting of AIME, St. Louis, Missouri, February, 1961 (29 W. 30th St., New York 18, N. Y.).

DISCUSSION

G. E. LAROCQUE (Canada):

With reference to Mr. Wantland's comment concerning the utilization of sledge hammer seismology in borehole velocity measurement, the Department of Mines, Ottawa, has always been interested in delineating the fracture zone around mine openings with this type of measurement. For this purpose a portable transistor unit was developed and has been in use for the last six months.

Velocity measurements are made between paralleled drill holes at a selected distance from the face. The receiving and transmitting units are identical; each consists of a hydraulic cylinder with a retractable side piston for wedging the unit into place. A barium titanate transducer is mounted on the back of each piston. Steel rods with couplers supply the mechanical connection between the units and the free face.

In operation, the pipe connected to the transmitter is hit with a hammer. The barium titanate transducer in the transmitter senses the arrival of the stress wave due to the blow and supplies a trigger that initiates transit time measurement. The identical barium titanate transducer in the other hydraulic cylinder acts as receiver. A small transistor preamplifier and line driver is included in the pipe behind each hydraulic cylinder. The whole sonic unit with two sets of rechargeable batteries (one set weighs 4 lb) can be used continuously for eight hours.

Figures 8 and 9 are photographs of the complete unit with the omission of connecting rods and hydraulic pump.

Fig. 8—Sonic console unit.

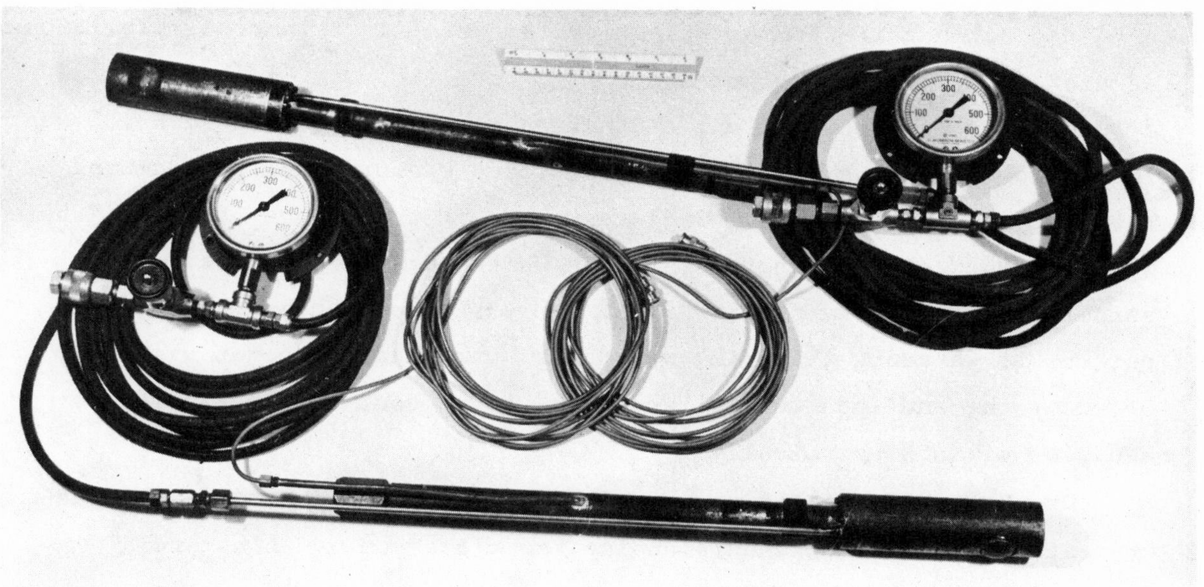

Fig. 9—Sonic receiving and transmitting units.

D. WANTLAND in reply:

Mr. Wiebenga* poses a question concerning my statement that a general range of Poisson's ratio is from 0.1 to less than 0.3, which I should emphasize is a general range. He notes that in certain mudstone in Tasmania, which has a wet density of 2.5, there was a Poisson's ratio of 0.38 plus or minus 0.005, measured with seismic methods. We all know that you can have various Poisson's ratios for various rocks, and I might comment that at the Bureau of Reclamation we had a negative Poisson's ratio one time, not measured with dynamic methods, which was a great embarrassment to all concerned.

A written comment was given to me by Dr. M. F. Bollo of Paris, France. He points out a number of things that I did not have an opportunity to present in my paper, some of which I was well aware of.

More particularly, he points out the importance of measuring the anelastic response of a rock. I did not mention it, but at the Bureau of Reclamation in our tests of dam sites we do measure that property, and we normally do it with jacks.

I was presenting numerical comparisons between the seismic determinations of elastic modulus and other forms of measuring that modulus, including tests of cores and jacking tests.

---

*Not in attendance at the conference.

Dr. Bollo calls to our attention the great importance of vibration studies, where you put vibration energy into the rock of the walls of a tunnel, or of a foundation, and note the response of the rock to such impressed vibrations. Among the quantities measured are the damping of the waves that you put in. Such tests are, I think, one of those "Where we are going" phases, that is, in the United States at least in our seismic rock property studies. Dr. Emery has pointed out to me that perhaps if we can get damping we can be smart enough to get viscosity of the rock, and then we are "knocking at the door" of the anelastic behavior of rock seismically.

Dr. Bollo also calls our attention to the fact that seismic measurements contribute to understanding how close you can blast to a structure or a proposed structure, and with what size of charges, and that such measurements are an aspect of seismic prospecting.

M. F. BOLLO (France):

I should like to thank Mr. Wantland for the abstract that he has made. I should like to call your attention to the fact that I have found a very interesting application of seismic methods for the detection of cracks in the rock foundation of very delicate constructions, such as nuclear reactors in Mont d'Ahrée or Chinon, France.

Another thing that I would like to call to your attention is the dynamic method such as indicated by Mr. Wantland. It is practically the only method that can give complete information on the compressed and decompressed zones of rock without giving too much trouble.

For the work in the site, for example in the Mount Blanc Tunnel, we have been able without hindering the work there to follow the decompression of the rock under the pressure of the mountain until 8000 ft above. We have followed this decompression up to a depth of some 30 ft. The log of the speed of the decompression propagation is very interesting.

In the general line for the carrying out of the studies on dam sites, we have been doing those at the present time at the rate of 1600 to 7000 seismograms for the site of a dam.

We are now at about the 250th dam, and only on about 15 per cent of the sites was it possible to make a direct correlation of the modulus of elasticity with the coefficient of static deformation. In all the other cases the results are translated taking into account the static results carried out by the available methods in each case.

The flatjack is one of the most useful methods in this case.

D. D. DICKEY (USA):*

In his paper Mr. Wantland has pointed out the invalidity of the assumption that rock is homogeneous and isotropic, and that this divergence will cause some error in the calculated elastic moduli. I would like to point out two other possible sources of error when comparing static with dynamic elastic moduli on the same section of rock.

1. If the rock is removed from the site and the measurements are made in the laboratory, the change in stress caused by removal of the rock may change its elastic properties. The magnitude of change of static Young's modulus with pressure is given in Fig. 1 of the paper by W. F. Brace.

2. Presence of a zone or layer of high velocity material near the seismic line will provide a path through which the waves that arrive first will travel, but this will not be representative or an average velocity for the rock mass as a whole. For example, a limestone bed in a shale sequence or a silicified fault zone in an altered granite may give misleading results if adequate care isn't taken in interpretation of seismic results. Just as this seismic method aids in the evaluation of the geology, a knowledge of the geology aids in the interpretation of the seismic data.

R. E. GOODMAN (USA):

Mr. Wantland suggested the use of a weighted average of the elastic moduli of the individual layers for calculating an equivalent over-all modulus for a layered rock mass. I would interpret the term "weighted average" to signify the operation specified by Eq. (1). (See Fig. 10.)

If one would compress a mass of layered rock in the direction of the layers, he could correctly calculate the deformation of the loaded rock mass using an equivalent modulus given exactly by the weighted

*Mr. Dickey's discussion was submitted after the conference.

average, as shown in Eq. (2). If, however, one applies a pressure to a layered rock mass in a direction perpendicular to the layers, he would calculate the deformation correctly if he used an equivalent modulus given by Eq. (3), which is quite different from the weighted average. These formulas are derived by assuming the layers to act ideally, and not particularly as rocks. For compression across the layers, the stress is considered to be the same in every layer, while in compression parallel to the layers, the strain is the same for each layer. This, of course, may not be true for layered rocks.

As an example, for a rock mass consisting of two 10-ft layers, one having a modulus of 5 million psi, and the other of 1 million psi, an equivalent modulus for the rock mass when compressed parallel to the layers would be 3 million psi, but only 1.7 million psi if compressed across the layers.

Thus the use of a weighted average modulus of elasticity for layered rock is justifiable only when the applied loading is in the direction of the layers.

1\. Average means $\dfrac{E_1 \ell_1 + E_2 \ell_2 + E_3 \ell_3 \cdots}{\ell_1 + \ell_2 + \ell_3 \cdots}$,

2\. E_1 E_2 E_3; ℓ_1 ℓ_2 ℓ_3 $\longrightarrow E_{eq} = E_{av}$,

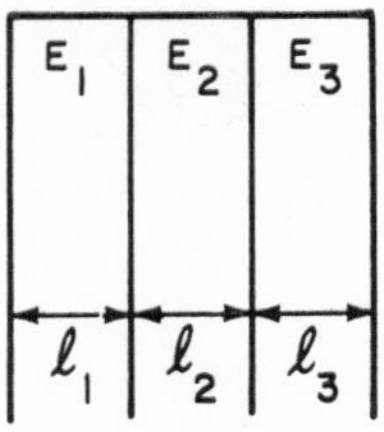

3\. E_1 ℓ_1; E_2 ℓ_2; E_3 ℓ_3 $\longrightarrow E_{eq} = \dfrac{\ell_1 + \ell_2 + \ell_3 \cdots}{\dfrac{\ell_1}{E_1} + \dfrac{\ell_2}{E_2} + \dfrac{\ell_3}{E_3} \cdots}$.

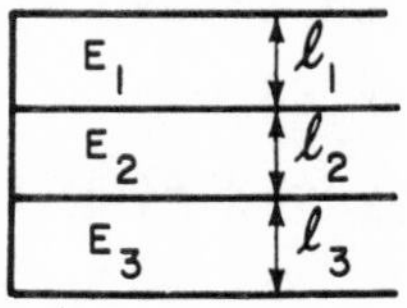

Fig. 10—Equivalent modulus of elasticity of layered rock.

ABSTRACT

Petrofabrics is the study of all the structural and textural elements of a rock ranging from the configuration of the crystal lattices of the individual mineral grains up to and including large-scale features that require field investigation. Accordingly, petrofabrics deals with the recognition, measurement, and illustration of the spatial arrangement of fabric elements, and with the kinematic and dynamic interpretations of these data.

The ability to map the principal stress directions in rocks at the time of deformation by use of petrofabric techniques is a relatively new development. It stems from the dynamic interpretation of fabric data that is based on an understanding of the mechanisms of rock deformation as gained through experimental and theoretical studies. Fracture and fault, gliding flow, rotation, and recrystallization phenomena provide fabric elements from which one may be able to determine the principal stress directions in rocks. Examples from the literature will be discussed to illustrate the utilization of petrofabrics in this area. These will include the derivation of stresses associated with geologic structures by study of fractures and faults, calcite and dolomite twin lamellae, quartz deformation lamellae, intragranular rotation phenomena, and the equilibrium orientation of grains.

RÉSUMÉ

La pétrographie structurale, ou "petrofabrics" est l'étude de tous les éléments structuraux et texturaux d'une roche, depuis la configuration des structures cristallographiques des grains minéraux individuels jusqu'aux caractéristiques à grande échelle qui requièrent des recherches sur les lieux. Par conséquent, l'analyse "petrofabric" comporte la reconnaissance, la mesure et l'illustration de l'arrangement dans l'espace des éléments "fabric" et les interprétations cinématiques et dynamiques de ces données.

La possibilité de faire la carte des directions des contraintes principales dans les roches au moment de leur déformation en utilisant les méthodes "petrofabric" est un progrès relativement récent. Il provient de l'interprétation dynamique des données "fabric," basées sur une compréhension des mécanismes de déformation des roches, fournie par des études expérimentales et théoriques. Les phénomènes de rupture et de faille, de flux de glissement, de rotation et de recristallisation fournissent des éléments "fabric" qui permettent la détermination des directions des contraintes principales dans les roches.

Des examples publiés seront discutés pour illustrer l'utilisation de l'analyse "petrofabric" dans ce domaine. Celle-ci comprendra la déduction des contraintes correspondant aux structures géologiques par l'étude des ruptures et des failles, des lamelles jumelles de calcite et de dolomite, des lamelles de déformation de quartz, des phénomènes de rotation intergranulaires, et d'orientation d'équilbre des grains.

AUSZUG

Die Gefügekunde umfasst alle Struktur- und Texturelemente eines Gesteines, angefangen vom Kristallgitter der einzelnen Mineralkörner bis zu den makroskopischen Erscheinungen, welche Felduntersuchungen erfordern. Sie beschäftigt sich infolgedessen mit der Erkennung, Messung und Beschreibung der räumlichen Verteilung der Gefügeelemente, sowie mit der kinematischen und dynamischen Deutung der Daten.

Das Vermögen, die Hauptspannungsrichtungen in Gesteinen und Gebirgen zur Zeit der Deformation durch gefügekundliche Untersuchungen zu bestimmen, stellt eine verhältnismässig neue Entwicklung dar. Sie wurde durch die dynamische Auswertung der Gefügedaten ermöglicht. Diese Deutung beruht ihrerseits auf dem Verständnis der Mechanismen der Gesteinsdeformation, das durch neuere, experimentelle und theoretische Studien gewonnen wurde. Durchklüftung und Verschiebung, laminare Gleitung, Rotation und Rekristallisierung ergeben Gefügeelemente, welche erlauben, die Hauptspannungsrichtungen in Gesteinen zur Zeit der Deformation festzulegen. Die Auswertung solcher Gefügedaten wird anhand von Beispielen aus der Literatur erläutert. Diese umfassen die Ableitung von Spannungen, die bestimmte geologische Gefüge verursacht haben, durch das Studium der Klüfte und Störungen der Zwillingslamellen in Kalzit- und Dolomitkristallen, der Deformationslamellen in Quarzkristallen, sowie der Gleichgewichtsorientierung der Körner.

PETROFABRIC TECHNIQUES FOR THE DETERMINATION OF PRINCIPAL STRESS DIRECTIONS IN ROCKS

Melvin Friedman*

Introduction to Petrofabrics

PETROFABRICS (from Gefügekunde der Gesteine[(1)]) is an important geological discipline which can provide knowledge on the state of stress associated with naturally deformed rocks. According to Turner's interpretation (Ref. 2, p. 149) of "Gefüge" (fabric), petrofabrics is the study of all structural and textural features of a rock as manifested in every recognizable rock element from the configuration of the crystal lattices of the individual mineral grains up to and including large-scale features which require field investigation. The fabric of a rock, therefore, is extremely complex, is seldom completely specified, and is developed throughout the entire history of the rock. An undeformed sedimentary rock, for example, has a fabric which relates to its depositional and diagenetic histories. For instance, the long axes of detrital grains, flute casts on bedding planes, and cross beds are fabric elements related to current directions during sedimentation. Permanent deformation of this rock (Ref. 3, p. 3) modifies the initial fabric and introduces many new fabric elements. These may range in size over at least 15 orders of magnitude from changes in the crystal lattices (10^{-9} m = 10 Å) to deformation of the rock into great folds and faults of mountain ranges and basins (10^{6} m). Although many kinds of elements can be distinguished, in practice a few easily measured ones are chosen which are thought or are known to be critical to the problem at hand. Those recognized as criteria of deformation include a variety of crystallographic parameters, deformation lamellae, kink

*Shell Development Company, Houston, Texas.

bands, lineations, fold axes, fractures, faults, and other planes of mechanical discontinuity. The spatial array of any one of these elements is called a "subfabric" (Ref. 4, p. 863), and the measurements that specify their orientation and distribution are termed "fabric data." Petrofabrics consists, therefore, of a descriptive phase in which the fabric elements are recognized, measured, and illustrated, and an interpretive phase in which the rock fabric serves as a basis for explaining the deformation history either kinematically or dynamically (Fig. 1).

Tectonically significant fabric elements have been recognized and the techniques for their study perfected primarily through the efforts of Professors Sander and Schmidt, their co-workers, and students in Germany and Austria, and Professors Knopf, Turner, Ingerson, Phillips, and Weiss, their co-workers, and students in the United States and England. The subject has evolved from the study of complexly deformed terrains for which normal field mapping methods did not provide structural resolution. In recent years several commonly occurring fabric

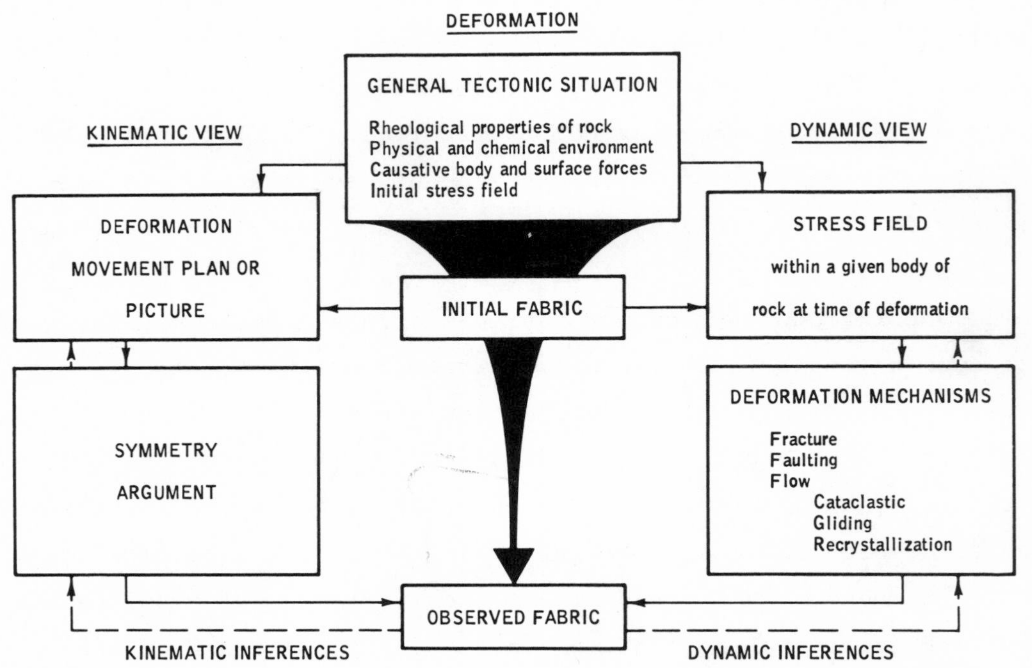

Fig. 1—Geologic deformation from the viewpoints of kinematic and dynamic petrofabrics.

elements have been given dynamic significance by comparisons between experimentally and naturally deformed rocks.

In the interpretive phase, the fabric elements are used to specify the nature of the deformation from two viewpoints, kinematic and dynamic (Fig. 1). Kinematic inferences concern the displacements that have transformed the initial fabric into the observed fabric. Dynamic inferences concern the nature of the stresses in the rocks at the time of deformation. In principle, all fabric elements that are significant criteria of deformation should lead to the same dynamic or kinematic interpretations even though they may be of different origin or scale.

The kinematic approach is favored by Sander and the Innsbruck school because "to correlate fabric with internal movements is less doubtful than is the more tenuous correlation with forces responsible for such movements" (Ref. 5, p. 2).* Symmetry is the basic criterion for correlating fabric with movement.** It is assumed that the symmetry of the rock fabric reflects the symmetry of the movement responsible for the evolution of that fabric. The principle is illustrated by the bending of wheat stalks and the rippling of the surface of water by wind. There is support for the validity of the principle in physics[4] and with regard to experimentally deformed rocks as discussed by Turner.[5] A second aspect of kinematic analysis involves unrolling ("Rückformungen") and levelling ("Horizontierung") proposed by Sander (Ref. 7, pp. 170-184) to derive an observed structure from an assumed earlier structure by the minimum amount and simplest possible kind of displacement consistent with the movement picture.

The kinematic viewpoint will not be considered further because it does not lead to the determination of the state of stress in rocks.***

---

*The forces referred to here are those of the "general tectonic situation" (Fig. 1).

**See Refs. 1, 2, and 4-14.

***There are vast amounts of excellent descriptive data on a variety of fabric elements, which to date have been only kinematically interpreted. These may eventually be amenable to dynamic analysis as the genetic relationships between the fabric elements and stresses in the rocks become known.

Moreover, usually only intensely deformed rocks exhibit fabrics with sufficiently clear symmetry for kinematic analysis. This excludes slightly and moderately deformed rocks from this form of petrofabric analysis.

The dynamic approach is an outgrowth of experimental and theoretical studies of deformation. The mechanisms by which common fabric elements are formed are systematically worked out by study of experimentally deformed single crystals, of monomineralic aggregates of these, and finally of polymineralic rocks. In addition, the relationships between these elements and the known stresses across the boundaries of the laboratory specimens are established. These specimens are then compared with their naturally deformed counterparts to determine statistically the orientation and relative magnitudes of the principal stresses in rocks at the time of deformation. This procedure is based on the assumption that the mechanisms of deformation observed in the laboratory are identical to those in nature. The dynamic approach is applicable to all deformed rocks irrespective of the intensity of deformation.

The limitations of the dynamic approach are as follows: (1) At present the mechanisms of deformation are known for only a relatively few fabric elements. (2) The orientations and relative magnitudes of the principal stresses can be determined statistically in a given volume of rock, but absolute magnitudes cannot be estimated from data of short-time laboratory experiments. (3) Although in principle it is possible to distinguish between two or more stress systems reflected in a given fabric, in practice it is difficult and often ambiguous.

Certain difficulties arise in petrofabric analysis because of the cumulative aspect of rock fabrics, the heterogeneity of stress and strain in rocks, and the scale on which a given fabric element may be sampled. Since the fabric develops throughout the history of a rock, distinction between the initial fabric and the modifications superposed by subsequent deformation(s) is a serious problem. For example, in deformed sedimentary rock it may be difficult to determine whether a given dimensional orientation of the grains is relict from the initial sedimentation or is a stable configuration in the strain field of the deformed rock. The states of stress and strain within different parts of a rock are rarely homogeneous with respect to direction and/or magnitude. Although these tend to be statistically homogeneous, stress sensitive fabric elements sometimes show large variations in their orientations and/or frequency. Accordingly descriptions of orientation patterns of the elements are necessarily statistical, as are any inferences drawn from them. Finally, it is important to keep in balance the scale of the fabric element and that of the field on which it is sampled in order to correctly evaluate the fabric. It is helpful, here, to consider the penetrative nature of a fabric element. For a fabric element to be penetrative the feature must be repeated statistically so that it effectively pervades the body and is present in the same average orientation in every sample. If the body is sampled on a scale smaller than the average spacing of the element, or if it is sampled on a scale larger than that pervaded by the element, the element becomes nonpenetrative (Ref. 4, p. 861). Hence, the fabric can be correctly evaluated only if penetrative fabric elements for a given size domain are studied.

Scope of the Paper

The primary purpose of this paper is to review dynamic petrofabric techniques now available for mapping principal stress directions in naturally deformed rocks. It is necessary to begin with the descriptive methodology employed in petrofabric analysis. This is discussed in some detail for the convenience of those readers not familiar with this subject. Included are sections on sampling and measuring, stereographic and equal-area projection, and the construction and statistical evaluation of petrofabric diagrams. This is followed by discussions of five important processes: fracturing, faulting, gliding flow, rotation, and recrystallization. As each process is discussed, the associated fabric elements are described and the literature is cited to demonstrate their usefulness in dynamic analysis.

Descriptive Methodology

Sampling and Measuring

Generalizations from observations on samples of the whole are merely statistical inferences, and the confidence within which a given statement can be made is related to the representative significance of the observations.[15,16] Systems are designed to facilitate and ensure sampling that will be representative of the features to be examined. The systems include consideration of the number and location of stations where the rock is sampled, as well as the number of measurements and the manner in which they are made at each station. In petrofabrics, such schemes are a function of the type and scale of the fabric element under consideration, the degrees of homogeneity and development* of the subfabric, and the purpose of the investigation. Sampling of macroscopic elements is obtained by standard field mapping techniques. Any microscopic petrofabric analysis begins with the collection of a geographically oriented specimen, i.e., one that is related to some fixed three-dimensional coordinate system. Oriented thin sections are cut from the hand specimen, so that the final results can be placed back into the geographic and geologic framework (Fig. 2). Various plans for sampling thin sections are discussed by Chayes.[17]

Poor sampling schemes can influence the observed orientation pattern for a fabric element. For example, one type of problem is the preferential sampling of a given fabric element because it is more easily measured in certain orientations than in others. If macrofractures are being measured along a road cut, for example, those fractures intersecting the long trend of the outcrop at high angles will tend to be preferentially sampled over fractures that trend more nearly parallel to the outcrop surface. Or on the microscopic level, a common bias in universal-stage work is to preferentially sample

---

*The orientation pattern of any fabric element varies between two end points: randomness and total alignment. The relative strength of a fabric increases as the state of total alignment is approached; and, conversely, the relative weakness of a pattern increases as it approaches randomness.

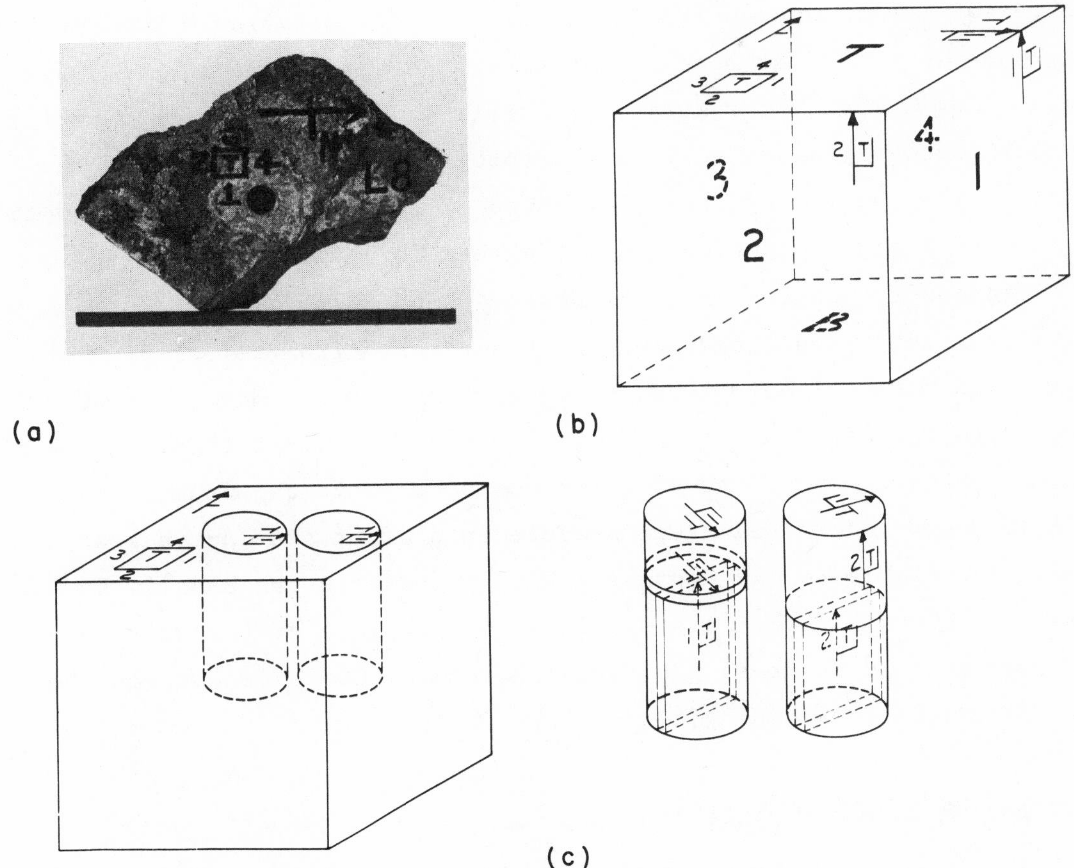

Fig. 2—Diagrams illustrating orientation convention commonly used to mark hand specimens and thin sections. (a) Hand specimen shows strike and dip directions marked on T (bedding) surface. Faces 1, 2, 3, and 4 are arranged in clockwise order with the 1 surface in the dip direction. (b) Cube shows relationships between different faces and the use of the orientation symbols. (c) Thin sections are taken from cylinders cored from the T surface of the block.

planar features inclined at high angles to the plane of a thin section at the expense of those inclined at lower angles. One can eliminate errors of this type by recognizing that they may exist and by designing a sampling plan that will eliminate or at least test for the effect.

Most measurements deal with the determination of the orientations of lines and planes in space. The choice of instrument one might use depends upon the scale of the fabric element, e.g., aerial photographs, a variety of telescopic instruments such as the theodolite and range finder, pocket transit (Brunton compass), petrographic microscope and universal stage, or the X-ray diffractometer. Detailed discussion of microscopic and X-ray measurement techniques are found in Refs. 2, 7-13, and 18-20. The accuracy of strike and dip measurements on aerial photographs varies with the topography and the scale of the photographs, but, in general, azimuths of lines can be determined to ± 3 degrees and dip angles to ± 5 degrees for angles > 60 degrees, ± 3 degrees for angles 45 to 60 degrees, and ± 1.5 degrees for angles < 45 degrees. Linear and angular measurements from telescopic instruments are very reliable (± 0.1 per cent and < ± 0.1 degree, respectively). Errors in the measurement of macroscopic fabric elements or in the marking of oriented specimens with the pocket transit can usually be held to ± 3 degrees. Universal-stage measurements of crystal optic axes are usually reliable to ± 2 degrees. Planar features can usually be located to ± 1 degree when they are inclined to the plane of the section at angles greater than 70 degrees. For inclinations of 30 to 70 degrees the error may be ± 3 degrees.*

Stereographic and Equal-area Projection

Petrofabric analysis involves spacial relationships between lines and planes and their illustration. The stereographic projection represents the surface of a sphere on a plane surface and provides an effectual means for both analytical and illustrative purposes.** Only those aspects needed to comprehend fabric orientation diagrams are reviewed here.

Consider a reference sphere cut by meridional or equatorial planes (Fig. 3), both passing through the center of the sphere. Either can serve as the reference plane of the projection, but the meridional is

---

*Recently, Kamb[21] has discussed the nature of the corrections needed in universal-stage work to compensate for differences in index of refraction between the mineral under observation and the glass hemispheres of the stage. He points out that the true corrections are negligible for cases where the ratio of the mineral refractive index to that of the hemispheres is between 0.95 and 1.05. As this condition is usually obtained by use of suitable hemispheres, refraction corrections are seldom applied to fabric data.

**See Refs. 11, 18, 22, and 23.

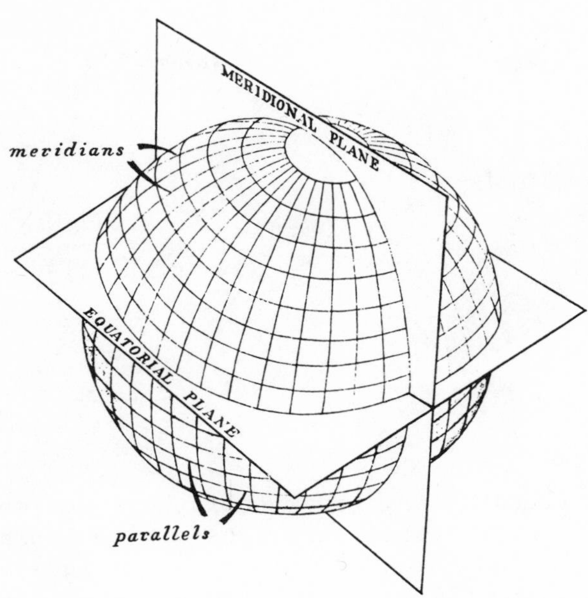

Fig. 3—A sphere or globe on which are drawn meridians or longitude circles separated by equal angles and parallels or latitude circles separated by equal angles (from Higgs and Tunell, Ref. 23, Fig. 2b).

used in petrofabrics. Next, consider any other plane of given attitude which is made to pass through the center of the sphere (Fig. 4(a)). A perpendicular to the plane is projected from the center of the reference sphere to the lower hemisphere of the sphere (P), in accord with petrofabric convention. From point P a line is projected to the zenith of the reference sphere. This line intersects the meridional plane at point P′, which is the lower hemisphere stereographic projection of point P--i.e., the normal to the shaded plane (Fig. 4(a)). If the viewpoint is changed such that the eye is at the zenith point and sights directly normal to the meridional plane, one sees the point P′ as in Fig. 4(b). The plane itself rather than its normal can be drawn by tracing onto the meridional plane the line that marks the intersection of the shaded plane and the meridional plane, and the great circle which marks the intersection of the shaded plane and the lower hemisphere.

Similarly, the azimuth and plunge of a line can be plotted by passing the line through the center and surface of the reference sphere (a point comparable to P). Then projection of P to the zenith intersects the meridional surface at another point (comparable to P′),

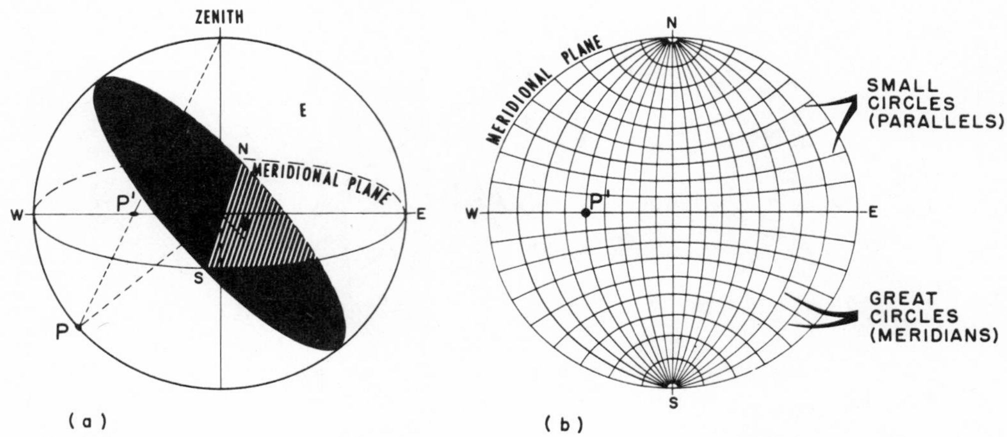

Fig. 4—Diagrams illustrating lower hemisphere stereographic projection. On the left is the reference sphere with a plane (strike N-S, dip 50°E) passing through center of sphere and intersecting the meridional plane along the N-S line. On the right, the normal to the plane (P′) is plotted in lower hemisphere stereographic projection.

which is the lower hemisphere stereographic projection of the line. Clearly, if the meridional surface is considered to be horizontal, a horizontal line will be represented by a point at a given azimuth on the periphery of the meridional plane, and a vertical line will appear as a point at the center of the meridional plane. Similarly, the normal to a vertical plane will appear on the periphery, and the normal to a horizontal plane will fall at the center.

The stereographic plotting of lines and planes is facilitated by the construction of polar or meridional nets (Fig. 5(a)). These represent the stereographic projection of points of different azimuths and vertical angles (circles of longitude and latitude). On 10-cm- and 20-cm-diameter nets the meridians and parallels are drawn at 2-degree intervals and on 40-cm nets at 1-degree intervals. The general formula for projection of any point stereographically is $r = R \tan \alpha/2$, where r is the distance along the equator from the center, R is the radius of the reference sphere, and α is the inclination from the zenith on the surface of the sphere. This generates a stereonet with true angular relationships (an equal-angle or Wulff net).

In plotting, a piece of tracing paper is usually pinned to the net so that it is free to rotate about the center of the net. A

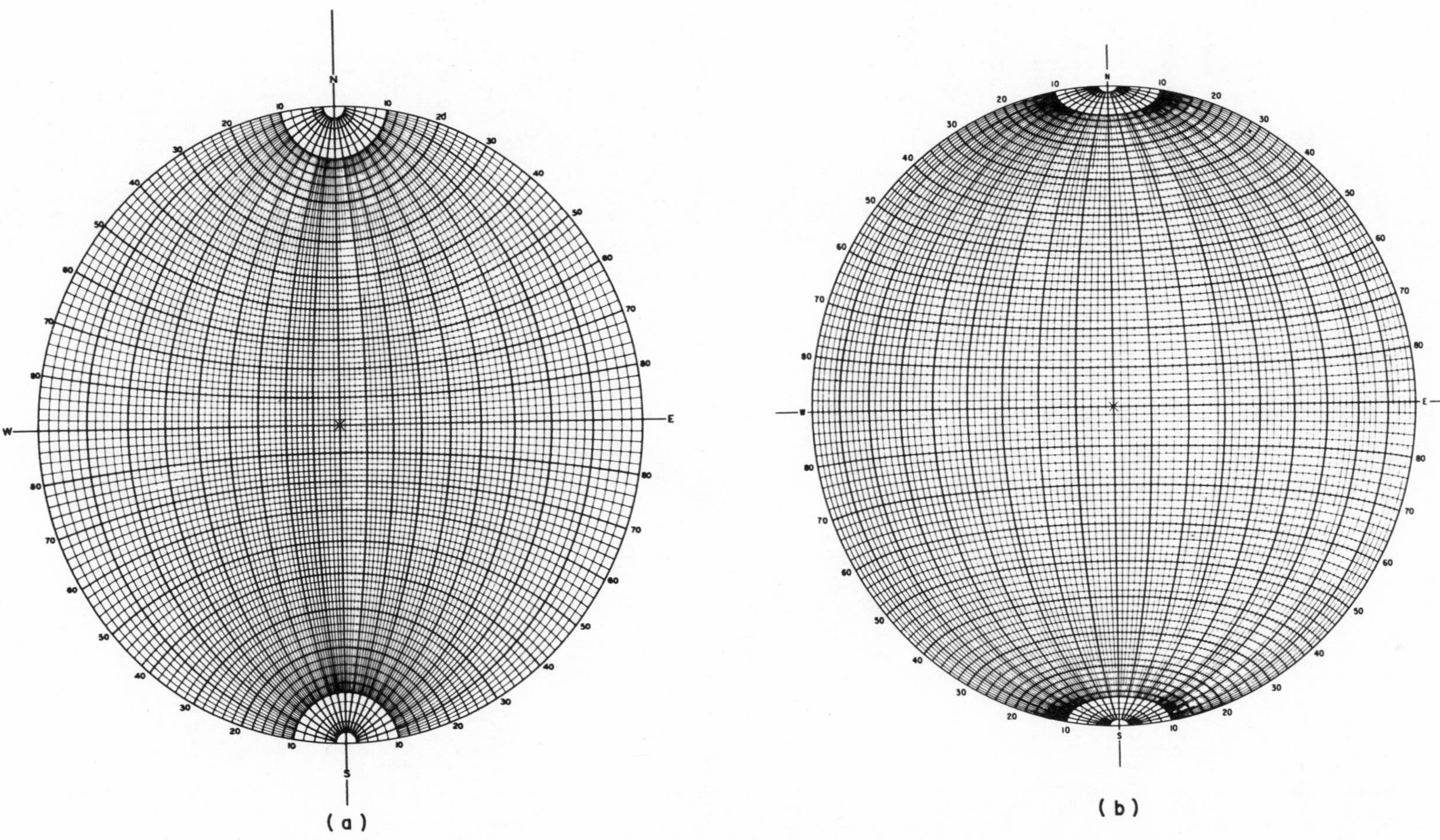

Fig. 5—Equal-angle or Wulff stereographic net on the left and an equal-area or Lambert-Schmidt net on the right.

reference point is marked on the paper over a corresponding point on the periphery of the net. Azimuths are read clockwise from north on the periphery, and angles from the vertical or horizontal are read along the equator. To plot the attitude of a line in space, the azimuth of the line is noted on the tracing paper. The paper is then rotated until the azimuth of the line coincides with the equator of the net. The plunge of the line from the horizontal or its deviation from the vertical is then marked off along the equator. Rotation of the paper back to its initial position permits one to visualize the attitude of the line in space. To plot the attitude of a plane, its strike line is noted on the tracing paper. The paper is rotated until the strike line coincides with the north-south diameter of the net. The dip is then plotted along the equator with respect to the lower hemisphere by either tracing the plane (a great circle) or plotting the normal to the plane. By working backward one can determine the azimuth and plunge of any line or the strike and dip of any planar element on a stereographic diagram.

Many other operations and graphical solutions are facilitated by use of the net--e.g., the determination of (1) the azimuth and plunge of the intersection of two planes, (2) the angle between two planes or two lines or a line and a plane (read along a great circle by rotating the tracing paper until the two corresponding points lie on the same great circle), and (3) the orientation patterns of many planes or lines that are plotted on the same diagram. Another important use of the net is to facilitate the rotation of fabric data from one plane of reference to another.[11]

Fabric orientation diagrams are amenable to statistical analysis only if the areal distribution of the data points at different locations on the diagram (different orientations) can be compared. The areal distortion inherent in the equal-angle or Wulff net is excessive. For example, a region bounded by 10 degrees of longitude and latitude near the center of the equal-angle net occupies a much smaller area than a 10-degree region near the periphery (Fig. 5(a)). To correct for this an equal-area or surface true net was devised by Lambert[24] for map projections and selected by Schmidt[25] for petrofabric analysis

(Fig. 5(b)). With this net, constructed from the formula $r = \sqrt{2}\ R \sin \alpha/2$, a unit area at any location on the projection corresponds to a unit area on the reference sphere, although somewhat distorted in shape. Equal subdivisions of the total area (e.g., 1 per cent area) can be chosen to illustrate concentrations of points projected onto the net from anywhere on the sphere.

The manipulation of the equal-area net in plotting lines or planes is identical to that previously described for the equal-angle net. Similarly, the equal-area net can be used to measure the angular relations between lines and planes. Moreover, rotation of data is more satisfactorily performed because the meridians are much more nearly equally spaced than on the equal-angle net. For these reasons the equal-area net is used almost exclusively in petrofabric work.

Petrofabric Diagrams

Petrofabric diagrams are the trademark of fabric studies. They illustrate as no other type of diagram can the three-dimensional orientations of fabric elements in a complete and concise manner. By convention these orientations are shown with respect to a specific plane of reference (the plane of the diagram) in lower hemisphere equal-area projection. Diagrams used in this review are of three types, as follows: (a) point or scatter diagrams (Fig. 6(a)) contain data from a number of measurements of a given fabric element, (b) contoured diagrams (Fig. 6(b)) show the same type of information as (a) except that the orientation pattern is emphasized by density contouring, and (c) stereograms (Fig. 6(c)) show the angular relations among relatively few lines and/or planes.

A _partial_ petrofabric diagram illustrates data from one given field of observation or only part of the available data. _Composite diagrams_ contain data from more than one field of observation on the same sample, e.g., elements measured in three mutually perpendicular thin sections cut from the same sample. Data from two of the sections are rotated into the plane of the third, or data from all three are rotated into some fourth plane of reference. _Synoptic diagrams_ are summary in nature and show fabric data for a number of different

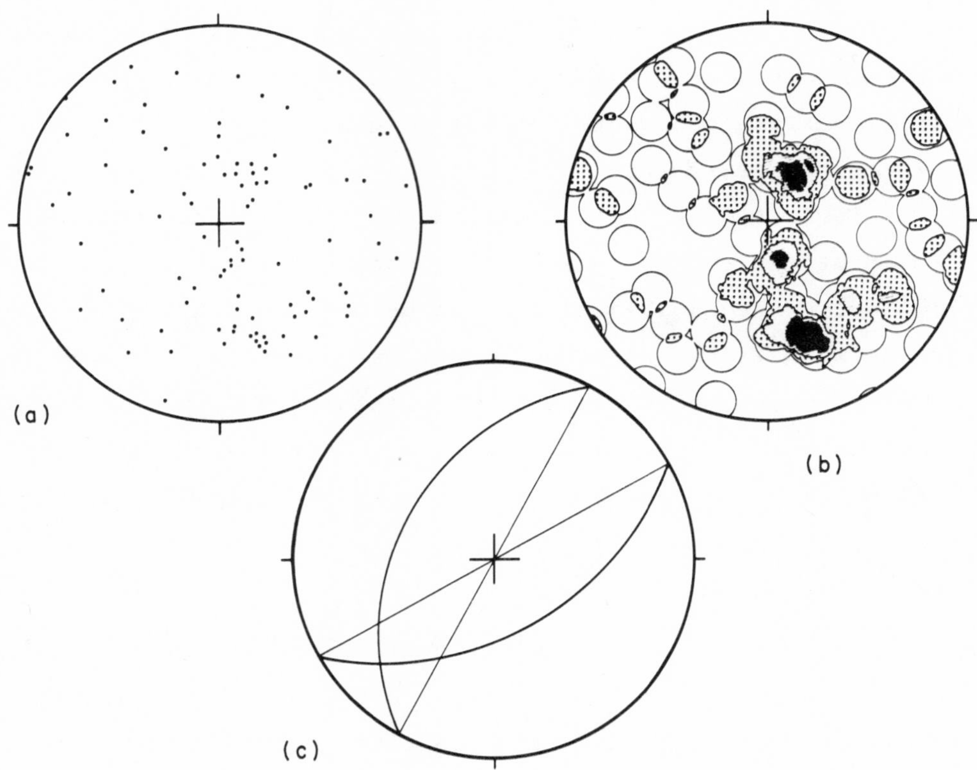

Fig. 6—Three types of petrofabric diagrams. (a) Point diagram illustrates the orientation of normals to 100 planar features or axes. (b) Contoured or density diagram of the points in (a); contours are at 1, 2, 4, and 6 per cent per 1 per cent area, 10 per cent maximum. (c) Stereogram shows two planes intersecting at 70 degrees. All diagrams are in lower hemisphere equal-area projection.

samples. In these, the data from the individual samples are rotated to a common plane of reference.

Petrofabric diagrams are easily read if the reader understands how lines and planes are plotted in lower hemisphere equal-area projection and if the writer has supplied a sufficiently informative caption. This should include at least the following: (a) sample designation, (b) type and number of data being illustrated, (c) orientation of the plane of the diagram, (d) location of geographic and/or geologic reference coordinates, and (e) nature of the contours and their values (if the diagram is contoured). The accuracy within which any point is located on a petrofabric diagram is a function of errors arising from original measurement and, for microscopic elements, from the collection of the hand specimen and the preparation of oriented thin sections. Small plotting errors included, this may be as large as ± 5 degrees, is usually ± 3 degrees, and with extreme care can be held to ± 1 degree.

Several techniques are used to contour diagrams in order to emphasize the orientation pattern. In general the contour lines are based on the number of points per unit area of the net, and the contour levels are selected with respect to the concentrations exhibited by the population of data points. A sufficiently accurate and rapid method, particularly suited to diagrams containing less than 200 points, is used by the author (Fig. 7). A contouring tool (Fig. 7(a)) for a 20-cm diagram is constructed with two circles scribed 20 cm apart. Each circle has a diameter of 2 cm so that its area is 1 per cent of the area of the diagram. Tapered holes are drilled at the center of each circle to accommodate a pencil point, and a slit is milled along the central portion of the tool to permit free rotation and translation about a pin through the center of the diagram.

To generate a contour line based, for example, on 3 points per 1 per cent area, the 1 per cent circle of the tool is placed on the diagram such that 2 points are <u>within</u> and 1 point is <u>on</u> the circle (Fig. 7(d)). The tool is then moved (pencil in contact with the paper through the center hole) such that 1 point is always on the circle and 2 are within. As new points are encountered by the circle, old ones are left behind. This is continued until the contour line is closed. Thus wherever one places the 1 per cent circle on or within the closed 3-point contour line, the circle will always be observed to contain <u>at least</u> 3 points. Points at the periphery are handled as illustrated in Fig. 7(e). The contour is expressed in terms of the percentages of the total number of points per 1 per cent area.

In practice, the contour line about 1 point is generated with a compass. In fact, the compass can be used throughout the procedure by drawing a 1 per cent circle about each point on the diagram and noting the areas of overlap.[26,27] Thus where two circles overlap, a 2 point per 1 per cent area is defined; and where three circles overlap, a 3

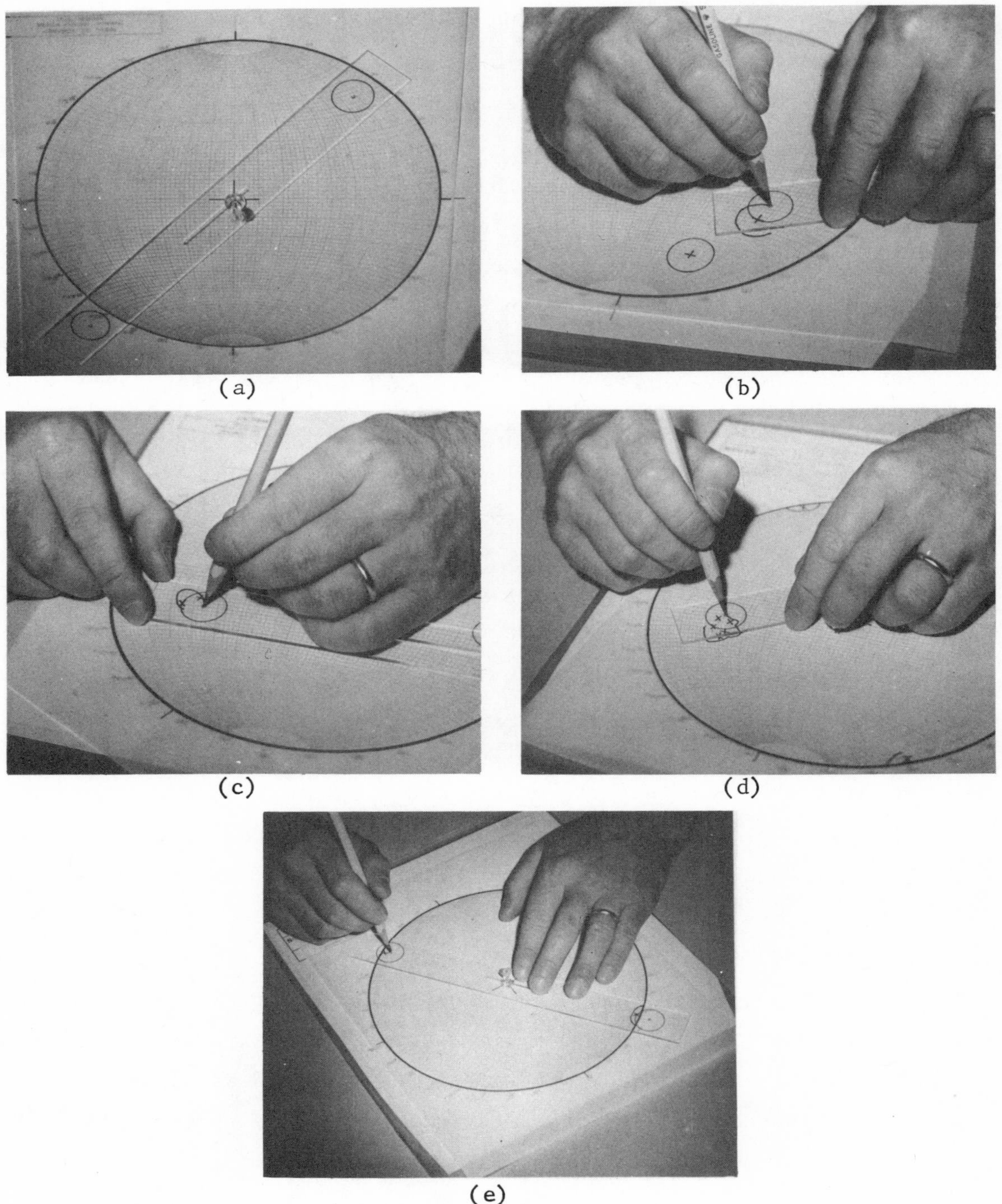

Fig. 7—Photographs illustrating a method for contouring points plotted in equal-area projection. Shown are (a) the contouring tool; (b, c, d) the method used to contour 1, 2, and 3 points per 1 per cent area, respectively; and (e) the procedure used to contour 2 points which lie on opposite sides of the diagram.

point per 1 per cent area is defined; etc. The author has found that searching for more than three overlaps consumes more time than contouring with the tool. The grid or Schmidt method is particularly useful in contouring large numbers of points.(11,18,25)

Kamb (Ref. 28, p. 1908) has prepared contoured diagrams by a novel procedure such that statistical inferences can be drawn directly from the diagrams. The area (A) of the contouring tool is so chosen that, if the population is randomly oriented, "the number of points (E) expected to fall within a given area (A) is three times the standard deviation of the number of points (n) that will actually fall within the area under random sampling of the population.... Observed densities that differ from E/A by more than two or three times the standard deviation σ (for random orientation) are then likely to be significant.... The observed densities are therefore contoured in intervals of 2σ at values 0, 2σ, 4σ, etc., the expected density = E/A for random orientation being 3σ."*

For a random population in which the distribution of n values for random samples of size N is binomial,

$$\frac{\sigma}{E} = \sqrt{\frac{(1 - A)}{NA}} ,$$

where E = NA, A is a given area expressed as its fraction of the total area of the diagram, and N is the total number of points on the scatter diagram. By setting $\sigma/E = 1/3$, one computes the appropriate area A of the counter to be used in preparing the diagram. Once A is obtained it is used to contour the point diagram by the Schmidt method. Kamb states that diagrams prepared in this way have a smoothed appearance in comparison with conventional contours because most of the irregular detail of the latter is of no statistical significance, the conventional A = 0.01 being usually much too small.**

---

*Here σ is used to designate both the standard deviation and stresses in accordance with conventions in statistics and stress analysis, respectively.

**All three contouring techniques contain a common source of error. Although the contoured points are plotted on an equal-area net,

Statistics

The orientation of fabric elements as illustrated on petrofabric diagrams can be treated in a purely statistical manner to determine whether the observed distribution significantly deviates from one which is randomly oriented. This implies statistical, but not necessarily geological, significance. Pincus[29] fully discussed the application of statistical methods to the analysis of aggregates of orientation data. Little can be added to his treatment here. Flinn[27] re-examined several of the more popular tests of significance as applied to petrofabric diagrams and found them generally unsuitable. He suggested that fabric diagrams should be compared to artificially prepared "random" diagrams and significant differences attributed to the rock fabric.

The contouring method of Kamb[28] allows statistical inferences to be made directly, but one possible drawback may be that the contours based on areas considerably larger than 1 per cent could tend to average or mask two or more closely spaced concentrations which may be geologically significant.

Another means of utilizing a contoured diagram as the basis for statistical inference has been employed by the author with some success. The probability of obtaining concentrations on a point diagram which deviate from a random distribution is approximated by the Poisson exponential binomial limit.[30] This has been confirmed by the author from tests of goodness of fit between this distribution and apparently random diagrams of fabric elements in rocks. The probability of obtaining at least a given number of points in any 1 per cent area of a fabric diagram is given by the following equation:

---

there is an unavoidable, progressive distortion of the equal areas on this net from the center outward. Fully accurate contouring, therefore, would require continual changes in the shape of the contour from circular at the center to elliptical at the periphery as recognized by Mellis and Strand (in Flinn, Ref. 27, p. 532). This error is considered to be negligible in light of the accuracy with which any given point is plotted and the nature of the interpretation of contoured fabric diagrams.

$$P = \sum_{x=x'}^{x=\infty} \frac{e^{-Np}Np^{x}}{x!},$$

where P is the probability, N is the total number of points in the sample, p is the probability that 1 point will occur in a given 1 per cent area (here, 0.01), and x equals the number of points per 1 per cent area. When Np = 1, the probabilities of finding at least x points in any 1 per cent area are as follows: 0 points, 1.00; 1 point, 0.63; 2 points, 0.26; 3 points, 0.08; 4 points, 0.02; 5 points, 0.004; 6 points, 0.0006; and 7 points, 0.0001. Thus in a sample of 100 points (Np = 1), the chances of obtaining a 6-point (6 per cent) concentration from a random distribution are 6 in 10,000, and a 7-point (7 per cent) concentration, 1 in 10,000.

Statistical tools should be used prudently--only to guide interpretations, not to dictate them. For example, none of the tests known to the author takes into account the locations (specific orientations) of the points on the diagram whose orientations are analyzed collectively. For instance, an orientation pattern may be characterized by points distributed within a band (girdle) along small or great circles. There is little doubt that this pattern would have geological significance, although statistical tests might show that the distribution was random. The role of statistics here should not be to dictate against reason that the pattern was random but rather to suggest that no significance should be attached to any concentrations within the girdle.

Deformation Mechanisms and Criteria for Dynamic Interpretations

Fracturing and Faulting

General. Many classifications of fracturing, faulting, and related phenomena are found in the engineering and metallurgical literature,[31] but none is generally applicable to geological problems. The types of failure pertinent here are extension fracturing and faulting (including shear fracturing).[32,33] Brace[34] has recently found that these types may be gradational under certain states of stress.

Fracturing is regarded as a process involving separation into two or more parts after total loss of cohesion and resistance to load, and release of stored elastic strain energy. Extension fracturing is separation of a body across a surface oriented normal to the least principal stress (σ_3).* There is no offset parallel to this surface. Macroscopically the least principal stress may either be negative (tensile) or positive (compressive). Tensile fracture (σ_3 negative) is regarded as a special case of extension fracture. Correlation between extension fractures (the feature) and the principal stresses follows from the criterion of no offset. The fracture surface is parallel to the plane of vanishing shear stress normal to σ_3, and contains σ_1 and σ_2 (Fig. 8(a)).

Faulting is defined as offset parallel to a more or less planar surface of nonvanishing shear stress. There is no restriction on the magnitude of offset. Faulting <u>may or may not</u> be accompanied by loss of cohesion and resistance to load, actual separation, or release of stored elastic strain energy. When these events do occur, it is proper to speak of shear fracturing. In the laboratory it is usually possible to distinguish between shear fracturing and faulting. In nature, however, it is rarely possible to observe the process and determine whether it is actually accompanied by a loss of cohesion, etc. Clearly, those features which have inherently maintained cohesion should be called faults. On the other hand, the fact that a feature exhibits no cohesion across its surface at present can not be used to infer that it formed as a result of shear fracturing. That is, the loss of cohesion may have occurred after the feature initially formed. Accordingly, there is a problem when dealing with naturally deformed rocks as to what to call the features that result from these processes. The usage adopted here is as follows: The term "shear fracture" will be used to designate features along which the shear displacement is less than an arbitrary amount (1 m), and the term "fault" will refer to features

---

*By the convention often adopted in geology, compressive stresses are positive, tensile stresses are negative. The greatest principal compressive stress is designated σ_1, the intermediate principal stress is σ_2, and the least principal stress is σ_3.

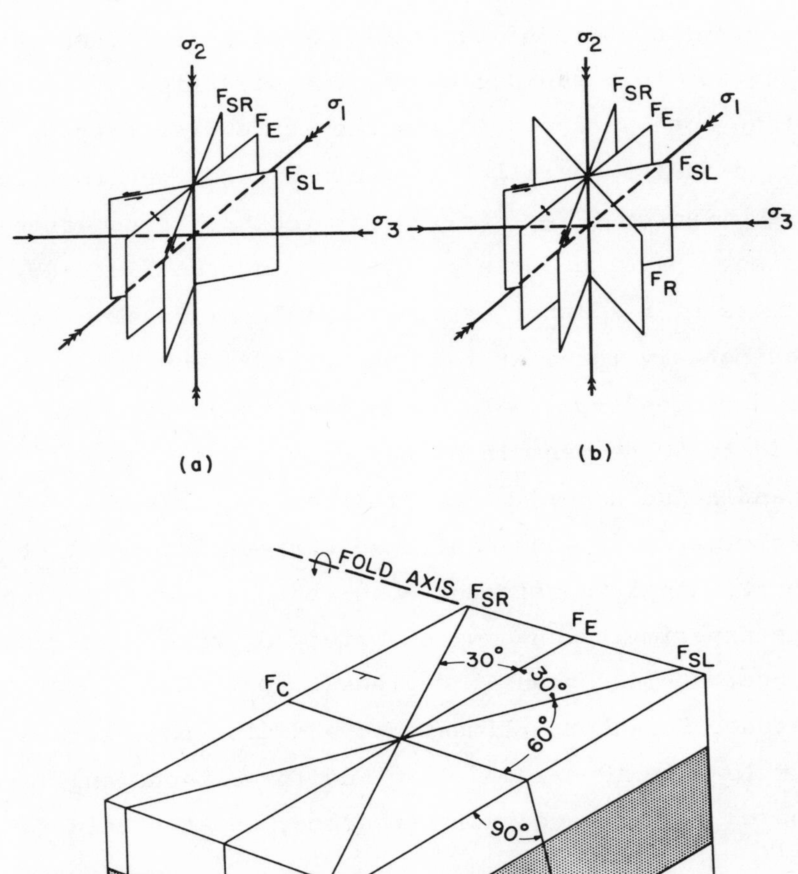

Fig. 8—Idealized geometric relationships between the principal stresses and fractures. (a) Two shear fractures (F_{SL} and F_{SR}) are illustrated with the enclosed extension fracture (F_E). (b) A relaxation fracture (F_R) is added to the configuration in (a). (c) An additional extension fracture (F_C) is added to the geometry in (a) in folded beds only.

along which the shear displacement is greater than about 1 m. These terms are then descriptive within the bounds of shear failure and do not necessarily imply distinction between the shear fracturing and the faulting processes.

The correlation between faults (including shear fractures) and the principal stresses has been established by a wealth of empirical data. In homogeneous, isotropic materials in which $\sigma_1 > \sigma_2 > \sigma_3$, a fault may occur along one or both of two equipotential surfaces, each inclined from 45 degrees to a few degrees to the direction of σ_1. When faulting occurs along both surfaces, σ_1 is the acute bisector; σ_2 lies in the plane of the faults (parallel to their line of intersection); and σ_3 is the obtuse bisector (Fig. 8(a)). For rocks that do not have pronounced planar anisotropy, the angle between σ_1 and the fault (θ) varies within narrow limits. In 70 short-time triaxial compression tests on a variety of dry sedimentary rocks at room temperature and 0 to 2-kb confining pressure, θ ranges from 25 to 35 degrees in 65 per cent of the cases and from 20 to 40 degrees in 95 per cent of the cases.[3] Subsequent work by Handin and Hager[35] at elevated temperatures and in the presence of pore water[36] has confirmed the earlier results. In fact, one of the outstandingly consistent observations from all properly designed short-time experiments on a wide variety of rock types is that faults tend to occur at less than 45 degrees to σ_1.[3,33-59] This holds also for sandstone, Solenhofen limestone, and diorite deformed at strain rates from 10^{-1} to 10^{-5} sec^{-1},[60] and for Solenhofen limestone, granite, diabase, dunite, and quartzite deformed at strain rates to 10^{-7} sec^{-1}.[61]

In rocks with strong planar anisotropy the value of θ is dependent upon the orientation of the foliation, schistosity, or cleavage (s-planes) with respect to the load axis (Fig. 9).[62-64] For rocks experimentally deformed at room temperature and under confining pressures up to 2000 bars, faults tend to develop parallel to the s-planes for inclinations up to 60 degrees to the direction of maximum principal stress. When the s-planes are at 60 and at 75 degrees to the load axis, θ tends to be between 40 and 60 degrees; and when the s-planes are at 90 degrees to the load axis, θ is again at about 30 degrees. Moreover, in all except the 90-degree orientation the strike of the fault tends to parallel that of the s-planes in the test specimens.[63]

No entirely satisfactory theory of shear fracture or faulting is yet available. This is not to say, however, that there are no criteria which qualitatively describe shear failure in rocks. The Coulomb-Mohr

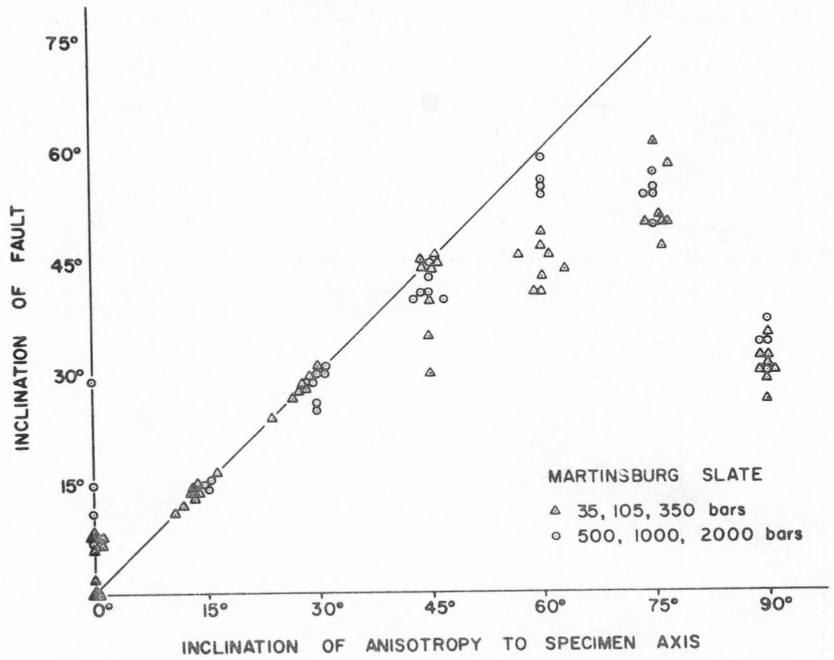

Fig. 9—Effect of cleavage on the angle of faulting, Martinsburg slate (Donath, Ref. 64).

and Griffith theories, for example, have been discussed by Brace earlier in these proceedings.(34) It is significant that all current theories and experimental observations show that θ is less than 45 degrees, and the relationships shown in Fig. 8(a) are essentially valid.

That naturally deformed rocks fault at less than 45 degrees to σ_1 is demonstrated by the attitude of normal faults in regions where it is logical to assume that σ_1 was nearly vertical at the time of faulting. Hubbert(65) points out that in areas of relatively uncomplicated structure normal faults occur with dips consistently greater than 45 degrees. He cites Sax's(66) study of the dip of 2102 individual underground faults in the coal measures of the Netherlands, which shows that 1651 of these were normal faults and had a well-defined preference to dip 63 degrees (the remaining were reverse faults with dip preference at 22 degrees). Moreover, the average dip of normal faults (from United States Geological Survey Folios published to 1913) is 78 degrees.(67) Anderson(68) has generalized these relationships by calling attention to the orientation of the principal stresses with respect to normal, reverse, and wrench faults (Fig. 10).

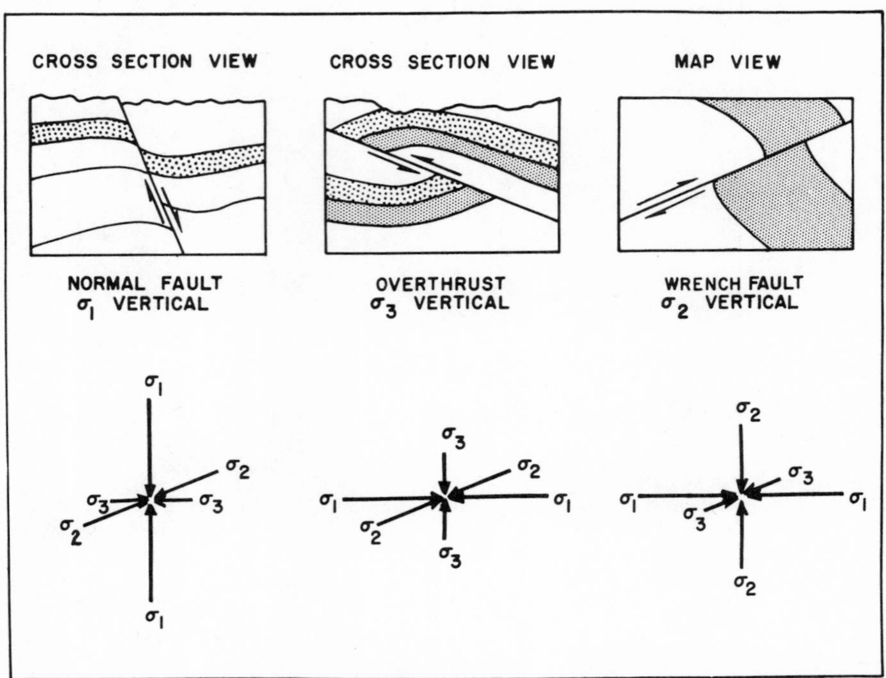

Fig. 10—The orientations of the principal stresses associated with the common fault types (after Anderson, Ref. 68).

Empirically the processes of faulting and of shear and extension fracturing are independent of scale down to at least the microscopic domain (10^{-5} m). Thus one may consider microfractures in individual grains of a rock as essentially identical genetically and geometrically to fractures and faults of outcrop scale and larger. This is a useful concept in petrofabrics as it leads to predictions of large-scale features from statistical inferences drawn from the study of smaller-scale features.

Dynamic Interpretation of Fractures and Faults. Suppose that a given domain contains extension and shear fractures which for the purpose of this discussion have no distinguishing characteristics other than their attitude in space. That is, extension fractures can not be distinguished by any physical feature from shear fractures. The fractures within a given size range are collectively regarded as a fabric element, their orientations are measured, and their subfabric is illustrated by the distribution of their normals on a petrofabric diagram.

This orientation pattern can be used to position the principal stresses if it is characterized by groups or concentrations of normals which define planes that can be related to patterns expected from previous experience.

For example, if θ is 30 degrees, the geometric relationships between shear and extension fractures for the general state of stress are shown in Fig. 8(a). By comparison to this configuration all three principal stresses can be located from the observed subfabric orientation pattern (a) if it exhibits three planar sets intersecting in a nearly common line and inclined at 30 and/or 60 degrees to each other, or (b) if it shows two sets of features intersecting at about 60 degrees (two conjugate faults). Frequently, partial geometries are encountered, perhaps in part because all elements of the basic geometry (Fig. 8(a)) do not always form. Accordingly, the pattern observed at any one locality may consist of any or all parts of the basic configuration. Indeterminate cases occur if the pattern consists of two elements intersecting at about 30 degrees or if a single concentration results. In the former case, it is not clear whether one is dealing with two conjugate shear fractures of small dihedral angle[69] or a set of extension fractures and a set of one of the potentially conjugate faults. At best, one can state only the orientation of σ_2 (parallel to the line of intersection) and, therefore, define the plane containing σ_1 and σ_3. In the case of a single concentration, one can not determine whether the concentration is related to a set of shear fractures or to a set of extension fractures. Ambiguities in the subfabric for a single locality can frequently be resolved by comparison with fabric data from other neighboring localities in similar structural positions where the pattern is unambiguous. Similar comparisons of observed fracture orientation patterns with expected patterns for the states of stress $\sigma_1 > \sigma_2 = \sigma_3$ and $\sigma_1 = \sigma_2 > \sigma_3$ lead to determinations of the unique principal stress only.

Along with the orientation data, information is sometimes available on the sense of shear along faults or shear fractures. Shear criteria permit reliable determination of principal stress directions provided the observed offset occurred at the time of faulting, the slip direction of the movement is known,* and the value of θ is known or can be determined experimentally. Given these facts one can determine (1) that σ_2 lies in the plane of the fault at right angles to the slip direction, (2) that σ_1 is inclined at θ degrees to the fault in the direction appropriate for the observed sense of shear and lies in the plane that is normal to the fault surface and contains the slip direction, and (3) that σ_3 lies in this same plane at 90 degrees to σ_1. Valid sense of shear criteria also permit resolution of the pattern consisting of two elements intersecting at a small dihedral angle. Observations of a consistent sense of shear displacement along one of the features and none along the other implies that the former is a fault and the latter an extension fracture set, provided the extension fracture set is disposed appropriately with respect to the observed sense of shear on the fault. Similarly, if consistent and opposite senses of shear are observed on each of two sets of features they can be recognized as two conjugate faults of low dihedral angle. Once the fractures and faults have been recognized, the derivation of the principal stresses follows from the basic relationship in Fig. 8(a).

Certain morphological features on the surfaces of macrofractures and faults have been investigated by a number of workers as possible criteria for the distinction between extension and shear failure. Three types of features have been recognized: slickensides, plumose structure, and conchoidal structure.[70-78] Unfortunately the genetic implications of these features are not well understood.

Clearly, slickensides (polished and striated surfaces) indicate offset parallel to the walls of the fracture or fault. They can not, however, be regarded as generally reliable criteria because they record only the last movement along the surface. Conceivably the surface of an extension fracture (initial movement normal to the fracture walls) may be slickensided as a result of later movements parallel to the walls. Plumose structure consists of grooves and ridges on a rock surface. There is a central axis into which barbs or plumes converge

---

*Direction of slip along a fault is needed only if one fault or one set of faults is present. For those cases involving intersecting conjugate elements, the line of intersection is parallel to σ_2 and normal to the slip direction.

in featherlike form. Conchoidal structure consists of concentric grooves and ridges which usually center about one or two points on the surface. It is not clear whether conchoidal structure is merely a part of a larger plumose structure or a separate phenomenon with distinct genetic implications. The only certain fact about these features is that where they occur there has been little shear movement along the surface in question. Hodgson (Ref. 77, p. 29) implies that plumose and conchoidal structure are not diagnostic criteria. On the other hand, Parker (Ref. 74, p. 397) and Roberts (Ref. 78, p. 486) find that plumose structures are apparently restricted to shear fractures or faults. This would imply at least some shear displacement at the time of formation. Further analysis of these features, perhaps experimentally, would seem justified as it might lead to diagnostic criteria.

Field Examples--Microfractures. To the writer's knowledge there are only a few published studies in which the subfabrics of microfractures have been used to derive the orientations of the principal stress in rocks at the time of deformation. On the microscopic scale, fractures occur in the individual grains or crystals of the rock. They may or may not cross grain contacts. Their size, therefore, is somewhat dependent upon grain size. Though they may be visible to the unaided eye, they are best studied in thin section with the aid of the petrographic microscope and universal stage. The microfractures are often essentially planar features such that their dip and strike can be measured by one setting of the universal stage. Commonly they are developed in sets of two or more parallel individual features. They are fresh clean breaks in all experimentally deformed rocks and are commonly healed or filled in naturally deformed rocks.

That microfractures can be valid dynamic fabric elements has been demonstrated by studies of experimentally deformed, dry, unconsolidated, quartz sand aggregates and calcite-cemented sandstones.[58,59] Results show statistically that the grains tend to fracture with respect to the principal stresses across the boundaries of the whole specimen rather than with respect to local stress concentrations at grain contacts (Fig. 11). That is, even though the stresses must be transmitted

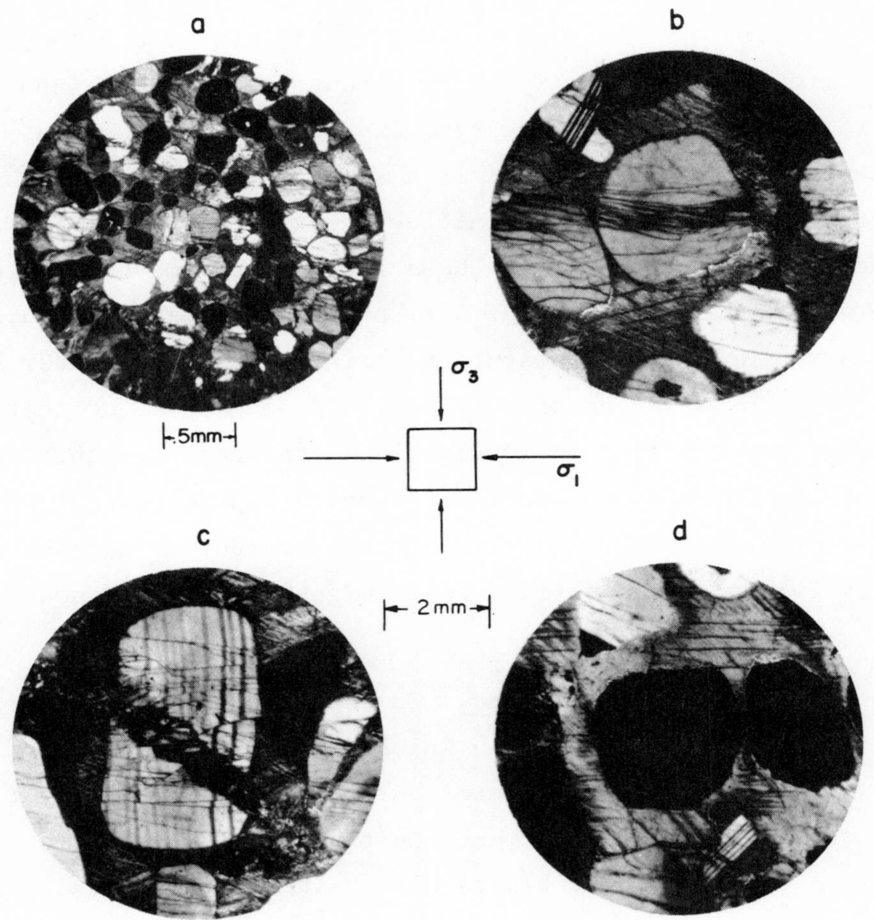

Fig. 11—Photomicrographs of microfractures in quartz and feldspar grains in an experimentally deformed calcite-sand crystal (from Friedman, Ref. 59, plate 2). The orientation of the principal stresses across the boundaries of the whole specimen is shown; extension fractures predominate.

through grain boundaries, the individual grains tend to fracture as if each grain were loaded in the same manner as the whole aggregate. This is a statistical statement, but reference to the photomicrographs in Fig. 11 shows that the phenomenon is quite pronounced. Moreover, the phenomenon in quartz grains is essentially independent of the crystal structure of quartz.*(59)

---

*This result for quartz sand aggregates is in marked contrast to recent data on quartz single crystals deformed at very high pressure. In such specimens the fracturing and faulting is controlled by the anisotropy of the crystal structure.

Bonham[79] studied both macrofabric and microfabric elements associated with the Pico anticline and syncline, which are located some 35 mi northwest of Los Angeles, California. The structure lies along the southern margin of the Ventura basin. Three Tertiary formations are encountered, the Modello (Miocene), the Pico (Pliocene), and the Saugus (Pliocene). These are composed of a poorly indurated turbidite sequence of interbedded arkose and graywacke sandstones, siltstones, and shales, with an aggregate thickness of nearly 15,000 ft. The anticline is 9 mi long and is folded tightly, showing an almost chevron cross section. The axial plane of the fold is nearly vertical and has a strike of about N-65°-W in the eastern and central parts and about N-75°-W in the western part. Dips on the flanks are commonly over 50 degrees with some overturning on the northern flank. The anticline plunges both east and west. Several normal faults, with as much as 500 ft of stratigraphic separation, cut the structure and tend to strike NE-SW and NW-SE. Bonham characterizes the anticline as a flexural slip fold with three mutually perpendicular axes of folding (Fig. 12).

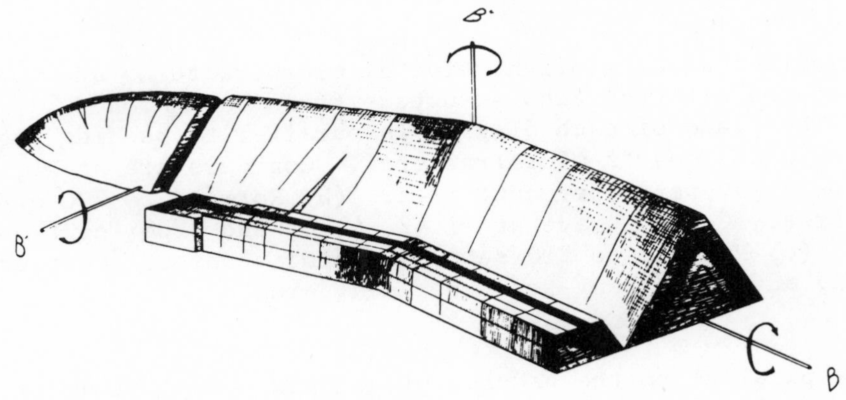

Fig. 12—Idealized block diagram of the Pico structure showing three axes of folding (from Bonham, Ref. 79, Fig. 3).

Bonham's data on orientation of microfractures and macrofractures (joints) are given in Fig. 13. For convenience, he refers his diagrams to three axes--a, b, and c. The b direction is parallel to the fold axis; the a axis lies in the bedding plane and is normal to the fold axis; and the c axis is normal to bedding. Accordingly, the

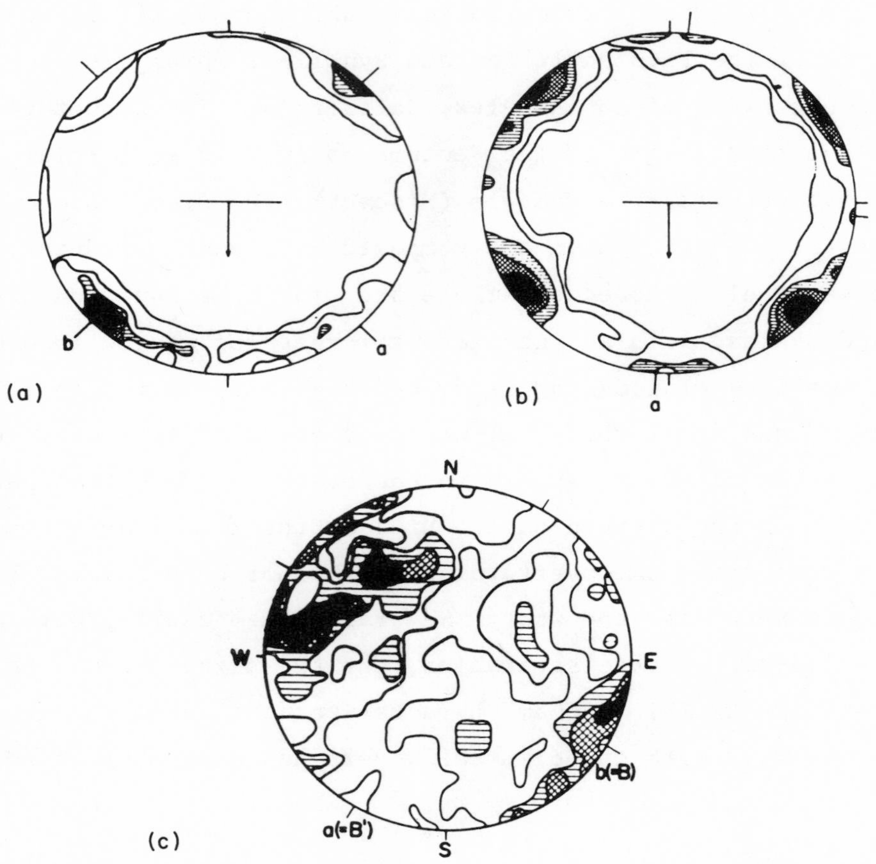

Fig. 13—Diagrams showing orientation of microfractures and macrofractures associated with the Pico structure (from Bonham, Ref. 79, Figs. 4 and 5). The plane of each diagram is parallel to bedding (ab). (a) Normals to 200 quartz microfractures. Contours are at >2, 4, 6, 8, and 10 per cent per 1 per cent area. (b) Normals to 200 quartz microfractures. Contours are at >1, 2, 4, 6, and 8 per cent per 1 per cent area. (c) Normals to 210 sets of macrofractures. Contours are at >1, 2, 3, and 4 per cent per 1 per cent area.

ab plane is parallel to the bedding plane, and the ac plane is normal to the fold axis (b = B). In diagrams with a single concentration (Fig. 13(a)), the microfractures are preferentially oriented in the ac plane of the fold. In diagrams with two concentrations (Fig. 13(b)), the microfractures are oriented in two planes which intersect in a line normal to bedding. The acute angle between these planes is bisected by the ac fabric plane. Macrofractures (Fig. 13(c)) also tend to lie in the ac plane. Stereograms showing the major plane(s) defined by the microfractures at each of 28 stations on the fold (Fig. 14) indicate

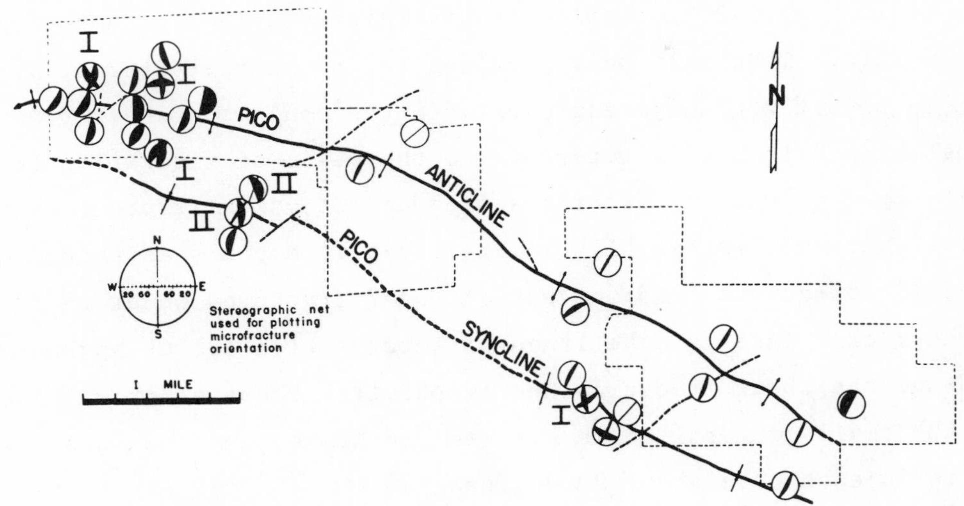

Fig. 14—Stereograms showing the orientation of major sets of microfractures associated with the Pico structure (from Bonham, Ref. 79, Fig. 7). The reference plane of each stereogram is horizontal.

that most are vertical and lie in the ac plane of the fold. Bonham points out that those near the western end of the structure reflect the westward plunge of the fold about the B′-axis.

Bonham does draw a dynamic inference from the ac fractures. He attributes them to tension parallel to the fold axis during deformation. Further inferences are possible from examination of the microfracture subfabrics. Consider first the diagram illustrating two concentrations (Fig. 12(b)). These define planes which intersect at about 75 degrees along a line nearly normal to the bedding plane. This configuration suggests that the fabric maxima define two sets of conjugate shear fractures (θ = 38 degrees) for which σ_1 is parallel to the bedding plane and normal to the fold axis, σ_2 is normal to the bedding plane, and σ_3 is in the bedding plane and parallel to the fold axis. In Fig. 14, there are four stereograms (marked "I") that contain two sets of planar features intersecting at 50 to 80 degrees. By correlating the line of intersection of the two sets with the probable dip of the beds (see fold axes), one concludes that at each station, sets comprise a pair of conjugate shear fractures that intersect in a line normal to bedding. In each case the orientations of the derived stresses are as stated above. Next, consider the many stereograms in

Fig. 14 that contain a single set of features oriented in the ac plane of the fold. These not only parallel the ac macrofractures, but also tend to bisect the acute angle between the conjugate shear fractures at nearby stations. By reference to the basic configuration (Fig. 8(a)), the ac features are recognized as extension fractures. Finally, at two stations (marked "II" in Fig. 14), it appears as though one set of ac microfractures and one set of shear fractures occur in the grains of the rock. That is, the lines of intersection are at high angles to bedding, the shear fractures nearly parallel those at other stations, and the small dihedral angle between the ac set and the shear fracture set is in each case about 20 degrees.

Thus, Bonham has mapped microfractures and macrofractures that (1) exhibit consistent orientations throughout the fold, (2) are geometrically related to the fold in a meaningful way, and (3) can be interpreted as genetically related to the same state of stress. These principal stresses appear to be uniformly oriented with respect to the bedding planes and to be independent of the dip of the beds. Unfolding brings principal stresses mapped at each station throughout the structure into congruency. Accordingly, one can conclude that the fractures occurred either throughout the folding process or early in the history of folding in response to a horizontal greatest principal stress oriented roughly N-25°-E, a vertical intermediate principal stress, and a horizontal least principal stress trending roughly N-65°-W. The first alternative requires that σ_1 remain nearly parallel to the bedding throughout folding.

Field Examples--Macrofractures and Faults. The literature is replete with purely geometric descriptions of macrofractures and faults. In most cases, the authors have referred to these features as joints (10^{-1} to 10^3 m in observed length) and related them to larger-scale folds and faults. Some workers have attempted to associate the joints genetically with the regional state of stress, but few have had an adequate understanding of the problem. Good examples of enlightened investigations are available, however. Several of these are reviewed by De Sitter (Ref. 80, pp. 122-142), and Schmidt (Ref. 81, pp. 10-17) gives a comprehensive review of the literature. Additional significant dynamic interpretations of fracture and fault assemblages are to be found in Cloos,[82] Dawson-Grove,[83] Price,[84] Harris et al.,[85] Muehlberger,[69] and Donath.[86]

Melton's reconnaissance study of the fracture systems in the Ouachita Mountains and Central Plains of Oklahoma[87] can be used to illustrate how dynamic inferences are made from geometric data. The geology of the area is sketched in Fig. 15. In general the intensity of the deformation decreases northwestward from the Ouachita Mountains through the Open Fold zone to the nearly flat-lying strata of the Central Plains. Melton measured the attitudes of fracture sets in outcrops distributed throughout this region (Fig. 16). He concluded

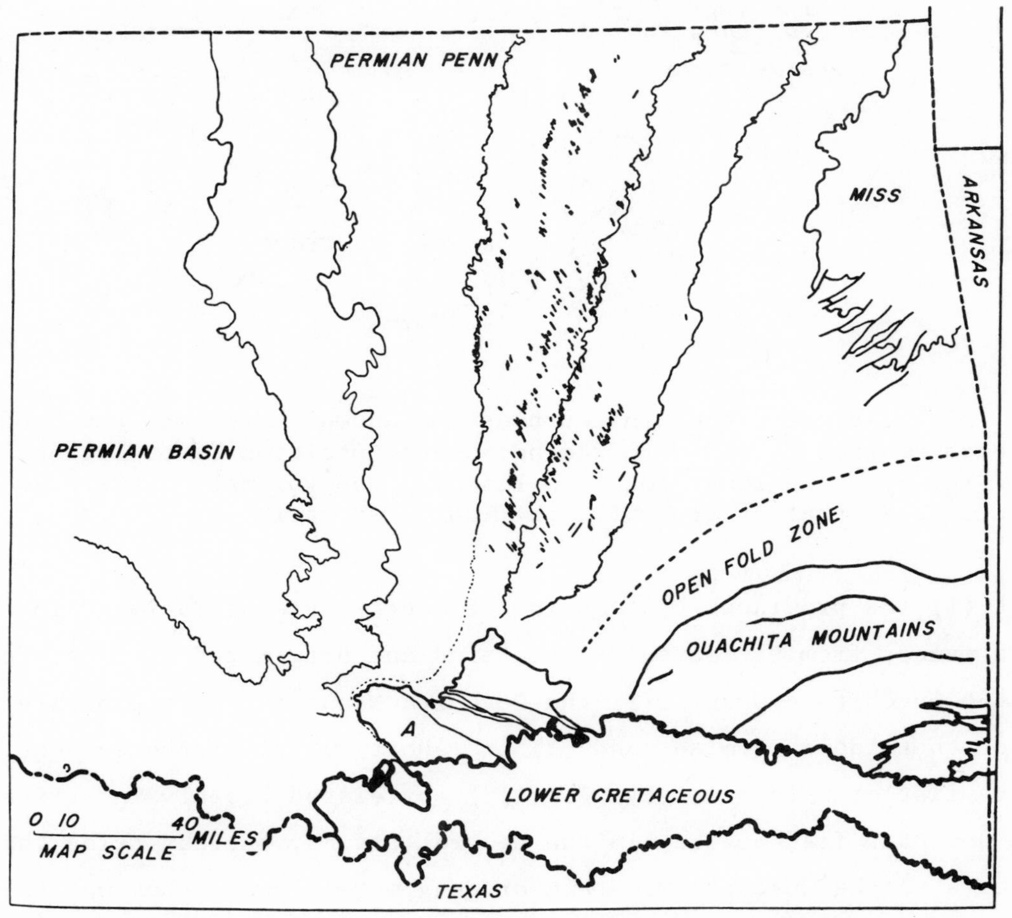

Fig. 15—The major structural units of Oklahoma. The Arbuckle Mountains (A), the Ouachita Mountains, the Permian (Anadarko) Basin, the Mississippian rocks of the southwestern part of the Ozark Dome, and the "belts" of en échelon faults in the Central Plains are shown (from Melton, Ref. 87, Fig. 1).

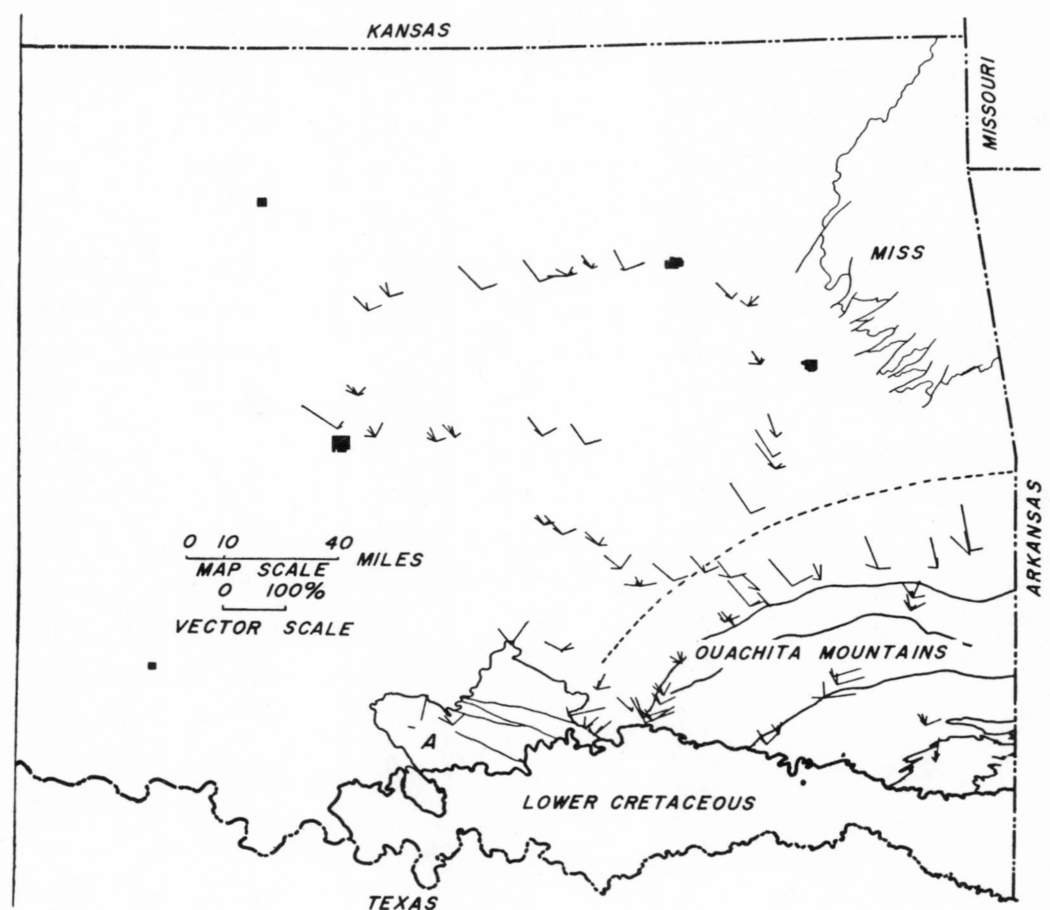

Fig. 16—Fracture trends in southeast Oklahoma shown with the bedding at each station unfolded to the horizontal position (from Melton, Ref. 87, Fig. 3). The length of each line is proportional to the number of fractures at that station with the indicated strike.

that (1) the prominent systems in the Central Plains radiated in a fan-like manner from the Ouachita Mountains and originated from the forces of the Ouachita orogeny, (2) the Ouachita Mountains were probably formed after the Middle Permian, and (3) the short faults of the <u>en échelon</u> belts east of Oklahoma City (Fig. 15) correlated closely in strike with the dominant fracture set in the Central Plains, thereby tying their genesis to the Ouachita Mountain orogeny more closely than was theretofore recognized.

If the fracture array at each station (Fig. 16) is examined closely, one can distinguish individual elements of a four-set pattern (Fig. 17). This pattern is repeated throughout the region, even though

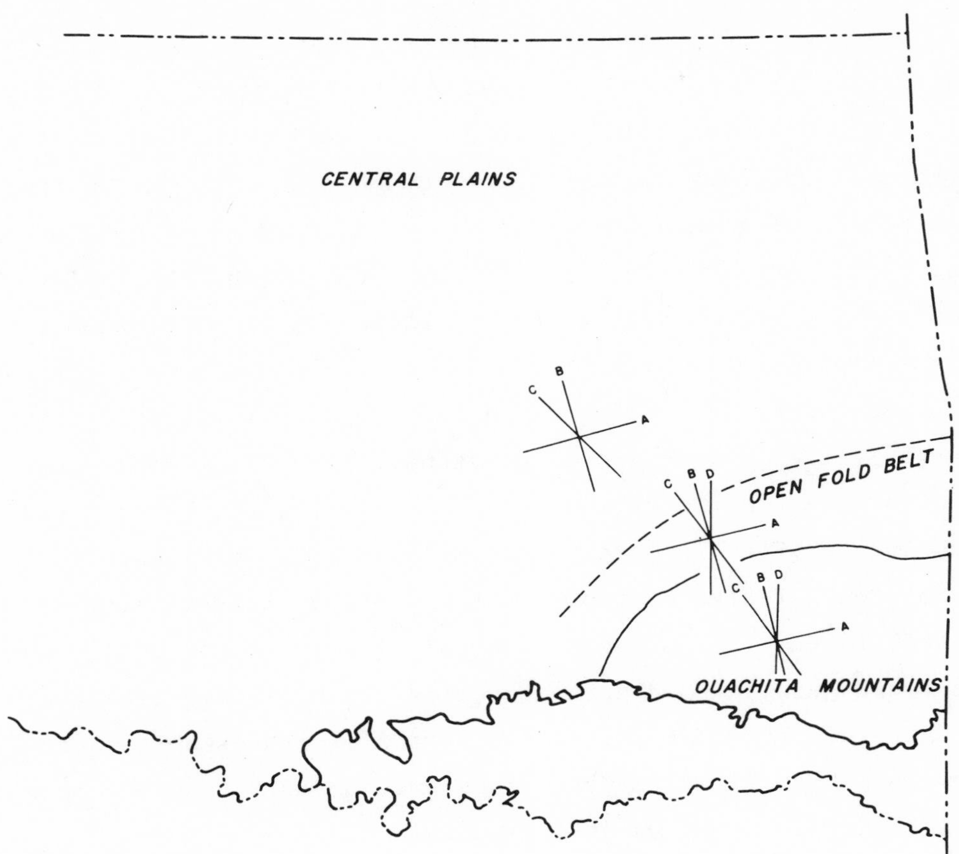

Fig. 17—The average fracture patterns in the Central Plains, Open Fold zone, and Ouachita Mountains (obtained from Melton, Ref. 87, Figs. 6, 8, and 10, respectively).

all four sets are rarely developed at any one station. Moreover, the pattern tends to rotate with the change in strike of the beds in the Ouachita Mountains and Open Fold zone. Each set is oriented normal to bedding. When beds in the Ouachita Mountains and Open Fold zone are unfolded, the fracture-fault sets become congruent with those in the Central Plains. Sets A and B (Fig. 17) intersect at 90 degrees. Sets C and D intersect at about 40 degrees and are about equally disposed on either side of set B. Set C is the best developed of the four. The configurations of sets B, C, and D suggest that they represent an extension fracture and two conjugate shear fractures, respectively. From this pattern, σ_1 is placed horizontal and everywhere normal to the strike of the Ouachita Mountain trend, σ_2 is vertical (normal to bedding), and σ_3 is horizontal and normal to σ_1.* The fourth element of the pattern (set A) needs further explanation. In the Ouachita Mountains and Open Fold zone it tends to parallel the strike of the beds (Ref. 87, p. 741) and therefore occupies an orientation similar to that of the often mentioned "tension" fractures that develop during folding (Fig. 8(c), and Ref. 80, p. 100; Ref. 82, p. 185; Ref. 88, p. 150; Ref. 89, p. 102; and Ref. 90, p. 118). It is unreasonable to extend this explanation for set A into the unfolded strata of the Central Plains. In the writer's view set A is there a relaxation fracture (Fig. 8(b)), i.e., an extension (or tensile) fracture formed upon release of stored elastic strain energy. Finally, the belts of en echelon faults (Fig. 15) are interpreted as near surface features related to wrench faults at depth, which would be left-lateral and would trend N-10-15°-E. The principal stress directions derived from these wrench faults are in good agreement with those derived from the regional joint pattern. It is significant that the fracture-fault pattern giving rise to these dynamic interpretations extends more than 100 mi into the Central Plains from the Ouachita Mountains.

An even-larger-scale example is afforded by the fault trends in the Great Basin of the western United States. Donath[86] compiled fault-strike frequency diagrams (Figs. 18(a) and 18(b)) for a 420-sq-mi area in south-central Oregon. The faults tend to strike in two main directions (N-35°-W and N-20°-E). The evidence indicates that the first movement on these faults was strike-slip and that this was followed by dip-slip displacement. Donath (Ref. 86, p. 1) states "... the intersection angle of approximately 55° and the nearly vertical dips indicate that the faults originally developed as conjugate strike-slip shears in a stress system characterized by a north-south maximum principal stress and an east-west minimum principal stress." The subsequent dip-slip movements on these planes reflect a redistribution of the surface forces acting on the individual fault blocks.

This same fabric is even more strikingly demonstrated by range-edge trends in a 50,000-sq-mi area in Nevada. On the assumption that

---

*Dynamic inferences by Friedman.

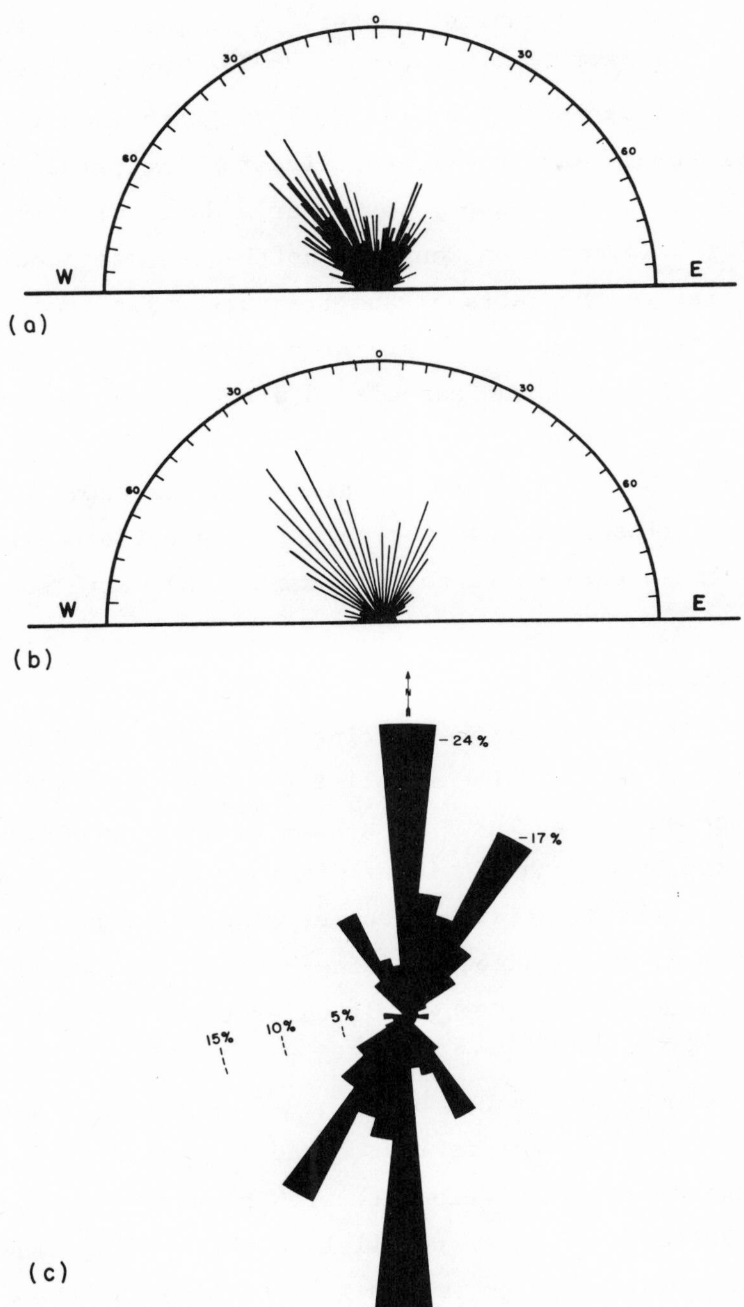

Fig. 18—Fault-strike frequency diagrams for areas in the Great Basin of western United States. (a) Strikes of 625 faults or fault segments plotted at 1-degree intervals. Radius of diagram equals 18 faults or fault segments. (b) Same data replotted for 5-degree class intervals. Radius of diagram equals 62 faults or fault segments (from Donath, Ref. 86, Fig. 4). (c) Strikes of 410 range-edges in over 50,000 sq mi in Nevada, as measured from 1:250,000 USGS topographic sheets (from Conger, Ref. 91).

these are delineated by faults, Conger[91] compiled the trends of 410 segments (Fig. 18(c)). North-south, N-30°-E, and N-30°-W trends are prominent, and comparison with Fig. 8(a) suggests a north-south greatest principal stress and an east-west least principal stress. Donath's explanation would seem to apply to Conger's data, with the exception that the initial movement on Conger's north-south set would have to have been normal to the walls of the fracture. The same stress orientations derived from these faults are also appropriate for the right-lateral strike-slip movement associated with the San Andreas wrench fault system in California.

The consistent fracture fault patterns in Oklahoma and in the Great Basin serve to illustrate that macroscopically uniform states of stress can be transmitted through large portions of the earth's crust.

Intracrystalline Gliding

General. Intracrystalline gliding flow (the "plastic" flow of metallurgy) takes place by the relative displacement of atomic or ionic layers over one another. The true nature of gliding seems to have been first recognized by Reusch[92] in halite and calcite. During the late nineteenth and early twentieth centuries mineralogists worked out the morphology of most of the known gliding systems. Since then metallurgists have much improved our physical understanding of the gliding (= slip) process.

In gliding, the strain can be regarded as a simple shear with no volume change. Displacement is restricted to a gliding (or slip) plane (T), a definite gliding direction (t) within that plane, and sometimes to a particular sense of shear parallel to the gliding line. These constitute the gliding system, which is determined by the crystal structure and is independent of the loading condition. Gliding is initiated when the shear stress along t exceeds some critical value (τ_c), which is essentially independent of the normal stress across the gliding plane[93-95] and of the orientation of the load relative to the gliding system. Accordingly in an aggregate or in a single crystal, gliding takes place most readily for systems of low τ_c and high resolved shear stress coefficient (S_o). (See Fig. 19.)

Translation gliding involves displacement through an integral number of interionic distances so that after slip the configuration of the crystal lattice is unchanged across the slip plane (Fig. 20). The slip need not be equally distributed throughout the deformed crystal, and the shear strain is not fixed. Although no general theory relates the translation gliding system to a particular crystal structure, in metals the gliding plane is usually one of high atomic

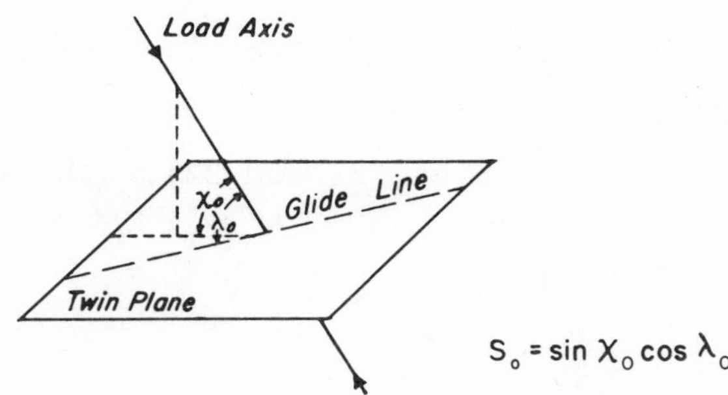

Fig. 19—Diagram showing nature of the resolved shear stress coefficient (S_o).

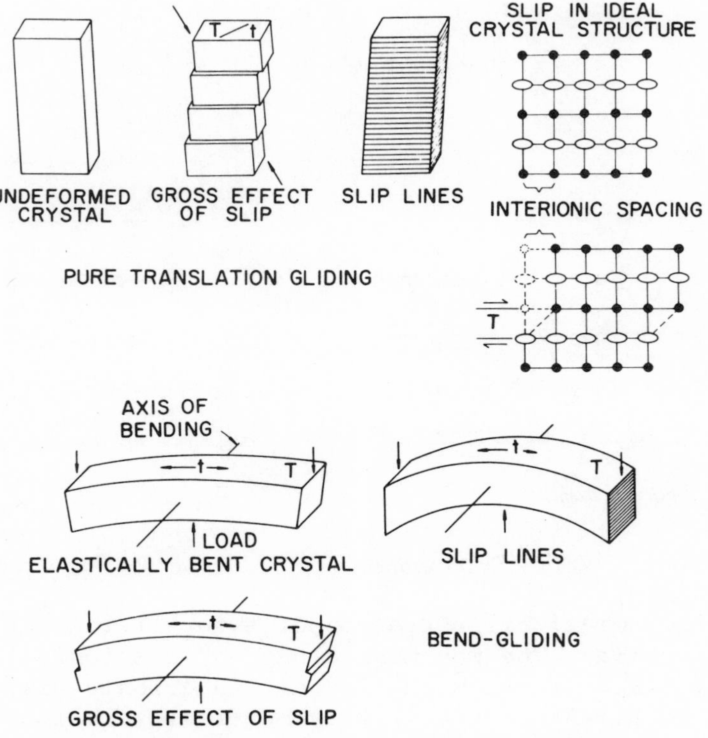

Fig. 20—Models of translation gliding.

or ionic density and of simple crystallographic index. The gliding direction is usually the densest atomic row (Ref. 93, p. 86). Bend gliding (Fig. 20) is a special form which occurs when planes are initially oriented parallel or normal to the load axes and are bent elastically before slip occurs. A compilation of translation gliding systems in 80 crystals is given by Handin.[96]

In mechanical twin gliding each ionic layer moves through a fixed fraction of the interionic distance so that the shear is fixed and the twinned portion of the crystal is in the proper symmetrical relationship to the original untwinned structure. The physical discontinuity between the twinned and untwinned portions of a crystal makes twin lamellae conspicuous in thin section (Figs. 21 and 22). The morphology of

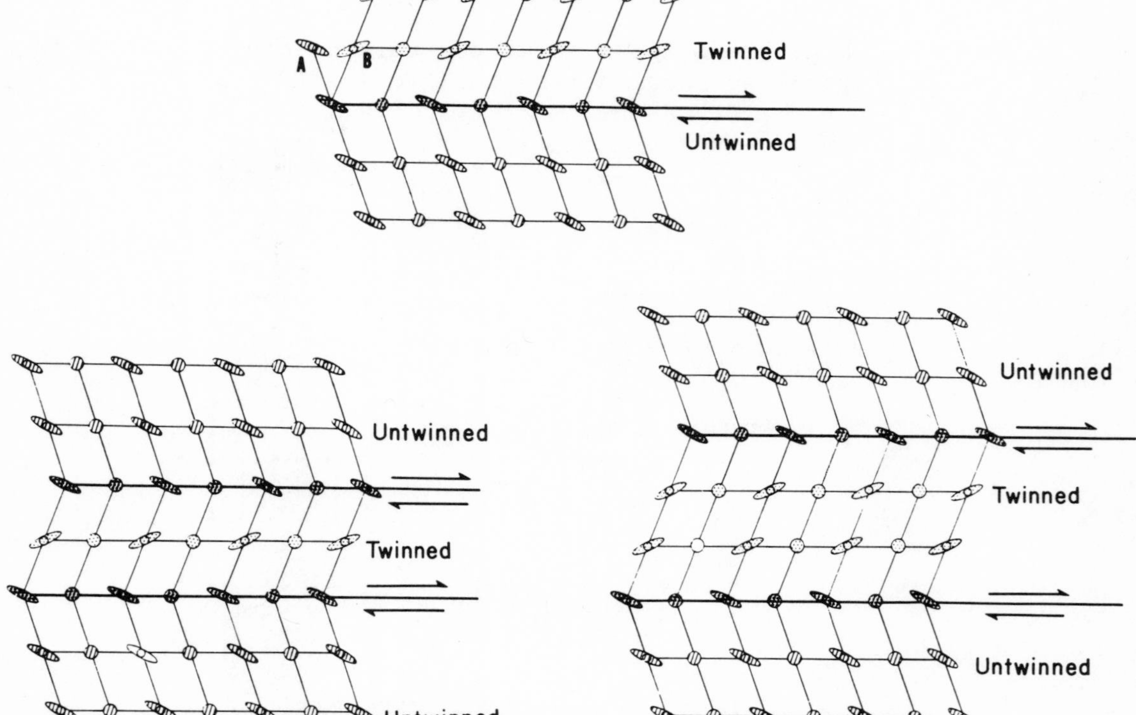

Fig. 21—Diagrammatic illustration of twin gliding and the development of a twin lamella. The movement along any one ionic layer (gliding plane) is a fixed fraction of the unit interionic distance, e.g., ion at A moves to B. As a result, a symmetrical relationship exists across the twin plane. A twin lamella is formed if twinned material is bounded on both sides by untwinned structure.

twinning has been discussed by Bell[97] and Pabst[98] and the metallurgical literature includes reviews by Cahn[99] and Hall.[100]

It is adequate for our present purpose to consider the model of twin gliding illustrated by the hypothetical structure shown in Fig. 23. If the relative displacement (sense of shear) of the upper layers is

Fig. 22—Photomicrograph of e{01$\bar{1}$2} twin lamellae in calcite grains of a Pre-Cambrian (?) marble, Schell Creek Mountains, near Ely, Nevada. Diameter of the specimen is 1/2 in.

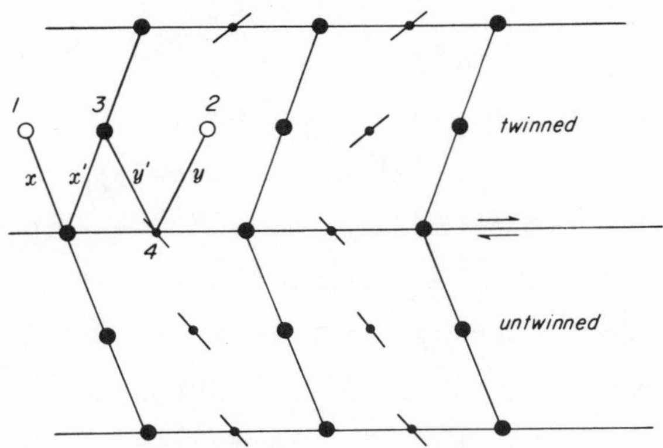

Fig. 23—Section through a hypothetical lattice illustrating different movements of the ions for different senses of shear in twin gliding (after Higgs and Handin, Ref. 55, Fig. 4). The plane of the paper is normal to the gliding plane and contains the gliding direction.

from left to right, the ion at position 1 (initial position) moves to position 3. The angle between lines x and x′ is the angle of shear (ψ), and the shear (s) is given by

$$s = 2 \tan \frac{\psi}{2} .$$

If the displacement of the upper layers is from right to left, the ion at position 2 (initial position) moves to position 3 and the angle of shear ψ' is between y and y′. Actually, twin gliding on a particular gliding plane is restricted to movement in only one direction, presumably that requiring the least energy. The proper sense of shear is indicated by the arrows. This qualitative picture of twin gliding involving only initial and final states of the ions is useful even though the actual paths of the ions may be unknown. Twin gliding systems for some 70 minerals have been compiled by Higgs.[(96)]

Dynamic Inferences from Gliding Systems. Twin and translation gliding is initiated for a given system when the resolved shear stress along the gliding direction and in the correct sense of shear exceeds a critical value τc. As gliding is largely independent of the normal stress, τc is reached most effectively when the resolved shear stress coefficient (S_o) is maximum (0.5). Accordingly, the most favorable state of stress in the crystal is characterized as follows: (1) σ_2 is parallel to the gliding plane (T) and normal to the gliding direction (t); (2) σ_1 is inclined at 45 degrees to T in the plane normal to T that contains t and is oriented so as to produce the correct sense of shear; and (3) σ_3 is inclined at 45 degrees to T in the plane containing t and σ_1. Clearly, if the gliding system(s) for a given crystal is known and if T and t can be recognized and measured, then one can derive the orientations of the principal stresses within the crystal that would best produce gliding. Petrofabric techniques are employed to locate these stresses in a number of individual crystals in polycrystalline aggregates, to plot them in fabric diagrams, and then to evaluate the average local state of stress in the rock at the time of gliding.

Dynamic Interpretation of Twin Lamellae in Calcite and Dolomite. Although in principle it is possible to make dynamic inferences from any known gliding system, in nature only a few common rock-forming minerals show the diagnostic features required in practice. Twinning in calcite and dolomite has received most attention in the laboratory and in the field. Knowledge of the deformation mechanisms in calcite has evolved from Brewster's observations(101) of mechanical twins in 1826 to the comprehensive experimental studies of deformed calcite single crystals and marbles, together with their petrofabric analyses.* As a result, gliding flow in calcite can be adequately accounted for by three gliding systems (Fig. 24):

(1) Twin gliding parallel to $e\{01\bar{1}2\}$ with $[e_1:r_2]$ as the gliding direction, and with a positive sense of shear,** effective throughout the temperature range of 20° to 800°C.

(2) Translation gliding on $r\{10\bar{1}1\}$ with $[r_1:f_2]$ as the gliding direction, sense of shear negative, effective over the temperature range 20° to 800°C.***

(3) Translation gliding on $f\{02\bar{2}1\}$ with $[f_1:r_3]$ as the glide direction, sense of shear negative, effective at 20°C and at 500° to 800°C, where it predominates over r translation.

Turner(111) has developed a technique for dynamic interpretations of twin lamellae in naturally deformed rocks by locating the mutually perpendicular directions of compression and extension that most favored development of the observed twin lamellae. The geometry of these relationships was initially set forth by Handin and Griggs (Ref. 105, pp. 866-869). If a maximum S_o value (0.5) for twinning is assumed,

---

*See Refs. 43, 95, and 102-110.

**Arbitrarily, relative displacement of the upper layers of the lattice upward toward the optic axis (or c axis, or c_v as used here) is called gliding in the positive sense; relative displacement of the upper layers downward from the upper end of the c_v is called gliding in the negative sense.

***Direct visual evidence of translation (e.g., slip lines) is rare. Accordingly, translation gliding systems will not be utilized here, but will be discussed in the section on intragranular rotation phenomena.

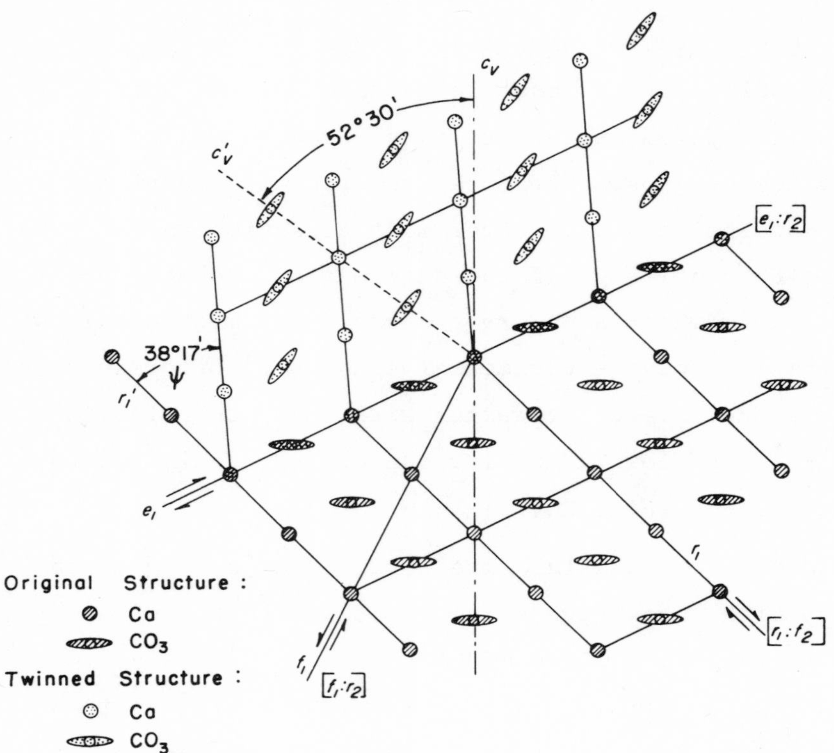

Fig. 24—Diagrammatic representation of the calcite structure. Section is drawn normal to the a_2 axis. The structure is twinned on e_1, with the gliding direction and sense of shear for the twin gliding indicated. Translation gliding systems along r_1 and f_1 are also shown.

the position of the load axis can be uniquely defined, because χ_0 and λ_0 must be 45 degrees (Fig. 19). Accordingly, σ_1' and σ_3' are fixed for twin gliding when $S_o = 0.5$ (Fig. 25(a)).* The compression axis σ_1' is inclined 45 degrees to e_1 or to the normal to e_1, and 71 degrees to c_v (the c axis).** The extension axis σ_3' is inclined 45 degrees to e_1 or the normal to e_1, and 19 degrees to c_v. For any calcite grain,

---

*Primes are used to denote principal stress axes derived from any one crystal.

**By convention, the three twin planes in each calcite crystal are designated as e_1, e_2, and e_3; e_1 is identified as the plane along which the twin lamellae are best developed, i.e., most densely spaced or widest, and e_3 is identified as the plane along which they are the most poorly developed. In a calcite crystal in which at least one set of twin lamellae is developed (e_1), the positions of the other potential sets can be determined from the crystallography of calcite.

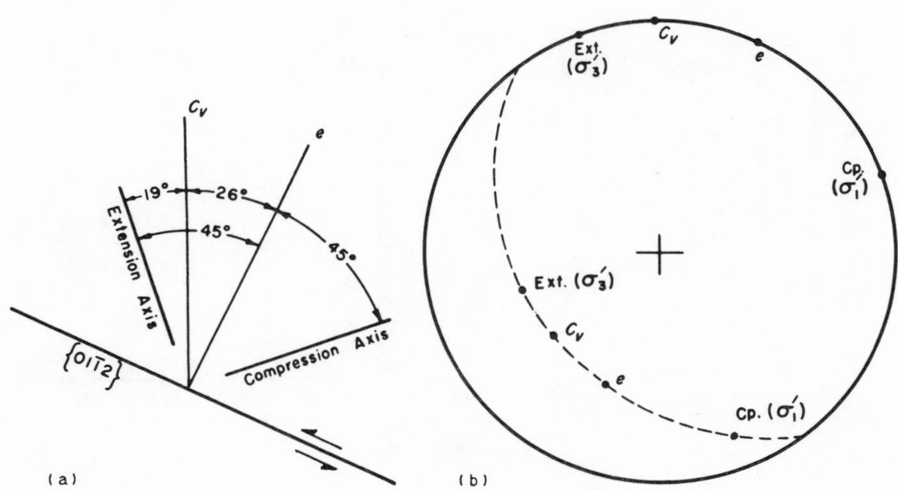

Fig. 25—Diagrams illustrating the orientation of the axes of compression (σ_1') and extension (σ_3') that would be most effective in causing twin gliding on e{01$\bar{1}$2} in calcite. (a) The plane of the diagram is perpendicular to the gliding plane and contains the gliding direction, the c_v, and the normal to the gliding plane (e). (b) Lower hemisphere, equal-area projection of the relationships in (a) are shown for two differently oriented cases.

therefore, the positions of σ_1' and σ_3' for $S_o = 0.5$ can be determined (Fig. 25(b)) by measuring and plotting e_1 and c_v.

Twinning can, of course, be initiated for S_o values less than 0.5 as long as τ_c is exceeded and the correct sense is maintained. For example, the average S_o value is 0.27 for the twinned calcite cement of an experimentally deformed sandstone.(59) Turner's technique utilizes $S_o = 0.5$, however, because this value allows unique location of the principal stresses. (There are an infinite number of possible orientations for the stress axes for $S_o < 0.5$.) Moreover, there is a good empirical relationship between the amount of twinning and S_o values. In experimentally deformed Yule marble and Hasmark dolomite, the greatest amount of twinning occurs on that set of twin gliding planes for which the resolved shear-stress coefficient was highest (Fig. 26).(52,106)

The technique using calcite twin lamellae to derive the orientations of the principal stresses in a rock consists of the following steps. (1) The orientation of the best developed set of twin lamellae and the host c_v in each grain are determined by universal-stage

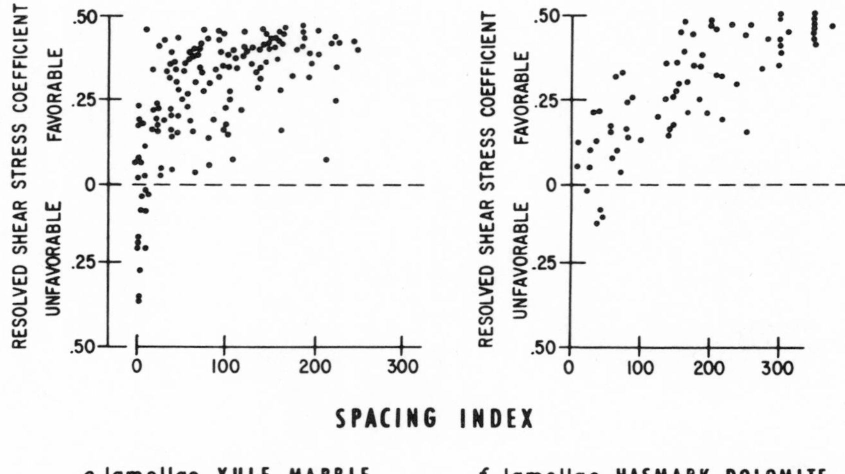

Fig. 26—Plots illustrating the relationship between e_1 and f_1 twin-lamellae spacing indices and S_o values for these planes as calculated from the known stress orientations in experimentally deformed Yule marble and Hasmark dolomite, respectively (from Turner and Ch'ih, Ref. 106, Fig. 6; and Handin and Fairbairn, Ref. 52, Fig. 6, respectively). Twin-lamellae spacing index is defined as the number of lamellae per millimeter when viewed on edge.

measurements. (2) These data are plotted in equal-area projection, and the compression and extension axes are located for each grain as outlined above. (3) The data measured in two or more mutually perpendicular thin sections are combined into a composite diagram, and the resulting orientation pattern is interpreted as reflecting the average orientation of σ_1 and σ_3 in the rock at the time twinning took place. Friedman[59] has found good agreement between the derived position of σ_1 and the known orientation of σ_1 in experimentally deformed calcite-cemented sandstones (Fig. 27).

A readily visible example of this technique is described by Friedman and Conger.[112] Calcite crystals within the walls of a naturally deformed fossil shell (Fig. 28) exhibit in thin section a systematic development of e twin lamellae. In transverse section the walls are elliptical and the crystals along two opposite sides of the ellipse (each encompassing 120 degrees of arc) are profusely twinned, whereas those along the two remaining 60-degree arcs are sparsely twinned. Calcite c_v are oriented radially. Fig. 29(a) shows that the preferential development of twin lamellae probably depends on favorable

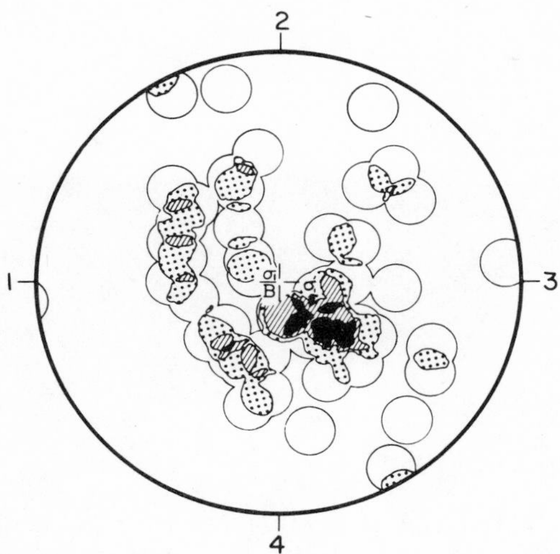

Fig. 27—Diagram of 50 compression axes derived from sets of e_1 twin lamellae in 50 calcite-cement crystals from an experimentally deformed calcite-cemented sandstone (from Friedman, Ref. 59, Fig. 16c). Specimen was shortened 9.2 per cent under 2-kb confining pressure at 300°C. Plane of the diagram is normal to the long axis of the deformed cylinder with the known position of σ_1 at the center of the diagram. Contours are at 2, 4, 6, and 8 per cent per 1 per cent area. The derived position of σ_1 is 10 to 15 degrees SE of the center.

or unfavorable orientation for twinning of a given crystal with respect to an assumed east-west σ_1. Moreover, compression and extension axes derived from c_v and e_1 lamellae in each grain are strongly oriented parallel to bedding and trend N-75°-W and N-15°-E, respectively, when the beds are unfolded (Figs. 29(b) and 29(c)). These results are in good agreement with the geologic framework because these directions are within 15 degrees of being perpendicular and parallel, respectively, to a series of nearly parallel north-south fold axes located where the sample was collected some 7 mi south of Drummond, Montana.

Twin lamellae in complexly deformed metamorphic rocks have been studied in considerable detail. Turner(111) applied his technique to the study of three marbles and concluded that the visibly twinned e lamellae developed during the last stages of deformation. McIntyre and Turner(113) employed the same methods in a study of three different marbles from Scotland and also concluded that twinning in calcite was the expression of minor postcrystallization deformation, probably compression transverse to the regional fold axis. Gilmour and

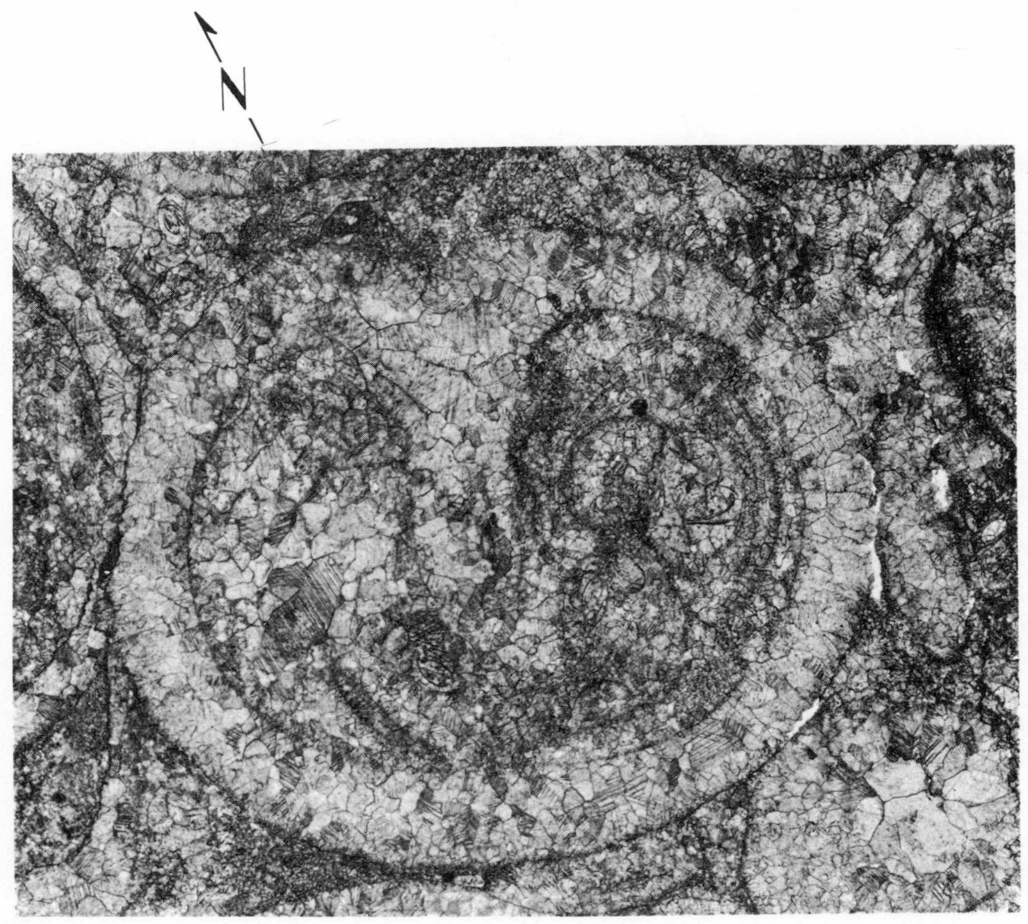

Fig. 28—Photomicrograph of a transverse section through a gastropod. Plane of the photomicrograph is also parallel to bedding. North is as indicated after the bedding is rotated to the horizontal. East-west diameter of the gastropod is approximately 4 mm (from Friedman and Conger, Ref. 112, Plate 1).

Carman,[114] taking the same approach, found that the compression axis deduced from postcrystallization e twin lamellae confirmed the direction of movement deduced from megascopic structures. Clark[115] studied the calcite twinning in still other marbles from the Scottish Highlands and found a consistent stress pattern over the 5 sq mi investigated. He concluded that the calcite twinning could best be explained by a compression oriented normal to the trend of the regional fold axis, followed by a mild "squeeze" at right angles. Weiss[116] investigated the dynamic significance of visibly twinned e lamellae in a marble-quartzite complex in southern California. He found that only

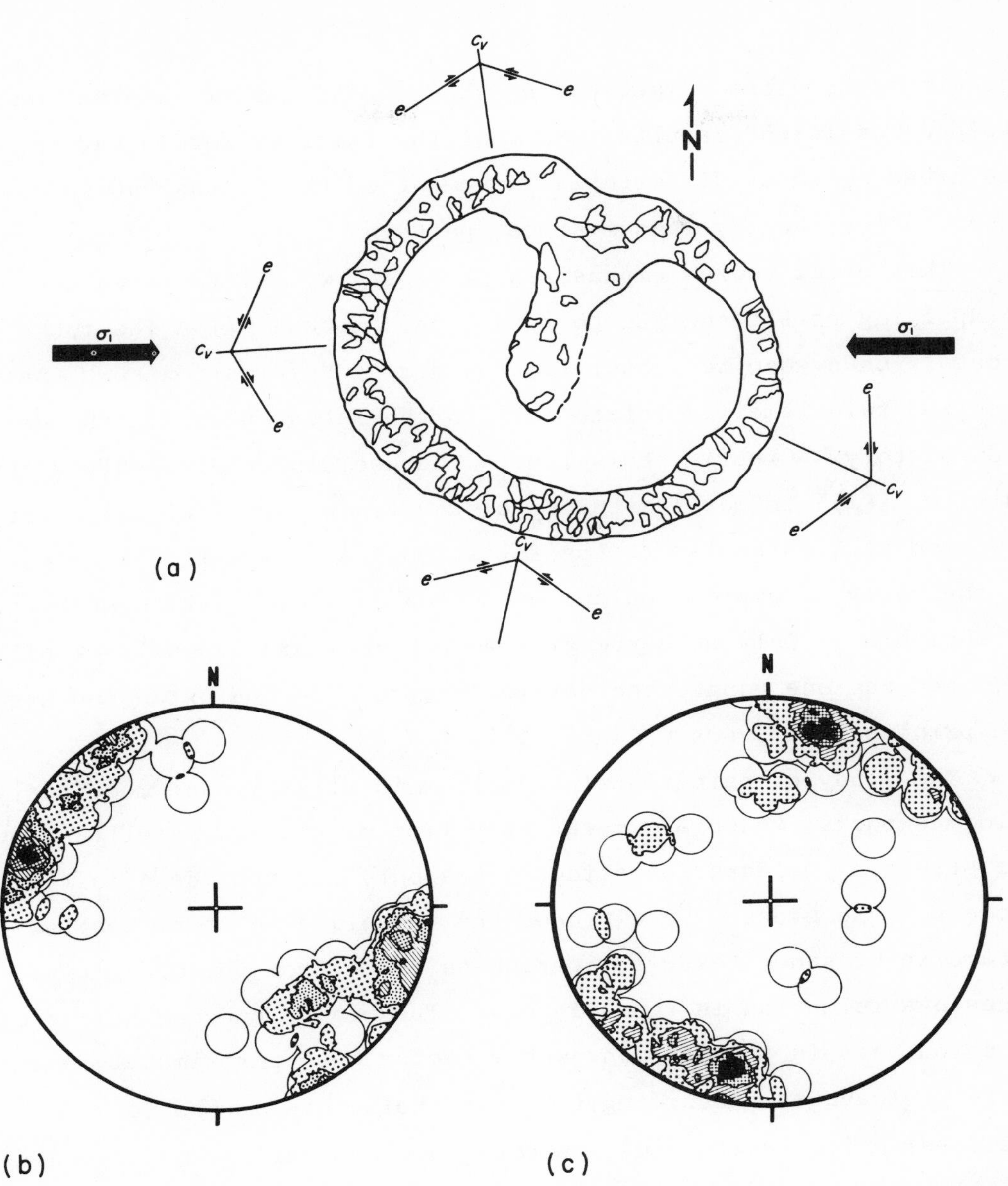

Fig. 29—Petrofabric analysis of calcite twin lamellae in the gastropod shell. (a) Approximate orientations of two e twin planes and the c_v in calcite crystals at four points within the gastropod shell. With respect to an assumed east-west greatest principal compression, the grains along the north and south sides of the shell are favorably oriented for twinning because of the sense of shear on the twin planes, whereas those on the east and west sides are unfavorably oriented (from Friedman and Conger, Ref. 112, Fig. 4). (b) Diagram illustrating the orientation of 70 compression axes derived from grains with high e_1 lamellae spacing index. Plane of the diagram is parallel to bedding with north as indicated for bedding unfolded. Contours are at 1.4, 2.9, 5.7, 8.6, 11.4, and 14.3 per cent per 1 per cent area, 15.7 per cent maximum (from Friedman and Conger, Ref. 112, Fig. 8b). (c) Diagram illustrating the orientations of 70 extension axes derived from same grains as in (b). Diagram oriented same as in (b). Contours are at 1.4, 2.9, 5.7, 8.6, and 11.4 per cent per 1 per cent area, 14.3 per cent maximum (from Friedman and Conger, Ref. 112, Fig. 9b).

those twin lamellae formed during the last stages of deformation yielded consistent results, and that the lamellae formed earlier were disturbed by later differential movement of the grains and so gave inconclusive, nearly random stress patterns.

This brief review emphasizes at least two difficulties encountered in applying this technique to metamorphic rocks. (1) The rocks have probably undergone recrystallization during deformation such that the observed twin lamellae relate only to the latest phase in the deformation history. Clearly, this limits the usefulness of highly deformed calcite. (2) In marbles the c_v of the grains are typically strongly oriented at high angles to the foliation. As a result the orientations of the derived compression and extension axes are restricted by the c_v subfabric. Only when the grains in a rock are randomly or diffusely oriented can one equate the derived compression and extension axes to the principal stresses σ_1 and σ_3.[117]

The study of calcite twin lamellae in slightly and moderately deformed rocks (which are relatively free of the recrystallization and the strong c_v orientation effects) has only recently been initiated. Nickelsen and Gross[118] extended the technique to study two low-grade metamorphic, sandy textured, carbonate rocks from the Ordovician Conestoga formation in Pennsylvania. They found concentrations of compression axes in a broad zone whose center was approximately normal to a slaty cleavage. According to the authors this agreed very well with the observation that grains, pebbles, and boulders were flattened parallel to the cleavage. In addition, the authors positioned the axes that corresponded to the bisectors of the acute angles between the two gliding lines for grains in which two sets of lamellae were developed. This gave a similar compression axis orientation pattern.

Conel[119] determined compression and extension axes in two specimens within a single bed of folded limestone (Fig. 30). His specimens were collected from the trough portion of a syncline--one near the top of the bed and the other near the bottom. In the former, compression axes are strongly grouped normal to the fold axis, whereas in the latter, the compression axes are strongly grouped normal to bedding. Conel concluded that these orientations were analogous to those expected from elastic analyses of bent plates.

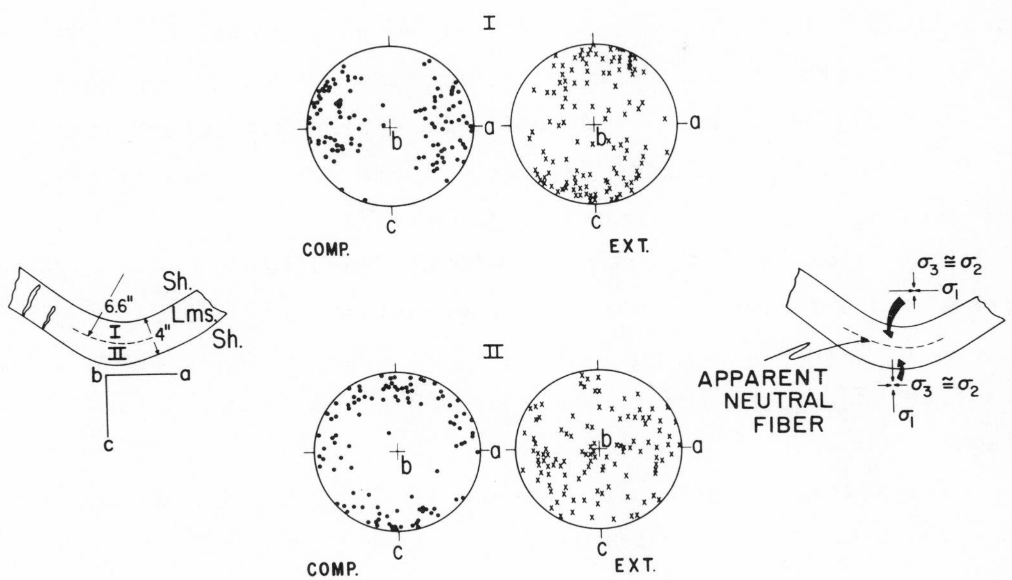

Fig. 30—Diagrams of compression and extension axes determined from calcite twin lamellae in regions I and II of a folded limestone bed in the Silurian McKenzie Creek formation of western Maryland (Conel, Ref. 119, Figs. 10a and 14). In region I, 114 compression axes are grouped about the a reference axis, and the 117 extension axes form a wide bc girdle with a tendency to be grouped about the c reference axis. In region II, 105 compression axes tend to be grouped about the c reference axis, while the 112 extension axes are diffusely oriented but show a tendency to lie in an ab girdle. Dynamic interpretation of these patterns is idealized at the right.

Hansen and Borg[120] studied deformed calcite cement in three specimens of folded Oriskany sandstone from eastern Pennsylvania. They found that the derived compression axes are oriented parallel to bedding and normal to the fold axis, while the extension axes are concentrated normal to bedding. This study will be discussed in more detail later in connection with quartz deformation lamellae.

Nissen[121] recently described a naturally deformed crinoidal limestone in which many of the individual calcite (crinoid) crystals exhibit two equally well developed sets of e twin lamellae. Nissen modified the Turner technique to locate the compression and extension axes that would produce equal S_o values on each of the two sets of twin planes. Thus, his compression axis is parallel to the crystallographic a axis cozonal with e_1 and e_2, and his extension axis is normal to the undeveloped or poorly developed third set of twin planes (e_3) in each grain. In Nissen's specimens these gave essentially the same stress pattern as that derived through use of the Turner technique.

Gliding mechanisms in dolomite have been determined from studies of experimentally deformed dolomite single crystals and rocks.* Only two gliding systems are known (Fig. 31):

(1) Translation gliding on c{0001} parallel to an a axis in either sense is the dominant flow mechanism below 400°C.

(2) Twin gliding on f{02$\bar{2}$1} in a negative sense along a line of the type [f_2:a_3] begins to occur at 400°C and is the major flow mechanism at 500°C.[55]

The writer is aware of only two studies in which dolomite twin lamellae have been dynamically interpreted by the Turner technique. This is in part due to the fact that twin gliding in dolomite is

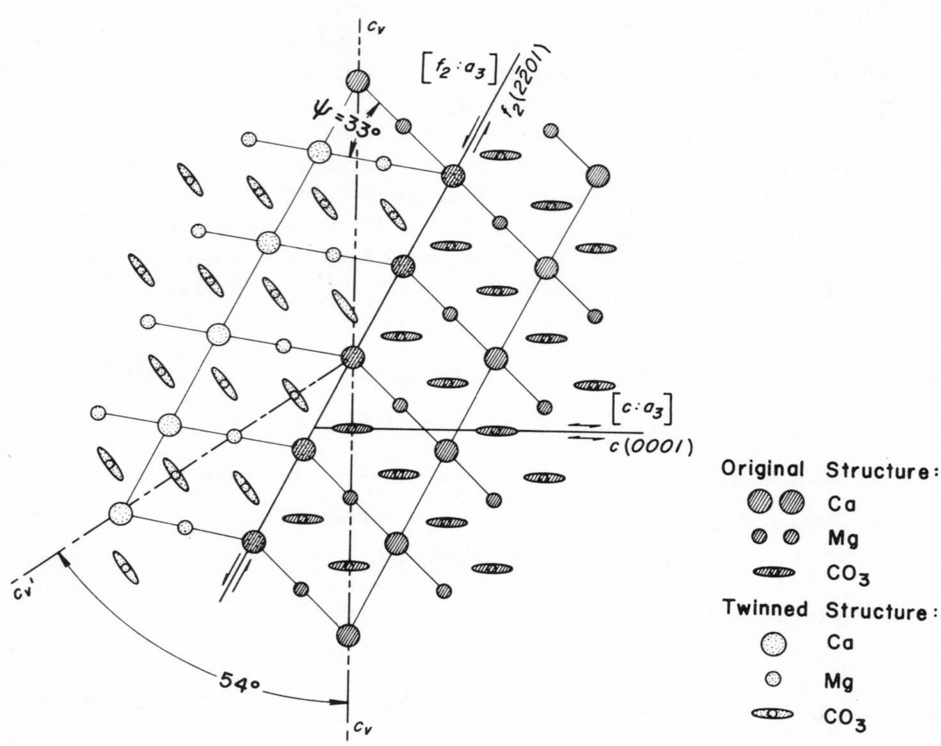

Fig. 31—Diagrammatic representation of the dolomite structure. Plane of the section is parallel to a_3(11$\bar{2}$0). Gliding direction for translation on {0001} is not in the plane of the section, but is parallel to any of the three a axes. The system for twin gliding parallel to f is also illustrated.

---

*See Refs. 52, 55, 122, and 123.

restricted to deformation at high temperatures, which is relatively rare in all but intensely deformed metamorphic rocks. Crampton[124] studied Cambro-Ordovician dolomite and calcite marbles from the northwest Highlands of Scotland. He determined compression and extension axes from f twin lamellae in dolomite as well as from e twin lamellae in calcite. Both gave similar results--compression axes oriented at high angles to the foliation. Christie[125] studied deformed dolomite from the Moine thrust zone. He concluded that the compression and extension axes inferred from twinned f lamellae reflect the final stage of deformation. He found the derived compression axes to be oriented at high angles to the foliation in the Moine thrust block.

In general, there are certain inherent limitations in the use of twin lamellae to derive principal stress directions. (1) Derived compression and extension axes are fixed by the crystallographic orientation of the grains in the rock. Accordingly, only when the grains are nearly randomly oriented can one equate the derived stress axes to the principal stresses σ_1 and σ_3.[117] (2) Some of the spread in the fabric diagrams of derived compression and extension axes results from the assumption that S_o is always 0.5. In reality twinning can occur when the load axes are inclined at angles far from 45 degrees to the gliding plane and direction, provided τ_c is exceeded in the proper sense. The scatter in these patterns can be reduced by constructing partial diagrams containing compression and extension axes from only the best developed e_1 lamellae.[112] (3) Designation of host and twinned lattices (characterized by c_v and c_v', respectively) is difficult in intensely deformed grains. By necessity the host is defined as the dominant lattice of the crystal. As twinning on e_1 progresses, however, the host lattice gives way to the twinned lattice as the predominant structure. In grains which are in reality more than half twinned, an observer must select the now dominant lattice as the host.* Compression and extension axes located with respect to e_1 and this new "host" (c_v') will depart 90 degrees from the true axes associated with the twinning (Fig. 32). (4) Use of false e_1 lamellae to

*Crystals more than half twinned are usually elongated by the shear strain of twinning. The dominant lattice of nearly equidimensional grains, therefore, is very probably the "true" host.

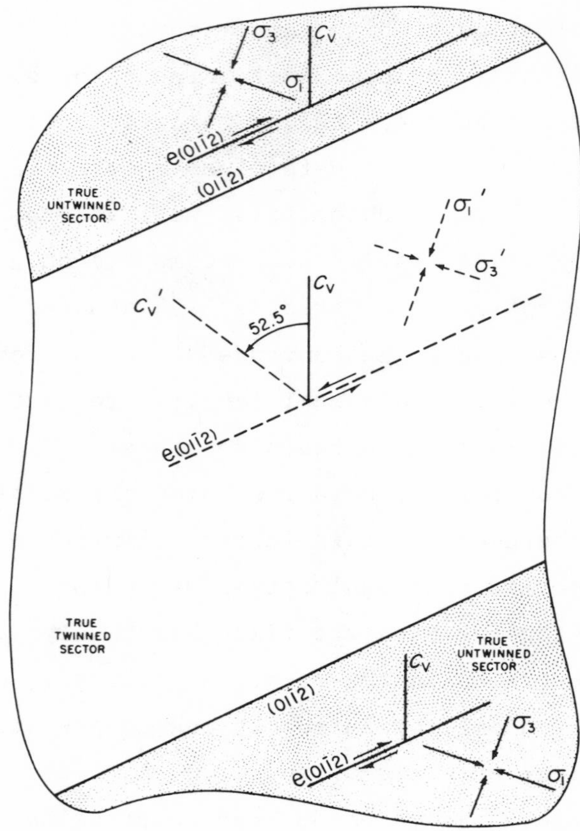

Fig. 32—Diagrammatic section through a calcite grain which is more than half twinned by gliding on e(01$\bar{1}$2). Plane of the section is normal to the twin plane and contains the gliding direction. (01$\bar{1}$2) is common to both the twinned and untwinned structures, which are designated by c_v' and c_v, respectively. Orientations of σ_1 and σ_3 that would best cause the twin gliding are shown at the top and bottom of the grain with respect to the true host structure. An observer, however, would mistake the dominant portion of the grain (the real twinned portion) to be the "untwinned host." That is, he would measure c_v' and the observed twin lamellae (01$\bar{1}$2), and from these measurements would position σ_1' and σ_3' as indicated in the center of the grain. These orientations for the principal stresses are 90 degrees out of phase with those actually best oriented to produce the twinning.

position compression and extension axes results in misleading interpretations. For example, as twinning nears completion on e_1, the "true" e_2 lamellae may appear to be the best developed set in the grain and incorrectly designated e_1; e_2 or e_3 sets may appear to be the best developed in a grain because the true e_1 lamellae are inclined at too low an angle to the plane of the thin section to be correctly evaluated;

or in highly twinned grains one can easily mistake well developed e_2 lamellae for e_1 lamellae. The angle between compression axes derived from e_1 and e_2 lamellae, respectively, is 70 degrees, while that for the extension axes is 32 degrees. Clearly, Turner's technique can be expected to give the most meaningful results when applied to slightly and moderately deformed rocks (strain less than about 20 per cent), for in such rocks, the effects of limitations 1, 3, and 4 are usually negligible.

Rotation Phenomena

Intragranular Rotations. Gliding in a constrained crystal is accompanied by an external rotation of the gliding planes toward the axis of extension and away from the axis of compression. Turner[(95)] recognized that this would cause an internal rotation of any intersecting pre-existing plane to an irrational position within the lattice.* He was then able to develop an important new technique to identify gliding systems and to compute shear strain (e.g., for calcite, Ref. 95; for dolomite, see Ref. 55) and to utilize translation gliding systems for dynamic analysis.[(125)]

Consider, for example, a lamella P in existence prior to deformation (Fig. 33). This will be rotated to a new position by gliding on the set of parallel planes T. In an unconstrained crystal, the lamella P has been rotated internally to the position L in the deformed sector. In a constrained crystal the gliding plane T and the lamella P in the deformed sector have both been rotated externally relative to the load axis to T′ and P′, and P′ has also been internally rotated to L. Internal rotation is always opposite in sense to that of external rotation.

The concept of intragranular rotation is applicable to any crystal that can be examined in thin section with the petrographic microscope

---

*External rotation is defined as rotation of crystallographic elements relative to external coordinates (e.g., rotation of the optic axis relative to the load axis). Internal rotation is defined as rotation of visible elements relative to internal coordinates (e.g., rotation of twin lamellae relative to the c_v of the crystal in the immediate vicinity of the lamellae).

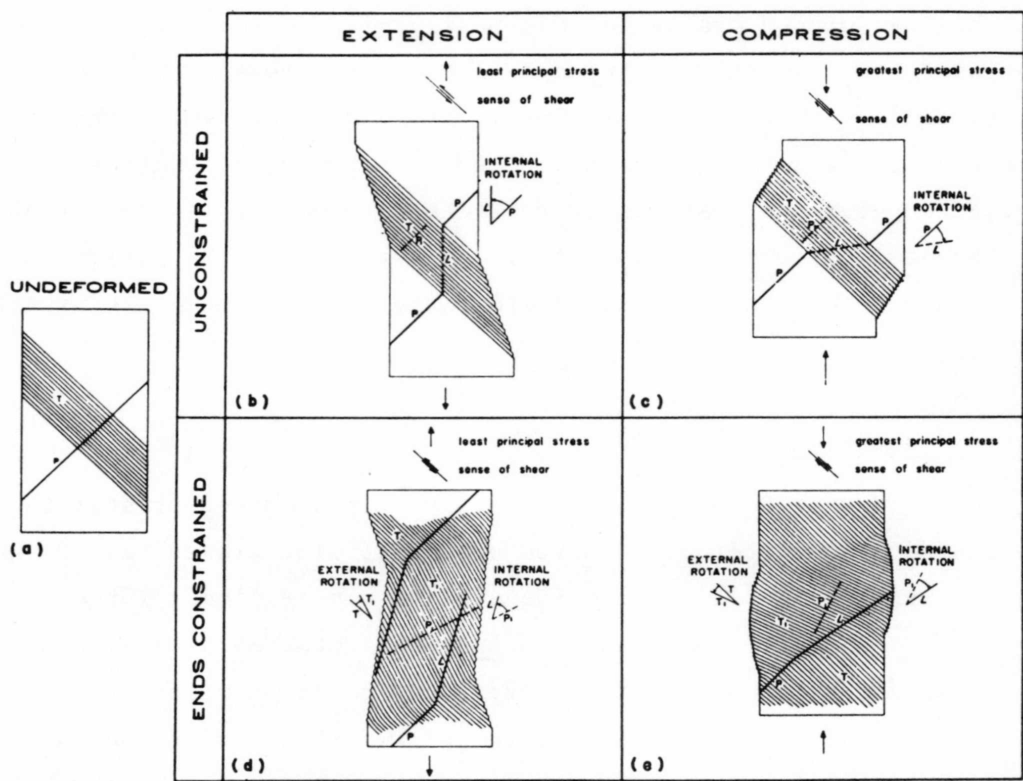

Fig. 33—Schematic illustration of internal and external rotations (from Higgs and Handin, Ref. 55, Fig. 1).

and universal stage. Visible rotations (Fig. 34) can be utilized even though the mechanism by which a pre-existing lamella is rotated through a crystal structure is not yet understood. Of importance here is that internal rotation of a visible feature is evidence that translation gliding has occurred even though there may be no visible trace parallel to the translation gliding plane itself. Once the translation gliding system is established, the directions of compression and extension that would best cause the gliding can then be determined in a manner similar to that employed for twin gliding. Christie[125] has utilized this approach and found good agreement between the principal stresses derived from internal rotation phenomena and those derived from f twin lamellae in dolomite.

Intragranular rotation phenomena in an experimentally deformed dolomite single crystal[55] serve to illustrate how a translation gliding system can be recognized and how the stresses that would most favorably produce the translation can be derived. A specimen was

Fig. 34—Photomicrographs illustrating features characteristic of internal rotation in an experimentally deformed dolomite single crystal. The internally rotated lamellae (L) are typically coarse and dark, and occasionally serrated. The rational lamellae (r) are straight, narrow, and sharply defined lines. X 175.

extended 12 per cent under 5-kb confining pressure, 24°C, at a constant strain rate of 1 per cent per minute. The specimen was extended normal to r_1 (cleavage plane), an orientation favorable for basal (c{0001}) translation. In thin section, the specimen is characterized by a central deformed sector bounded on either side by relatively undeformed sectors. The orientations of the c_v and cleavage are measured in the undeformed areas (c_v and r_1) and in the deformed sector (c_v' and r_1'). These are plotted in lower hemisphere equal-area projection (Fig. 35). It is apparent from their positions that there has been a clockwise external rotation of 12 degrees between the undeformed and deformed sectors. Detailed examination of the deformed sector reveals a coarse dark set of planar features ($L^{c'}_{r_1'}$) that make a small angle with r_1'. The normals to r_1' and $L^{c'}_{r_1'}$ lie on the same great circle as c_v', which indicates that c′{0001} is the active gliding plane.* Moreover, the

*The active gliding plane and a rotated plane in all positions of rotation must be cozonal, i.e., share a common axis of intersection or rotation. Hence, the normals to these planes must lie on the same great circle.

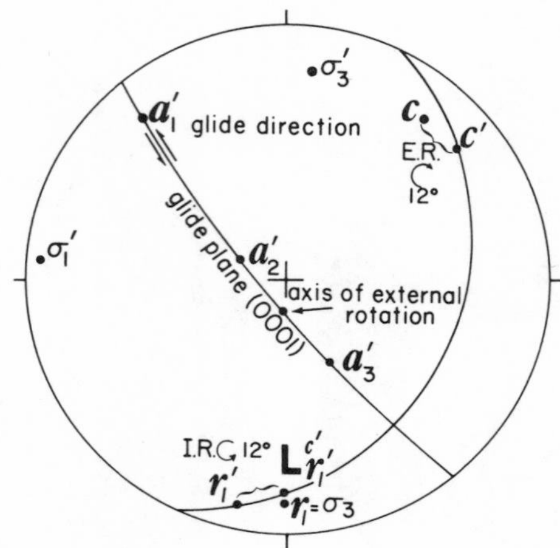

Fig. 35—Stereogram showing the internal and external rotation phenomena in experimentally deformed dolomite single crystal specimen No. 161 (from Higgs and Handin, Ref. 55). The specimen was extended 7.5 per cent under 5-kb confining pressure at 24°C. The least principal stress of the experiment (σ_3) was oriented normal to r_1 as indicated. External rotation of c to c' and r_1 to r_1' and the internal rotation of r_1' to $L^{c'}/r_1'$ are shown. σ_1' and σ_3' are the stresses derived from translation gliding on c' along the a_1' gliding direction in the sense indicated. The angle between σ_3 and σ_3' is 31 degrees.

angle between r_1' and $L^{c'}_{r_1'}$ indicates that a counterclockwise internal rotation of 12 degrees has occurred. Although c' is established as the active gliding plane, dynamic inferences still require identification of the direction and sense of shear. Basal translation can occur along any one of the three crystallographic a axes in either sense. As a first step, draw the great circle representing c', i.e., the great circle normal to c_v' (Fig. 35). Because the rational position of r_1' is known, we can locate a_1', a_2', and a_3' along the active gliding plane. One of these must be the gliding direction. This direction can be established from the axis of external rotation which must lie in the gliding plane and perpendicular to the gliding direction. Accordingly, the normal to the great circle through c_v and c_v' and to that through r_1 and r_1' defines the axis of external rotation. Since this lies on the active gliding plane midway between a_2' and a_3', a_1' at 90 degrees to the axis of external rotation must be the gliding direction. Finally, the sense of shear is fixed by the sense of internal rotation. To bring r_1' to $L^{c'}_{r_1'}$ requires a left-lateral sense of shear along c' (Fig. 35).

The complete gliding system is now established, the gliding plane is c′, the gliding direction is parallel to a_1', and the sense of shear is consistent with the known principal stress directions. The compression and extension axes that would best cause this gliding (S_o = 0.5) are in good agreement with the known stresses across the boundaries of the specimen during the experiment (Fig. 35).

Kink Bands. Kink bands are deformation features in crystals (Fig. 36) and in aggregates with pronounced planar anisotropy (Fig. 37). They were first recognized by Orowan[126] in compressed cadmium crystals, and have since become well known in metallurgy.* Current understanding

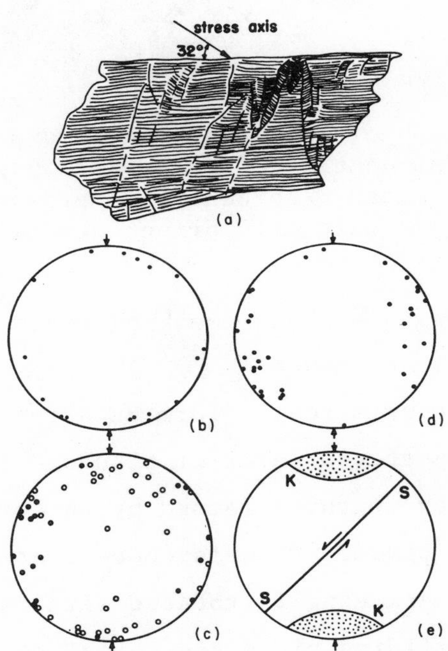

Fig. 36—Kink band development in biotite grains in experimentally deformed Westerly granite (from Griggs, Turner, and Heard, Ref. 43, Fig. 11). (a) Sketch of kink bands in a single grain of deformed biotite. Arrow shows orientation of compression axis. (b) Normals to kink band boundaries in 11 biotite grains. Arrows show axis of compression. (c) Normals to (001) in biotite grains with kink bands (solid circles) and lacking kink bands (open circles). Arrows show axis of compression. (d) Distribution of normals to (001) in 42 biotite grains with kink bands, as measured in 5 thin sections. Arrows show the axis of compression, which is also the common axis of reference for the 5 sections. (e) Relation of macroscopic surfaces of shear (SS), maximum concentration of normals to kink band boundaries in biotite (K), and sense of shear (arrows).

*See Refs. 93, 94, and 127-129.

Fig. 37—Photomicrograph of kink bands in an experimentally deformed specimen of slate (from Handin and Borg, Ref. 53). Specimen was shortened 27 per cent under 5-kb confining pressure at 500°C. Slate cleavage (east-west) was initially oriented parallel to σ_1.

of their formation suggests that some kink bands may be dynamically significant petrofabric elements.

Kink bands characteristically form as a result of gliding flow along a set of closely spaced parallel planes. The boundaries of the bands are nearly planar features formed by an abrupt change in attitude of the active gliding plane. Flow is concentrated within the band where the structure is externally rotated with respect to the host about an axis in the gliding plane and normal to the gliding direction.(94) Kink band formation in a single crystal differs from that in an aggregate in that the gliding direction is fixed in the active gliding plane of the crystal but is apparently unrestricted in the s-plane of the aggregate.

Kink banding can result from pure translation gliding in a constrained crystal (e.g., for calcite, see Ref. 95, Plate 3, and Figs. 33(d) and 33(e) of this paper). However, kink bands are best developed in those crystals or aggregates whose active gliding plane is oriented subparallel to the axis of compression (Ref. 126, p. 644; Ref. 127, p. 192)--i.e., initially parallel to the plane of vanishing shear stress.

Bend gliding (Fig. 20) is presumably initiated after some elastic distortion, and the kink band is formed as the material buckles. According to Barrett (Ref. 94, p. 375), the kink band boundaries are initially planes that are normal to the gliding direction and the gliding plane. It follows, therefore, that well-developed kink band boundaries are initially oriented at high angles to the greatest principal stress.

Kink bands developed in biotite crystals of experimentally deformed Westerly granite[43] demonstrate their potential usefulness as a dynamic petrofabric element. Cylinders of the granite were compressed 15 to 23 per cent under 5-kb confining pressure at 500°C. Most of the shortening was achieved by faulting, localized along a zone of mylonitization. Biotite crystals near the shear zone were deformed by kink banding. The individual kink bands are described as being narrow and sharply bounded by nearly planar surfaces trending at high angles to the basal cleavage (001). They are conspicuous because the cleavage is sharply deflected through angles of 20 to 50 degrees. Opposite boundaries of a kink band typically are nonparallel, so that the band is wedge- or lens-shaped. In some grains several subparallel kink bands are present. In others there are two sets, symmetrically inclined to (001) of the host grain, giving a chevronlike pattern (Fig. 36(a)). Normals to the kink band boundaries tend to group around the known compression axis, σ_1 (Fig. 36(b)). Moreover, kink bands tend to develop preferentially in grains whose [001] axes (normals to (001)) are steeply inclined to σ_1 (Figs. 36(c) and 36(d)). That is, the (001) cleavages (the active gliding plane) are subparallel to σ_1. Subsequent work by Handin and Borg[53] on Fordham gneiss has substantiated these findings.

Accordingly, if biotite crystals in a naturally deformed rock are nearly randomly oriented, it is reasonable to conclude that the center of concentration of normals to kink band boundaries can be equated to σ_1 in the rock at the time of deformation. On the other hand, as biotite in most rocks exhibits a nonrandom orientation, this technique can position the greatest compression axis only somewhere within the s-plane defined by the basal cleavages.

Quartz Deformation Lamellae

General. Quartz deformation lamellae are here treated independently of any one mode of origin because the process by which they are formed in naturally deformed rocks is still somewhat in doubt. Sufficient information is available, however, to utilize these features in dynamic analyses. They occur in deformed quartz in a variety of rock types. Usually grains with deformation lamellae also show undulatory extinction. Their physical appearance is variable (Fig. 38 and Fig. 42 on p. 521). Most workers agree, however, that their tendency to lie at low angles to the {0001} plane in the quartz host is diagnostic (Fig. 39).

Origin. The origin of quartz deformation lamellae has been the subject of much speculation since they were first reported 85 years ago. According to Griggs and Bell[(42)] and Ingerson and Tuttle,[(130)] Kalkowsky[(131)] first described the lamellae; Boehm[(132)] identified them as planes of liquid inclusions; Judd[(133)] thought they were secondary twin lamellae; and Becke[(134)] identified them as healed fractures. Mügge,[(135)] Sander,[(1)] Hietanen,[(136)] Fairbairn,[(137,138)]

Fig. 38—Photomicrographs of quartz deformation lamellae in grains of a naturally deformed calcite-cemented sandstone from the Jurassic Piper formation, Park County, Montana. Note two sets of lamellae in one grain. Crossed Nicols, X 100.

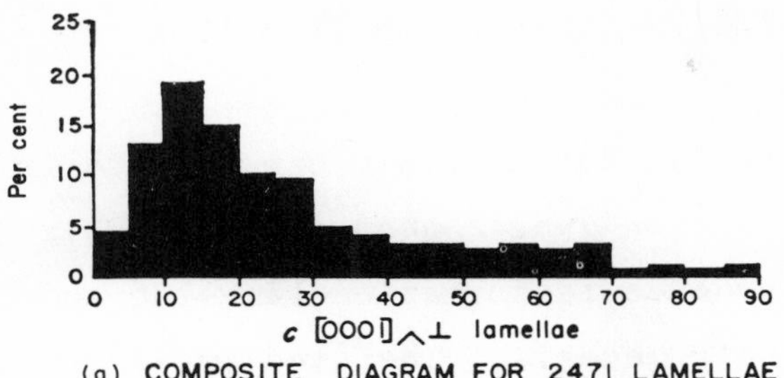

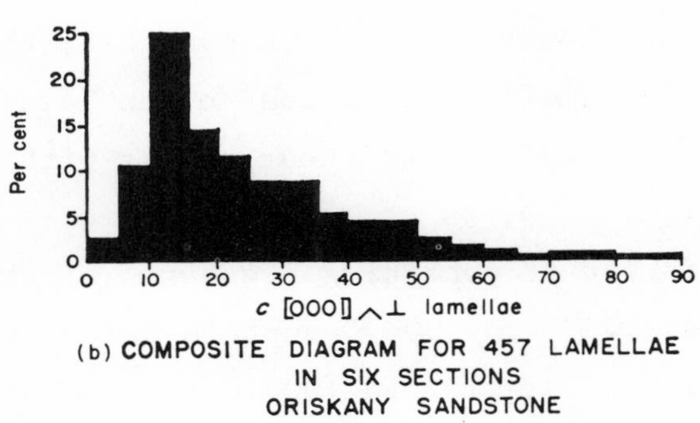

Fig. 39—Histograms showing the orientation of quartz deformation lamellae with respect to the c_V in host grains (from Hansen and Borg, Ref. 120, Fig. 1). (a) Composite histogram compiled from previous literature: 775, Christie and Raleigh (1959); 885, De (1958); 102, Saha (1955); 373, Ingerson and Tuttle (1945); 336, Fairbairn (1941). Sources cited by Hansen and Borg, Ref. 120. (b) Histogram for the Oriskany sandstone studied by Hansen and Borg.

Brace,[139] and Christie and Raleigh[140] have suggested that the lamellae are a result of gliding mechanisms. Ingerson and Tuttle (Ref. 130, p. 105), on the other hand, concluded that the lamellae are not controlled by definite crystallographic planes or zones in the quartz structure and that they are "apparently controlled almost entirely by the stress pattern which determined the fabric of the quartz in the rocks." Riley,[141] Turner,[2] and Weiss[116] suggested that the lamellae are only partially controlled by the structure of quartz and that they are formed late in the deformation history. Bailey _et al_.[142] found that grains with deformation lamellae have Laue photograph patterns that are somewhat more disturbed than those from grains with no lamellae, but they found no evidence to establish whether or not the lamellae represent gliding planes. The controversy seems to center mainly on whether the lamellae are caused by intracrystalline gliding or by fracture.

Attempts to produce this feature experimentally have only recently been successful. Work on quartz sand and on quartz single crystals[143-146] has produced deformation lamellae and provided a better understanding of their genesis. The sand specimens, compressed at confining pressures from 12 to 50 kb and temperatures from 25° to 700°C, contain abundant undulatory extinction, deformation bands, and deformation lamellae. Lamellae occur in over half the grains and at low angles to {0001}. Their inclination to the load axis indicates that they formed in planes of high shear stress with about as many inclined at < 45 degrees to σ, as at > 45 degrees to σ.[146] All the single crystals deformed at confining pressures of about 15 kb and at temperatures between 300° and 1500°C contain deformation lamellae. The lamellae are almost parallel to {0001} in those crystals compressed so that there was a high shear stress on {0001}. In crystals compressed along a line parallel or perpendicular to {0001}, the lamellae developed at angles from 30 to 60 degrees to {0001}, but always in planes of high resolved shear stress. Subsequent studies show conclusively that these artificial deformation lamellae result from translation gliding on {0001} with an a axis as the glide direction.[145]

Dynamic Interpretation of Quartz Deformation Lamellae. Even if natural deformation lamellae result from translation gliding, they can not be dynamically interpreted as, for example, calcite twin lamellae because the gliding direction (an a axis) cannot be located optically in quartz. Also there would be difficulties in establishing the sense of shear.* It is possible, however, to draw dynamic inferences from their orientation pattern in naturally deformed rocks based on their experimentally determined formation in planes of high shear stress

*Recently, Carter, Christie, and Griggs[146] have shown experimentally that the more deformed parts of kink bands and zones of undulatory extinction in quartz contain more abundant and closely spaced near-basal lamellae (the active slip plane) than the less deformed parts. They point out that a sense of shear can be established for the lamellae from consideration of the sense of external rotation within the kink bands or within zones of undulatory extinction.

provided that (a) for situations in which $\sigma_1 > \sigma_2 > \sigma_3$, one dihedral angle between two conjugate sets of lamellae is less than 90 degrees, and σ_1 is unambiguously the acute bisector; or (b) for cases in which $\sigma_1 > \sigma_2 = \sigma_3$ or $\sigma_1 = \sigma_2 > \sigma_3$, the lamellae lie along conical surfaces with half-angles of less than 45 degrees and greater than 45 degrees, respectively. In this hypothesis they are interpreted in the same manner as shear fractures or faults. Some examples of the angle between sets of quartz deformation lamellae and the attitude of the acute bisector for these sets in discrete samples are listed in the table on page 516. This compilation shows that the sets of lamellae intersect at an average acute dihedral angle of 74 degrees, and that they, therefore, probably formed in planes of high resolved shear stress. Moreover, the compilation indicates that in most cases the acute bisector, which the writer has equated to σ_1 as a working hypothesis, is related meaningfully to geologic reference lines and planes. These relationships, however, do not by any means establish unambiguously that the acute bisector parallels σ_1 in the rock at the time of lamellae formation. The hypothesis is strengthened, however, when other fabric elements are studied along with the deformation lamellae.

From studies of the Baraboo quartzite, Riley[141] showed that the character and orientation of lamellae could be used as a qualitative measure of the intensity of deformation. In addition he differentiated between microfractures and deformation lamellae, and found both strongly oriented and geometrically related to the major structure. It is instructive to examine Riley's Figs. 11(a) and 11(b) (see Fig. 40) which show the orientation of these features in his specimen. Two sets of deformation lamellae are defined which intersect at 60-80 degrees, and the normals to the microfractures define a single concentration. If the orientation diagrams for microfractures and lamellae are superimposed, it is apparent that the microfractures bisect the angle between the two sets of deformation lamellae. This geometrical relationship is identical with that of two sets of shear fractures and the enclosed extension fracture, i.e., two sets of lamellae are bisected by σ_1. In this case, σ_1 is oriented normal to bedding and to the fold axis (Figs. 40(c) and 40(d)).*

---

*Dynamic inferences by Friedman.

SOME EXAMPLES OF THE ANGLE BETWEEN SETS OF QUARTZ DEFORMATION LAMELLAE AND THE ATTITUDE OF THE ACUTE BISECTOR BETWEEN THE LAMELLAE PLANES

Reference	Rock Type	Approximate Angle[a] between Best Developed Sets of Lamellae (degrees)	Orientation of Acute Bisector[b] (σ_1)
Fairbairn, Ref. 138	Ajibik quartzite	80	Normal to s-plane
Ingerson and Tuttle, Ref. 130	Ajibik quartzite	76	Normal to s-plane
	Biotite-gneiss	80	Parallel to s-plane
Riley, Ref. 141	Baraboo quartzite		
	Specimen No. 9	60	Horizontal and at high angles to regional fold axis
	Specimen No. 18	75	Nearly horizontal and subparallel to axial plane foliation and regional fold axis
	Specimen No. 38	65	Nearly horizontal, normal to bedding, and at high angles to regional fold axis
	Specimen No. 47	75	Nearly horizontal, subparallel to bedding, and subparallel to regional fold axis
	Specimen No. 55	85	Horizontal, subparallel to bedding, and at high angles to regional fold axis
	Specimen No. 58	65	Horizontal, at high angles to bedding and to regional fold axis
Mackie, Ref. 147	Quartz-piedmontite schist	60	Normal to fold axis and to s-plane

Preston, Ref. 148	Kinahmi and Kuopio quartzites (3 specimens)	67 76 81	High angles to s-planes
Naha, Ref. 149	Quartz-mica schist	78	Microfractures in garnet grains bisect acute angle between sets of deformation lamellae
Christie and Raleigh, Ref. 140	Quartzites (Orocopia Mts., southern California)		
	Specimen No. 1	90	. . . .
	Specimen No. 2	88	Subnormal to first generation fold axis and axial plane
	Specimen No. 3	76	Parallel to first generation fold axis
	Quartzite (Moine thrust zone, Scotland)	76	High angles to foliation
Hara, Ref. 150	Calcite-quartz vein (Sangun formation, western Japan)	72	Subparallel to c fabric axis and normal to σ_1 derived from calcite twin lamellae
Hara, Ref. 151	Quartz vein (Sangun formation, western Japan)	67	Subparallel to c fabric axis
Hansen and Borg, Ref. 120	Calcite-cemented sandstone		
	Specimen E2	74	Parallel to bedding, normal to fold axis, and subparallel to σ_1 derived from calcite twin lamellae
	Specimen E4	76	Subparallel to bedding, normal to fold axis, and subparallel to σ_1 derived from calcite twin lamellae

[continued on following page]

Reference	Rock Type	Approximate Angle[a] between Best Developed Sets of Lamellae (degrees)	Orientation of Acute Bisector[b] (σ_1)
Hansen and Borg (continued)	Specimen E6	60	Parallel to bedding, subnormal to fold axis, and parallel to σ_1 derived from calcite twin lamellae

[a]Determined by Friedman if not provided by the author.

[b]Attitude of acute bisector was determined by Friedman if not provided by the author. This bisector is equated to σ_1 by Friedman as a working hypothesis.

The writer disregards the influence of pre-existing preferred orientations of quartz c_v on the deformation lamellae pattern. He agrees with Turner and Weiss (Ref. 13, pp. 433-434), who find that the quartz lamellae pattern, with its characteristic orthorhombic symmetry pattern, can be used to reconstruct the stress system (also commonly orthorhombic in symmetry) "because [even] in tectonites there is a wide range of orientation of c axes, and because crystallographic control of lamellae in any crystal is not rigorous, the influence of the initial orientation pattern of quartz axes upon the lamellae pattern is commonly negligible."

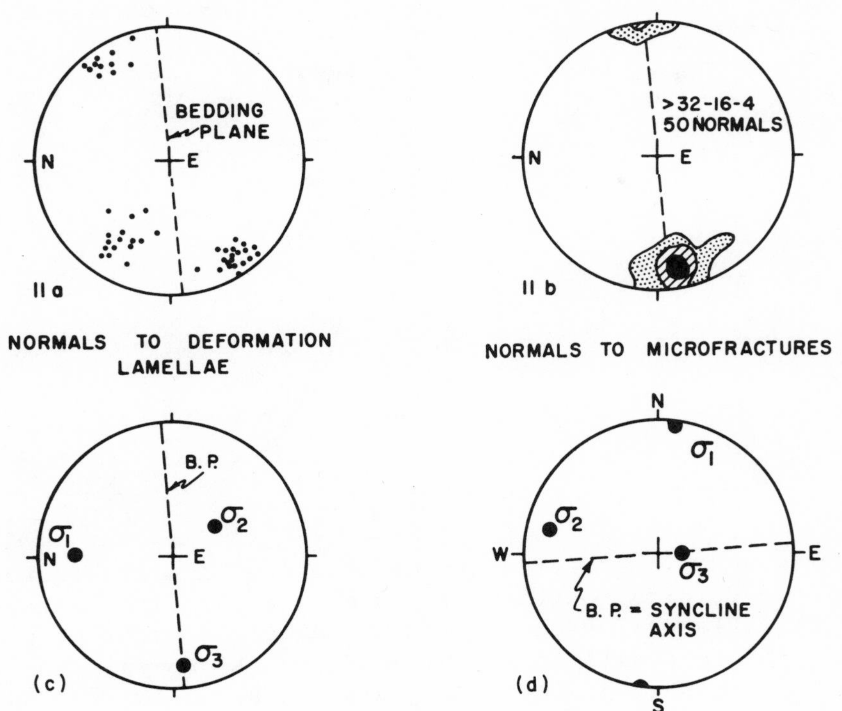

Fig. 40—Diagrams (11a and 11b) illustrating the orientation of 50 sets of quartz deformation lamellae and of microfractures, respectively, in a Baraboo quartzite specimen (from Riley, Ref. 141, Fig. 11). (c) Stereogram shows the orientation of the principal stresses deduced from the deformation lamellae and the microfractures, assuming the latter are extension fractures and the former are in planes of high shear stress and inclined at < 45 degrees to σ_1. (d) Stereogram shows same data rotated to horizontal plane, with north as indicated. σ_1 is nearly perpendicular to both the synclinal axis and the bedding plane.

From his studies of quartz deformation lamellae and microfractures in garnet grains of a quartz-mica schist, Naha (Ref. 149, p. 120) concluded that the normals to "deformation lamellae in quartz ... show two maxima in an incomplete girdle normal to the fold axis [i.e., the lamellae intersect in the fold axis] and are symmetrically situated with reference to late tension cracks in garnet." Thus the lamellae "have formed parallel to the two planes of maximum shearing strain...." His diagrams (see Fig. 41) show that the two sets of lamellae intersect at 78 degrees and that the set of microfractures acutely bisects this angle. This combined geometry is similar to that obtained by Riley. Here the derived σ_1 and σ_3 axes are oriented normal to the fold axis.

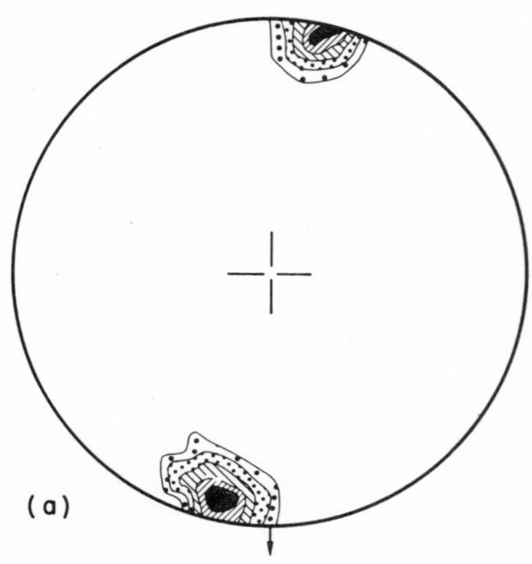

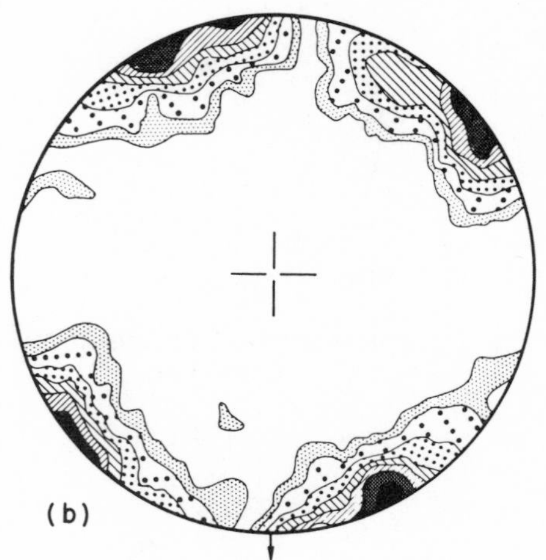

Fig. 41—Diagrams illustrating the orientation of microfractures in garnet (a) and deformation lamellae in quartz (b) (from Naha, Ref. 149, Fig. 2). Plane of each diagram is parallel to the ac plane of a fold, i.e., the fold axis is at the center. (a) Normals to 54 microfractures in 14 garnet grains. Contours are at 5, 10, 20, 30, 40, and 45 per cent per 1 per cent area, 52 per cent maximum. (b) Normals to 158 deformation lamellae in quartz. Contours are at 0.66, 1.3, 3.5, 7, 9, and 11 per cent per 1 per cent area, 13.7 per cent maximum.

Additional confirmation of these interpretations is afforded by Hansen and Borg[120] who studied both quartz deformation lamellae (Fig. 42) and calcite twin lamellae in three oriented samples of Devonian Oriskany calcite-cemented sandstone from an Appalachian fold in eastern Pennsylvania. Compression and extension axes derived from the deformed calcite cement are shown in Figs. 43(a), 43(b), 44(a), 44(b), 45(a), and 45(b). In each case the derived σ_1 is within 10 degrees of the bedding and is normal to the fold axis, and σ_3 is normal to the bedding,* so that σ_2 is subparallel to the bedding and to the fold axis. The orientation patterns of normals to the quartz deformation lamellae (Figs. 43(c), 44(c), and 45(c)) correspondingly show two distinct concentrations connected by an incomplete small circle girdle of 52 to 60 degrees half-angle, i.e., the lamellae lie along the surfaces of cones with half-angles of 30 to 38 degrees.

---

*The girdle of extension axes in specimen E2 is an exception. It indicates that $\sigma_2 \simeq \sigma_3$ in magnitude for this specimen.

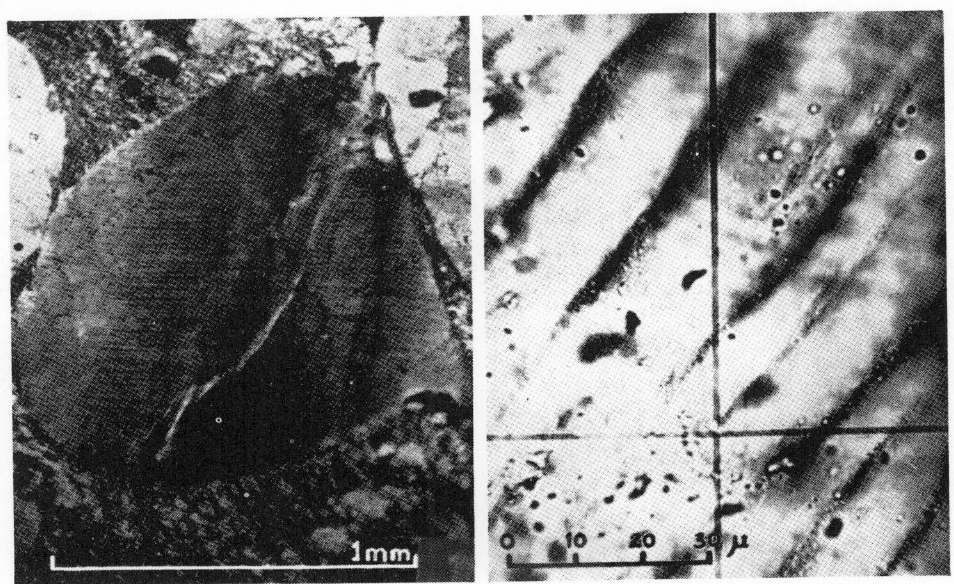

Fig. 42—Photomicrographs of quartz deformation lamellae in grains of Oriskany sandstone (from Hansen and Borg, Ref. 120, Plate 1). On the left is a grain with well-developed lamellae subnormal to zones of undulatory extinction. Crossed Nicols. On the right are lamellae at high magnification inclined at 75 to 80 degrees to the plane of the paper. Crossed Nicols.

Those corresponding to the major concentrations intersect at 60 to 76 degrees in lines parallel to the fold axis. The small circle girdle is best developed in specimen E2, in which the calcite extension axes also fall in a girdle. The acute bisector between the two major sets of deformation lamellae is within 10 degrees of the position of the corresponding derived σ_1 from the twinned calcite; the obtuse bisector also agrees with σ_3 from the calcite. This supports the conclusions that the deformation lamellae form in planes of high shear stress at less than 45 degrees to σ_1 and so are the significant dynamic criteria.

A different view of the dynamic interpretation of quartz deformation lamellae is presented by Christie and Raleigh.[140] They

> showed that the poles of lamellae in four quartzite specimens are distributed along small circles containing maxima. Three of these rocks are from a metamorphic terrain with a complex tectonic history, in which two major deformations are recognizable [Christie, personal communication]. The axes of the small-circles (A_1 in Fig. 46) are almost parallel in the three specimens and are interpreted as the axes of maximum principal (compressive) stress in the deformation

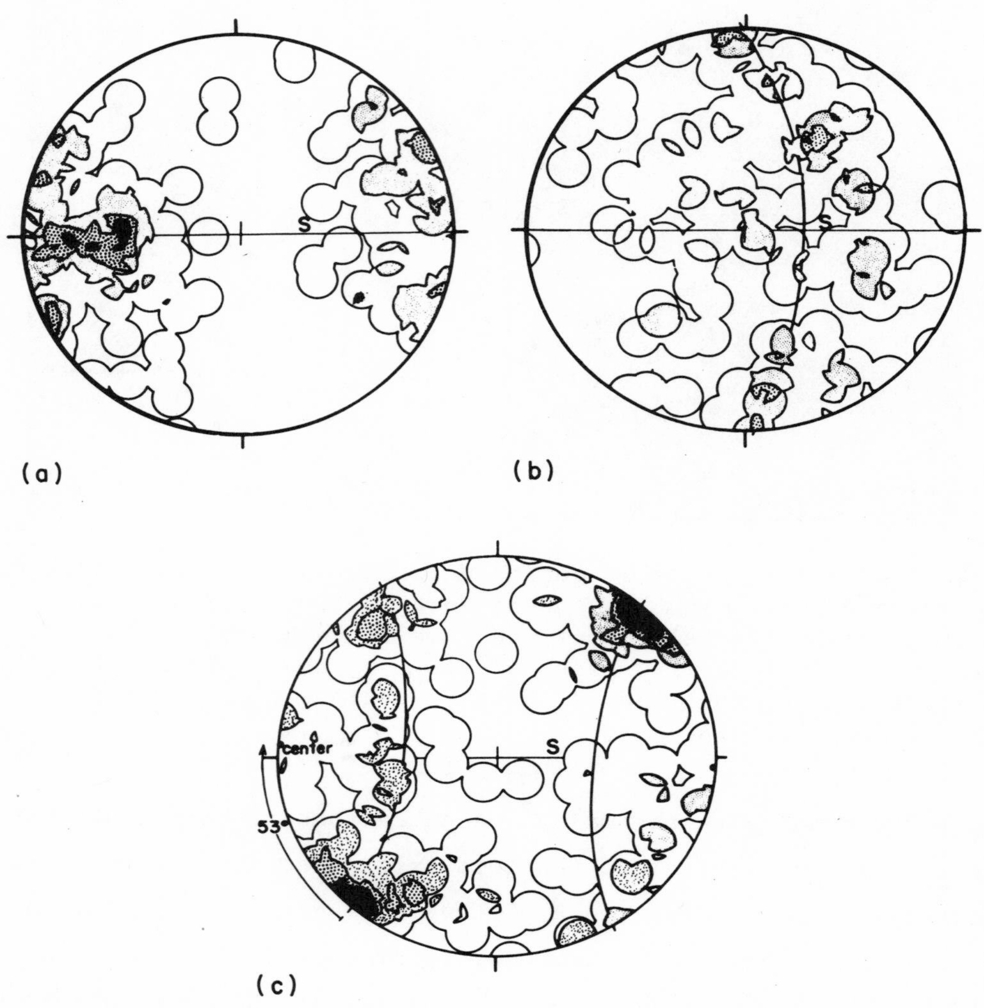

Fig. 43—Diagrams showing orientations of compression and extension axes (a and b) derived from calcite-cement twin lamellae and of quartz deformation lamellae (c) in Oriskany sandstone specimen E2 (from Hansen and Borg, Ref. 120, Figs. 4 and 5). The plane of each diagram is parallel to the ac plane of the fold with the fold axis near the center; s is the bedding plane. (a) 118 compression axes derived from well-developed sets of e twin lamellae in 200 grains. Contours are at 0.9, 2.6, 4.3, and 6.0 per cent per 1 per cent area, 6.8 per cent maximum. (b) 118 extension axes. Contours are at 0.9, 2.6, 4.3, and 6.0 per cent per 1 per cent area. Great circle indicates the trend of the plane containing the extension axes. (c) Normals to 147 deformation lamellae in 400 grains. Contours are at 0.7, 2.1, 3.5, 4.9, and 6.3 per cent per 1 per cent area, 8.4 per cent maximum. The center of the small circle defined by the normals to the lamellae is marked by a dot.

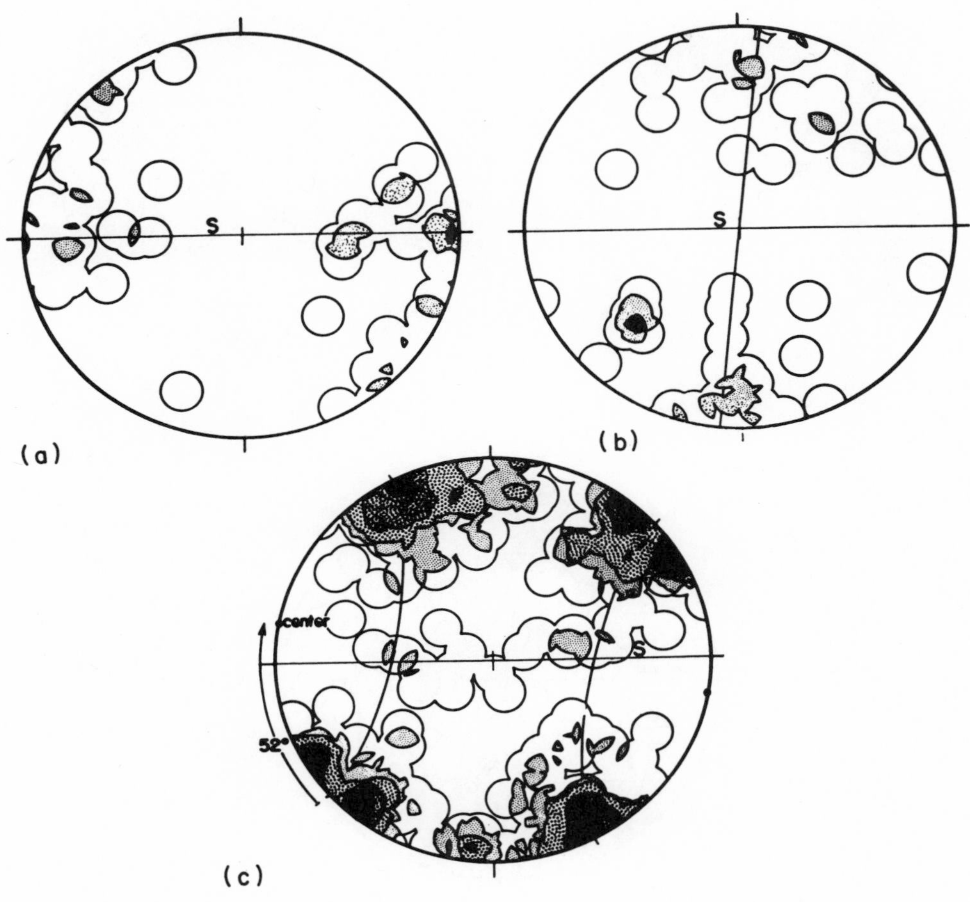

Fig. 44—Diagrams showing orientations of compression and extension axes (a and b) derived from calcite-cement twin lamellae and of quartz deformation lamellae (c) in Oriskany sandstone specimen E4 (from Hansen and Borg, Ref. 120, Figs. 6 and 7). Diagrams are oriented similar to those in Fig. 43. (a) 49 compression axes derived from the well-developed sets of e twin lamellae in 100 grains. Contours are at 2, 6, and 10 per cent per 1 per cent area. (b) 49 extension axes. Contours are at 2, 6, and 10 per cent per 1 per cent area. The great circle defines the plane normal to the major concentration of compression axes. (c) Normals to 220 deformation lamellae in 412 grains. Contours are at 0.5, 1.4, 2.3, 3.6, and 5.0 per cent per 1 per cent area, 7.3 per cent maximum. The center of the small circle defined by the normals to the lamellae is marked by a dot.

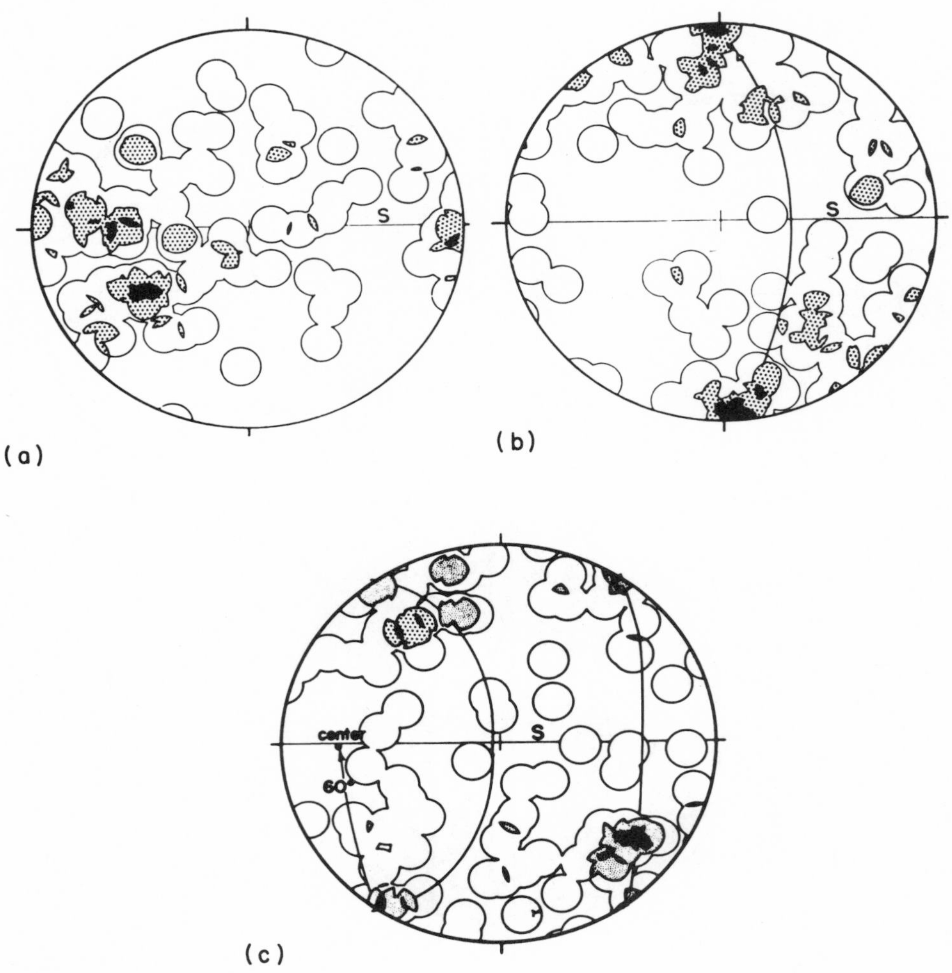

Fig. 45—Diagrams showing orientations of compression and extension axes (a and b) derived from calcite-cement twin lamellae and of quartz deformation lamellae (c) in Oriskany sandstone specimen E6 (from Hansen and Borg, Ref. 120, Figs. 8 and 9). Diagrams are oriented similar to those in Figs. 43 and 44. (a) 92 compression axes derived from the well-developed sets of e twin lamellae in 134 grains. Contours are at 1.1, 3.3, 5.5 per cent per 1 per cent area, 7.6 per cent maximum. (b) 92 extension axes. Contours are at 1.1, 3.3, and 5.5 per cent per 1 per cent area, 8.7 per cent maximum. Great circle defines the plane normal to the major concentration of compression axes. (c) Normals to 92 deformation lamellae in 210 grains. Contours are 1.1, 3.3, and 5.5 per cent per 1 per cent area, 5.9 per cent maximum. The center of the small circle is marked by a dot.

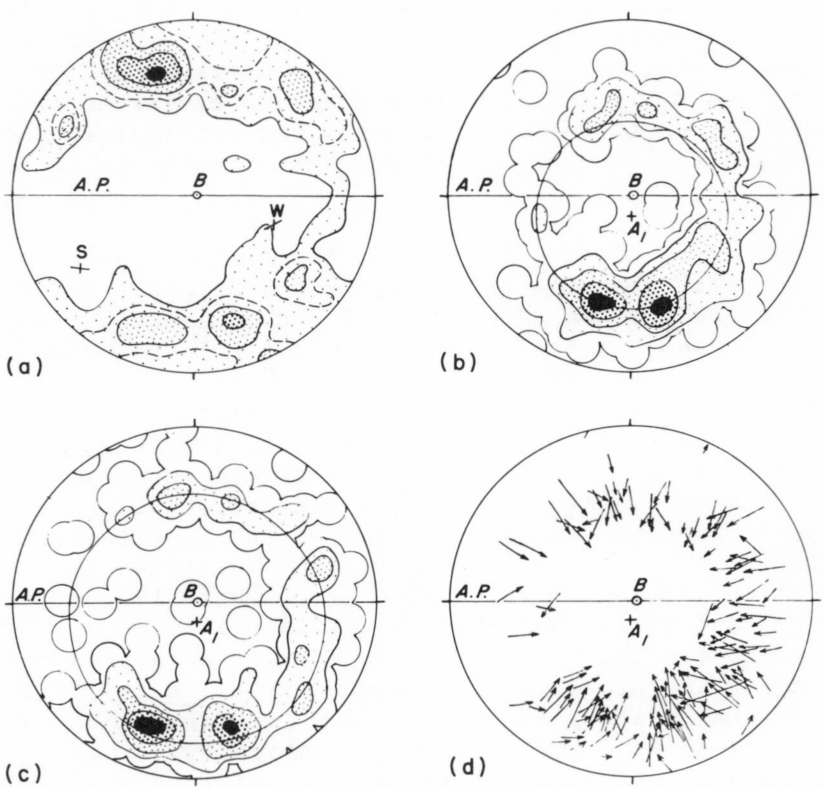

Fig. 46—Orientation diagrams for c_V and deformation lamellae in quartz of Specimen I, a quartzite from the Orocopia Mountains in southern California (from Christie and Raleigh, Ref. 140, Fig. 2). (a) 817 quartz c_V; contours at 1, 1.5, 2, 3, and 4 per cent per 1 per cent area. (b) Normals to deformation lamellae in 195 grains (195 sets of lamellae); contours at 1, 1.5, 3, 5, and 8 per cent per 1 per cent area. (c) c_V in same 195 grains containing deformation lamellae; contours at 0.5, 1.5, 3, 5, and 7 per cent per 1 per cent area. (d) Normals to deformation lamellae (point of arrow) and c_V (end of arrow) in a representative number of grains from each section. B is the first generation fold axis and A.P. is the axial plane of the first generation fold. A_1 is the axis of the small circle defined by the normals of the lamellae and c_V in grains containing lamellae. All four diagrams have the same orientation, shown by south (S) and west (W) directions in (a), and are plotted in lower hemisphere equal-area projection.

which produced the lamellae. This is consistent with the stress-field during the second deformation, as inferred from macroscopic folds and conjugate shear-surfaces. It is worth noting that in these specimens the radius of the small-circles of poles of lamellae are 44°, 45°, and 52°, a fact inconsistent with the claim that the lamellae in any sample should be inclined, on the average, at less than 45° to σ_1. The rocks have moderate preferred orientations of quartz dating from the first deformation and this anisotropy appears to affect the orientation of the lamellae.

Carter, Christie, and Griggs [Ref. 146] find no support in their experimental work for the hypothesis that the lamellae are consistently inclined at less than 45° to σ_1. In sand samples deformed in a simple squeezer the lamellae are inclined at angles from 0° to 85° to the compression axis and in quartzite samples at angles from 15° to 85° to the compression axis [Ref. 146, Fig. 10]; in both types of samples most lamellae are inclined at slightly more or less than 45° to the compression axis. In some of their samples the poles of the lamellae lie consistently closer to the compression axis than the c-axes of the grains in which they occur [Ref. 146, Figs. 7 and 9] and it is suggested that this criterion might be used to distinguish between σ_1 and σ_3 in fabrics consisting of two planes of lamellae. Carter *et al*. also suggest that a study of the rotations in kink bands and undulatory zones might resolve the ambiguity, since lamellae are better developed in more deformed zones and the c-axes in such zones are rotated towards σ_1.*

Clearly, a unique resolution to the dynamic interpretation of quartz deformation lamellae is not as yet established. Support for the experimental findings of Christie, Griggs, and Carter can be found in the field studies of Christie and Raleigh,(140) Riley,(141) and Hara.(151) While exceptions to their findings and support for the acute bisector equals σ_1 hypothesis can be found in the work of Hansen and Borg (Ref. 120, Fig. 5(d)) and in Naha,(149) it is the writer's opinion that a unique solution will be forthcoming from a study of the sense of shear along the lamellae** in slightly and moderately deformed

*The preceding two paragraphs were written by Dr. J. M. Christie at the request of the author.

**See Ref. 146, p. 10-89 n.

rocks in which the tectonic history is simple and interfering problems of recrystallization and preferred orientations of c_v are minimal.

Recrystallization

General. Recrystallization of a given mineral species in a polycrystalline aggregate can occur below the melting temperature by solution and redeposition and/or solid diffusion. There is a large literature on the subject, especially in metallurgy, ceramic engineering, and glass technology.[94,152-158] Of concern here is the fact that the petrofabric literature abounds with descriptions of preferred crystallographic orientations of crystals, which for textural reasons can not be explained on the basis of cataclastic or gliding flow. That is, there is no visual evidence of grain breakage or intragranular gliding. Presumably these orientations have resulted from recrystallization during deformation (the paratectonic or syntectonic recrystallization of Sander[1]), i.e., under conditions of nonhydrostatic stress. Most of them have been interpreted only kinematically by means of the symmetry argument. The same crystallographic orientations, however, might be amenable to dynamic analysis if one understood the relationships between the orientations of the recrystallized grains and the principal stresses. Significant contributions to the problem have been made through thermodynamic and experimental investigations. Initial results are encouraging and suggest the potential usefulness of these approaches.[159]

Thermodynamic Approach. The thermodynamic principles of the behavior of elastically strained solids under nonhydrostatic stress in contact with fluids were laid down by Gibbs.[160] Significant departures from the Gibbs treatment were given by Goranson,[161-163] Verhoogen,[164] and MacDonald.[165] Kamb[166] has reviewed these in detail and finds that all are essentially identical with respect to prediction of the most stable crystal orientation. Kamb[167] believes that only the Gibbs theory is valid and applies it to the simplest possible model of the recrystallization process. Recently, Kamb[168] has demonstrated experimentally the validity of the Gibbs approach and the inadequacy of all other theories so far advanced. It is instructive to compare the MacDonald and the Gibbs-Kamb approaches and the crystal orientations they predict.

MacDonald[165] assumed that the most stable orientation for a mineral would be the one for which the potential energy of the external forces plus the potential energy of strain was minimized. If the deformation is isothermal and obeys a linear elastic stress-strain law, then the most stable orientation will be the one in which the Helmholtz free energy is a maximum. This neglects any permanent strain energy, and the nature of the path from the initial to final states of the mineral, i.e., the final orientation, is supposed to be independent of the orienting mechanisms.

Brace,[169] using MacDonald's prediction, calculated the most stable orientations for calcite, high and low quartz, and ice for a uniaxial stress. His results show that the following crystallographic planes should be nearly normal to the load axis: $\{10\bar{1}1\}$ in ice and calcite, $\{10\bar{1}2\}$ in high quartz, and $\{02\bar{2}1\}$ in low quartz. Their respective c_v will then tend to lie along small circles or girdles of specific half-angle about the unique stress axis (Fig. 47). Brace also examined the equilibrium orientation in a stress field of three different, nonzero, principal stresses. He found that the position of most stable orientation is a function of the confining pressure as well as the stress difference. Accordingly, different orientation patterns of grains might occur in rocks for which the stress difference was similar but the depth of burial different.

Very different results have been obtained by Kamb[167] from the Gibbs theory. He considers that the orientation of a mineral depends only on the stress deviators and is therefore independent of changes in hydrostatic pressure. When recrystallization takes place by solution and redeposition, the most stable orientation of the crystal is that which minimizes the chemical potential across the plane normal to the greatest principal pressure. Accordingly, the axis of least elastic modulus (e.g., the c_v of calcite) tends to align itself parallel to the greatest principal stress axis (axes), and the axis of greatest elastic modulus (e.g., the c_v of quartz) tends to become

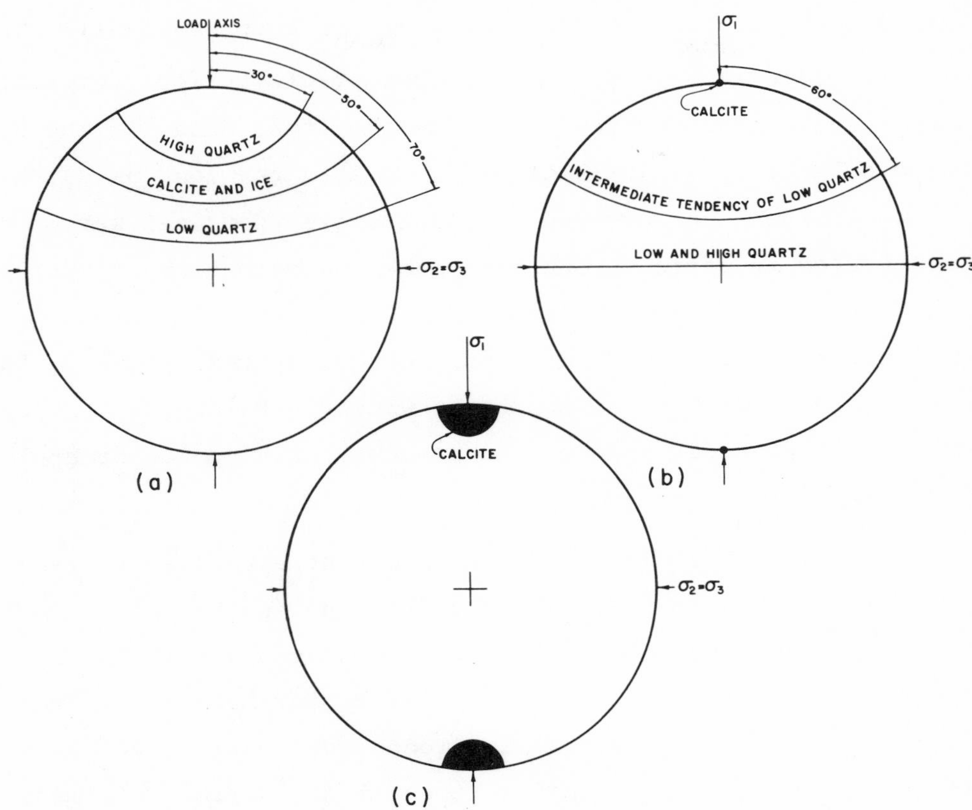

Fig. 47—Stereograms showing the predicted (a and b) and the experimentally determined (c) orientations of c_v in minerals recrystallized under a uniaxial state of stress (stereographic projection). (a) Orientations of c_v predicted by MacDonald[165] and calculated by Brace (Ref. 169, Fig. 4) for calcite, ice, high and low quartz. (b) Orientations of quartz and calcite c_v predicted by Kamb[167] for recrystallization in the presence of a fluid phase. (c) Orientation of calcite c_v produced in experiments as a result of syntectonic, dry recrystallization.

normal to σ_1 (Fig. 47(b)). Moreover, low quartz is considered by Kamb to be unique among the common hexagonal and rhombohedral minerals in that the theory predicts "intermediate" orientation positions. For the case $\sigma_1 > \sigma_2 = \sigma_3$, the c_v will tend to lie along a small girdle at about 60 degrees to σ_1; and for $\sigma_1 = \sigma_2 > \sigma_3$, the c_v will tend to develop a small girdle at about 29 degrees to σ_3. Kamb also extends Gibbs' theory to predict orientations under conditions of dry recrystallization. He finds that if the initial grain shapes are equant, the most stable orientation for most hexagonal or rhombohedral crystals is such that their axes of least elastic modulus tend to lie normal to the unique stress axis (whether tensional or compressional). This orientation is directly opposite to that predicted for recrystallization by solution and redeposition. On the other hand, if the dry grains are initially much flattened normal to the load axis, the orientations would be reversed, and therefore identical wet or dry. If Kamb's view is correct, one must know the mechanism of recrystallization before applying his theory.

Although the thermodynamic approach holds great promise, the predictions of MacDonald and Kamb differ widely. Furthermore, they are based on infinitesimal strain. At present, they cannot be applied to geologic problems.

Experimental Approach. A most important experimental result has been the dry recrystallization of calcite and quartz under simulated geological conditions, since previously most geologists had seemed to regard the agency of solutions as necessary for metamorphic recrystallization.* Syntectonic recrystallization occurs during the course of triaxial compression or extension tests, i.e., during deformation. Annealing recrystallization (familiar in metallurgy) occurs when an aggregate is first deformed at low temperature (cold-worked) and then held at an elevated temperature to accelerate the process. Presumably there is a critical temperature below which annealing recrystallization does not take place regardless of the duration of heating.[(171)] This temperature decreases as the initial strain is increased. Syntectonic recrystallization appears to occur not only above some critical temperature, but also below some temperature above which strain energy is annealed out faster than it can be stored.

Syntectonic recrystallization of calcite was first observed in specimens of Yule marble deformed at 300°C and at 5-kb confining pressure.[(108)] The evidence is intergranular flow and development of textures closely resembling those of naturally deformed marble. In specimens deformed at 400° and 500°C[(110)] the evidence is augmented by the development of small lobes and lensoid streaks in which new optic axes, differing from those of the host grains, tend to parallel the

*See Refs. 43, 108, 110, 143, and 170.

direction of maximum principal stress in both compression and extension regardless of the orientation of the host crystals. Syntectonic recrystallization reaches a maximum at 600°C (Fig. 48) in short-time tests (strain rate about 10^{-4} per second).[43] At lower strain rates the maximum seems to occur at a lower temperature, since the phenomenon is common in specimens deformed at 350° to 500°C and at 10^{-7} sec^{-1}.[170] In these, recrystallization may be the major mechanism of steady-state flow. What is important here is that the recrystallized grains are always strongly oriented with their c_v aligned parallel to the axis of maximum principal compressive stress (Fig. 47(c)), a result which agrees with the Gibbs-Kamb[168] prediction (but in the presence of solutions).

Annealing recrystallization of calcite crystals and aggregates has also been produced experimentally.[172] The critical temperature of annealing of Yule marble is about 500°C, and the annealing time (15-120 min) did not influence the recrystallization. In contrast to the syntectonic process, the annealing tends to produce a random orientation, although traces of the inherited fabric of the original material remain. An interesting exception was noted in specimens of powdered Yule marble (fragments 1 to 2 μ) that were compressed between steel pistons at 1000°C for 30 min. After this treatment the material consists of groundmass, nearly uniform grains 10 to 20 μ in diameter, and porphyroblasts, about 0.2 mm in diameter. The c_v of the porphyroblasts tend to parallel the axis of compression, i.e., tend to lie normal to the piston faces (Fig. 49).

Crystallographic Orientations as Criteria for Dynamic Interpretations. The experimental fact that syntectonically recrystallized calcite crystals are oriented with their c_v parallel to the greatest principal compressive stress serves as a working hypothesis for the dynamic interpretation of natural calcite fabrics.

In calcite marbles c_v characteristically lie at high angles to the major foliation plane (e.g., Yule marble, Fig. 50).* According to the

*For those marbles in which the grains are not highly elongated, one can reasonably be sure that the c_v subfabric has resulted from

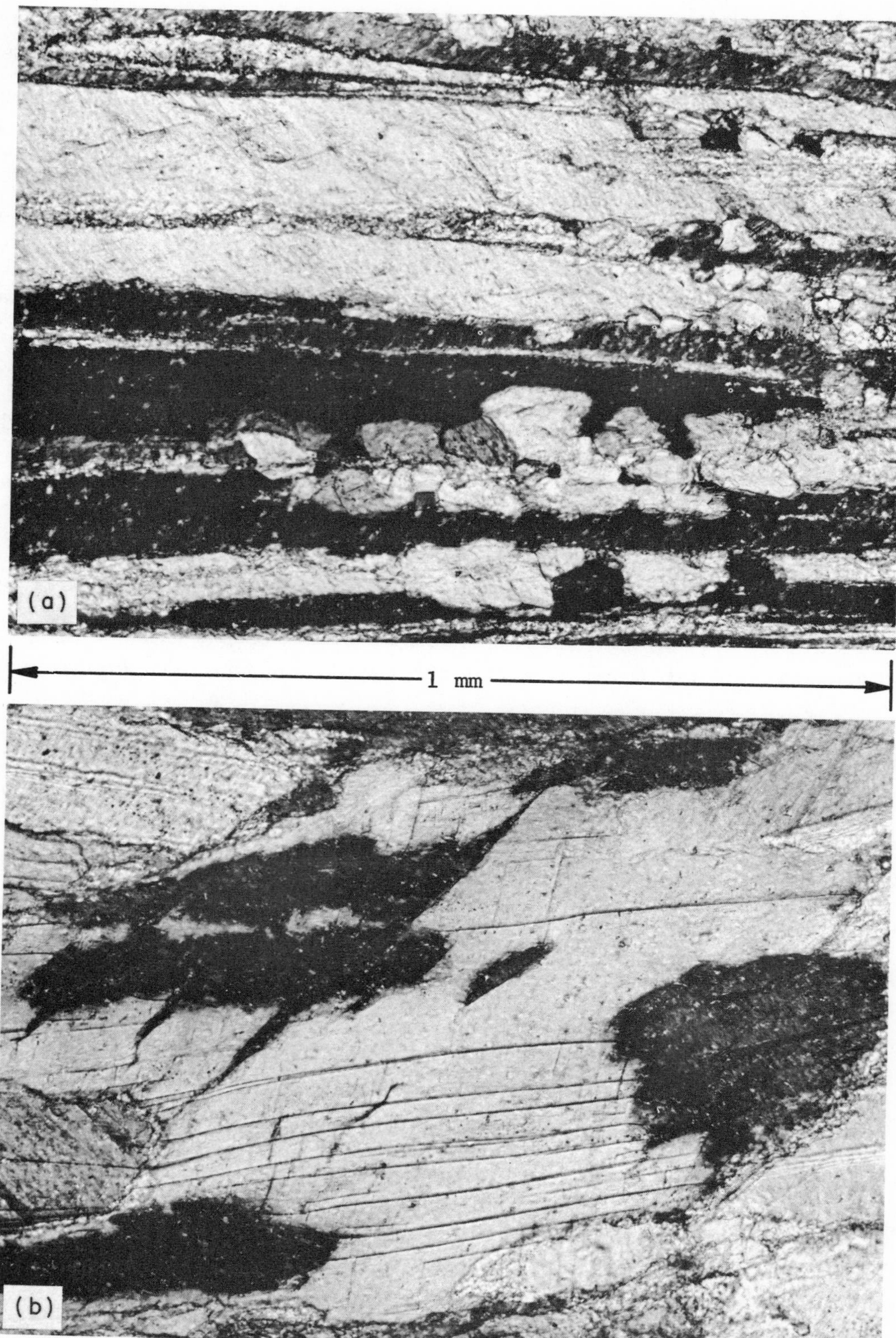

Fig. 48—Photomicrographs of Yule marble showing effects of syntectonic recrystallization (from Griggs, Turner, and Heard, Ref. 43, Plate 12). (a) New grains developed in marginal clusters in specimen locally elongated 590 per cent at 3-kb confining pressure and 600°C. Crossed Nicols. (b) New grains (dark) developed at three centers in single host crystal in specimen elongated 50 per cent at 3-kb confining pressure and 600°C. Crossed Nicols.

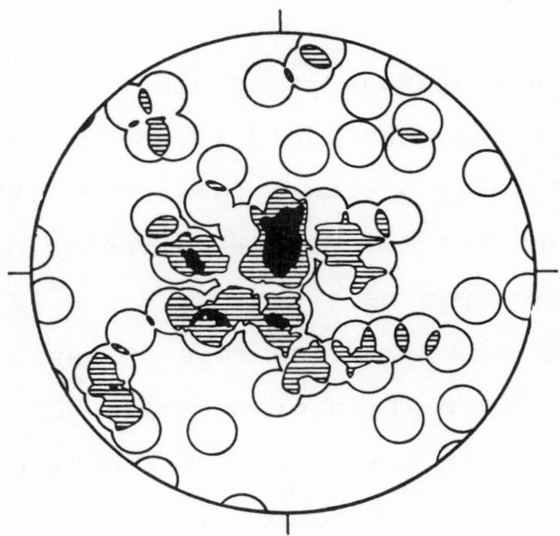

Fig. 49—Diagram illustrating orientation of calcite c_V in 72 porphyroblasts that resulted from the annealing recrystallization of powdered Yule marble. Contours are at 1.3, 2.7, and 5.5 per cent per 1 per cent area. Plane of the diagram is oriented normal to the axis of compression.

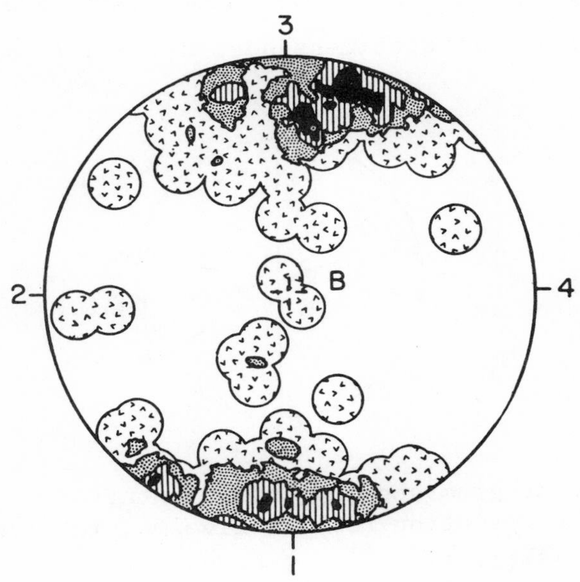

Fig. 50—Diagram showing orientation of 100 calcite c_V in Yule marble (from Handin, Higgs, and O'Brien, Ref. 175, Fig. 12). Contours are at 1, 3, 5, and 7 per cent per 1 per cent area. The foliation plane is parallel to the 2-B-4 plane.

hypothesis, the greatest principal compressive stress during the latest stages of recrystallization should be oriented nearly normal to the foliation. An interesting example is afforded by marble fabrics in five specimens from a recumbent fold in the Mojave Desert of southern California.[116] In three of these (Nos. 278, 161, and 70) the grains are nearly equant, and there is little evidence of postcrystallization strain. The other two specimens (Nos. 298 and 72) are conspicuously deformed and markedly elongated (average dimensional ratio is 1:2:7 with long and intermediate grain diameters in or near the plane of the foliation). Despite these textural differences the c_v maxima in all five specimens are similarly oriented at high angles to the foliation (Fig. 51). Weiss (Ref. 116, p. 77) concluded that "the preferred orientation of c_v in the calcite marbles dates from the main deformation, and the direction of maximum concentration in each specimen is thought to coincide with the axis of maximum compressive stress immediately before cessation of movements." This statement seems to recognize the ability of calcite grains to become reoriented by gliding flow. Although the textures of specimens Nos. 298 and 72 suggest that this

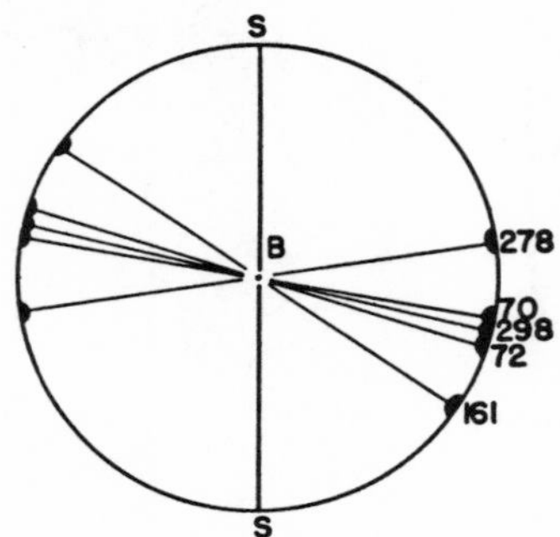

Fig. 51—Synoptic diagram showing the orientation of calcite c_v maxima in relation to the foliation (SS) in five marbles from southern California (from Weiss, Ref. 116, Fig. 27).

---

recrystallization rather than from mechanical reorientation of the grains by gliding flow. A nearly similar c_v subfabric can be produced by complete twinning and external rotation of grains. This, however, produces markedly elongated grains, remanent internally rotated lamellae, and a tendency for the c_v to lie along small circles rather than in a distinct maximum (Ref. 43, p. 91).

might be so, the equant grains in the other three specimens suggest that the orienting mechanism is the recrystallization. Weiss may have been dealing with a situation in which the effects of recrystallization and of gliding flow have both produced essentially the same c_v subfabric.

One could cite examples in which recrystallized grains have their c_v distributed in nearly complete girdles, which are related to $\sigma_1 > \sigma_2 = \sigma_3$ or $\sigma_1 = \sigma_2 > \sigma_3$ states of stress, but the technique should now be clear. It is important to point out, however, that if preferred orientations of calcite c_v arise from recrystallization under nonhydrostatic conditions, then nearly random patterns would probably imply recrystallization under essentially hydrostatic pressures. Although published accounts of random calcite c_v subfabrics are rare, random orientations are in fact common in recrystallized sedimentary rocks.[59,120,173] In undeformed rocks this might be evidence of an essentially hydrostatic state of stress due to simple overburden pressure.

Summary and Conclusions

Petrofabrics is the study of fabric elements that may range in size over 15 orders of magnitude from the crystal lattices to mountain ranges. It consists of a descriptive phase in which fabric elements are recognized, measured, and illustrated, and an interpretive phase in which the rock fabric serves as a basis of inference to the kinematic or dynamic aspects of the deformation. The kinematic approach, based on the symmetry argument of Sander,[1] provides no knowledge of the state of stress. The dynamic approach utilizes fabric elements to derive the orientations and relative magnitudes of the principal stresses in the rocks at the time of deformation. It is based on an understanding of the mechanisms of rock deformation gained largely through experiments.

The current physical understanding of fracturing and faulting, gliding flow, rotation, and recrystallization and of the corresponding fabric elements often enables one to determine the principal stress directions in rocks at the time of deformation. A review of the literature shows that laboratory observations on the fabric elements are compatible with those from the field and illustrates the types of dynamic interpretations that can be made from the fabric data.

Faults, shear fractures, and extension fractures are viewed as phenomena that are independent of scale (down to the microscopic field, i.e., >0.01 mm) and that exhibit predictable orientations to the principal stresses in the rock at the time of failure. In naturally deformed rocks, fractures and faults are distinguished and identified primarily from their combined orientation pattern. Derivation of the principal stress directions follows from the genetic relationships--extension fractures are normal to the least principal stress, and faults and shear fractures are inclined at less than 45 degrees to the greatest principal compressive stress. Studies of microfractures in individual grains of folded sandstones and of fracture and fault systems in the Ouachita Mountains and Central Plains of Oklahoma and in the Great Basin of the western United States demonstrate the dynamic interpretations of these elements and emphasize that the elements are independent of scale. Consistent fracture-fault trends over large portions of the earth's crust suggest that the stress pattern is homogeneous on a regional scale.

Intracrystalline gliding involves mechanical twinning and translation parallel to a definite gliding plane along a fixed direction with or without restricted sense of shear. As gliding flow is essentially independent of normal stress, the most favorable state of stress for gliding is that which gives the maximum shear stress along the gliding line (in the proper sense). That is, the greatest and least principal stresses are each oriented at 45 degrees to both the gliding plane and direction so as to yield the known sense of shear. Accordingly, if the gliding systems are known, the principal stresses in each crystal can be derived from the gliding evidence (e.g., twin lamellae or internally rotated lamellae). From laboratory experience it is reasonable to suppose that the stress orientation pattern determined from many grains will correspond to the orientation of the principal stresses in the rock at the time of deformation. The dynamic interpretations based on calcite and dolomite twin lamellae and on phenomena of internal rotation give the most reliable and meaningful results when studied in slightly and moderately deformed rocks.

Experimental kink banding in biotite crystals provides another criterion for dynamic inferences. In biotite the kink band boundaries are initially oriented at high angles to the greatest principal compressive stress.

In experimentally deformed quartz crystals and sands, deformation lamellae which resemble the natural counterparts lie only in planes of high resolved shear stress. This laboratory observation is valid in the field as well where the angle between the deformation lamellae and the greatest principal stress is evidently less than 45 degrees. Principal stress directions derived from the quartz deformation lamellae on this basis agree well with those located from extension fractures and calcite twin lamellae.

Experimental syntectonic recrystallization of dry calcite suggests that in the most stable orientation the c_v tend to parallel the greatest principal compressive stress. This agrees with the Gibbs-Kamb[167] thermodynamic prediction for recrystallization in the presence of solutions, but the theory is still controversial. Application of the experimental results to marbles implies that the greatest principal stress is characteristically oriented at high angles to the observed foliation during recrystallization. The random calcite c_v subfabrics common in recrystallized sedimentary rocks may be evidence that the state of stress due to simple overburden pressure is nearly hydrostatic.

It has been emphasized repeatedly that some petrofabric techniques serve to map the principal stresses in rocks at the time of their deformation. Future research will have to determine the relationship, if any, between these stresses and the present state of stress in the rocks.

Acknowledgments

Dr. D. V. Higgs and Dr. J. W. Handin of the Shell Development Company, through their stimulating and enthusiastic approach to rock deformation, are largely responsible for my interest in petrofabrics. Thanks are due these gentlemen and Mr. D. W. Stearns and Dr. D. J. Atkinson, also with the Shell Development Company, for many helpful discussions and critical review of the manuscript.

REFERENCES

1. Sander, B., Gefügekunde der Gesteine, Springer, Wien, 1930.
2. Turner, F. J., "Mineralogical and Structural Evolution of the Metamorphic Rocks," Geol. Soc. Am. Mem. 30, 1948.
3. Handin, J. W., and R. V. Hager, Jr., "Experimental Deformation of Sedimentary Rocks under Confining Pressure: Tests at Room Temperature on Dry Samples," Bull. Am. Assoc. Petrol. Geologists, Vol. 41, 1957, pp. 1-50.
4. Paterson, M. S., and L. E. Weiss, "Symmetry Concepts in the Structural Analysis of Deformed Rocks," Bull. Geol. Soc. Am., Vol. 72, 1961, pp. 841-882.
5. Turner, F. J., "Lineation, Symmetry, and Internal Movement in Monoclinic Tectonite Fabrics," Bull. Geol. Soc. Am., Vol. 68, 1957, pp. 1-17.
6. Sander, B., "Über Zusammenhänge zwischen Teilbewegung und Gefüge in Gesteinen," Mineral. Petrog. Mitt., Vol. 30, 1911, pp. 281-315.
7. Sander, B., Einführung in die Gefügekunde der geologischen Körper, 1, Springer, Wien, 1948.
8. Sander, B., Einführung in die Gefügekunde der geologischen Körper, 2, Springer, Wien, 1950.
9. Schmidt, W., "Gefügesymmetrie und Tektonik," Jahrb. Geol. Bundesanstalt, Vol. 76, Part 1, 1926, pp. 407-430.
10. Schmidt, W., Tektonik und Verformungslehre, Bornträger, Berlin, 1932.
11. Knopf, E. B., and F. E. Ingerson, "Structural Petrology," Geol. Soc. Am. Mem. 6, 1938.
12. Turner, F. J., and J. Verhoogen, Igneous and Metamorphic Petrology, McGraw-Hill Book Company, Inc., New York, 1951.

13. Turner, F. J., and L. E. Weiss, *Structural Analysis of Metamorphic Tectonites*, McGraw-Hill Book Company, Inc., New York, 1963.

14. Weiss, L. E., "Structural Analysis of the Basement System at Turoka, Kenya," *Overseas Geological Surveys*, Her Majesty's Stationery Office, London, 1959, pp. 1-65.

15. Krumbein, W. C., "Some Problems in Applying Statistics to Geology," *Appl. Statist*., Vol. 9, 1960, pp. 82-91.

16. Krumbein, W. C., "The 'Geological Population' as a Framework for Analyzing Numerical Data in Geology," *Liverpool Manchester Geol. J*., Vol. 2, 1960, pp. 341-368.

17. Chayes, F., *Petrographic Modal Analysis*, John Wiley & Sons, Inc., New York, 1956.

18. Fairbairn, H. W., *Structural Petrology of Deformed Rocks*, 2d ed., Addison-Wesley Publishing Company, Inc., Reading, Massachusetts, 1949.

19. Higgs, D. V., M. Friedman, and J. E. Gebhart, "Petrofabric Analysis by Means of the X-ray Diffractometer," in "Rock Deformation," *Geol. Soc. Am. Mem*. 79, 1960, pp. 275-292.

20. Gehlen, K. V., "Die röntgenographische und optische Gefügeanalyse von Erzen, insbesondere mit dem Zahlrohr-Texturgoniometer," *Beitr. Mineral. Petrog*., Vol. 7, 1960, pp. 340-388.

21. Kamb, W. G., "Refraction Corrections for Universal-stage Measurements, I. Uniaxial Crystals," *Am. Mineralogist*, Vol. 47, 1962, pp. 227-245.

22. Phillips, F. C., *The Use of the Stereographic Projection in Structural Geology*, Edward Arnold, Ltd., London, 1954.

23. Higgs, D. V., and G. Tunell, *Angular Relations of Lines and Planes, with Application to Geological Problems*, Wm. C. Brown Co., Dubuque, Iowa, 1959.

24. Lambert, J. H., "Anmerkungen und Zusätze zur Entwerfung der Land- und Himmelscharten," *Ostwald's Klassiker der Exakten Wissenschaften*, Nr. 54, Wilhelm Engelman, Leipzig.

25. Schmidt, W., "Gefügestatistik," *Mineral. Petrog. Mitt*., Vol. 38, 1925, pp. 392-423.

26. Mellis, O., "Gefügediagramme in Stereographischer Projektion," *Z. Mineral. Petrog. Mitt*., Vol. 53, 1942, pp. 330-353.

27. Flinn, D., "On Tests of Significance of Preferred Orientation in Three-dimensional Fabric Diagrams," J. Geol., Vol. 66, 1958, pp. 526-539.

28. Kamb, W. B., "Ice Petrofabric Observations from Blue Glacier, Washington, in Relation to Theory and Experiment," J. Geophys. Res., Vol. 64, 1959, pp. 1891-1909.

29. Pincus, H. J., "The Analysis of Aggregates of Orientation Data in the Earth Sciences," J. Geol., Vol. 61, 1953, pp. 482-509.

30. Spencer, E. W., "Geologic Evolution of the Beartooth Mountains, Montana and Wyoming, Part 2, Fracture Patterns," Bull. Geol. Soc. Am., Vol. 70, 1959, pp. 467-508.

31. Orowan, E., "Fracture and Strength of Solids," Phys. Soc. London Report Prog. in Phys., Vol. 12, 1949, pp. 185-232.

32. Handin, J. W., "An Application of High Pressure in Geophysics: Experimental Rock Deformation," Trans. ASME, Vol. 75, 1953, pp. 315-324.

33. Griggs, D. T., and J. W. Handin, "Observations on Fracture and a Hypothesis of Earthquakes," in "Rock Deformation," Geol. Soc. Am. Mem. 79, 1960, pp. 347-364.

34. Brace, William F., "Brittle Fracture of Rocks," these Proceedings, pp. 111-174.

35. Handin, J. W., and R. V. Hager, Jr., "Experimental Deformation of Sedimentary Rocks under Confining Pressure: Tests at High Temperature," Bull. Am. Assoc. Petrol. Geologists, Vol. 42, 1958, pp. 2892-2934.

36. Handin, J. W., R. V. Hager, Jr., M. Friedman, and J. N. Feather, "Experimental Deformation of Sedimentary Rocks under Confining Pressure, Part III--Tests on Fluid-saturated Samples," Bull. Am. Assoc. Petrol. Geologists, Vol. 47, 1963, pp. 717-755.

37. Daubrée, A., "Application de la Méthode Experimentale a l'Étude des Déformations et des Cassures Terrestres," Bull. Soc. Geol. France, 1878 à 1879, pp. 108-141.

38. Kármán, Th. von, "Festigkeitsversuche unter allseitigem Druck," Z. Ver. Deutsch. Ingenieure, Vol. 55, 1911, pp. 1749-1757.

39. Ros, H. C., and A. Eichinger, "Experimental Attempt To Solve the Problem of Failure in Materials--Non-metallic Materials," Zurich Federal Materials Testing Lab., Report 28, 1928, Trans. in U.S. Bur. Reclamation Tech., Memoir 635.

40. Prandtl, L., and F. Rinne, in Plasticity, ed. by A. Nádai, McGraw-Hill Book Company, Inc., New York, 1931.

41. Griggs, D. T., "Deformation of Rocks under High Confining Pressure," J. Geol., Vol. 44, 1936, pp. 541-577.

42. Griggs, D. T., and J. F. Bell, "Experiments Bearing on the Orientation of Quartz in Deformed Rocks," Bull. Geol. Soc. Am., Vol. 49, 1938, pp. 1723-1746.

43. Griggs, D. T., F. J. Turner, and H. C. Heard, "Deformation of Rocks at 500° to 800°C.," in "Rock Deformation," Geol. Soc. Am. Mem. 79, 1960, pp. 39-104.

44. Balsley, J. R., "Deformation of Marble under Tension at High Pressure," Trans. Am. Geophys. Union, Part 2, 1941, pp. 519-525.

45. Jones, V., "Tensile and Triaxial Compression Tests of Rock Cores from the Passageway to Penstock Tunnel N-4 at Boulder Dam," U.S. Bur. Reclamation, Basic Structural Research Report SP-6, 1946, pp. 1-9.

46. Goguel, J., Introduction à l'Étude Mécanique des Déformations de l'Ecorce Terrestre, Imprimerie Nationale, Paris, 1948.

47. McHenry, D., "The Effect of Uplift Pressure on the Shearing Strength of Concrete," Intern. Congr. on Large Dams, 1948, pp. 1-31.

48. Bridgman, P. W., Studies in Large Plastic Flow and Fracture, McGraw-Hill Book Company, Inc., New York, 1952.

49. Balmer, G. G., "A Revised Method of Interpretation of Triaxial Compression Tests for the Determination of Shearing Strength," U.S. Bur. Reclamation, Basic Structural Research Report SP-9, 1946, pp. 1-20.

50. Balmer, G. G., "Physical Properties of Some Typical Foundations Rocks," U.S. Bur. Reclamation, Concrete Lab. Report SP-39, 1953, pp. 1-15.

51. Robertson, E. C., "Experimental Study of the Strength of Rocks," Bull. Geol Soc. Am., Vol. 66, 1955, pp. 1275-1314.

52. Handin, J. W., and H. W. Fairbairn, "Experimental Deformation of Hasmark Dolomite," Bull. Geol. Soc. Am., Vol. 66, 1955, pp. 1257-1273.

53. Informal communication from J. W. Handin and I. Y. Borg, Shell Development Co., Houston, Texas, 1962.

54. Handin, J. W., "Experimental Deformation of Rocks and Minerals," Quart. Colo. School Mines, Vol. 52, No. 3, 1957, pp. 75-98.

55. Higgs, D. V., and J. W. Handin, "Experimental Deformation of Dolomite Single Crystals," Bull. Geol. Soc. Am., Vol. 70, 1959, pp. 245-277.

56. Robinson, L. H., Jr., "Effects of Pore and Confining Pressures on Failure Characteristics of Sedimentary Rocks," Quart. Colo. School Mines, Vol. 54, 1959, pp. 177-199.

57. Heard, H. C., "Transition from Brittle Fracture to Ductile Flow in Solenhofen Limestone as a Function of Temperature, Confining Pressure, and Interstitial Fluid Pressure," in "Rock Deformation," Geol. Soc. Am. Mem. 79, 1960, pp. 193-226.

58. Borg, I., M. Friedman, J. Handin, and D. V. Higgs, "Experimental Deformation of St. Peter Sand: A Study of Cataclastic Flow," in "Rock Deformation," Geol. Soc. Am. Mem. 79, 1960, pp. 133-191.

59. Friedman, M., "Petrofabric Analysis of Experimentally Deformed Calcite-cemented Sandstones," J. Geol., Vol. 71, 1963, pp. 12-37.

60. Serdengecti, S., and G. D. Boozer, "The Effect of Strain Rate and Temperature on the Behavior of Rocks Subjected to Triaxial Compression," Proceedings of the Fourth Symposium on Rock Mechanics, Pennsylvania State University, 1961, pp. 83-97.

61. Heard, H. C., "The Effect of Large Changes in Strain Rate in the Experimental Deformation of Rocks," Ph.D. thesis, University of California at Los Angeles, 1962.

62. Jaeger, J. C., "Shear Failure of Anisotropic Rocks," Geol. Mag., Vol. 97, 1960, pp. 65-72.

63. Donath, F. A., "Experimental Study of Shear Failure in Anisotropic Rocks," Bull. Geol. Soc. Am., Vol. 72, 1961, pp. 985-990.

64. Donath, Fred A., "Strength Variation and Deformational Behavior in Anisotropic Rock," these Proceedings, pp. 281-297.

65. Hubbert, M. K., "Mechanical Basis for Certain Familiar Geologic Structures," Bull. Geol. Soc. Am., Vol. 62, 1951, pp. 355-372.

66. Sax, H. G. J., "De Tectoniek van het Carboon in het Zuid-Limburgsche Mijngebied," Mededel. Geol. Sticht., Ser. C-I-I, No. 3, 1946, pp. 1-77.

67. Leith, C. K., Structural Geology, 1st ed., Henry Holt and Co., New York, 1913.

68. Anderson, E. M., The Dynamics of Faulting, Oliver and Boyd, London, 1951.

69. Muehlberger, W. R., "Conjugate Joint Sets of Small Dihedral Angle," J. Geol., Vol. 69, 1961, pp. 211-219.

70. Woodworth, J. B., "Some Features of Joints," Science, Vol. 2, 1895, pp. 903-904.

71. Woodworth, J. B., "On the Fracture System of Joints, with Remarks on Certain Great Fractures," Proc. Boston Soc. Nat. History, Vol. 27, 1896, pp. 163-184.

72. Dale, T. N., and H. E. Gregory, "The Granites of Connecticut," U.S. Geol. Surv., Bull. No. 484, 1911, pp. 1-137.

73. Sheldon, P. G., "Some Observations and Experiments on Joint Planes," J. Geol., Vol. 20, 1912, pp. 53-79, 164-190.

74. Parker, J. M., III, "Regional Systematic Jointing in Slightly Deformed Sedimentary Rocks," Bull. Geol. Soc. Am., Vol. 53, 1942, pp. 381-408.

75. Raggat, H. G., "Markings on Joint Surfaces in Angelsea Member of Demon's Bluff Formation, Angelsea, Victoria," Bull. Am. Assoc. Petrol. Geologists, Vol. 38, 1954, pp. 1808-1810.

76. Hodgson, R. A., "Regional Study of Jointing in Comb-Ridge-Navajo Mountain Area, Arizona and Utah," Bull. Am. Assoc. Petrol. Geologists, Vol. 45, 1961, pp. 1-38.

77. Hodgson, R. A., "Classification of Structures on Joint Surfaces," Am. J. Sci., Vol. 259, 1961, pp. 493-502.

78. Roberts, J. C., "Feather-fracture, and the Mechanics of Rock-jointing," Am. J. Sci., Vol. 259, 1961, pp. 481-492.

79. Bonham, L. C., "Structural Petrology of the Pico Anticline, Los Angeles County, California," J. Sediment. Petrol., Vol. 27, 1957, pp. 251-264.

80. De Sitter, L. U., Structural Geology, McGraw-Hill Book Company, Inc., New York, 1956.

81. Schmidt, R. G., Joint Patterns in Relation to Regional and Local Structures in the Central Foothills Belt of the Rocky Mountains of Alberta, Ph.D. thesis, University of Cincinnati, 1957, Doctoral Dissertation Series, Pub. No. 23,242.

82. Cloos, H., Einführung in die Geologie, Gebrüder Bornträger, Berlin, 1936.

83. Dawson-Grove, G. E., "Analysis of Minor Structures near Ardmore, County Waterford, Eire," Quart. J. Geol. Soc. London, Vol. 111, 1955, pp. 1-22.

84. Price, N. J., "Mechanics of Jointing in Rocks," Geol. Mag., Vol. 96, 1959, pp. 149-167.

85. Harris, J. F., G. L. Taylor, and J. L. Walper, "Relation of Deformational Fractures in Sedimentary Rocks to Regional and Local Structure," Bull. Am. Assoc. Petrol. Geologists, Vol. 44, 1960, pp. 1853-1873.

86. Donath, F. A., "Analysis of Basin-range Structure, South-central Oregon," Bull. Geol. Soc. Am., Vol. 73, 1962, pp. 1-16.

87. Melton, F. A., "A Reconnaissance of the Joint Systems in the Ouachita Mountains and Central Plains of Oklahoma," J. Geol., Vol. 37, 1929, pp. 729, 746.

88. Nevin, C. M., Principles of Structural Geology, 4th ed., John Wiley & Sons, Inc., New York, 1949.

89. Hills, E. S., Outlines of Structural Geology, 3d ed., Methuen and Co., Ltd., London, 1953.

90. Billings, M. P., Structural Geology, Prentice-Hall, Inc., Englewood Cliffs, New Jersey, 1954.

91. Informal communication from F. B. Conger, Shell Oil Company, Los Angeles, California, 1961.

92. Reusch, E., "Über eine besondere Gattung von Durchgängen im Steinsalz und Kalkspath," Poggendorffs Ann. Physik., Vol. 132, 1867, pp. 441-451.

93. Schmid, E., and W. Boas, Plasticity in Crystals, F. A. Hughes and Co., Ltd., London, 1950.

94. Barrett, C. S., The Structure of Metals, McGraw-Hill Book Company, Inc., New York, 1952.

95. Turner, F. J., D. T. Griggs, and H. Heard, "Experimental Deformation of Calcite Crystals," Bull. Geol. Soc. Am., Vol. 65, 1954, pp. 883-934.

96. Handin, J. W., "Strength and Ductility," in Handbook of Physical Constants, 2d ed., Geological Society of America (in preparation).

97. Bell, J. F., "Morphology of Mechanical Twinning in Crystals," Am. Mineralogist, Vol. 26, 1941, pp. 247-261.

98. Pabst, A., "Transformation of Indices in Twin Gliding," Bull. Geol. Soc. Am., Vol. 66, 1955, pp. 897-912.

99. Cahn, R. W., "Twinned Crystals," Advan. Phys., Vol. 3, 1954, pp. 363-445.

100. Hall, E. D., Twinning and Diffusionless Transformations in Metals, Butterworth & Co. (Publishers), Ltd., London, 1954.

101. Brewster, D., "On a New Cleavage in Calcareous Spar, with a Notice of a Method of Detecting Secondary Cleavages in Minerals," Edin. J. Sci., Vol. 9, 1826, pp. 311-314.

102. Knopf, E. B., "Fabric Changes in Yule Marble after Deformation in Compression," Am. J. Sci., Vol. 247, 1949, pp. 433-461, 537-569.

103. Turner, F. J., "Preferred Orientation of Calcite in Yule Marble," Am. J. Sci., Vol. 247, 1949, pp. 593-621.

104. Griggs, D. T., and W. B. Miller, "Deformation of Yule Marble, Part I--Compression and Extension Experiments on Dry Yule Marble at 10,000 Atmospheres Confining Pressure, Room Temperature," Bull. Geol. Soc. Am., Vol. 62, 1951, pp. 853-862.

105. Handin, J. W., and D. T. Griggs, "Deformation of Yule Marble, Part II--Predicted Fabric Changes," Bull. Geol. Soc. Am., Vol. 62, 1951, pp. 863-885.

106. Turner, F. J., and C. S. Ch'ih, "Deformation of Yule Marble, Part III--Observed Fabric Changes," Bull. Geol. Soc. Am., Vol. 62, 1951, pp. 887-905.

107. Griggs, D. T., F. J. Turner, I. Borg, and J. Sosoka, "Deformation of Yule Marble, Part IV--Effects at 150°C," Bull. Geol. Soc. Am., Vol. 62, 1951, pp. 1386-1406.

108. Griggs, D. T., F. J. Turner, I. Borg, and J. Sosoka, "Deformation of Yule Marble, Part V--Effects at 300°C," Bull. Geol. Soc. Am., Vol. 64, 1953, pp. 1327-1342.

109. Borg, I., and F. J. Turner, "Deformation of Yule Marble, Part VI--Identity and Significance of Deformation Lamellae and Partings in Calcite Grains," Bull. Geol. Soc. Am., Vol. 64, 1953, pp. 1343-1352.

110. Turner, F. J., D. T. Griggs, R. H. Clark, and R. H. Dixon, "Deformation of Yule Marble, Part VII--Development of Oriental Fabrics at 300° to 500°C," Bull. Geol. Soc. Am., Vol. 67, 1956, pp. 1259-1294.

111. Turner, F. J., "Nature and Dynamic Interpretation of Deformation Lamellae in Calcite of Three Marbles," Am. J. Sci., Vol. 251, 1953, pp. 276-298.

112. Friedman, M., and F. B. Conger, "Dynamic Interpretation of Calcite Twin Lamellae in a Naturally Deformed Fossil," J. Geol., Vol. 71, 1963.

113. McIntyre, D. B., and F. J. Turner, "Petrofabric Analysis of Marbles from Mid-Strathspey and Strathavon," Geol. Mag., Vol. 90, 1953, pp. 225-240.

114. Gilmour, P., and M. Carman, "Petrofabric Analysis of the Loch Tay Limestone from Strachur, Argyll," Geol. Mag., Vol. 91, 1954, pp. 41-60.

115. Clark, R. H., "A Study of Calcite Twinning in the Strathavon Marble, Banffshire," Geol. Mag., Vol. 91, 1954, pp. 121-128.

116. Weiss, L. W., "A Study of Tectonic Style," Univ. Calif. (Berkeley) Publ. Geol. Sci., Vol. 30, 1954, pp. 1-102.

117. Turner, F. J., "'Compression' and 'Tension' Axes Deduced from $\{01\bar{1}2\}$ Twinning in Calcite," J. Geophys. Res., Vol. 67, 1962, p. 1660.

118. Nickelsen, R. P., and G. W. Gross, "Petrofabric Study of Conestoga Limestone from Hanover, Pennsylvania," Am. J. Sci., Vol. 257, 1959, pp. 276-286.

119. Conel, J. E., "Studies of the Development of Fabrics in Naturally Deformed Limestones," Ph.D. thesis, California Institute of Technology, 1962.

120. Hansen, E., and I. Y. Borg, "The Dynamic Significance of Deformation Lamellae in Quartz of a Calcite-cemented Sandstone," Am. J. Sci., Vol. 260, 1962, pp. 321-336.

121. Nissen, H. U., "Dynamic and Kinematic Analysis of Deformed Crinoid Stems in a Quartz Graywacke," J. Geol., Vol. 71, 1963.

122. Johnsen, A., "Biegungen und Translationen," Neues Jahrb. Mineral. Geol. Palaeont., Teil II, 1902, pp. 133-153.

123. Turner, F. J., D. T. Griggs, H. Heard, and L. W. Weiss, "Plastic Deformation of Dolomite Rock at 380°C," Am. J. Sci., Vol. 252, 1954, pp. 477-488.

124. Crampton, C. B., "Structural Petrology of Cambro-Ordovician Limestones of the North-west Highlands of Scotland," Am. J. Sci., Vol. 256, 1958, pp. 145-158.

125. Christie, J. M., "Dynamic Interpretation of the Fabric of a Dolomite from the Moine Thrust-zone in Northwest Scotland," Am. J. Sci., Vol. 256, 1958, pp. 159-170.

126. Orowan, E., "A Type of Plastic Deformation New in Metals," Nature, Vol. 149, 1942, pp. 643-644.

127. Crussard, Ch., "Les Deformation des Cristaux Metalliques," Bull. Soc. Franc. Mineral., Vol. 68, 1945, pp. 187-197.

128. Hess, J. B., and C. S. Barrett, "The Structure and Nature of Kink Bands in Zinc," Trans. AIME, Vol. 185, 1949, pp. 599-606.

129. Washburn, J., and E. R. Parker, "Kinking in Zinc Single Crystal Tension Specimens," J. Metals, Vol. 4, 1952, pp. 1076-1078.

130. Ingerson, E., and O. Tuttle, "Relations of Lamellae and Crystallography of Quartz," Trans. Am. Geophys. Union, Vol. 26, 1945, pp. 95-105.

131. Kalkowsky, E., "Die Gneissformation des Eulengebirges," Habilitationsschrift, Leipzig, 1878.

132. Boehm, A., "Über Gesteine des Wechsels," Mineral. Petrog. Mitt., Vol. 5, 1883, p. 204.

133. Judd, J. W., "The Development of a Lamellar Structure in Quartz Crystals," Mineral. Mag., Vol. 36, 1888, pp. 1-8.

134. Becke, F., "Petrographische Studien am Tonalit der Rieserferner," Mineral. Petrog. Mitt., Vol. 13, 1892, p. 447.

135. Mügge, O., "Der Quarzporphyr der Bruchhäuser-Steine in Westfalen," Neues Jahrb. Mineral., Vol. 10, 1896, p. 757.

136. Hietanen, A., "Petrology of the Finnish Quartzites," Bull. Comm. Geol. Finlande, No. 122, 1938, pp. 1-118.

137. Fairbairn, H. W., "Correlation of Quartz Deformation with Its Crystal Structure," Am. Mineralogist, Vol. 24, 1939, pp. 351-368.

138. Fairbairn, H. W., "Deformation Lamellae in Quartz from the Ajibik Formation," Bull. Geol. Soc. Am., Vol. 52, 1941, pp. 1265-1278.

139. Brace, W. F., "Quartzite Pebble Deformation in Central Vermont," Am. J. Sci., Vol. 253, 1955, pp. 129-145.

140. Christie, J. M., and C. B. Raleigh, "The Origin of Deformation Lamellae in Quartz," Am. J. Sci., Vol. 257, 1959, pp. 385-407.

141. Riley, N. A., "Structural Petrology of the Baraboo Quartzite," J. Geol., Vol. 55, 1947, pp. 453-475.

142. Bailey, S. W., R. A. Bell, and C. J. Peng, "Plastic Deformation of Quartz in Nature," Bull. Geol. Soc. Am., Vol. 69, 1958, pp. 1443-1466.

143. Carter, N. L., J. M. Christie, and D. T. Griggs, "Experimentally Produced Deformation Lamellae and Other Structures in Quartz Sand," J. Geophys. Res., Vol. 66, 1961, pp. 2518-2519.

144. Christie, J. M., N. L. Carter, and D. T. Griggs, "Plastic Deformation of Single Crystals of Quartz," J. Geophys. Res., Vol. 67, 1962, pp. 3549-3550.

145. Christie, J. M., D. T. Griggs, and N. L. Carter, "Experimental Evidence of Basal Slip in Quartz," J. Geol., Vol. 72, 1964 (in press).

146. Carter, N. L., J. M. Christie, and D. T. Griggs, "Experimental Deformation and Recrystallization of Quartz," J. Geol., Vol. 72, 1964 (in press).

147. Mackie, J. B., "Petrofabric Analyses of Two Quartz-Piedmontite-Schists from Northwest Otago," Trans. Roy. Soc. New Zealand, Vol. 76, 1947, pp. 362-368.

148. Preston, J., "Quartz Lamellae in Some Finnish Quartzites," Bull. Comm. Geol. Finlande, Vol. 180, 1958, pp. 65-78.

149. Naha, K., "Time of Formation and Kinematic Significance of Deformation Lamellae in Quartz," J. Geol., Vol. 67, 1959, pp. 120-124.

150. Hara, I., "Dynamic Interpretation of the Simple Type of Calcite and Quartz Fabrics in the Naturally Deformed Calcite Quartz Vein," J. Sci. Hiroshima Univ., Ser. C., Vol. 4, 1961, pp. 35-53.

151. Hara, I., "Petrofabric Study of the Lamellar Structures in Quartz," J. Sci. Hiroshima Univ., Ser. C., Vol. 4, 1961, pp. 55-70.

152. Buckley, H. E., Crystal Growth, John Wiley & Sons, Inc., New York, 1951.

153. Smoluchowski, R. (ed.), Phase Transformations in Solids, John Wiley & Sons, Inc., New York, 1951.

154. Shockley, W. (ed.), Imperfections in Nearly Perfect Crystals, John Wiley & Sons, Inc., New York, 1952.

155. Cottrell, A. H., Dislocations and Plastic Flow in Crystals, Clarendon Press, Oxford, 1953.

156. Read, W. T., Jr., Dislocations in Crystals, McGraw-Hill Book Company, Inc., New York, 1953.

157. Fisher, J. C. (ed.), Dislocations and Mechanical Properties of Crystals, John Wiley & Sons, Inc., New York, 1957.

158. Creep and Recovery, American Society for Metals, Cleveland, Ohio, 1957.

159. Voll, G., "New Work on Petrofabrics," Liverpool Manchester Geol. J., Vol. 2, 1960, pp. 503-567.

160. Gibbs, J. W., "On the Equilibrium of Heterogeneous Substances," in Collected Works of J. Willard Gibbs, Yale University Press, New Haven, Connecticut, 1906.

161. Goranson, R. W., "Thermodynamic Relations in Multicomponent Systems," Carnegie Institution of Washington, Pub. No. 408, 1930, pp. 1-329.

162. Goranson, R. W., "'Flow' in Stressed Solids: An Interpretation," Bull. Geol. Soc. Am., Vol. 51, 1940, pp. 1023-1033.

163. Goranson, R. W., "Physics of Stressed Solids," J. Chem. Phys., Vol. 8, 1940, pp. 323-334.

164. Verhoogen, J., "The Chemical Potential of a Stressed Solid," Trans. Am. Geophys. Union, Vol. 32, 1951, pp. 251-258.

165. MacDonald, G. J. F., "Orientation of Anisotropic Minerals in a Stress Field," in "Rock Deformation," Geol. Soc. Am. Mem. 79, 1960, pp. 1-8.

166. Kamb, W. B., "The Thermodynamic Theory of Nonhydrostatically Stressed Solids," J. Geophys. Res., Vol. 66, 1961, pp. 259-271.

167. Kamb, W. B., "Theory of Preferred Crystal Orientation Developed by Crystallization under Stress," J. Geol., Vol. 67, 1959, pp. 153-170.

168. Kamb, W. B., "An Experimental Test of Theories of Nonhydrostatic Thermodynamics," abstract, J. Geophys. Res., Vol. 67, 1962, p. 1642.

169. Brace, W. F., "Orientation of Anisotropic Minerals in a Stress Field (Discussion)," in "Rock Deformation," Geol. Soc. Am. Mem. 79, 1960, pp. 9-20.

170. Heard, H. C., "The Effect of Large Changes in Strain Rate in the Experimental Deformation of Yule Marble," J. Geol., Vol. 71, 1963, pp. 162-195.

171. Buerger, M. J., and E. Washken, "Metamorphism of Minerals," Am. Mineralogist, Vol. 32, 1947, pp. 296-308.

172. Griggs, D. T., M. S. Paterson, H. C. Heard, and F. J. Turner, "Annealing Recrystallization in Calcite Crystals and Aggregates," in "Rock Deformation," Geol. Soc. Am. Mem. 79, 1960, pp. 21-37.

173. Sander, B., Contributions to the Study of Depositional Fabrics, trans. by E. B. Knopf, American Association of Petroleum Geologists, Tulsa, Oklahoma, 1951.

174. Christie, J. M., H. C. Heard, and P. N. La Mori, "Experimental Deformation of Quartz Single Crystals at 27-30 Kb Confining Pressure and 24°C," abstract, Bull. Geol. Soc. Am., Vol. 71, 1960, p. 1842.

175. Handin, J. W., D. V. Higgs, and J. K. O'Brien, "Torsion of Yule Marble under Confining Pressure," in "Rock Deformation," Geol. Soc. Am. Mem. 79, 1960, pp. 245-274.

176. Paterson, M. S., and L. E. Weiss, "Experimental Folding in Rocks," Nature, Vol. 195, 1962, pp. 1046-1048.

177. Turner, F. J., "Rotation of the Crystal Lattice in Kink Bands, Deformation Bands, and Twin Lamellae of Strained Crystals," Proc. Natl. Acad. Sci. U.S., Vol. 48, 1962, pp. 955-963.

DISCUSSION

D. COATES (Canada):

Can you suggest a definition and description of joints so that they can be identified on a rock face? It is often difficult to distinguish a joint plane from a plane which has been formed by blasting. Some material property aside from bedding provides a preferential surface for such breaking.

M. FRIEDMAN (in reply):

Certain trivial cases can easily be recognized. For example, there is little doubt about the origin of fractures that radiate from blast holes or of fractures that are filled with natural vein material. However, consider two situations where such obvious features are not developed. In the first, let us assume it is possible to move away from the blasted face to some structurally similar location where only the natural fractures can be observed. In the second, we will assume it is impossible to observe an unblasted exposure.

In the former situation, it is possible to map the natural fracture geometry, the average spacing for each of the fracture sets, and any markings that might exist on the natural fracture surfaces (e.g., plumose or conchoidal structures) in the region unaffected by blasting. These can then be traced back into the blasted area and compared with the observed fracture array. At the blasted face, one could then detect the development of new fracture sets (recognized by their geometry or perhaps by the presence or absence of surface markings) or changes in the spacing between natural fracture sets.

In the second situation, where one cannot move away from the blasted area, my first inclination is to say that one could not differentiate between the natural and the induced fractures. After all, they are fractures in rock, and the only difference between them is the energy source used to initiate them. However, if you can measure the residual strain gradient away from the fracture surfaces you may find that the gradients associated with paleo-fractures are different from those adjacent to induced fracture surfaces. The nature of this difference, if any, would have to be determined under controlled conditions. As far as

I am aware, the measurement of such gradients has not been attempted, but it should be possible by use of resistance foil gages, photostress methods, or X-ray diffractometry.

R. P. TRUMP (USA):

I have a question for Dr. Friedman. In your talk on the fold, you show a rather high amplitude fold having the neutral axis within the fold. If this fold is developed by thrusting or buckling, the neutral axis is not contained in the fold in the early stages of the process. Now, the question is, do you believe that your correlation with the neutral axis means that what you are seeing is a late stage deformation after the wave length has already been determined and, in turn, that the lateral pressure was relatively low at the time of initial fold development?

M. FRIEDMAN (in reply):

I do not think this question can be answered from study of the fractures alone. However, in the course of studying fractures on folds we have also investigated the deformed calcite in limestones. In many cases the same orientations for the principal stresses are derived from the calcite twin lamellae as from the fracture geometry. As the critical resolved shear stress to initiate twin gliding in calcite is low (less than 100 bars), I visualize the calcite as beginning to deform rather early in the development of the fold. Accordingly, as both the calcite and the fractures reflect the same principal stress orientations, I conclude that some fractures also were initiated early in the folding history.

GENERAL DISCUSSION TO PART III

K. F. DALLMUS (Venezuela):*

The validity of the elastic theory with respect to rocks was openly questioned at the meeting. In the first place the terms used to put the theory into numbers for mathematical statement were set up to judge the suitability of certain materials for construction purposes, and not as a general theory for the behavior for all materials under stress. What is needed is a number to describe the mechanical behavior of material, independent of the size and shape of the test specimen.

Fundamentally, there are only two stress patterns that can be applied to any material: (1) balanced stress (hydrostatic) and (2) unbalanced stress.

In the same manner all materials can be classified with respect to the manner in which they react to stress: (1) isotropic with respect to stress, and (2) anisotropic with respect to stress.

These two general classes can be broken down further as follows:

1. Isotropic:
   a. Impermeable with respect to hydrostatic confining stress.
   b. Permeable with respect to hydrostatic confining stress.
2. Anisotropic:
   a. Impermeable with respect to hydrostatic confining stress.
   b. Permeable with respect to hydrostatic confining stress.

Some of the procedures used in testing materials are an attempt to make a specific material react in the same manner as another material that is fundamentally different in its reaction to stress. Such a procedure is the jacketing of specimens to prevent internal deformation

---

*Mr. Dallmus' discussion was submitted after the conference.

by hydrostatic confining pressure in materials that are permeable.

There is a physical measurement that can be made that is applicable to all materials and that will describe their mechanical behavior under specific conditions. This measurement is based on a very simple geometrical fact, and not on a postulated theory; namely, a sphere has the least surface area for a given volume.

If the sphere is made of material that is isotropic and impermeable, hydrostatic confining stress (radial balanced stress) will produce a reduction in volume and a corresponding proportional decrease of surface area, but there is no change of shape. If the material is isotropic and permeable, there will be a volume decrease, but the decrease in surface area may not be proportional. In neither case will the deformation cause an increase in surface area with respect to the original amount.

Under unbalanced stress the sphere composed of any material whatsoever will change its shape, and there will be an increase of surface area for the given volume of the sphere. This axiom can be extended to any three-dimensional geometric figure all of whose internal angles are 180 degrees or less. Any three-dimensional figure of this description under unbalanced stress will increase its surface area and the increase is independent of the shape of the test specimen, but the site of the increase of area will depend on how the specimen is loaded. The barreling effect before failure of cylindrical test specimens under uniaxial compression, which is supposed to be objectionable, is the expression of the increase of surface area of the same volume.

The increase of volume after deformation is mentioned in most textbooks; however, I can find no mention of the increase of surface area before failure, nor any reference to anyone who has ever tried to measure it. I understand that some private investigations are going on at the present time, but nothing has been published.

If the ratio of surface area just before failure to an original area before loading can be determined, it will be a unique number describing the mechanical behavior of the material, provided there was no strain energy (latent stress) present in the test specimen. The rate at which any material can extend its surface area will depend on the rate at which the smallest units composing the material can move with respect to one another and on the bonding forces between them. These units may be atoms, molecules, crystals, or grains. A material may be composed of any of these or two combinations of them. For all materials where the rate of loading is significant, the number will be qualified by another number, which represents the work necessary to break the bonds between the constituent units.

Since an increase of surface area means extension, the stresses involved are necessarily tensile stresses. The mechanical behavior of any material can, therefore, be expressed in terms of tensile strength, and from this it follows that any geometric figure, all of whose internal angles are 180 degrees or less, may be deformed (change of shape) by unbalanced stress, either compression or tension, but the failure will always be by tension.

The same terminology can be extended to liquids and gases. Since gases do not have a shape of their own, the number will automatically be infinity. Contrary to the popular definition of liquids that they conform to the containing vessel, all liquids have one form in common, and that is a sphere. Everyone is familiar with the perfect spheres of mercury on glass. All liquids will form perfect spheres, and the size of the perfect sphere depends on the surface tension that exactly counterbalances the weight of the sphere when resting on a nonwettable material. The size of the sphere can be calculated if the surface tension is known.

All liquids can, at least theoretically, be spread out in a layer one atom or one molecule thick. If the diameter of the constituent units is known, then the area of a layer one unit thick can be calculated, and the ratio of this area with respect to the surface area of the sphere will be analogous to the number used for materials that can maintain any shape under their own weight. In the case of the liquids the qualifying number will be the surface tension.

Using a number such as described, arbitrary limits can be set for the various kinds of liquids and solids now described by names, and much confusion can be eliminated. Also, since the bonding forces between the constituent units are probably constant and independent of time, it may well be that the total amount of work done between the so-called elastic limit and failure is also a constant independent of time.

Another condition that may be investigated in materials that are composed of crystals that twin under unbalanced stress is the maximum amount of increase of surface area that can be obtained in this manner, and the work necessary to break the bonding forces after this maximum is reached.

Rocks are very often composed of a mixture of all four kinds of materials, and the failure of the whole will depend on the percentage of material with the lowest ratio of increase of surface area. As pointed out by Professor Emery, grain or crystal boundaries are preferred sites of failure. The sum of the volume of the void spaces so created is the increased volume caused by the deformation. In petroleum geology the void spaces (fractures) are called secondary porosity, and frequently provide the means of escape of the contained fluid in rocks with very low primary porosity.

One principle often overlooked in deformation phenomena, either natural or artificial, is the principle of least work. Since practically all materials are weaker in tension than in compression, it appears logical to assume that failure by any deformation involving change of shape with an increase of surface area will be initiated by tension. The reasoning above, therefore, involves the principle of least work.

J. L. SERAFIM (Portugal):

We have seen here results of measurements of internal stresses, and so I would like to ask this question.

As most of the rocks are anisotropic, has any attempt been made to compute the stresses not taking into consideration the usual theory of elasticity but the theory of elasticity for anisotropic materials?

C. L. EMERY (Canada) in reply:

I would say that probably numbers of attempts have been made at this, and I make such attempts myself. I think Dr. Serafim is already aware that some of the people at Sheffield, England, are using methods somewhat similar to my own, doing this kind of thing.

I would rather not discuss that at this point, unless someone asks about it, but I would like to put this question to others of you who may be doing work in this field.

The obvious thing, of course, is that if the rock has directional properties, and you ignore them, the computations will be in error by the extent of the error of your assumptions.

The measurements in most cases are fairly well done and well recorded, and it is how they are used that is the important point. And, any of these calculations that you have heard about or any plans or designs based on them will only be as good as the assumptions that are made.

Given any set of assumptions, most of us can work out mathematical data that meet the needs of the assumptions. If the assumptions are wrong, the data are not too good and maybe you might as well have gone fishing.

G. P. EATON (USA):

If one wished for some reason to map the stress trajectories and measure the magnitudes of stress in a horizontal plane at depth in fairly flat terrain near a large mountain range (for example, eastern Colorado), what method might one practically use, and what technical problems could be encountered? Could he possibly use a modified borehole method in wells drilled during petroleum exploration?

C. L. EMERY (Canada) in reply:

I think I will start the discussion on this, and then possibly some others might like to speak.

First of all, the modified borehole method mentioned at the end of the question would be subject to the conditions required for using those kinds of borehole meters that are available, and this may require overcoring techniques; if the hole were a newly made hole, one might get some information by the deformation of the hole itself as it readjusted to the conditions. On the other hand, work such as Dr. Friedman has been doing and work such as Bob Merrill and Dr. Potts have discussed, where holes may be put down and instruments put in and loads applied, might be used if some modification of these might possibly be made.

In Dr. Friedman's work, samples may be taken and the principal stress directions derived from the samples. These are of known orientation and if plotted on a large plan, presumably some kind of a stress trajectory would be derived.

In fact, this sort of thing is done and our own techniques involving photoelasticity allow us to use a drill core, the only requirement being that the drill core be oriented. This is difficult in a small diameter core, although it can be done. However, it is not too difficult in a large diameter core, such as is sometimes available in oil well boring, where cores are taken at certain sections, so that it is possible to map stress trajectories in a horizon if one has a sufficient number of samples from that horizon.

Most of these techniques would allow you to map stress trajectories in three dimensions. Is there anyone who wants to comment?

M. FRIEDMAN (USA) in reply:

Another possible way to map stress trajectories in such a region would be through study of hydraulic fracturing in oil and gas wells. First, the fracturing pressure in the hole is apt to be fairly close to the magnitude of the least principal stress. This would be true if the induced fractures were extension or tensile fractures and if the tensile strength of the rocks was negligible. Second, it is possible to determine the orientation of the induced fractures either by studies of packer impressions, of communication between wells, or by use of down hole television cameras. From the orientation of the induced fractures, and assuming they are vertical extension or tensile fractures, one can position the least principal stress in the horizontal plane normal to the best developed set of fractures, and the greatest and intermediate principal stresses in the plane of the fracture set. Accordingly, it may be possible to map the least principal stress at each well and to construct a trajectory diagram by extrapolation between wells. One would have to further distinguish between the greatest and intermediate principal stresses in the plane of the fracture set before a complete trajectory diagram could be drawn.

E. L. J. POTTS (England) in reply:

Mr. Eaton has, in his question, emphasized the tremendous difficulties inherent in remote rock stress measurement *in situ*. The papers presented to the conference have described methods suitable for stress evaluation at a single location and at a specific time by relaxation techniques. Because of the mechanical overcoring technique required, the approach is limited to an evaluation close to the side of the opening or the collar of the borehole used. The question implies the use of deep holes whose initial purpose was for petroleum exploration. I do not consider that a satisfactory technique has been developed for any borehole deformation meter using an overcoring or relaxation approach, which could be used for the objective quoted in the conditions described. It should be appreciated by the conference that the instrumentation described in these papers has its limitations of use. The hydraulic stressmeter described in my own contribution was developed to measure, remote from mine openings, the stress changes resulting from mining development and the influence of load transfer as the support afforded by the rock or mineral mined is removed. In such cases, continuity of stress change evaluation is vitally important, and the "spot" evaluation of absolute stress obtained by a relaxation technique is not so important as the location of pressure abutments and their growth and change as mining development proceeds. An example of the use of this approach will clarify this amplification to Mr. Eaton's question, underline the limitations of relaxation techniques, and emphasize the difficulties that his question raises in any attempt to evaluate stress, at depth, in a from-surface borehole.

Figure 1 shows the measuring network of boreholes drilled into a strong hornblende schist, country rock, which forms the foot and hanging wall of a gold bearing, near vertical, quartz reef, at Kolar, South India.

The schistosity is parallel to the reef, the footwall drive and crosscuts providing access to the preformed reef drives. The wastes behind the inclined rill faces are solid packed with granite blocks brought from surface, the stoping width being about 60 in. The borehole extensometer described in my paper is used to measure strata movement

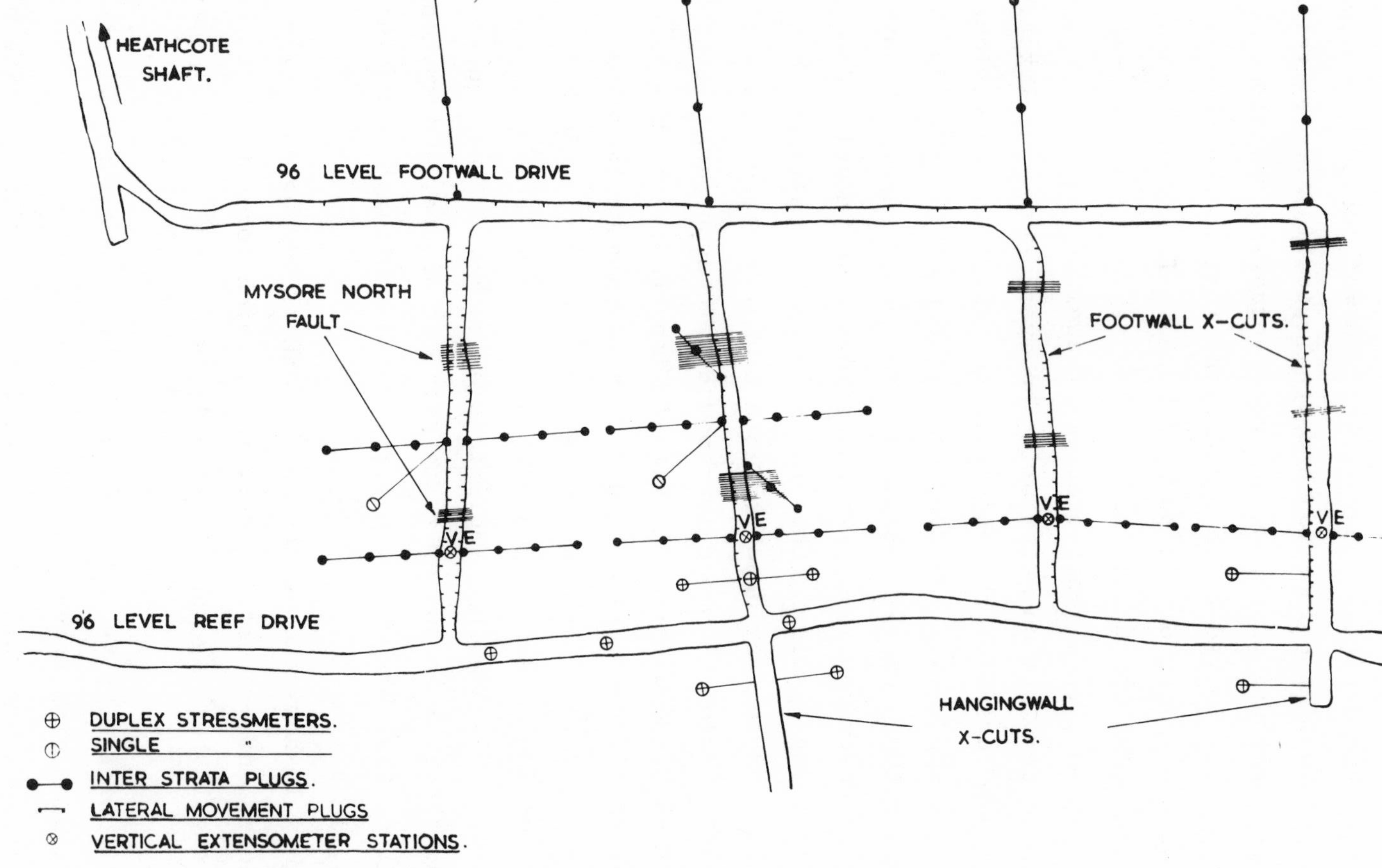

Fig. 1—Measuring network of boreholes in reef, Kolar, South India.

parallel to the reef, while the hydraulic stressmeters are located strategically to measure stress change from the initial prestress at installation as the rill faces, above and below the reef drive, advance. The installation was near completion in November 1962 when a major rockburst occurred. The analysis of stressmeter output, which was continuously recorded, has provided important data on the events leading up to this major burst. The displacement measurements both in boreholes and along drives and crosscuts have equally been demonstrative of the strata movement as mining proceeded.

It will be appreciated, therefore, that under the specific conditions suitable for the use of the hydraulic stressmeter, it will provide valuable data. Mr. Eaton has posed a problem that, in my opinion, is more likely to be answered at our present state of knowledge and instrumentation development by the approach suggested by Dr. Friedman, although I feel that even with this approach an emphatic stress evaluation will be difficult to obtain.

Y. OKA (Japan):*

What is the most suitable way of determining the stress state that existed before a drift or a borehole was made?

C. L. EMERY (Canada) in reply:

This is the $64 question, and you have heard some opinions on these things. Generally speaking, most of us require some kind of a borehole or a sample. A lot can be inferred from those naturally occurring surfaces that one sees around rocky places, but there are various ways of determining the relative stress state.

I don't know of anybody who can tell me what the stress is as far as absolute magnitudes are concerned; however, I would like to refer you to the conference papers and discussions by Dr. Jaeger, Mr. Merrill, Dr. Friedman, others, and myself. In general, with our present techniques, we can do a very good job on stress trajectories. In other words, directions are pretty good, but magnitudes are pretty awful. All we

---

*<u>In absentia</u>.

do with magnitudes is to record some sort of relative change that occurred during a measurement period.

I might just interject here that of the various techniques that you have heard about, involving in some cases short time tests and in other cases long time tests, you know that any of these things involve instantaneous elastic rebound and time-delay elasticity, and it depends on the kind of test you use as to what part of the deformation you are measuring. In all cases, however, the direction should be pretty good. The magnitude would depend on what part of the total energy you were involved with, and it would depend on the kind of test.

I now wish to throw this question open to the audience.

L. OBERT (USA) in reply:

In answer to the question about possible ways of measuring stress without a borehole, I think that the seismic method was used sometime ago, and is still a possible method. It is being used quite extensively in Germany today in salt deposits.

Of course, this is a presumption that the velocity of the wave varies with the stress level, and that there is some known relationship between the stress level and the velocity.

The two papers that I call to mind are both, I believe, published by Springer (Vienna, Austria). They are dissertations that were turned out at the University of Freiberg.

E. HOEK (South Africa) in reply:

I think that all of us will agree that the necessity to measure stress is a very definite one. However we do it, I am not in a position to argue on the merits or demerits of various systems, but I should like to introduce, on behalf of my colleague, Mr. E. R. Leeman, a recent development in rock-stress measuring tools for which he has been responsible. This development resulted from the desire to determine the absolute stress in rock at distances up to and exceeding, say, 100 ft inside the rock mass on a more or less routine basis. The stress relieving technique is applied in the following manner to achieve this.

A borehole is drilled into the rock to the depth at which it is desired to determine the stress, as shown in Fig. 2. The end of the borehole is ground flat and smooth, and one or more electrical resistance wire strain gauges are cemented upon it.

The length of the borehole is next extended by means of a suitable coring tool as shown in Fig. 3. The strain gauges cemented upon the rock at the end of the borehole are thereby trepanned. The stresses on the rock face on which the strain gauges are cemented are relieved and this will be exhibited as a change in strain by the strain gauges.

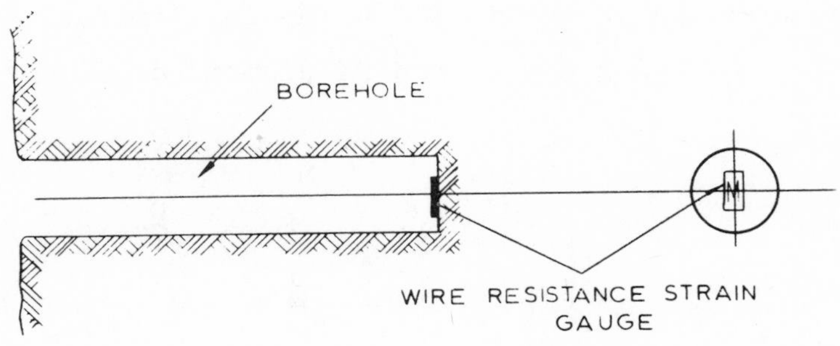

Fig. 2—Before stress relieving.

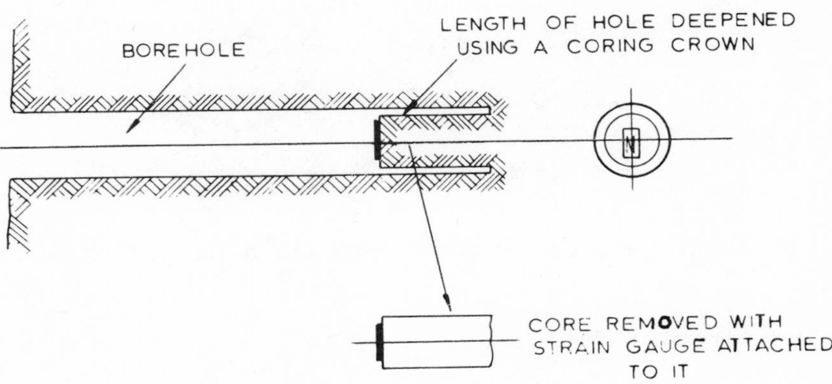

Fig. 3—After stress relieving.

The core is removed from the end of the borehole and strain readings are again taken from the strain gauges. The difference between the strain readings before and after coring is a measure of the stress in the rock surrounding the end of the borehole.

One of the difficulties associated with the use of wire resistance strain gauges is their sensitivity to water.

Mr. Leeman has developed a strain cell incorporating wire resistance strain gauges, and a cementing-on tool, by which it is possible to use the trepanning stress relieving technique described above under wet or dry drilling conditions and at almost any depth inside the rock mass.

As will be seen from the diagrammatic sketch of one of the strain cells in Fig. 4, a wire resistance strain gauge is cemented upon a circular piece of shim 1 3/8 in. in diameter. The diameter of the strain cell was chosen so that it could be used in a standard BX diamond drilled borehole, which is 2 3/4 in. in diameter. Trepanning is therefore effected using a standard BX diamond drill coring crown.

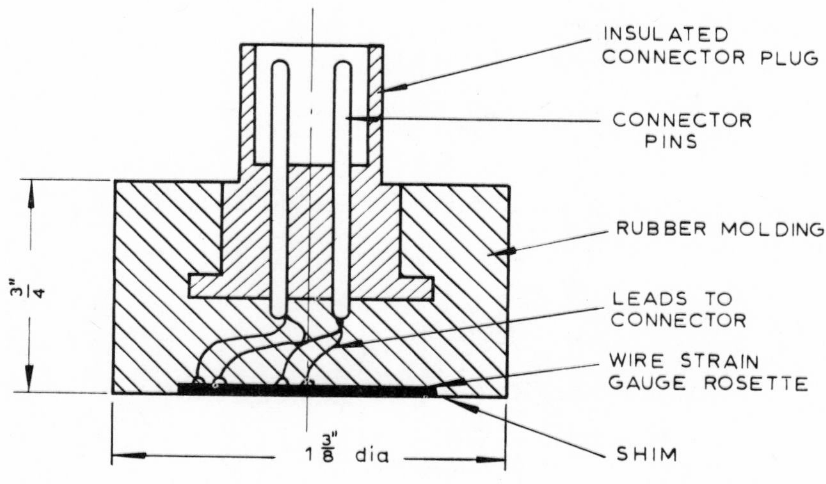

Fig. 4—Cross section through a strain cell.

The material from which the shim is made depends on the modulus of elasticity of the rock in which the stresses are to be determined. Steel shim 0.008 in. thick has been found to be suitable for hard rocks such as quartzite. Epoxy resin shim, 0.025 in. thick, has proved satisfactory for both hard and soft rocks including coal. When the strain cell is cemented upon a smooth surface, changes in strain on the surface are transmitted via the shim to the strain gauge. Tests have shown that the presence of the shim has an almost negligible effect on the strain readings.

The wire resistance strain gauge incorporated in the strain cell may take the form of a single gauge or of a rosette of two or more gauges. In the latest form of this strain cell, a rectangular rosette of three gauges is used, oriented to measure strains in the vertical, horizontal, and 45-degree directions.

The leads from the strain gauges are connected to four pins in an insulated connector plug. Both the shim and the plug are molded into a silicone rubber "plug" as shown in Fig. 4. The rubber acts as a protection for the strain gauges and prevents moisture from reaching it.

To use the strain cell, the end of the borehole is ground smooth with a square-faced diamond impregnated bit and a flat-faced bit. This gives a sufficiently smooth surface upon which to glue the cell. The cell is inserted into the borehole with a special cementing-on tool and glued on the end of the borehole.

Strain readings are taken during the period that the glue is hardening. Once the glue has set, the strain indicator readings become constant and the tool may be removed from the borehole.

The strain cell is then drilled out of the borehole by means of a standard BX diamond drilling crown and is recovered adhering to a piece of core. It is plugged into the strain indicator and the stress relieved strain reading taken. The principal stresses in the rock on the flat face at the end of the borehole are calculated from the strain readings. From three-dimensional photoelastic analysis, the relationship between the stresses on the face at the end of a borehole and those in the surrounding rock are known, and the stress condition in the rock can be determined.

This technique has been successfully used for distances up to 40 ft inside the rock.

The cores upon which the strain cells are glued are taken back to the laboratory and compression specimens prepared. From these the elastic constants of the rock are determined. A disk, 1/2 in. thick, is also cut off the end of the core to which the strain cell is cemented. This is subjected to a diametral compressive loading test during which readings from the strain gauges in the strain cell are taken. Load-strain curves are plotted from these readings and if these show no irregularities it is accepted that the strain cell had been satisfactorily glued upon the end of the borehole.

L. MÜLLER (Austria) in reply:

There is an interesting question by Professor Oka that we haven't quite understood yet, as it seems. The question is: Can we measure the state of the original stresses or changes in the stress state only?

Here is the problem. To know what kind of stresses there are in the rock, normally we drill boreholes or excavations. Executing these holes and measuring the stresses we are late each time, because what we want to measure has already happened by 80 per cent, and we can measure only the remaining 20 per cent of the event. We reach the theater during the last act.

In order to bring about a change in this situation we tried in three experiments, which are being conducted now in Germany and Austria, to be there earlier than the miner with our observations.

Before excavating the adit for the measurements (see Fig. 5), we introduce measuring devices into four boreholes. In these boreholes we can then measure the deflections of the rock around the adit by means of new devices that we call deformation indicators, and by means of which we can measure the deflection of a borehole perpendicular to its axis with an accuracy of one tenth of a millimeter. Moreover, in

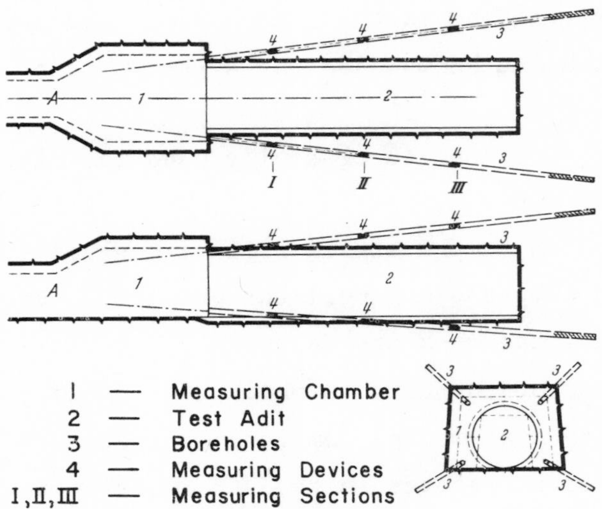

Fig. 5—Stress measuring devices in test tunnel.

boreholes we can determine the direction of stress according to the method of Hiramatsu and Oka by introducing photoelastic material into the boreholes.

Taking first the measuring devices into the boreholes and only then excavating the adit, we can observe along several cross sections the deformation of the rock. The measuring chamber is situated at the entrance of the adit.

Now, taking several cross sections and comparing them, we arrive at a four-dimensional result that has three dimensions in space and one dimension in time.

I hope that within a year I will be able to give you more details on the results of these tests.

C. L. EMERY (Canada):

Thank you very much, gentlemen. This is most interesting. As I think my remarks have more or less been borne out, we are pretty good on directions; we hope to do better on magnitudes.

We haven't gotten down to absolutes yet. It will be, however, through endeavors, such as Dr. Müller has pointed out, and others of course, that we will approach better confidence on our absolute stresses in the future.

PART IV--DESIGN

Don U. Deere, Chairman

Presentation of the principles of rock mechanics as applied to specific design problems--with special emphasis on the influence of residual rock pressures on the design, construction, and performance of arch-dam abutments, open-pit mines, rock slopes, and underground openings.

PROJETS

Présentation des principes de la Mécanique des Roches appliquée à des problèmes bien définis--en insistant particulièrement sur l'influence des pressions résiduelles des roches sur le calcul, l'exécution et la bonne tenue des butées des barrages à voûte, des mines à puits ouverts, des talus de roches et des ouvertures souterraines.

ENTWURF

Darstellung der Grundsätze der Felsmechanik im Hinblick auf bestimmte Entwurfsprobleme--mit besonderer Betonung des Einflusses des Gebirgsdruckes auf Entwurf, Bau und Verhalten von Talsperrenwiderlagern, bergbaulichen Tagebauten, Felsböschungen und unterirdischen Hohlräumen.

CONTENTS

INTRODUCTION

Don U. Deere*

THE TITLE of this part is "Design." It could more properly be called "The Significance of the State of Stress in Rock Engineering." In the part on fundamentals, concepts regarding fracture and flow of rocks were discussed, as well as some of the fundamental geological causes of the stresses which lead to these phenomena in the earth's crust.

In the third part, entitled "Measurements," various methods were given for field determination of the state of stress of *in situ* rock at the surface and in deep underground workings. By state of stress is meant the orientation and magnitudes of the major, minor, and intermediate principal stresses. In this part it will be our goal to focus our attention on problems of rock engineering and to see in what manner the state of stress may influence our design and construction procedures in the fields of mining engineering, civil engineering, and petroleum engineering.

For underground openings the concept of "heavy ground" or stressed rock is not a new one, and the engineers who have the responsibility for the stability of such openings are aware of the occasional disastrous effects as these stresses make themselves known. The effects may include the slow, squeezing, working action of the rock as it moves inward into the opening with crushing of timber supports, distortion of steel supports, and snapping of rock bolts--or they may manifest themselves by rock bursts of explosive violence. In all cases work is done by the rock, energy is released, and the stresses surrounding

*Department of Geology and Civil Engineering, University of Illinois, Urbana, Illinois.

the opening are redistributed. Of course, the strength and the stress-strain characteristics of the rock mass and the type of rock support artifically supplied--as well as the character of the initial stress field--will influence the type of behavior.

On the other hand, in surface workings the possible occurrence of high horizontal stresses and their probable detrimental effects have not generally been recognized. However, in the past decade or so there have been several examples where apparently rather large horizontal stresses have been encountered in shallow excavations for quarries, open-pit mines, rock cuts for highway, railway, or canal location, and excavations for spillways and power plants at dam sites; in these instances horizontal displacements or vertical uplift have resulted in certain engineering problems.

In the case of natural excavations, such as river valleys carved into rock masses, the stresses may have been relieved to a certain extent; however, in the process joints parallel to the valley wall and dipping toward the stream may have formed. These joints are called "sheeting joints" and are often the cause of concern in the design and construction of dams and associated works in rock valleys. If the stresses are relieved by a man-made cut, the horizontal displacements may lead to general instability and slope failure of the excavation walls.

Thus it is seen that we may be concerned with high rock stresses in two ways: first, in dealing with joints already formed by geological stress relief, and second, the stability problems encountered when we excavate into a highly stressed rock medium.

In the following papers some of these points will be developed which later will be enhanced by discussion from the panel and the audience.

ABSTRACT

Residual stresses influence the stability of high slopes considerably. If their magnitude and orientation is determined by *in situ* measurement they can be utilized in the geostatic analysis.

The theoretical analysis of the stability of rock slopes is based on the knowledge of both the mechanical states and processes within the slope system and the technological properties of the rock mass.

The rock substance is differentiated from rock masses by the different order of magnitude of the discontinuities and by the morphologic anisotropy developed by tectonic fracturing. The mechanical anisotropy results from this morphologic anisotropy. Rock masses have a principally different type of bond between their constituents than soil masses.

The strength of rock masses can not be determined in the laboratory but only by means of *in situ* testing.

Observations of natural phenomena and model tests provide information on the states of stress and deformation within a sloping region. Static considerations deal with momentary states only; however, different states, which are the phases of a flow process within the rock mass are to be investigated. This flow process is not limited to the immediate vicinity of the slope surface, but extends to greater depth and width than normally assumed. Within this greater region not only shear strains but also tensile strains can be observed. In most cases, fissured material can not resist this tensile straining; the respective sensitivity depends on the orientation of the joints in respect to the direction of flow.

Seismic action, joint-water pressure, and orientation of the prevailing joint systems have to be considered in a rock slope stability analysis.

RÉSUMÉ

Les contraintes résiduelles influencent considérablement la stabilité des fortes pentes. Si on connaît leur grandeur et leur orientation, par mesurage sur place, on peut en tenir compte dans l'analyse géostatique.

L'analyse théorique de la stabilité des pentes rocheuses se fonde sur une connaissance à la fois des états et processus mécaniques dans un système de pentes, et des propriétés techniques de la masse rocheuse.

La masse rocheuse se distingue de la substance rocheuse par la différence de taille des discontinuités et par l'anisotropie morphologique résultant de la fracturation tectonique. L'anisotropie morphologique cause l'anisotropie mécanique. La masse rocheuse se distingue du sol aussi par la différence de type de lien entre ses composants.

La résistance des masses rocheuses ne peut pas être déterminée en laboratoire mais seulement au moyen d'essais sur les lieux.

Des informations sur l'état des contraintes et des déformations dans le système peuvent être tirées de l'observation des phénomènes naturels et d'essais sur modèles.

L'observation statique a trait à des états provisoires seulement; il est cependant nécessaire d'observer plusieurs états, ceux-ci étant les phases d'un processus d'écoulement à l'intérieur de la masse rocheuse. Ce processus n'est pas limité au voisinage immédiat de la surface du talus, mais se propage plus loin et plus profondément qu'on ne présume normalement. Dans cette zone élargie apparaissent non seulement des contraintes de cisaillement, mais aussi des contraintes de tension. Dans la plupart des cas, les matériaux fissurés ne peuvent pas résister à ces contraintes de tension. Les sensibilités respectives dépendent de l'orientation des joints par rapport à la direction d'écoulement.

Les tremlements de terre, la pression de l'eau dans les joints et l'orientation des systèmes à joints prédominants doivent être considérés dans une analyse de la stabilité d'une pente rocheuse.

AUSZUG

Krustenspannungen beeinflussen die Standsicherheit hoher Böschungen erheblich. Wenn man ihre Grösse und Richtung aus Messungen *in situ* kennt, kann man sie in der geostatischen Berechnung berücksichtigen.

Die theoretische Behandlung der Standsicherheit von Felsböschungen setzt sowohl die Kenntnis der mechanischen Zustände und Vorgänge im Böschungssystem als auch die Kenntnis der technischen Eigenschaften des Materials Fels voraus.

Das Material Gebirge (Fels) unterscheidet sich vom Material Gestein durch die andere Grössenordnung der Diskontinuitäten und die bei der tektonischen Zerbrechung erworbene morphologische Anisotropie; diese ist die Ursache einer mechanischen Anisotropie. Vom Lockergestein unterscheidet sich das Gebirge durch die grundverschiedene Art des Verbandes der Einzelteile.

Gebirgsfestigkeit kann nicht im Laboratorium, sondern nur in Grossversuchen *in situ* bestimmt werden.

Über den Spannungs- und Formänderungszustand in Böschungen sagen Naturbeobachtungen und Modellversuche aus. Statische Betrachtung liefert nur einen augenblicklichen Zustand, man hat aber verschiedene Zustände zu betrachten, welche nichts anderes sind als Phasen eines Fliessvorganges. Dieser Vorgang beschränkt sich nicht auf die unmittelbare Nachbarschaft der Böschungsfläche. Dabei treten nicht nur Scherbeanspruchungen auf, sondern auch Zugbeanspruchungen, denen gegenüber zerklüftetes Gebirge besonders empfindlich ist. Diese Empfindlichkeit hängt hochgradig von der Orientierung der Klüfte ab.

Erdbebenkräfte, Kluftwasserschub und die Orientierung der vorherrschenden Kluftsysteme müssen in einer Böschungsberechnung berücksichtigt werden.

APPLICATION OF ROCK MECHANICS IN THE DESIGN OF ROCK SLOPES

Leopold Müller*

Introduction

THE DESIGN of rock slopes should always start from three essential elements: (1) the properties of the particular material of rock mass, (2) the states of stress existing in the rock mass, and (3) the conditions of the ground water. All three of these factors are essential for the stability of artificial and natural slopes. Unfortunately it is still very difficult to take into consideration the stresses existing in the earth's crust; there are, however, onsets in this direction. In order to judge their influence it is appropriate to regard the static and dynamic attitude of rock slopes from the point of view of rock mechanics generally.

So far there is no generally acknowledged theory on the formation of mountains and continents; and there is no agreement about the origin of the forces which have operated in the earth's crust and are still operating. Only the one point--that during and after the lithogenesis significant processes of formation have taken place--is generally acknowledged. Observations of nature show that these processes of formation were connected with strains which overstepped the flow limit of the rocks and reached their rupture limit everywhere. The theory of strength proves that in such a case bodies do not turn stressless again but that they keep residual stresses, unless they recrystallize under high temperatures. Recrystallization and pressure alone, creeping, and earthquake vibrations cannot cause complete removal of stress.

---

*Consultant, Ingenieurbüro für Geologie und Bauwesen, Internationale Versuchsanstalt für Fels, Salzburg.

Stresses of Tectonical Origin

Thus we have to take into account at least residual stresses in the earth's crust, often even active tectonic stresses, or, in a great number of places, stresses as a cause of unloading owing to erosion or melting of the ice.[1] In many cases the actual existence of such stresses is confirmed.

The rock slopes of the Vajont Valley, cutting into the earth more than 1000 ft, in which Carlo Semenza[2,3] has erected the highest dome dam in the world (900 ft), are subjected to stresses to such an extent that, during the excavation of the rock, owing to the stress relief large rock parts were detached from the walls with a sound like a gun-shot and rock plates in the shape of shells and with a thickness of 4 in. burst away from the bottom of the excavation. If they were knocked at, these shells of massive Jura limestone sounded like phonolite and cracked like porcelain. At a depth of 30 to 60 ft regular rock burst occurred during the excavation of the bottom.

Valley Creep and Joints Parallel to the Surface

The slopes of a great number of mountain valleys creep downward under the influence of a disordered equilibrium such as valley erosion. This was already shown by Stini's investigations in the thirties.[4-10] The rock zone that takes part in this movement extends (Fig. 1) in steep slopes as far to the depth and about twice as far behind the crest of the slope as the slope is high. At the 13th Colloquium of the International Society of Rock Mechanics in Salzburg, Müller[11] and Scheiblauer[12] reported on recent investigations and model tests referring to the problem of steep slopes.

Whereas valley creeps of this kind can be understood and treated mathematically as rheological flow movements of an elastical viscous medium, the similar scaling phenomenon (joints parallel to the surface) can be understood statically. The joints are caused by stress relief, and they are the result of states of deformation running discontinuously in a rhythm of space and time in the interior of the rock mass (Fig. 2).

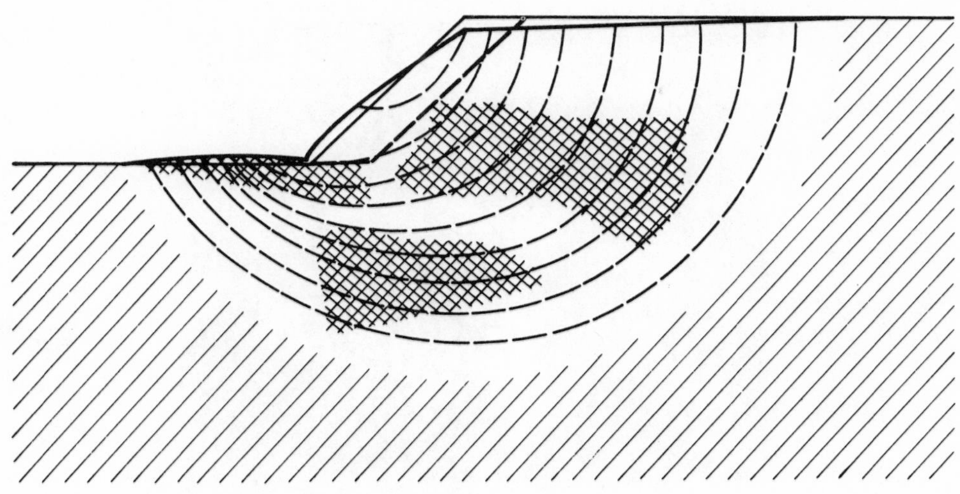

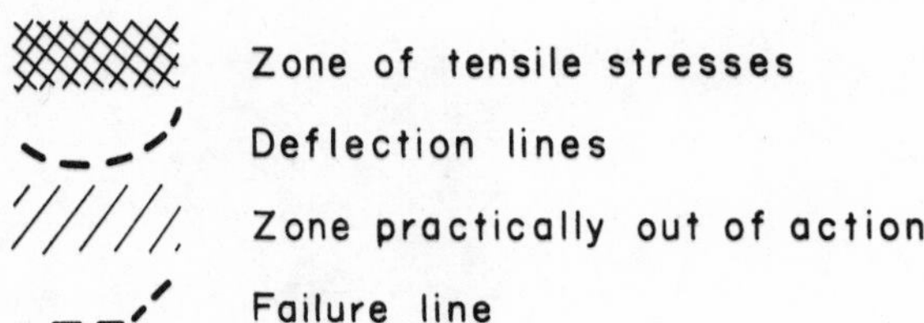

Fig. 1—Zone of most important deformations around a steep rock slope.

Kieslinger[13] described this phenomenon as an effect of residual stresses and loosening stresses at the 9th Colloquium of Geomechanics in Salzburg; he has shown in a convincing number of examples that it takes place in all kinds of material, even in sand, and in all mountains of the world. At a Spanish dam site the author has measured such a scaling even below the bottom of the valley (Fig. 3); it corresponded well to the above mentioned model tests.

At the Vajont Dam (which now is named "Diga Carlo Semenza") this phenomenon of the joints being parallel to the slopes caused the disconnection of the rock mass into large, almost vertical rock tables (Fig. 2) having an average thickness of 30 ft. This regular disconnection of the rock mass rendered very difficult the spreading, and thus the derivation, of the induced dam forces.

Because of a geostatical calculation the author therefore recommended reinforcement of larger rock zones (Fig. 4) with the purpose of

Fig. 2—Major joints parallel to surface in the abutment of Vajont Dam, Italy.

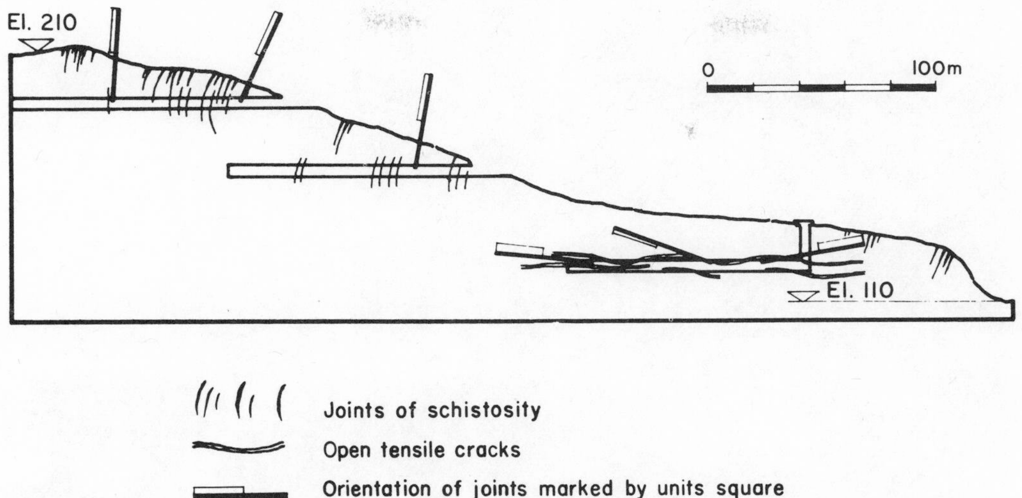

Fig. 3—Valley creep in the Tajo Valley, Spain.

preventing the large lateral expansions to be expected, and of increasing the effective resistance against the shearing away on the zones of the major joints.

Engineering Properties of the Rock Mass

The scaling also supports the valley creep. It contributes to the erosion of the slope because it makes the rock mass loose and mobile. Whereas mass creeping usually appears only in intensely and regularly jointed rock mass, however, slope slides occur chiefly in banked rock mass, and scaling for the most part in massive rock mass. Nature always has several ways to reach the same aim.[14-16] Which way it will take depends on the engineering properties of the material.

By this we mean the properties of the material rock mass, not of the rock substance; the properties of these two materials differ considerably. While the strength of rock is a strength of the substance, the strength of rock mass is predominantly bonding strength, characterized by the planes of fabric and only to a small extent by the properties of the substance.

Technological tests in the jointed granite under a Japanese dam, performed on a really large scale probably for the first time (by the

Fig. 4—Rock reinforcement by means of tie rods (100-ton capacity) (system **PZ**) in the right abutment of Vajont Dam, Italy.

International Research Institute for Rock, Salzburg), confirmed this considerable difference between the properties of rock and rock mass. So far this difference could only be inferred from observations of nature. Only since these tests we really know how great the difference can be; it can amount up to 1:300. Hall stated at the last International Meeting of the International Bureau for Rock Mechanics at Leipzig that the shearing resistance of rock mass of a granite of 35,000 psi pressure strength as a prism amounted in a certain case to only 85 psi.

The tests conducted in Japan (Fig. 5) showed above all that the Poisson's ratio μ of the jointed rock mass, which is already a fractured material, can be much higher than 0.5, and that the influence of the confining pressure (Fig. 6) on the compressive strength is much higher in jointed rock masses than in rock tests performed in the laboratory. The efficiency of prestressed reinforcements, which often is surprisingly high, is due to this as yet not widely known fact.

These few remarks already show quite clearly that rock engineering must be based on the properties of the material rock mass. Likewise we start out from the properties of the building material in the other branches of engineering. This was the way taken by Terzaghi,[17] when he originated soil mechanics and separated it from the older views of foundation engineering which had not regarded sufficiently the properties of the substance soil. The statics of a steel bridge, a reinforced concrete bridge, and a wood bridge are similar; but the technology of steel, concrete, and wood is different. This difference has such an influence on the design that it was necessary to develop consequently the different branches of wood engineering, steel engineering, and concrete engineering, and to train specialists for the different branches.

The properties of rock mass also differ considerably from those of soil; consequently a special rock technology would be required and also the development of rock engineering with special methods and purposes. Rock technology and rock engineering have to be based on the mechanics of fractured media, i.e., discontinua, as well as on rheology. There are already onsets in both branches. It would be a

Fig. 5—Test chamber in the granite of Kurobe IV Dam, Japan.

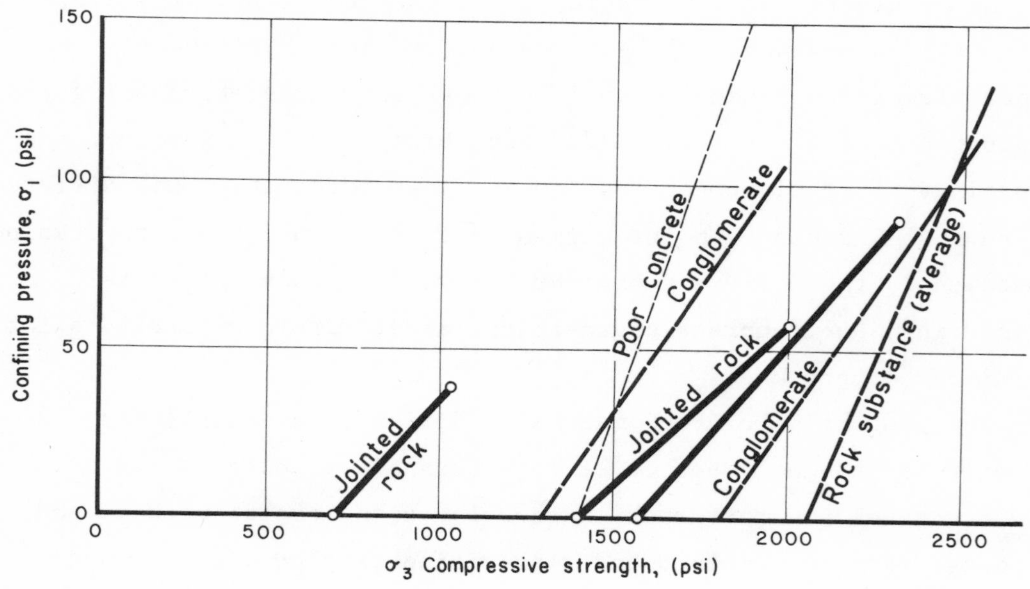

Fig. 6—Influence of confining pressure on strength.

vain attempt to develop the mechanics of the solid earth crust (called geomechanics in Europe) by means of an extrapolation from the mechanics of aggregates (soils).

Petrofabrics

The interior kinematics of the fractured masses, their rheological properties, and their strength properties are determined by plane fabrics and petrofabrics if long-duration tectonic processes, high temperature, or high pressure are effective. If processes of short duration (e.g., seismic loads), normal temperature, and low pressure are the case, however, they are determined practically only by the plane fabrics. This view, advocated by the author for 30 years, was confirmed recently by statistic model tests made by **INTERFELS**. Therefore the science of petrofabrics[18,19] is one of the most significant foundations of rock mechanics. It does not only give us an exact description of all data on geological bodies, it teaches us above all to correlate the mechanic anistropy and the relative order of magnitude (the "Bereich" as Sander meant it) to the physical or kinematical system. It also discloses previous formations and their temporal succession, i.e., the history of the material.

Strength, elasticity, lateral expansion, viscosity, time of relaxation, even permeability, transmission of sounds, etc.--in other words all physical properties and the anistropy of these properties--are thoroughly different in the very same rock, according to the particular size: if we look at a hand piece, a boulder, an abutment, or a range. Thus it depends upon the size of the physical system whether we have to treat a problem of rock engineering rheologically, by means of the mechanics of the continuum, or by means of the mechanics of the discontinuum.

The science of petrofabrics teaches us to distinguish homogeneous zones, i.e., to define areas of different interior configuration.[20] It indicates, for example, for which part of the rock mass a conducted large-scale test is valid.

The science of petrofabrics teaches us also that the axes of the interior kinematics (Fig. 7) in a zone of fabrics frequently show a higher degree of persistence than the individual data on fabrics, joints, planes of strata, etc. Consequently the technically significant fact appears that the mechanical homogeneous zones for certain kinds of strain are often much larger than the homogeneous zones defined morphologically. This fact was also confirmed by the tests conducted in Japan.

Only through the exact methods of the describing science of petrofabrics a number of objections can be refuted that are made again and again against large-scale tests *in situ* and against the methods of analysis applied in the mechanics of discontinuous media. Terzaghi[21,22] objects, for instance, that all engineering calculations may be overthrown if, in spite of an ever so thorough analysis of petrofabrics, one single joint did not appear in the joint statistics; which of course may easily happen.

Against this objection we can quite rightly argue that it is indeed impossible to measure all joints, but that nevertheless an exact survey, conducted according to the science of petrofabrics, guarantees admissible statistics. These statistics indicate sufficiently the distribution of the directions and the zone of their spreading. It becomes also clear which orientations might be taken by joints not observed and in which orientations such joints are not likely to exist. The probability that no hidden joints exist naturally is considerably lower in strictly regular plane fabrics (Figs. 8a and 8c) than in an irregularly jointed rock mass (Figs. 8b and 8d).

This can be judged rightly not by theoretical reflections but only by practical experience. Only he who has measured thousands of joints himself with the compass and who has statistically evaluated the results himself knows that good joint statistics make it possible to state exactly the degree of the probability of individual exceptional values (Fig. 9) being beyond the maxima of accumulation. It is a prior condition, however, that minor joints, major joints, and giant joints be treated separately in the survey as well as in the geomechanical evaluation.

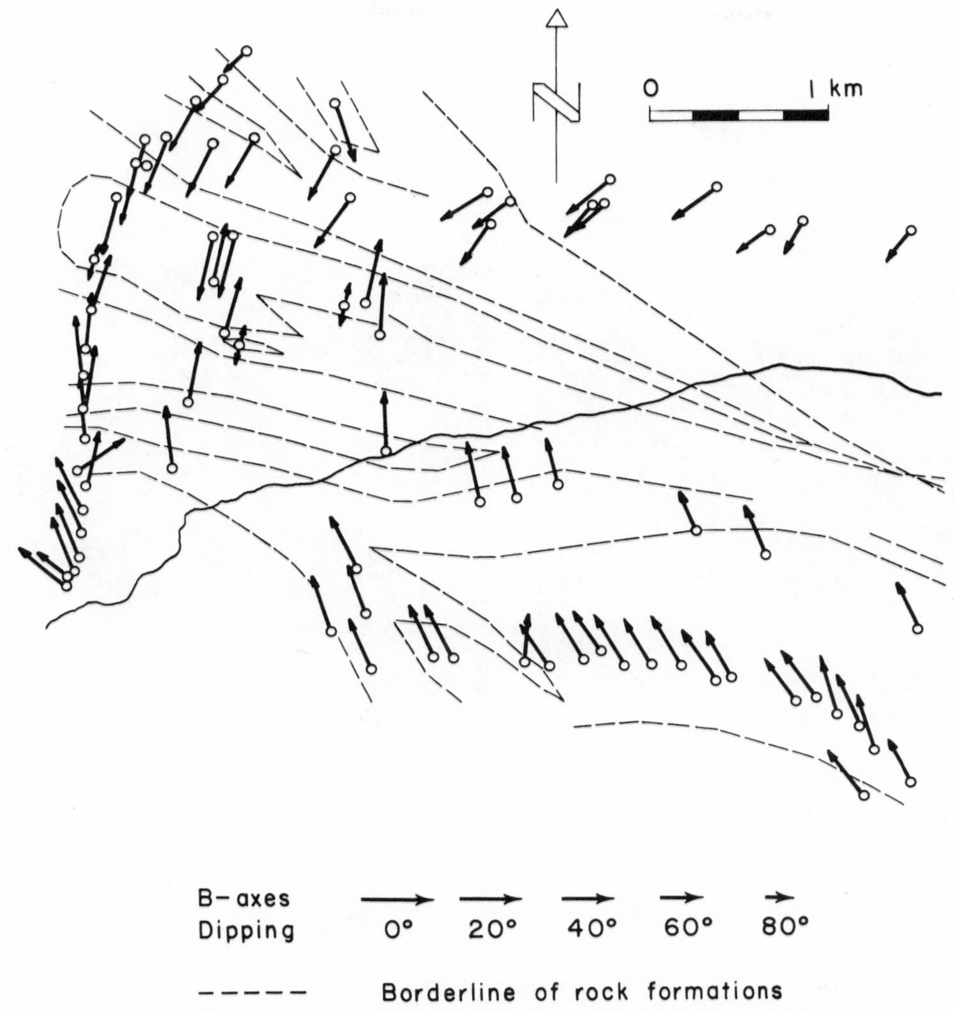

Fig. 7—B-axes in the area of a syncline (after Beck-Managette).

In designing rock slopes or in judging rock abutments of a concrete dam our calculations are based not only on the rock fabrics observed in the outcrop. We also investigate the interior of the cosupporting zone of fabrics by means of the optical borehole sounding[23,24] and by a sufficient number of exploratory tunnels. Above all, we put into the calculation not the average values of the statistics but the most unfavorable orientations of the statistical spreading so we can be sure that, according to the strict law of accident, even more unfavorable orientations of the joints may not occur.

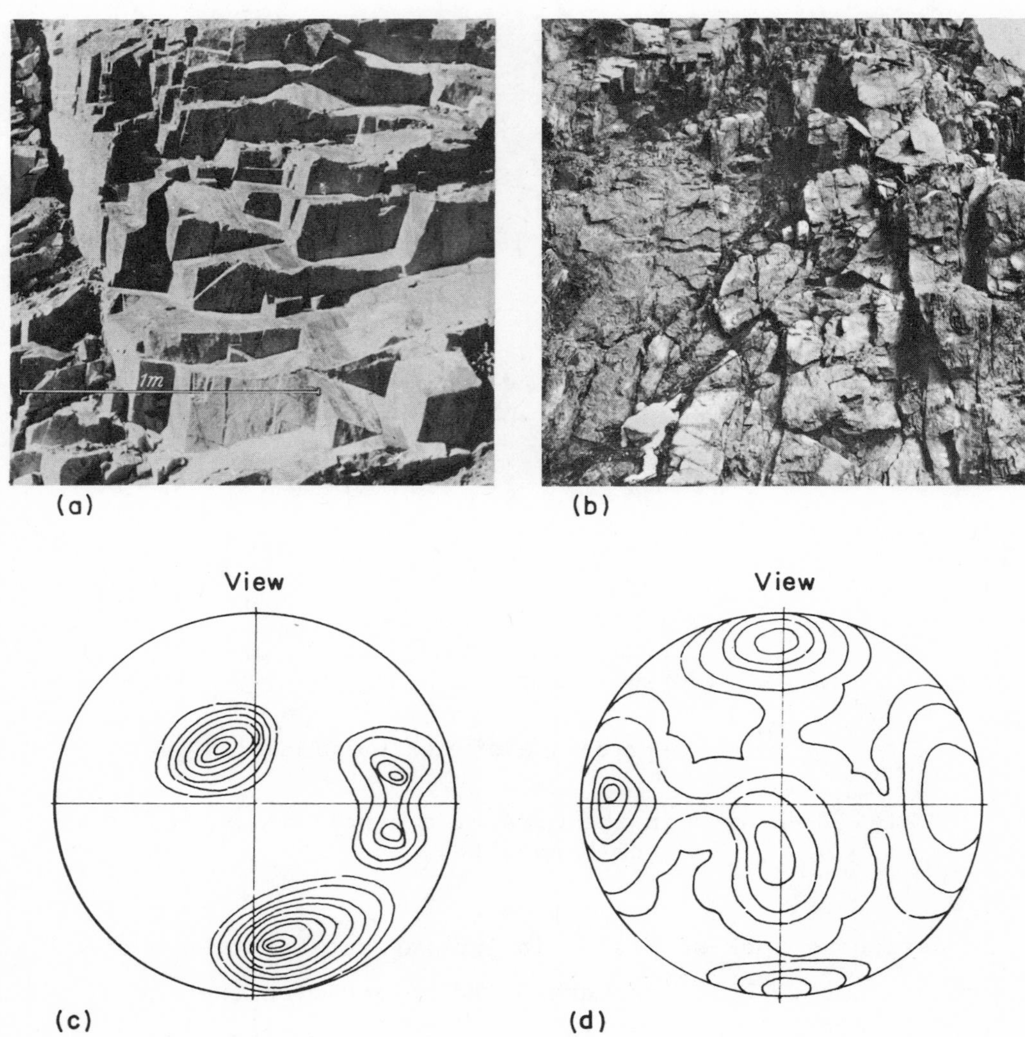

Fig. 8—Example of a highly regulated (a) and a slightly regulated (b) rock mass, with referring pole diagrams.

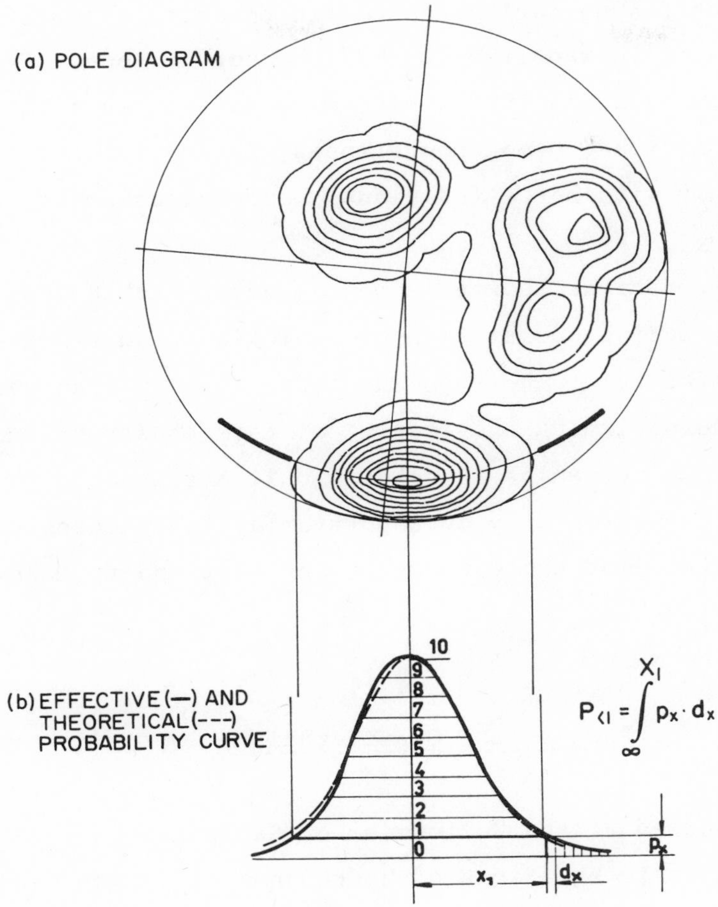

Fig. 9—Probability of a gradually changing orientation in joint statistic diagrams.

It is a matter of experience that such exceptional values occur only very rarely. In fact, the theory of probability argues that they do not occur. This discrepancy is explainable if we realize that each single plane of fabrics was formed according to the statistically valid laws of the formation of rupture planes. Every expert in tectonics is well aware, for example, that in the area of a wide folding or of a shield the basic law of the stress distribution (and therefore of ruptures) can not change irregularly on any spot, but that it can only have local deviations on inhomogeneous zones.

Thus another objection can be refuted which was frequently made and discussed, recently expressed by Terzaghi[21,25,26]: It was objected that fabrics measurings conducted on certain spots could not be extrapolized and intrapolized beyond the survey zone. Experience contradicts this objection; the observer, however, has to define precisely the individual homogeneous zones; he can superpose his diagrams statistically only within the individual homogeneous zones, not in the total survey zone.

These considerations show that we can never tackle rock mechanical problems without active cooperation of geologists and experts in the science of petrofabrics. Thorough knowledge in both branches is necessary for the designing and judging of rock constructions. That is the reason why rock mechanics is so difficult a science. No expert should deliver an opinion only on laboratorial and theoretical grounds; it is necessary to spend whole weeks at the site and to measure and judge repeatedly.

Rheological Considerations

Torre(27) showed at the Colloquium of Geomechanics in Salzburg in 1952 that the basic equations of hydrodynamics, which usually are applied only at high speeds and low viscosity of the flowing media, are likewise valid for high viscosity and accordingly low speeds. He estimated the coefficient of viscosity of jointed rock mass as $\mu = 10^{10}$ to 10^{12} kg sec/cm^2. If we know this coefficient and the flow net we can, according to his "Hydrodynamic Theory of Firm Substances," calculate in the case of creeping slopes the stresses in a flowing medium. On account of the normal stress-velocity relation of hydrodynamics the stresses in any point of the flowing medium can be calculated.

For an underground hall in Austria (Fig. 10) Torre calculated in 1949 the flow pressure of the rock mass against the cavity; he received suitable values. He had defined the period after which a steady state will occur at about 19 years. The size of the thus calculated flow pressure corresponded well to large-scale tests in lined galleries, conducted in the last years by INTERFELS and evaluated by Rabcewicz.(28,29) The period calculated for the decrease of the flow is in good accord

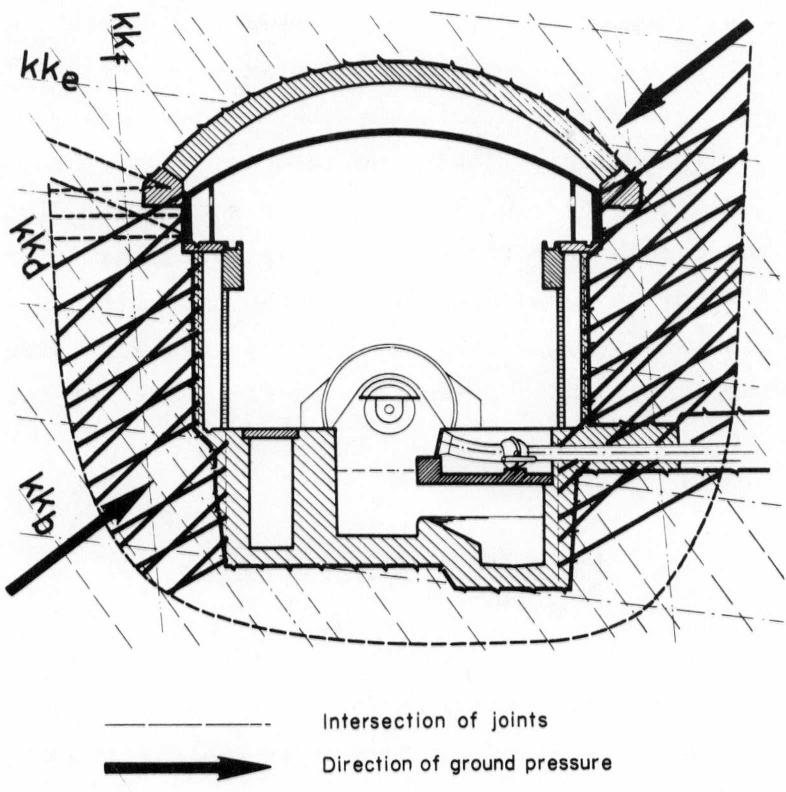

Fig. 10—Section through the underground power station at Braz, Austria.

with practical experiences. The example of this hall is very instructive: because of an analysis of petrofabrics, an inclined pressure gradient of rock mass (Fig. 10) had to be considered, which led to a steel reinforcement of underground excavations in 1949--the first one, to our knowledge.

It has already been stated that the rheological method can be used only in areas that are very large in proportion to the block units. It can easily be understood that the calculation of a geological body consisting of merely some hundred interjoint block units cannot be based on the physical conception of a viscous liquid. Such a conception can only be applied if the geological body consists of hundreds of thousands or millions of such block units. If that is the case, it does not matter whether the block units are completely separate as in dry masonry or whether they maintain a residual bonding where both friction on the joints and the material bridges support a connection.

Even then we have to consider that a viscous medium of this kind is anisotropic to a great extent. Hydrodynamical treatment cannot yet be applied for anisotropic flowing media. It is possible, however, to use statical equations in the case of anisotropic media; on the grounds of these equations the results of a hydrodynamic calculation can be modified or at least discussed if the circumstances are favorable. There is also the possibility of treatment according to the theory of elasticity, which extends to simple states of anisotropy, e.g., orthotropy.

Mechanics of the Continuum

In recent years Russian scientists (Panow and Ruppeneit,[30] Awerschin,[31,32] and Protodjakonow[33]) have dealt with the problem of whether and to what extent the theory of elasticity and the theory of plasticity can be applied in the solution of rock mechanical problems. Unfortunately their research did not consider that the particular size is a decisive factor for this application, and the strength of the substance, the degree of jointing, and the state of stress are also factors of great significance. In small orders of magnitude the mechanics of continuum can be applied only in a limited number of cases: either if the strength of substance is very low (clay rocks, weak marl, etc.) or if the disconnection of the rock is unimportant (quasimonoliths), or if the pressures are so high that the joints remain tightly closed and the friction on them reaches the size of the shearing strength. In all other cases the mechanics of continuum can be applied only in areas of large size.

Practice has shown that treatment according to the mechanics of the continuum may well be the basis for the investigation of stress distributions, if the results of this investigation are modified according to the anisotropy of rock mass. For the solution of strength problems the mechanics of the continuum are less suited.

Model Tests

For this reason rock mechanics makes use of model tests and also photoelastic tests. Model tests were used by Takano, for example (see Ref.(11)), when he studied the influence exerted by forces of earthquakes on slopes; in these tests he could, however, reproduce statically only the forces of earthquakes according to the principle of D'Alembert; thus the impacts were not regarded.

The model tests conducted by INTERFELS with the purpose of examining the states of stress in steep rock slopes were very instructive.[11,12] They explained the reason why high rock slopes very rarely are steeper than 45° to 55°, even in firm rocks. At present these tests are extended in order to examine the influence of residual crust stresses. They showed, at least qualitatively, that (a) the shearing stresses behind the slope plane are essentially decreased by horizontal crust stresses; (b) at the toe, however, a considerable concentration of stress and therefore a considerable increase of the shear stresses occurs; (c) the zone of extenuations in the middle third of the slope remains; and (d) it is joined by another zone of extenuations under the toe of the slope. (See Fig. 11.)

Model tests have a great number of advantages: They can reproduce complicated limit conditions, they make it possible to reproduce statistically discontinuous media, and they do not require similarly simplifying conditions as are required by most calculations. In model tests the conditions of compatibility are no problem, even in discontinuous media. Model tests conducted by Ismes and INTERFELS (Fig. 12) showed the enormous influence that the disconnection caused by the joints exerts on strength and safety of slopes and dam foundations. Today we are able to calculate, although roughly, the influence of the plane fabrics, which had played a decisive part in the catastrophe at Malpasset. Today we can pass more precise judgments concerning the safety of rock constructions than has so far been possible using only the methods of engineering geology. We can say confidently that the foundation technique, on which the life of men often depends, does not lag in this respect (see Ref. 25).

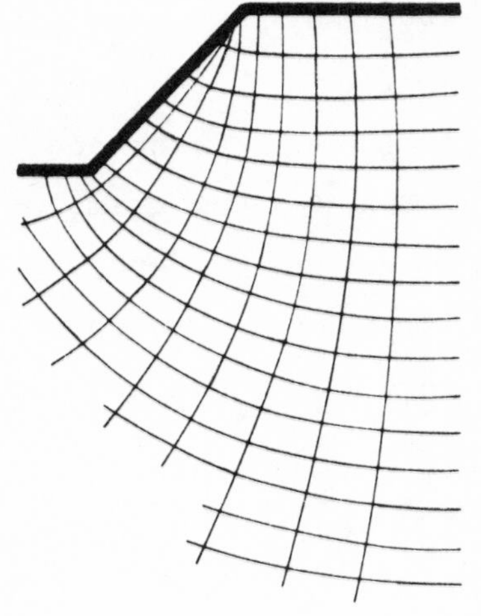
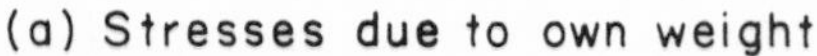

(a) Stresses due to own weight

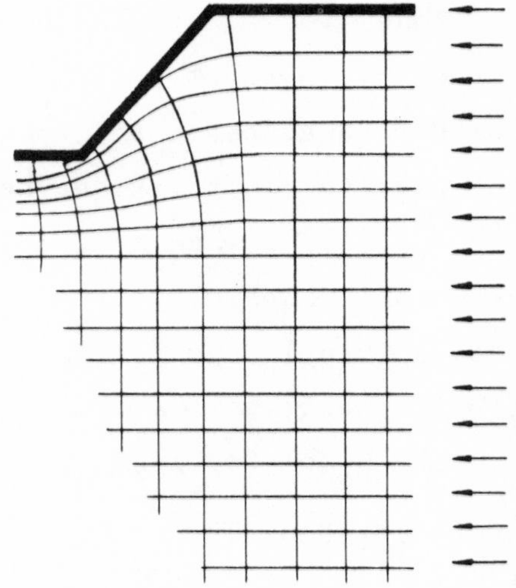

(b) Stresses due to horizontal residual stresses

Fig. 11—Trajectories of stresses in a steep rock slope.

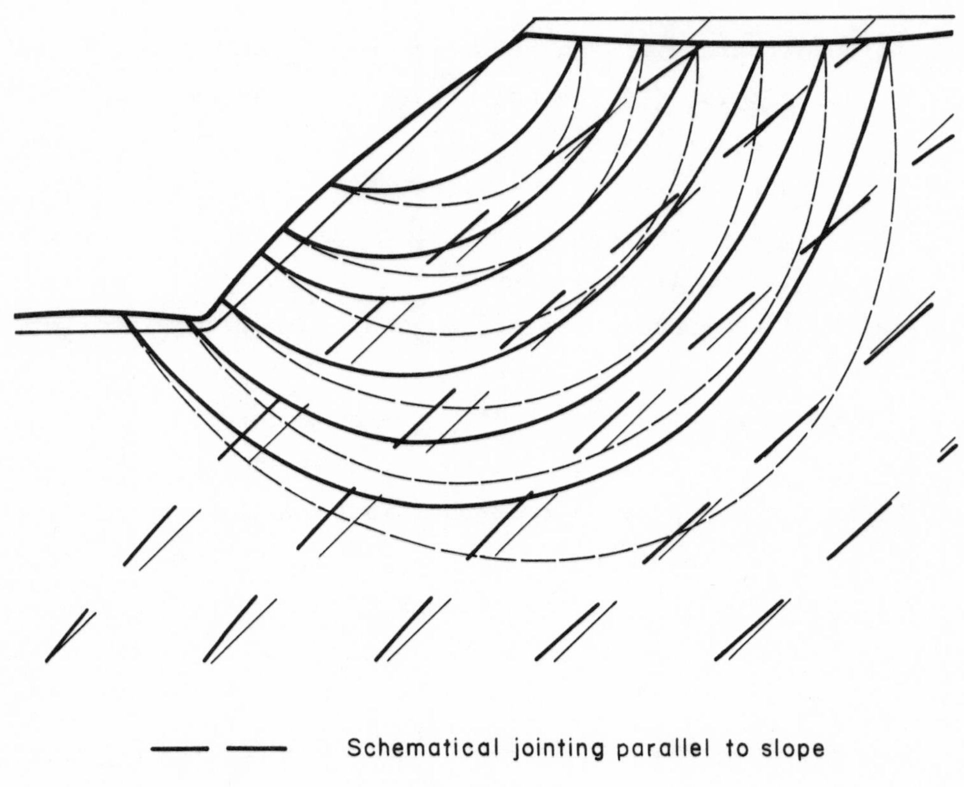

Fig. 12—Model test of a jointed rock slope charged by own weight.

Thus, for every proposed design, the tests concerning the influence of tectonic stresses on the stability of slopes and dam foundations should be conducted.

Joint-water Pressure and Buoyancy Effect

From the branch of soil mechanics, i.e., the mechanics of grain aggregates, rock mechanics can take over certain experiences as to buoyancy and neutral stress. This can not be done directly, however, for in jointed rock the effects of void water are joined by the effects of the free joint water, which are of a different kind.

The void water in the solid rock substance and the void water in the fillings of the joints behave similarly to the void water in concrete or in soils. The free joint water, however, causes directed pressures, not only if it seeps through the rock, but also if it is backed up by faults and loam-fill joints, which happens quite frequently. Therefore we regard the joint-water pressure in rock mechanics as predominantly a static force.

The buoyancy effect plays a part similar to that in soil mechanics. It can, however, be diminished according to the degree of separation of the joints and to the orientation of the sets of joints. Its direction can also deviate from the vertical.

At this point we have to mention an objection recently made by Terzaghi.[26] Terzaghi remarks that the definition of the permeability value k in jointed rock is problematic because of the questionable applicability of Darcy's law. He concludes that the effect of the stream flow pressure in rock can not be calculated. In rock mechanics however, e.g., at dam foundations, we do not calculate the (dynamic) flow pressure, but the (static) joint-water pressure; moreover, we do not calculate it on account of the k values and a seepage flow according to these values, but on account of the actual course that the ground water level takes in the mountain. This course can be measured by several methods. In his lectures in Vienna, Terzaghi was the first to advance an example for this view. Already in 1923 an Austrian physicist and philosopher had remarked after the collapse of an

Italian dam that such catastrophes can be caused by buoyancy and joint-water pressure in the rock.

Petzny(34) and Stini(35) published the circumstances of three hydraulic rock ground ruptures which occurred in 1954. An analysis showed that the cause of the ground ruptures was not the pressure exerted by the seepage flow, but the static difference of the pressures on the front side and the back side of a major joint backing up water. Although the strata were declined toward the mountain (Fig. 13) and the rock was completely sound (gneiss), the joint-water pressure was able to push upward on the strata planes rock masses 100,000 to 1,000,000 cu ft in size, and to cause a collapse of the slope.

Pressure impacts on the ground water have a considerable effect on the rock mass. Not only impacts on the void water, but also impacts

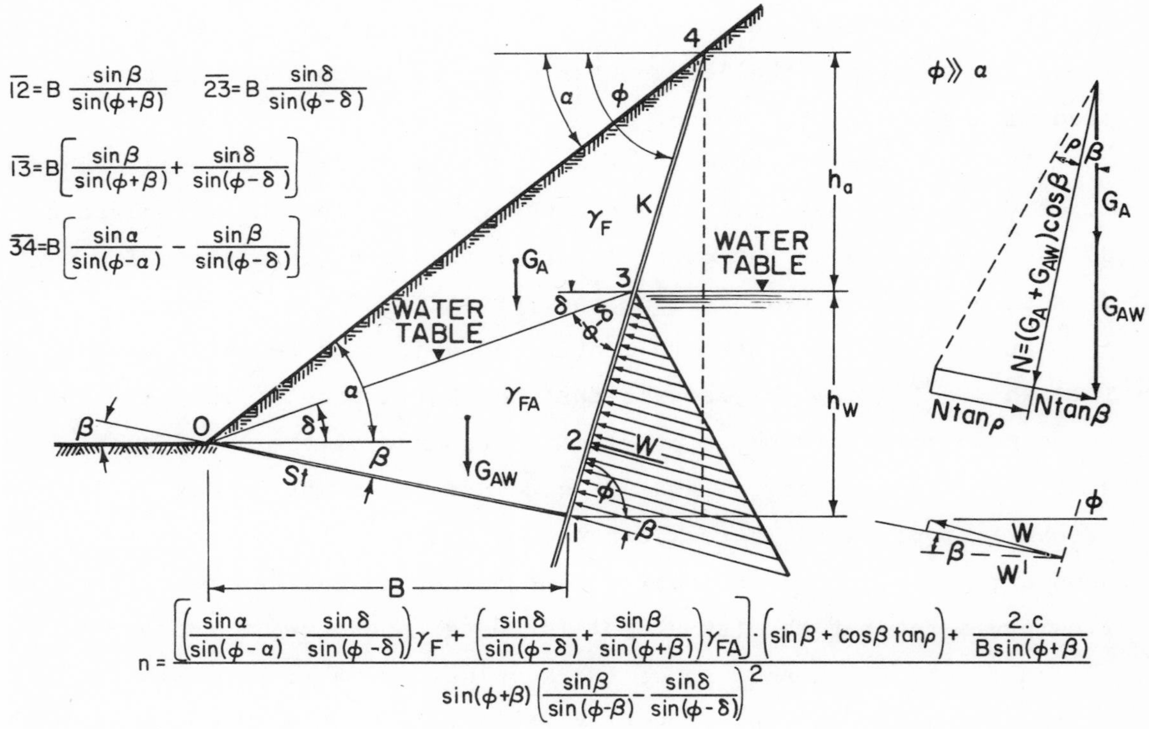

Fig. 13—Analysis of a simple ground failure caused by hydrostatic joint-water pressure beyond a major joint.

on the free joint water, were efficient, especially in minor joints. Therefore the loosening effect of explosive blasts frequently extends far deeper into the mountain than is generally expected, according to certain tests.

Slope Analysis

Detailed examples of the analysis of slopes were already given on another occasion (see Ref. 36). Thus only some test results of general interest are mentioned here:

1. The safety of rock slopes depends in a higher degree on the planes of fabrics, i.e., on the bond of the interjoint units, than on the substance strength of the material. Laboratory tests alone accordingly can not give sufficient information about the safety of rock slopes.

2. In the very same rock mass, slopes that are inclined equally differ in safety if the directions are different.

3. The natural slope angle of jointed rock mass, i.e., the angle of a slope with safety 1.0, can be smaller than the slope angle of the same rock material if it is blasted and heaped in a pile.

4. The safety of natural rock slopes frequently is not much higher than 1.0.

5. Rupture of high slopes is not always caused by an overstepped shearing strength at the toe of the slope; frequently extenuations in the interior of the slope body, which occur very often, are the cause.

Three phases at a slope rupture can be distinguished: (a) the phase of the first tensile cracks behind the slope head; (b) the phase of the extenuations in the interior of the slope body, in most cases accompanied by a bulge of the slope surface; and (c) the collapse of the slope toe.

6. The strength of the zone before the slope also has an influence on the safety of the slope.

7. Residual stresses exert great influence on the stability of slopes if the main direction of these stresses is chiefly in the crosscut plane.

8. Percussions diminish the stability of slopes. Earthquakes frequently split off advancing piers and ridges.

9. The influence of the mountain water as well as elevation and inclination of its level are important. Fluctuations of the mountain water level can be very dangerous.

10. The safety of the rock slopes is a function of time.

11. The safety of slopes can be increased by artificial means, e.g., by change of form, support of toe, anchoring, and tying, but not by bolting or grouting.

12. Berms generally do not increase the safety of rock slopes; occasionally they even decrease it.

13. Rock slopes with a concave toe line show a higher stability than those with a convex toe line.

It is generally known that rock mass is open to the influence of tensile stresses. The Japanese large-scale tests showed, however, that the lateral expansion also decreases the bond of the rock mass considerably and that it can become as dangerous as tensile stresses. Pacher (see Ref. 11) therefore proposed to include lateral expansions in the calculation with the value $E.\epsilon$; this value has the dimension of a tensile stress without necessarily being a tensile stress.

Several technological experiences made only recently seem to lead to a new rupture hypothesis for jointed media; it is still too early, however, to reveal details.

REFERENCES

1. Watznauer, Adolf, "Über die Möglichkeit des Auftretens einer tektonischen Komponente im Gebirgsdruck," Intern. Gebirgsdrucktagung Leipzig, Akademie-Verlag, Berlin, 1958.
2. Semenza, Carlo, "Einige praktische Überlegungen zum Problem der Gründung von Staumauern und Staudämmen," Geol. Bau., Jg. 24, H. 2, 1958.
3. Semenza, Carlo, "Die Staumauer Vajont und die Entwicklung der Wasserkraftanlagen der SADE im letzten Jahrzehnt," Schweizer Bauzeitung, H. 27, 1960.

4. Stini, Josef, "Unsere Täler wachsen zu," Geol. Bau., Jg. 13, H. 3, 1941.

5. Stini, Josef, "Nochmals der Talzuschub," Geol. Bau., Jg. 14, H. 1, 1942.

6. Stini, Josef, "Talzuschub und Bauwesen," Die Bautechnik, 1942.

7. Stini, Josef, "Neuere Ansichten über Bodenbewegungen und ihre Beherrschung durch den Ingenieur," Geol. Bau., Jg. 19, H. 1, 1952.

8. Stini, Josef, "Talzuschub und Wildbachverbauung," Geol. Bau., Jg. 19, H. 2, 1952.

9. Stini, Josef, "Ein 'Talzuschub' im Burgenlande," Geol. Bau., Jg. 19, H. 2, 1952.

10. Stini, Josef, "Der Gebirgsdruck und seine Berechnung," Geol. Bau., Jg. 19, H. 3, 1952.

11. Müller, Leopold, "Die Standfestigkeit von Felsböschungen als spezifisch geomechanische Aufgabe," Rock Mechanics and Engrg. Geol., Vol. I, No. 1, 1963.

12. Scheiblauer, Johann, "Modellversuche zur Klärung des Spannungszustandes in steilen Böschungen," Rock Mechanics and Engrg. Geol., Vol. I, No. 1, 1963.

13. Kieslinger, Alois, "Restspannung und Entspannung im Gestein," Geol. Bau., Jg. 24, H. 2, 1958.

14. Cloos, Hans, "Brüche und Falten," Naturwissenschaften, Jg. 19, H. 11, 1931.

15. Cloos, Hans, "Über Biegung und selektive Zerlegung," Geol. Rundschau, Bd. 24, H. 1/2, 1933.

16. Cloos, Hans, "Gang und Gehwerk einer Falte," Z. Deut. Geol. Ges., Bd. 100, 1948.

17. Terzaghi, Karl, Erdbaumechanik auf bodenphysikalischer Grundlage, Franz Deuticke, Leipzig, 1925.

18. Sander, Bruno, "Festigkeit und Gefügeregel am Beispiel eines Marmors," Neues Jahrb. Mineral., Beilageb. LIX, Ab.A. S. 1-26, 1929.

19. Sander, Bruno, "Gefügekunde und Baugeologie," Festschr. Geol. Ges., Bd. 48, Wien, 1955.

20. Sander, Bruno, Einführung in die Gefügekunde der geologischen Körper, 1, Springer, Wien, 1948.

21. Terzaghi, Karl, "Engineering Geology on the Job and in the Classroom," J. Boston Soc. Civil Engrs., April, 1961.

22. Terzaghi, Karl, "Past and Future of Applied Soil Mechanics," J. Boston Soc. Civil Engrs., April, 1961.

23. Burwell, E. B., and R. H. Nesbitt, "The 'NX' Bore-Hole Camera," U.S. Department of the Army, Corps of Engineers, January, 1954.

24. Müller, Leopold, "Geologische Erkundungsmethoden beim Bau des Pumpspeicherwerkes Tanzmühle," Kraftswerkgruppe Reisach-Rabenleite, München, 1961.

25. Terzaghi, Karl, "Does Foundation Technology Really Lag?" Engineering News-Record, February, 1962.

26. Terzaghi, Karl, Discussion of the paper by K.W. John, "An Approach to Rock Mechanics," Proc. Am. Soc. Civil Engrs., Vol. 88, S.M. 4, August, 1962.

27. Torre, Cosimo, "Hydrodynamische Theorie fester Stoffe," Oesterr. Ing.-Arch., Bd. 6, H. 5, 1952.

28. Rabcewicz, Ladislaus, "Aus der Praxis des Tunnelbaues," Geol. Bau., Jg. 27, H. 3-4, 1962.

29. Rabcewicz, Ladislaus, "Bemessung von Hohlraumbauten," Rock Mechanics and Engrg. Geol., Vol. I, No. 1, 1963.

30. Panow, A. D., and K. W. Ruppeneit, "Fragen der Gebirgsdruckforschung," Intern. Gebirgsdrucktagung Leipzig, Akademie-Verlag, Berlin, 1958.

31. Awerschin, S. G., "Erfahrungen aus der Gebirgsdruckforschung," Intern. Gebirgsdrucktagung Leipzig, Akademie-Verlag, Berlin, 1958.

32. Awerschin, S. G., Diskussion, Beitr. z. 4. Ländertreffen d. Intern. Büros f. Gebirgsmechanik, Leipzig, 1962.

33. Protodjakonow, M., Diskussion, Beitr. z. 4. Ländertreffen d. Intern. Büros f. Gebirgsmechanik, Leipzig, 1962.

34. Stini, Josef, and Hans Petzny, "Wassersprengung und Sprengwasser," Geol. Bau., Jg. 22, H. 2, 1956.

35. Stini, Josef, "Felsgrundbrüche im Baugelände von Wasserkraftanlagen," Geol. Bau., Jg. 22, H. 3-4, 1956.

36. Müller, Leopold, Der Felsbau, Ferd. Enke, Stuttgart, 1963.

DISCUSSION

R. T. LAIRD (USA):

Can the impact of ground water pressures be measured by field techniques before and after excavation? What is the influence of these pressures and changes in pressures on the distribution of stresses?

L. MÜLLER in reply:

Theoretically we have two kinds of influences to distinguish: static and dynamic effects in the ground water.

The ground water itself appears in threefold nature: as free joint-water, as pore water in the joint fillings, and as pore water in the rock substance. The latter is less important in this connection.

The dynamic influence--we call it "dynamic" in order to simplify the matter--is in the pore water of the joint fillings, as is known in soils. A sudden change in stress can change the internal equilibrium of the joint fillings in such a way that there will be a zero friction and a danger of sliding along a joint filled by clay or silt. This same danger exists if there is no filling material other than water in the joint.

Practical tests are not available at this time, only theoretical investigations based on assumptions.

The static influence of ground water is easily calculated. It can be seen in Fig. 14.

If there is a stress state σ_1 and σ_3 in the rock mass and the joints are filled with water at pressure p, this water pressure will attempt to press against both sides of a joint and to press it apart. Thus, we could say that a quasitensile stress is superimposed on σ_1 and σ_3. That is, both stresses are decreased by the value of the joint-water pressure, p. This p not only reduces the friction on the joint walls, but it also shifts the Mohr's circle much closer to the Mohr's envelope and thus into a more dangerous position than before. At the same time, occasionally the Mohr's envelope will change from H_t, which is true for dry rock, to a lower position H_n in the case of saturated rock.

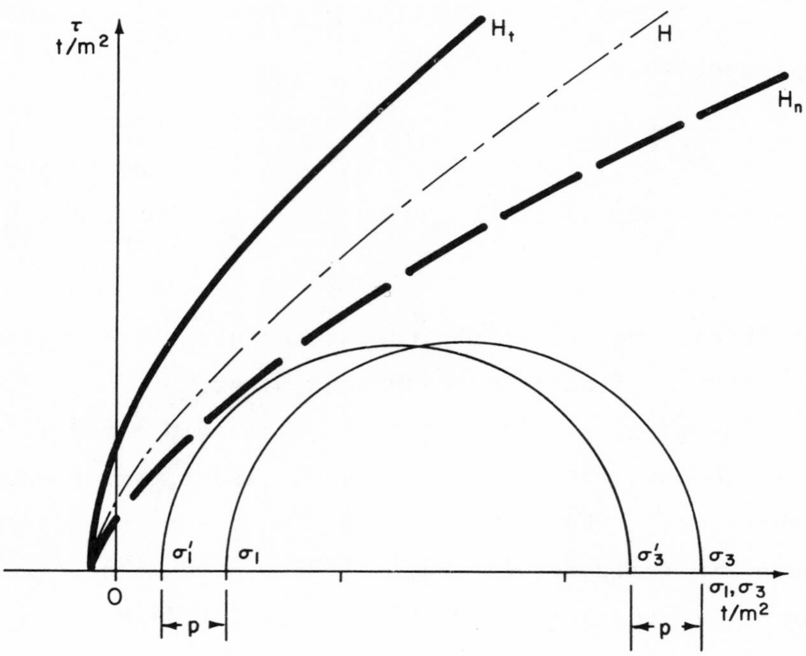

Fig. 14—Mohr's circle at superimposed joint-water pressure.

Measurements of the ground water pressure have been conducted and show good coincidence with the theoretical assumptions, but we still need more investigation to prove the danger of failure. The observations of ground water pressure normally are carried out either by means of piezometer boreholes or by means of pressure gauges. The measurement of pore pressure in the joint fillings is still a problem.

S. G. A. BERGMAN (Sweden):

Dr. Müller has made a series of conclusions in his paper. Conlusion 11 reads: "The safety of slopes can be increased by artificial means, for example, change of form, support of the anchoring, tying, but not by bolting or grouting."

I feel like challenging this conclusion, especially as regards grouting. I think that a lot of practical experience can be mobilized about the stabilizing effect of competent grouting by increasing friction and bondage in faults and joints, at least in hard rocks of granite and gneiss type. I would also refer to Dr. Serafim's tests, which show a very much higher deformation resistance for grouted rock than for natural ungrouted rock.

L. MÜLLER in reply:

Stabilizing the rock masses by grouting is possible in all cases where we can bring about a cementation of joints, particularly when the joints are not filled by clay. Bonding with cement, however, requires a fairly high grout pressure, which in turn necessitates a cover of rock or concrete. For example, we know that in dam construction we must first load the contact area by pouring a section of the dam to provide a counterweight against the grouting pressure.

Without such a concrete surcharge we can cement grout zones which lie 4, 5, or 6 m below the surface because at these depths we have sufficient weight of overburden. But if we try to stabilize the upper 2 m by grouting, we only loosen the rock more because the grout simply produces a hydraulic pressure in the joints. Unfortunately it is exactly this outer 2-m zone which is so decisive for the stability of the slope. Therefore, in order to bring about the stabilization by grouting we would have to add external confinement pressure. For example, we would have to bolt it in depth before grouting.

T. LANG (USA):

With respect to Dr. Müller's large-scale <u>in situ</u> triaxial test in Japan, how did he take into account the relaxation of the rock block which took place while he was preparing for the tests and before he actually applied his stress difference?

L. MÜLLER in reply:

We have considered these problems and have come to the conclusion that it was impossible to measure relaxation in these initial tests. We were optimistic in believing that there would not be too much relaxation effect before applying the stresses. We hoped that because of the size of the specimen there would be relaxation only in the outer envelope. However, we found that the rock reaction was greatly dependent on time, particularly when the stress was cycled.

W. WITTKE (Germany):

I should like to discuss the following question concerning the contribution of Dr. Müller.

I have thought about developing a method of calculating the stability of rock slopes by means of vector analysis. I shall briefly outline the simplifying assumptions which are necessary with regard to the rock.

The rock is jointed by several plane joints as, for instance, strata, planes of schistosity, fissures, etc. These joints cross throughout the whole rock mass and divide it into separate blocks so that relative movements of the blocks can only be prevented by frictional forces. Additional cohesion within the joints represents an additional amount in the factor of safety.

The strength of the rock mass has to be high in comparison to the shear strength within the joints so that in the calculations the rock can be considered rigid.

There are no crust stresses in the rock mass and no water pressures within the joints which affect the stability. Thus gravity is the only force which acts on the rock blocks. Possibly horizontal earthquake accelerations also can be taken into account.

Dr. Müller has indicated similar considerations in his book <u>Der Felsbau</u>. Now, I wonder whether cases to which these assumptions apply occur in practice and whether there is a use and need to try developing methods of calculating the stability of slopes in this way.

L. MÜLLER in reply:

I believe that it is indeed interesting to follow this question. A similar proposal was once made at a Geomechanic Colloquium in Salzburg by Dr. Tremmel; however, no one has taken the trouble so far to make a vector analysis, and it will be very good indeed if you would try it yourself.

The three assumptions put forth by you do not always exist, although they may in many cases. Your first assumption applies only if the joints are continuous--which we call a high degree of separations. Your second premise of describing the shear strength along joints in terms of cohesion and internal friction is indeed relevant. This was confirmed by model tests on a large scale as reported by Prof. Krsmanovic at the Rock Mechanics Congress in Belgrade, held at Easter in 1963. Your third premise, that only gravity is effective and there is no crust stress, must be proved in each individual case, either on a geological basis or by stress measurements after Potts, Hast, Talobre, or others.

Your additional question whether the shear strength along joints can be defined by friction only without cohesion, can be answered as follows: It is possible whenever the joint filler is sandy and not clayey.

J. A. TALOBRE (France):*

I think that every one among us may agree with Dr. Müller about most of the considerations that he developed concerning rock mechanics. We are familiar with these considerations and we know that they are quite general. The influence of the flaws in rocks, of confining stresses, and of pore pressure has been for decades successfully studied in the United States. Some very good papers concerning part of the above questions have been presented at this conference. In this country as elsewhere many valuable investigations leading to decisive conclusions have been conducted.

The difficulties encountered in applied geology do not come from the lack of geological theories. Similarly the difficulties presently encountered in rock mechanics do not come from theoretical shortcomings. The question is no longer of knowing what to do, but how to do it in the best way.

Dr. Müller emphasized the usefulness of petrofabrics graphs and of diagrams of joints. These graphs are now commonly employed everywhere. I also made use of them. I will not explain again (this has been done in my book on rock mechanics) why the number of principal directions of joints in rocks is generally reduced to three, or how it is possible, from diagrams of joints, to figure approximately what the characteristics of the rock were when the jointing took place.

*This discussion was submitted subsequent to the conference.

It is relatively easy to establish diagrams of joints when the joints are plane. It is hard to build such diagrams when the rock is slivered, when the jointing is irregular, and when the cracks, although dangerous,can not be detected visually. I know by personal experience many geological sites in which the use of diagrams of joints could not be established as regular practice. But the main difficulty is not there.

Petrofabrics graphs and diagrams of joints by no means represent the properties of the rock. They are only intermediaries, by the help of which test results can be compared. It is clear that the graphs cannot furnish answers to all questions and that the answers cannot be precise.

The roughness of the joints, the bond at contacts, the intensity of weathering and of minute cracking, and the amount and nature of gouge fillings vary from one joint to another, since their ages and history are widely different. Relations between the graphs and the properties of rocks may be reliable for some rocks, but not for all.

Dr. Müller has mentioned that no diagram of joints was made for the Malpasset dam. It would not have helped much to have done one. The dip and strike of the main joints of the abutment of this dam were very apparent, and the most dangerous joint plane was normal to the dam thrusts. I hope that some day the tests that I conducted on the site after the disaster, for the French Department of Agriculture, will be fully published and discussed. I can say that the safety factor of the foundation of this dam was found to be equal to one, and that the methods of soil mechanics, in this case, proved to be very helpful.

The usefulness of petrofabrics graphs has been questioned. According to the Griffith theory, the strength of a rock depends only upon the length of the cracks which affect it. It is independent of crack frequency and crack orientation. According to the Portuguese engineers, most of the properties of a rock depend upon a unique parameter: the alteration index. The above correlations are extraordinarily simple, and I doubt that they could be fully trustworthy. However, the correlations between the alteration index and the rock properties may be less wide of the mark than most of the qualitative correlations furnished by petrofabrics examinations. If so, on account of their lack of accuracy, petrofabrics graphs may lose much of their interest.

In fact, the value of any of the above correlations is now secondary for the applications of rock mechanics. I indicated in my paper that at the present time, new and more precise methods are available, such as a scientific process of sampling inspection as used in industry. The sampling inspection must be extended to the totality of the rock mass to be surveyed. The investigation must bear on the totality of the useful rock properties. The factors having some influence on the properties of rocks (time, chemical alteration, water content, temperature, etc.) have to be investigated in their totality. This program seems far-reaching; but it can be fulfilled with the help of the efficient, rapid, and economical equipment that I previously mentioned.

Furthermore, as I have pointed out, there are drawbacks to the Russian method of analyzing rock masses; and scale models also have very serious shortcomings. These drawbacks can be overcome by my method of joint analytical techniques. Let us only remember that these most recent enrichments of rock mechanics can contribute with full success to the safety of large dams.

ABSTRACT

The problem of the stability of rock slopes finds its most important applications on dam abutments and in open-pit mines. Conventional approaches to this problem, however, present no assured degree of safety. An additional fact, one which is rarely realized, is that the problem of open-pit mine stability is basically an economic one.

A new, statistical approach to this problem has been presented by the author. This approach permits a complete analysis of the slide risk to be made in advance, provided the necessary data are collected in the field in time. The author has patented new measuring devices, which can be placed at any depth in boreholes, in order to obtain these data.

The conclusion is that the cost of a cubic yard of ore must be based on the cost of a cubic yard of slide rock multiplied by the probability of failure.

RÉSUMÉ

Le problème de la stabilité des talus de roche trouve ses applications les plus importantes dans les butées de barrage et dans les mines à puits ouvert. Cependant, les façons habituelles d'aborder ce problème ne présentent pas un niveau de sécurité assuré. De plus--on ne s'en rend pas souvent compte--le problème de la stabilité d'une mine à puits ouvert est essentiellement un problème économique.

L'auteur a présenté une nouvelle méthode statistique. Cette méthode permet de faire à l'avance une analyse complète du risque de glissement, à condition de recueillir à temps les données nécessaires sur le terrain. L'auteur a breveté de nouveaux appareils de mesure que l'on peut placer à n'importe quelle profondeur dans des trous de sonde, afin d'obtenir ces données.

On en conclut que le prix du mètre cube de minerai doit se fonder sur le prix du mètre cube de roche susceptible de glisser multiplié par la probabilité de rupture.

AUSZUG

Die Frage der Standfestigkeit von Felsböschungen findet seine wichtigsten Anwendungen bei Widerlagern von Talsperren und bei Tagebauten des Bergbaues. Herkömmliche Methoden zur Lösung dieses Problemes ergeben jedoch keine eindeutigen Sicherheitsgrade. Die zusätzliche Tatsache, dass das Standsicherheitsproblem bei Tagebauten grundsätzlich ein wirtschaftliches ist, wird nur selten erkannt.

Eine neue, statistische Annäherung zu dem gegebenen Problem wurde von dem Verfasser bereits gegeben. Diese Methode erlaubt eine vollständige Untersuchung der Rutschgefahr im voraus, wenn die notwendigen Felddaten rechtzeitig ermittelt werden. Der Verfasser hat neue Messinstrumente patentiert, die in beliebiger Tiefe in Bohrlöcher eingebaut werden können, um diese Daten zu bestimmen.

Abschliessend wird dargelegt, dass die Kosten pro Kubikmeter Erz von den Kosten pro Kubikmeter Felsrutsch, mit der Wahrscheinlichkeit eines Rutsches multipliziert, abhängen.

THE PROBLEM OF OPEN-PIT MINE STABILITY

Joseph Antoine Talobre*

THE PROBLEM of the stability of rock slopes finds its most important applications on dam abutments and in open-pit mines. The topic of the stability of open-pit mines has long been discussed by many authors. Some years ago, a conventional approach to this problem was proposed by Mr. Stanley Wilson. His solution was based on the principles of soil mechanics. Mr. Wilson has given an example of computation of angle slope.

Of course, even if on the whole it can be accepted, such a conventional approach presents no assured degree of safety. The actual resistance of a rock to slides is much lower than the resistance which its average strength would warrant. This discrepancy is easy to explain. Manifestly, the surmise that equality of mean rock strength and destructive stresses may exist everywhere is quite unreal and can not be retained.

There is another fact whose importance has rarely been realized. It is that the problem of open-pit mine stability is basically an economic one. In order to make a decision on the usefulness of an excavation work, a mining engineer has to know beforehand what the cost of the extracted cubic yard of ore will be. The steeper the slope of the pit walls and the lower the stripping ratio, the larger the landslide risk will be. This risk is money too.

In a paper presented to the American Society of Civil Engineers (ASCE), the title of which is "Rock Mechanics and Statistical Methods,"**

---

*Consulting engineer, Paris, France.

**To be published by ASCE.

I have described a new and precise approach to this problem. It uses statistical methods. They are the only operative methods for the precise computation of the risk of landslide.

The evolution of landslides presents several distinctive steps. In the first phase, irreversible rock movements can be traced. In the next phase, cracks open and deflections increase. At last, the sliding zone collapses. There is no example of a landslide's stopping when fully unleashed. It follows that landslides must not be permitted to begin. The definition of the risk will be founded on the probability of the first irreversible movements of the rock.

A complete analysis of the slide risk can be made far in advance if the necessary data are collected in the field in time. As previously said, this analysis can be performed with the help of quite new and precise approaches which use largely statistical methods.

The scientific processes of sampling inspection used in industry are good examples of what can be done for geotechnical surveys. Any time a sampling inspection is used for rocks, it has to be extended to the totality of the rock mass to be surveyed, and the investigation must bear on the totality of the useful rock properties. The rock characteristics controlling the origin of landslides are numerous: specific weight of rock, cohesion, internal friction, stresses existing before excavation work, water table level, permeability of rock, etc. In order to render practicable, precise, and fast field measurement of these characteristics, I had to design patented new devices which can be placed at any depth in bore holes. A description of these devices will be given later.

The factors having some influence on the properties of rocks (time, chemical alteration, water content, temperature, etc.) have also to be investigated with care.

From sampling inspections, it is easy to represent statistically rock properties such as heterogeneity (or statistical scatter) and anisotropy (or directional scatter). Heterogeneity of rocks is no longer a problem.

When the behavior of the rock, under all modifications of stresses which seem to present some interest, has received a definition at each point, the analysis of the behavior of the whole rock mass is within reach. Mr. Müller has mentioned the attempts made in Russia for the analysis of dam abutments. In my opinion the drawbacks of the methods such as the ones that have been developed in Russia seem serious, and the efficiency of these methods seems very poor. These imperfect methods can not take into account most of the rock properties revealed by the tests, such as heterogeneity, anisotropy, plasticity, internal stressing, dependence on time, and rock swelling. Consequently they render comprehensive tests useless. I am also far from confident about scale models. These oversimplified devices have still more serious shortcomings. All these drawbacks are totally overcome by my quite new and precise method, which I called the "method of joint analytical techniques." It is specially adapted to the analysis of large dam abutments and of large dams; but it can also be used for the analysis of rock slopes. This method is as yet unpublished.

In any case, its description is not within the objective of this conference. Let us only remember that it is very versatile and that the only limitation to the perfection of its precision is the limit adopted for analysis expenses. Numerous trials with diverse values of the parameters can be made easily and rapidly. The new method can be simplified freely, and even then it presents overwhelming advantages over all known processes.

Whatever the method of analysis used, a sine qua non condition has to be fulfilled in order that the rock of the slope may tend to a steady equilibrium. This condition is the key to the problem of slope stability.

Finally, it can be demonstrated that, in any case, the cost of a cubic yard of ore has to include the cost of each cubic yard of earth in rock slides multiplied by the probability of failure.

ABSTRACT

In situ properties of the foundation rock must be fully investigated when designing concrete dams. The most important are deformability, shear strength, and permeability. Laboratory tests complement the information of *in situ* tests.

Jack tests in galleries for determining deformability are described and typical results are presented. The various factors affecting deformability are studied: (a) geological details such as faults, folds, stratification, dikes, and fissures; (b) petrographic structure; (c) degree of alteration of the rock; (d) direction of the load; (e) degree of consolidation, decompression, and fissures produced by excavation; and (f) internal stresses.

Shear strength of rocks was determined *in situ* by "box" tests and in the laboratory by triaxial tests. The factors previously described affect the strength of the rock mass. It is especially important to take into account the shear strength along faults, stratification plans, and altered rock near fissures.

The permeability of rock masses is especially important, as a rule, along faults and fissures. Comprehensive studies of percolation and interstitial water pressure conditions along them are often very important when studying the stability of rock abutments of dams.

Internal stresses in the rock, which have a decisive influence on the behavior of rock foundations, must be taken into consideration notably in the case of deep valleys and strong rocks.

RÉSUMÉ

Il faut étudier a fond les propriétés des roches de fondation sur place pour les projets de barrages en béton. Les propriétés les plus importantes sont la déformabilité, la résistance au cisaillement et la perméabilité. Des essais de laboratoire complètent les information fournies par les essais sur les lieux.

On décrit des essais de vérins en galeries pour déterminer la déformabilité et on presente des résultats typiques. On étudie les différents facteurs affectant la déformabilité: (a) détails géologiques tels que failles, plis, stratification, filons et diaclases; (b) structure pétrographique; (c) degré d'alteration de la roche; (d) direction de la charge; (e) degré de consolidation, décompression, et diaclases dues à l'excavation; et (f) contraintes internes.

La résistance des roches au cisaillement a été déterminée sur place par des essais de "boîte de cisaillement" et en laboratoire par des essais triaxiaux. Les facteurs décrits ci-dessus affectent la résistance de la masse rocheuse. Il est particulièrement important de tenir compte de la résistance au cisaillement le long des failles, des plans de stratification, et des roches altérées près des diaclases.

En général, la perméabilité des massifs rocheux est particulièrement importante le long des failles et des diaclases. Des études approfondies des conditions d'infiltration et de pression d'eau interstitielle le long de celles-ci ont, très souvent, une grande importance dans l'étude de la stabilité des appuis rocheux des barrages.

Les contraintes internes dans la roche ont une influence décisive sur le comportement des fondations rocheuses et il faut en tenir compte, surtout lorsqu'il s'agit de vallées profondes et de roches résistantes.

AUSZUG

Beim Entwurf einer Betonstaumauer müssen die *in situ* Eigenschaften des für die Gründung herangezogenen Gebirges gründlich untersucht werden. Am wichtigsten sind hierbei die Verformbarkeit, die Scherfestigkeit und die Durchlässigkeit. Die *in situ* gewonnen Erkenntnisse werden durch Untersuchungen im Laboratorium vervollständigt.

Es werden Belastungsversuche in Stollen zur Bestimmung der Verformbarkeit des Gebirges beschrieben. Typische Ergebnisse werden gegeben. Die verschiedenen Faktoren, die zur Verformbarkeit beitragen, werden untersucht: (a) Geologische Einzelheiten wie Störungen, Falten, Schichtungen, Gänge und Klüfte. (b) Petrographisches Gefüge. (c) Zersetzungsgrad des Gesteins. (d) Belastungsrichtung. (e) Verdichtungsgrad und durch den Aushub hervorgerufene Entspannungen und Klüfte. (f) Innere Spannungen.

Die Scherfestigkeit wird *in situ* durch Versuche mittels Schergerät und im Laboratorium durch Triaxialversuche bestimmt. Es ist sehr wichtig, dass die Scherfestigkeiten in Richtung von Störungen, Schichtungsflächen und von, von Zersatzbereichen umgebenen, Klüften in Rechnung gezogen werden.

Die Durchlässigkeit des Gebirges entlang Störungen und Klüften ist von grösster Bedeutung. Umfassende Untersuchungen der durch sie beeinflussten Durchsickerungsbedingungen und Porenwasserdrücke sind für Standsicherheitsuntersuchungen von Staumauern sehr wichtig.

Die inneren Spannungen im Gebirge haben einen entscheidenden Einfluss auf das Verhalten von Felsgründungen, besonders wenn es sich um tiefe Täler und um widerstandsfähige Gebirgsmassen handelt.

ROCK MECHANICS CONSIDERATIONS IN THE DESIGN OF CONCRETE DAMS

J. Laginha Serafim*

Introduction

THE BEHAVIOR of the foundations of concrete dams, although of primary importance for the safety of these structures, is a subject about which little has been discussed. In the design of dams it is assumed that the foundations are continuous elastic semi-infinite solids with a certain modulus of elasticity; strength considerations are often based on the compressive strength of rock samples, and, in a few cases of buttress and arch dams, on results of shear tests. It was only recently that elastic and rupture studies in models of dams began to be carried out taking into account mechanical properties of the foundations, which differ from region to region, and certain geological details.[1-3] Likewise the stability of the rock foundation along potential surfaces of rupture, such as faults, stratification planes, or joints of the rock mass, is only now a subject of study and discussion.[4]

From studying significant cases of failures of concrete dams, it is immediately clear that the weakness of the foundations is the most important cause of collapse, e.g., the St. Francis and Malpasset dams. In many of these cases either the strength of the rocks or the geological details which could affect the over-all stability of the dam were not properly investigated.

On the other hand, results of model studies have indicated that concrete dams, particularly some very economical types such as arch

*Consulting engineer, formerly at the Laboratório Nacional de Engenharia Civil (Portugal).

and large span multiple arch, can be built with safety on foundations previously considered to be of poor quality, provided appropriate studies and tests are carried out in the sites and consolidation and drainage of the rock mass are conveniently secured. The behavior of a large number of existing dams, some of them with important foundation problems, proves the same. It is to be remarked that the figures for the number and height (maximum = 65 m) of the dams that failed are very small when compared with the same figures for the existing ones.[5]

Observation of concrete dams has shown that foundations have a marked influence on their behavior; small settlements of rock foundations are often noticed, and movements along time due to the load, if not visible, are obvious in some instances when interpreting opening of joints.

Basic Problems in the Design of Dams

Rock masses being very complex media, they display peculiar mechanical responses to loads and varied conditions of percolation of water. Many problems of rock mechanics are raised now which require careful and long-term research in order properly to understand and to predict the behavior of the foundation of a large concrete dam. Even when considering simplified assumptions about that behavior, such as a perfectly elastic response and a shear strength based on Coulomb's theory, the analytical problems connected with foundation conditions are very difficult. Models can solve some of these problems.

Recent results have shown that, besides a marked influence of the deformation and settlements of the foundations on the state of stress in the dam, tensile stresses may develop in the concrete when these deformations are large. Such is the case of the stresses near the left bank of the arch dam indicated in Fig. 1. The low modulus of elasticity of the left bank gave rise to high tensile stresses along the downstream insertion of the arch. Studies on models also

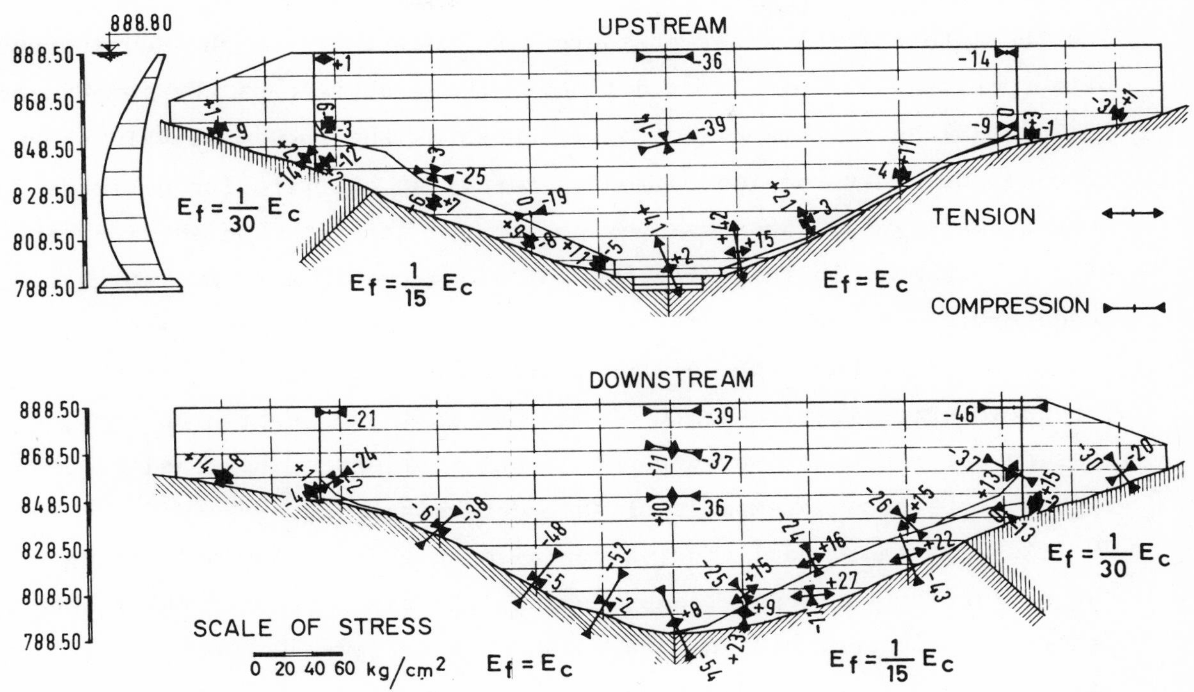

Fig. 1—Preliminary design of Alto Rabagão Dam: principal stresses in the prototype due to the hydrostatic pressure at level 888.80 m--model I.

indicate that such stresses tend to increase if a crack develops in the arch, but the problem can be solved by reinforcing the concrete along the foundation downstream. It is well known that the modulus of elasticity of the rock is of primary importance in the distribution of stresses at the base of buttress and gravity dams, as was shown by models(6) and analytical studies.(7) It is thus essential, for the design of a dam, to determine the deformability of the rock mass by methods which can indicate not only stress-strain diagrams but also possible settlements due to the compaction and the increase of deformations with time. These factors can be taken into consideration by assuming lower moduli of elasticity in the foundation or by making provisions for prestressing and forcing the opening of the joints from time to time.(8)

All types of concrete dams set up normal and tangential loads at the surface of the foundation due to the hydrostatic pressure on the dam and the weight of the concrete; furthermore, the weight of the rock and the interstitial water pressure also contribute to the development of a complex state of stress not only at the surface of the foundation but also along certain planes of weakness of the rock mass. In order to make sure that the foundation is safe, the distribution of loads must be well known, and the average bearing strength and shear strength of the rock mass in large regions and along such planes of weakness must be determined by appropriate tests in the laboratory and the site.

A complete knowledge of the loads acting on the foundation, although sometimes difficult to obtain, as in the case of arch dams, must be secured either by dependable calculations or by model studies. Figure 2 presents an example of such determinations.

As was said, model studies can also be helpful for predicting the rupture of the foundations, taking into consideration their heterogeneities, discontinuities, and weaknesses. Figure 3 concerns the model of an arch dam(2) founded on almost vertical strata of limestone, mudstone, and sandstone.

The permeability of rock masses, although they are generally considered as homogeneous media, is mainly due to the percolation of water along faults, fissures, and altered and crushed zones. Such paths of percolation must be studied comprehensively in order to design the grouting curtains and the drainage systems; otherwise, important and sometimes very dangerous uplift forces may develop in the foundations of dams.

When the rock is being excavated at the bottom of deep valleys, it is observed that the internal stresses sometimes give rise to exfoliation of the rock or to occasional rock bursts. Although in general they are not an important problem, such conditions can bring about difficulties in the impermeabilisation of the rock mass underneath the dam in the stressed zones.

All these considerations show the importance of the prospection of the petrographical, geological, and geotechnical characteristics of the site of a dam. It can now be considered essential, in addition to a careful geological study, to make and observe borings and galleries at the dam site, to carry out geophysical prospection, and to undertake comprehensive programs of <u>in situ</u> and laboratory tests and investigations of the deformability, strength, and conditions of percolation

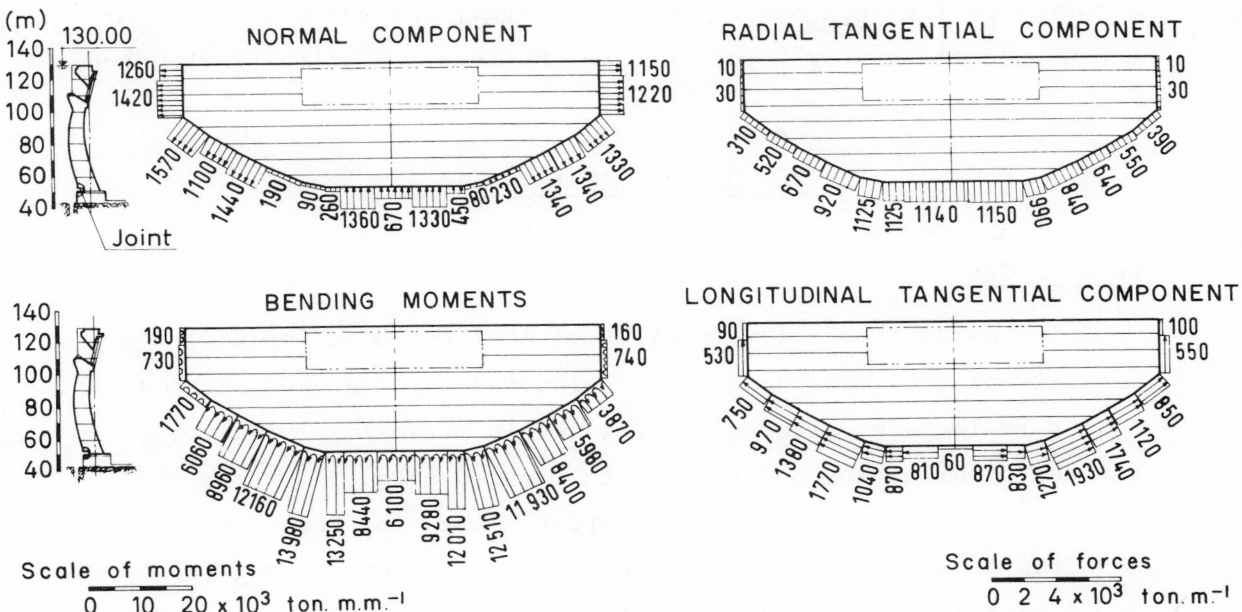

Fig. 2—Stability of the foundation of Cambambe Dam: applied forces and moments at the contour due to the hydrostatic pressure--model III.

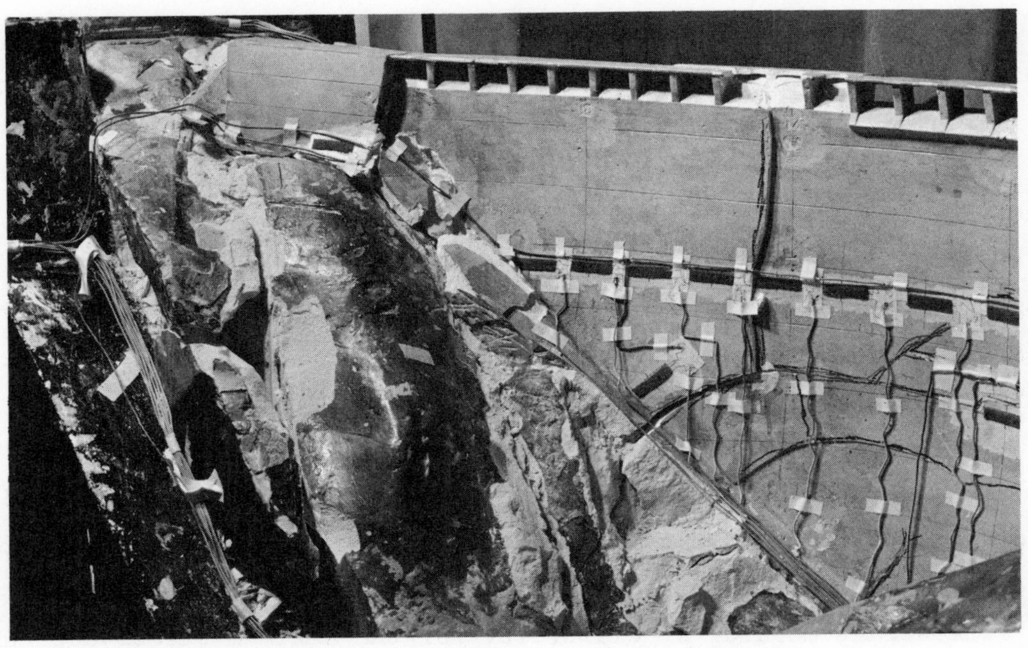

Fig. 3—Model of Tang-E Soleyman Dam after rupture tests.

of water in the rock mass. Such laboratory tests and research are considered to be very important for the interpretation of the studies.

The Deformability of Rocks

In Situ Tests

The methods usually employed for the determination of the deformability of rock masses consist in loading a given area, generally inside a gallery, by means of jacks, and in measuring the deformation at the center or at the boundary of the loaded area.

In Portugal the equipment currently used for tests inside galleries consists of metallic or rubber cushions which apply a uniform load on two opposite rock surfaces of 1 sq m, the force being supplied by two 300-ton hydraulic jacks. Two dial gauges are used which measure separately the deformations of the two loaded areas with respect to a fixed beam placed along the axis of the gallery. This is the only difference from the setup previously employed.(9)

Studies are under way in order to load larger areas, areas of, say, 6 or 10 sq m.

The test program usually includes cycles of loading and unloading up to a maximum value, two additional cycles being performed for the same value. When the maximum load and the zero load are reached in each cycle, they are maintained for a period of 30 min unless the gauges indicate no increase in deformation with time. A few days after these loading-unloading cycles, another test is carried out which consists in maintaining the maximum load for 2 or 3 days and reading deformations as a function of time.

This testing procedure has been in use for about 14 years, and more than 300 tests were carried out in galleries and trenches in dam sites. Results obtained up to 1954 have been reported elsewhere.(9)

The secant modulus of elasticity for the maximum stress during the second cycle at maximum load is currently taken as the modulus of elasticity of the rock. Other values such as the tangent modulus at the beginning and at the end of the loading and unloading curves are also determined.

Basically two types of diagrams are obtained in this kind of test.

The first type, shown in Figs. 4 and 5, indicates an initial settlement of the rock which generally increases with the load. The second and third cycles at the maximum load run parallel during both loading and unloading. It is to be noted that if a given load was previously applied twice, the next diagrams up to the same loading are parallel

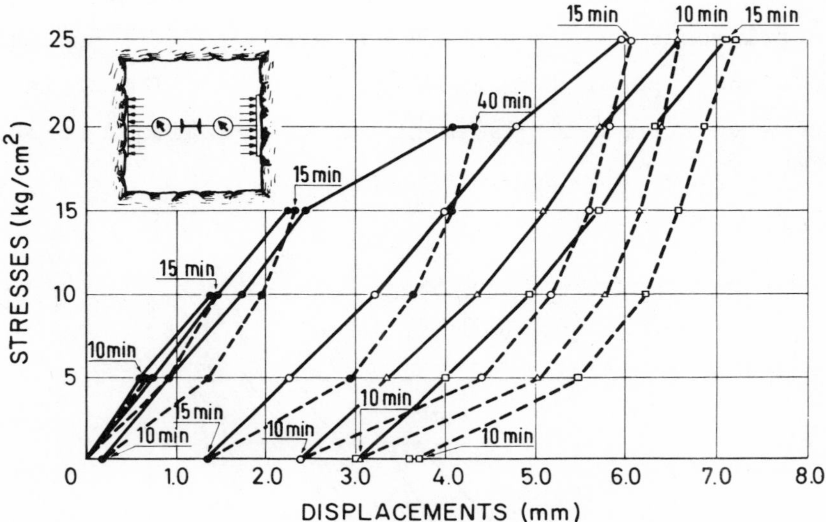

Fig. 4—Deformability tests in granites before grouting: Alto Rabagão Dam.

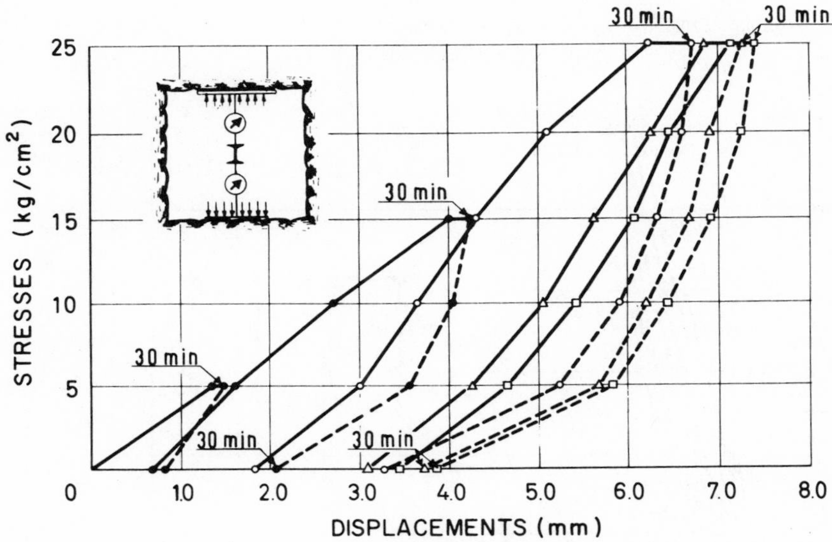

Fig. 5—Deformability tests in granites before grouting: Valdecañas Dam.

to the second. This shows that the structure of the rock mass became elastic after its compaction under the initial loading (Fig. 5); in other cases, elasticity is not achieved until a certain stress is reached, such as the 15 kg/cm^2 in Fig. 4. This is obviously the mechanical effect of macrojoints or cracks in the rock mass incompletely filled with compressed material. Generally in such cases the grouting of these joints, if properly carried out, succeeds in reducing the settlements observed in the diagrams (Fig. 6 and Ref. 9).

In the second type of diagram (Fig. 7), no settlement is observed, and the successive stress-displacement curves are almost coincident, as in the case of Fig. 6 after grouting.

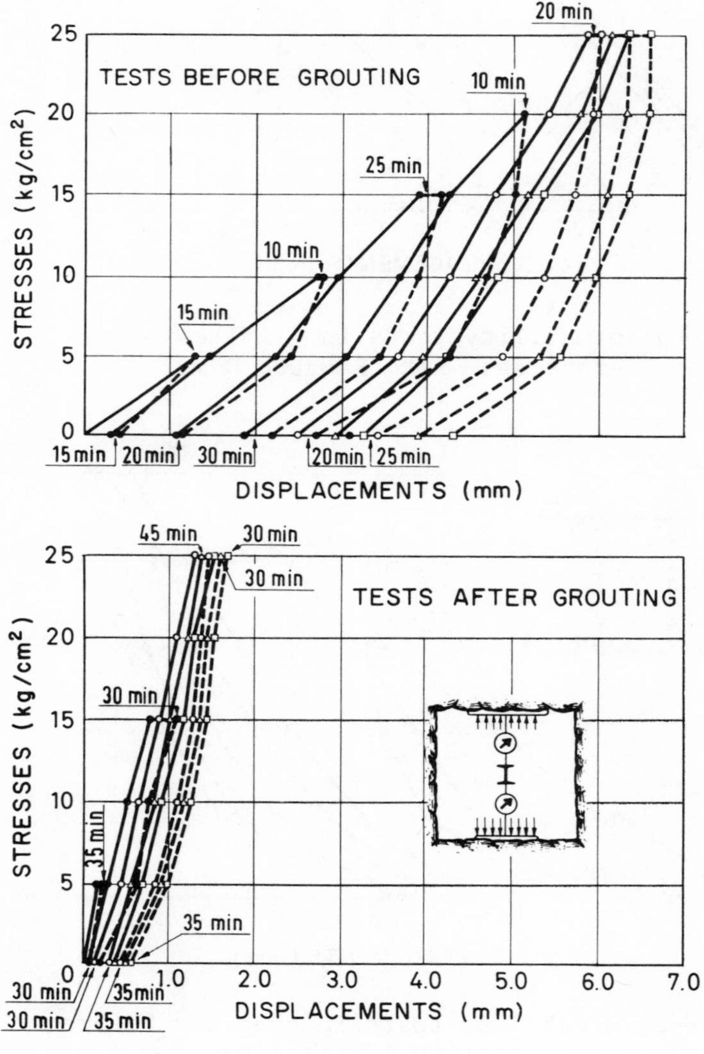

Fig. 6—Deformability tests in granites before and after grouting: Girabôlhos Dam.

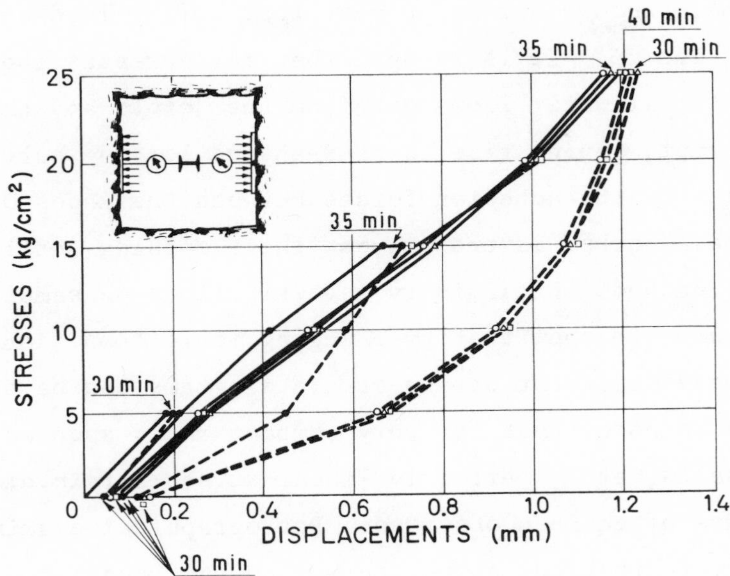

Fig. 7—Deformability tests in granites before grouting: Bemposta Dam.

Factors Affecting Deformability

The above results indicate the importance of the mechanical behavior of a rock mass of large joints and also of faults approximately parallel to the loaded surface. Sometimes when the joints are filled with soft clay, a certain amount of creep is observed when pressure is maintained at the end of each loading. If the clay is very compact, the effect of the joints is minimized.

Some deformability tests were also performed near important faults and inside their crushed zone. Obviously the importance of such accidents was quite decisive, and anomalous diagrams were obtained. It seems, however, that on a much larger scale the effect of large faults on the deformability of a rock mass is similar to the effect of joints.

Other important remarks regarding Figs. 4-7 are necessary. They show that during loading a noticeable increase in rigidity of the rock mass takes place and that during unloading a very small deformation occurs for the high stresses applied, with very large recuperation of the deformations being observed for stresses near zero. The same type of diagrams can be obtained in prisms. They are typical of rocks that are porous, altered, or in which a microtectonics can be observed in the microscope. Such diagrams suggest that during loading, the joints in the rock close up. It is thought that for stresses above a certain value, interstitial water flows out from the joints and the structure begins to disrupt, giving rise to an S-shaped loading curve. Once joints are closed, the adhesion forces between the faces of the fissures or between the adsorbed layers prevent their opening until a certain unloading is reached. Preliminary investigations on samples of rocks of this kind seem to show that by changing ionic conditions inside the rock pores, very different stress-strain diagrams can be obtained.

It is to be noted that for very compact rocks such as limestones, granites, basalts, etc., perfectly linear stress-strain diagrams without hysteresis have often been observed. Petrographical examinations can give important indications about the mechanical behavior of rocks and help in the interpretation of the tests. In very old rocks which were subject to high stresses, opening of new or old fissures in the crystals due to the blastings for excavation of a tunnel or the foundation of a dam can be expected. The stronger the blasts, the heavier the fissuration of the rocks.

This is one good reason for explaining why, in the observation of dams and their foundations, measurements with strain meters placed inside the rock generally indicate very large deformability[10,11] when compared either with the results of *in situ* tests in small galleries opened with care, or with geophysical measurements of the modulus of elasticity. Measurements on buried strain gauges of great length, which will be reported soon, also show compaction during the construction of the blocks of the dams.

Results of triaxial tests on rock cylinders with measurements of deformations indicated an increase of the modulus of elasticity with the hydrostatic pressure.[12] Figure 8 presents another example ($\sigma_1 = \sigma_2$ is the lateral pressure).

From all these results it can be concluded that the internal state of stress in the rock has a great influence on its deformability. If the rock was blasted and decompressed, its deformability during the loading can be quite high; but if the rock is in its natural state in the crust, its deformability can be much smaller. This has been proved for granites, gneisses, schists, quartzites, and graywackes.

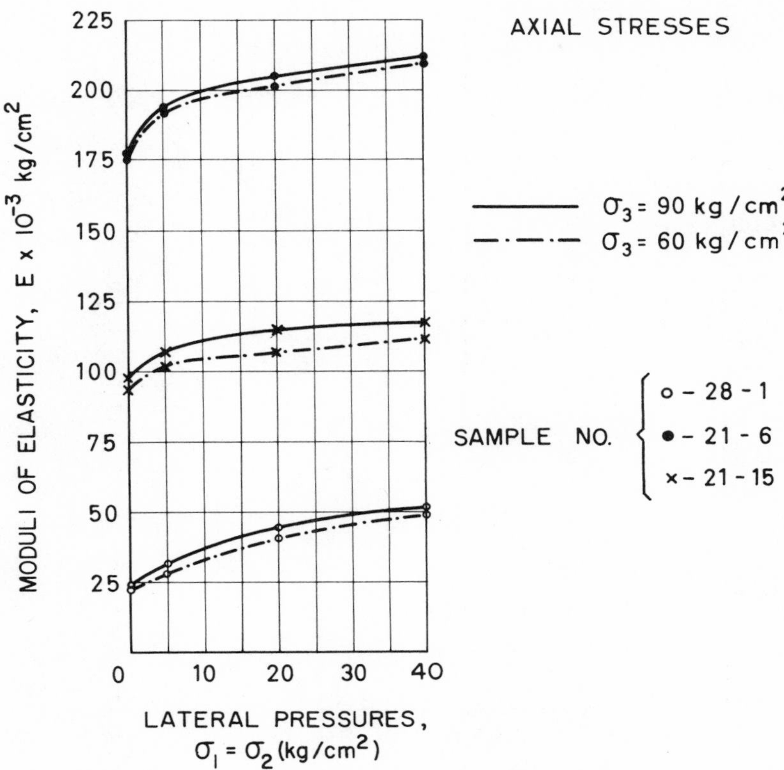

Fig. 8—Variation of the elastic characteristics of granite cores as a function of the lateral pressures for different axial stresses: Alto Rabagão Dam.

When trying to interpret the results of creep tests carried out at a number of galleries of a given site, it was also concluded that the degree of consolidation of the rocks and their internal stresses are the only factors that explain why similar rocks of the same site can display very different creep coefficients from place to place.[13]

Almost all these creep tests *in situ* have an asymptotic diagram. Such behavior can be due to the confinement of the loaded rock.

Measurements of the speed of propagation of shocks (geophysical method) were carried out at the sites of some Portuguese dams on which static tests inside galleries were also performed. The moduli of elasticity determined by this method were quite close to the tangent modulus E_u in the beginning of the unloading and much above the secant modulus E (Fig. 9). It is to be noted that this difference is more marked for rocks presenting greater deformability. Actually, this is also indicated

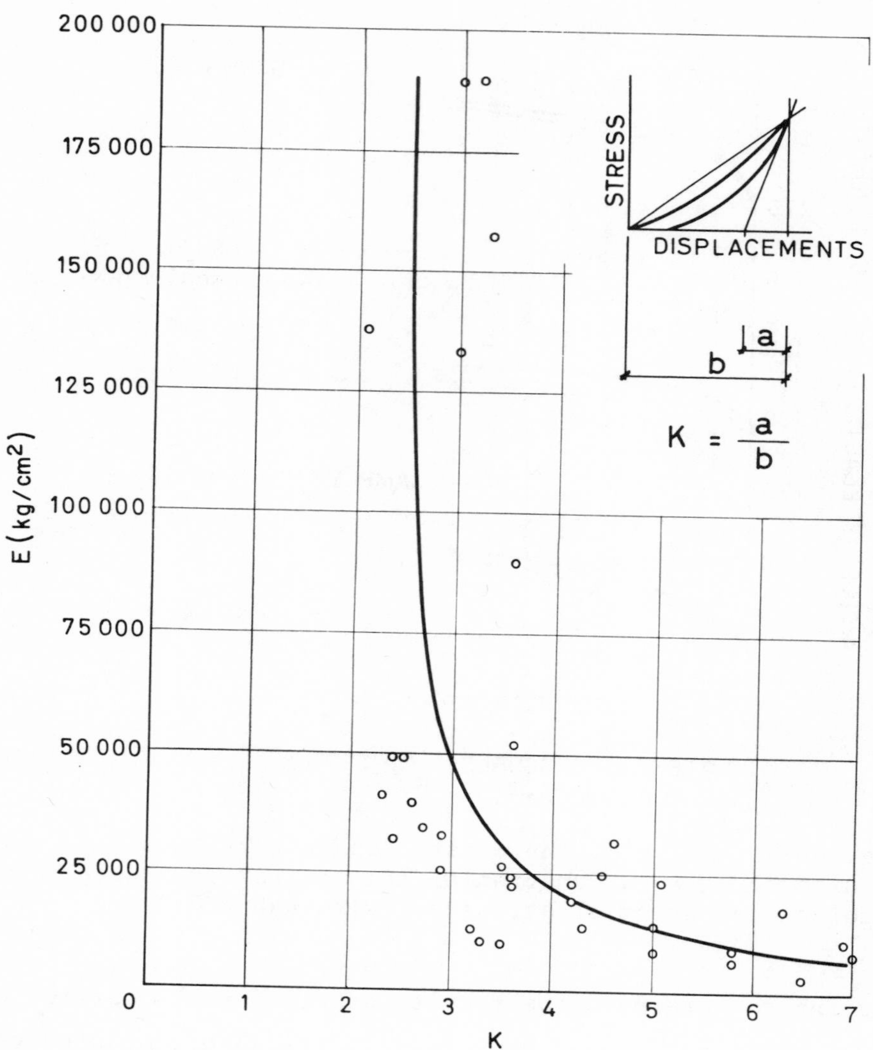

Fig. 9—Ratio between loading and unloading deformability: Bemposta Dam.

in Fig. 9, which shows the values of the ratio K between E_u and E obtained from the results of tests carried out in galleries in the Bemposta Dam site.

Deformability measurements of the rocks of a certain site yield as a rule very different results. This can be due to factors other than those mentioned before. One of these is the degree of alteration of the rock, another is the direction of the load. A property which can define the degree of alteration of a rock is its density; another is its porosity, which can be approximately expressed by the quantity of water in percentage of the total weight that a saturated sample loses when dried at 110°C. In this paper, the index of alteration, i, is taken at that percentage.(14)

Figures 10 and 11 show the results of various deformability tests carried out at the site of Alto Rabagão Dam in the horizontal and in the vertical directions as a function the index of alteration. It is noticed that the deformability along the vertical direction is on the average about twice that in the horizontal direction. The dispersion of the results shown in these figures seems to be from different conditions regarding the proximity of macrofissures, from different states of stress and fissuration of the rock, and, finally, from difficulty in obtaining a correct average value of the index of alteration of the loaded zone.

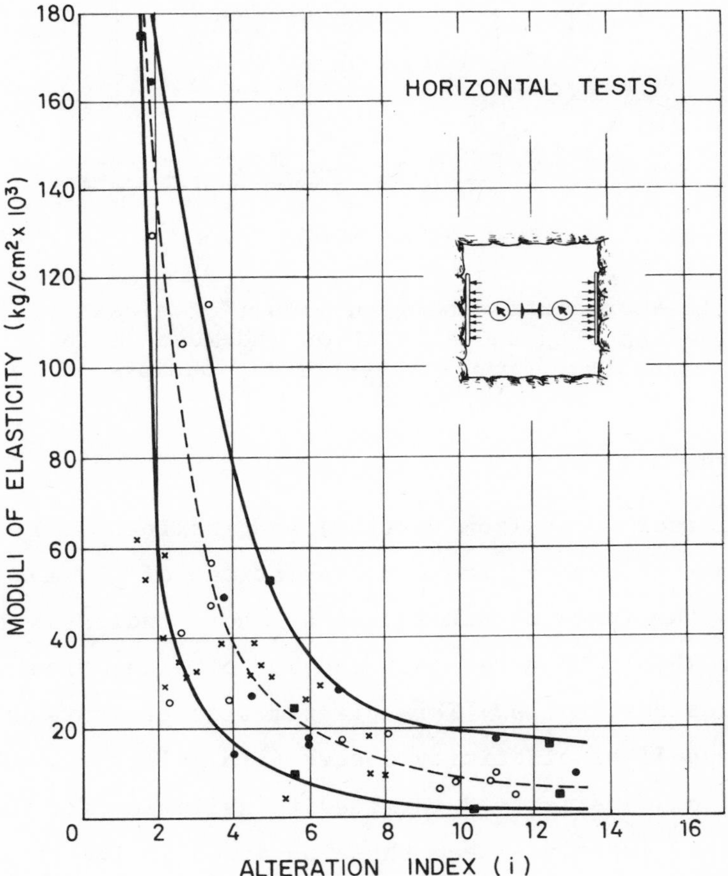

Fig. 10—Correlation between moduli of elasticity obtained *in situ* and alteration indexes: Alto Rabagão Dam (granite)--horizontal loading.

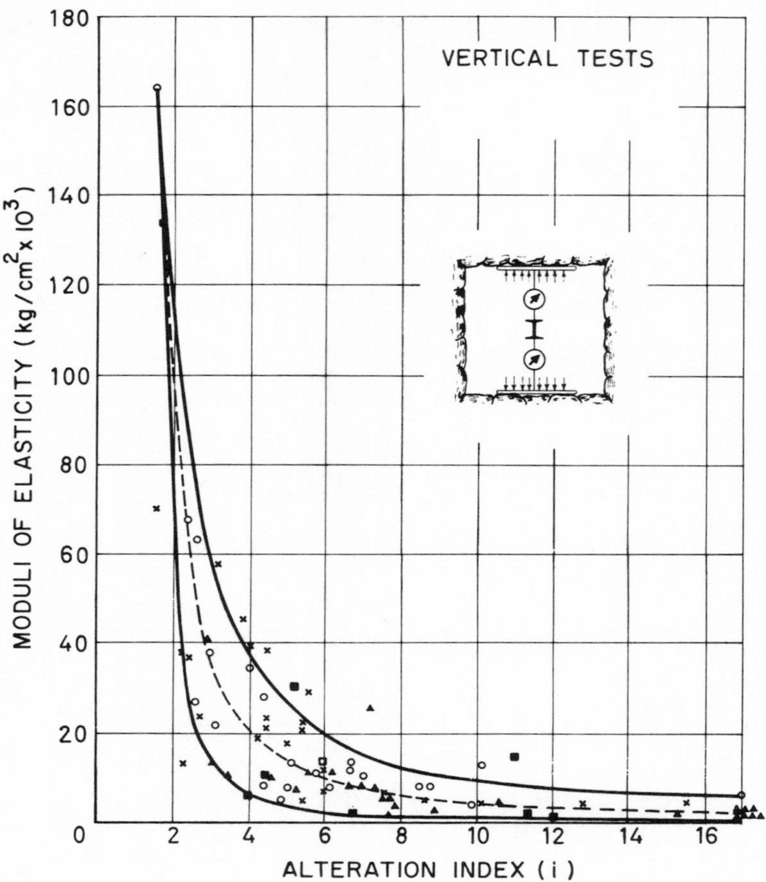

Fig. 11—Correlation between moduli of elasticity obtained *in situ* and alteration indexes: Alto Rabagão Dam (granite)--vertical loading.

Laboratory Tests

By testing rock cores from vertical and inclined drill holes and prisms cut from rock pieces, the same variations of the modulus of elasticity with the index of alteration as those indicated in Fig. 12 were found. It should be noted that indexes of alteration above 15 correspond to residual soils. This diagram also presents average values of the moduli of elasticity observed in galleries. The fact that the moduli of elasticity determined in galleries for indexes of alteration above 7 largely exceed those obtained in the samples can be explained by the fact that, in the case of tests in galleries, the load is applied to an infinite body with lateral restraint which develops

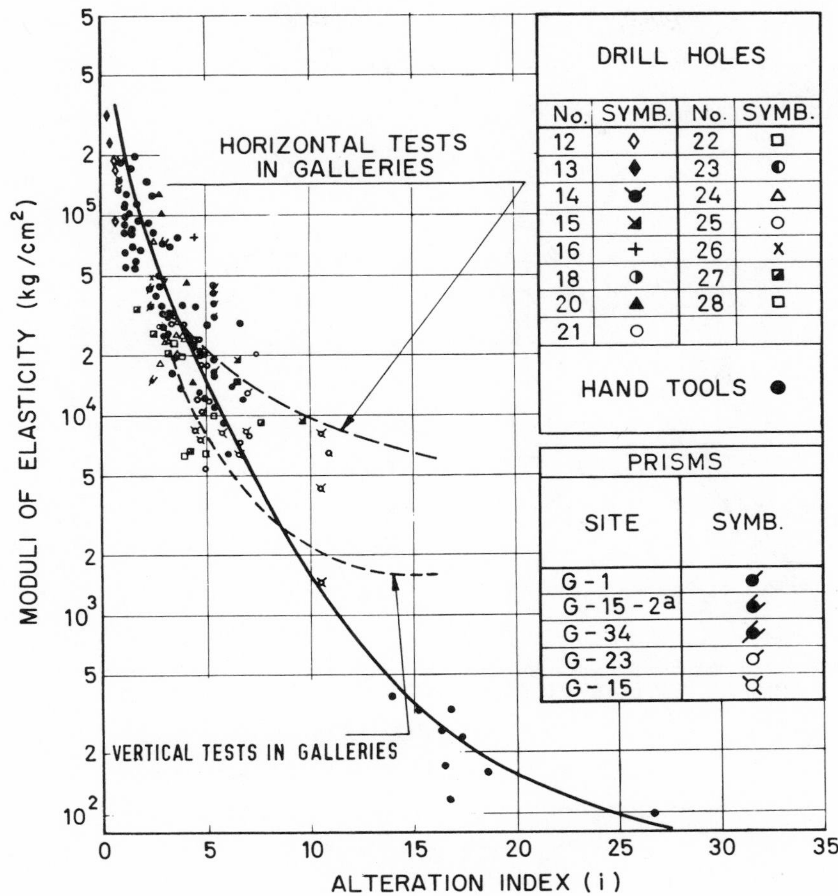

Fig. 12—Correlation between moduli of elasticity of core samples and prisms and alteration indexes: Alto Rabagão Dam (granite).

a triaxial state of compression, thus increasing the modulus of elasticity.

A criticism of the results indicated in Fig. 12 shows that the dispersion is due in part to the fact that the direction of the axes of the prisms and of the cylinders was not the same. By taking only those of the vertical direction, dispersion is reduced. Another cause of dispersion is the presence of incipient cracks in the samples from cutting.

The anisotropy of granites was studied in a number of cases by cutting three prisms along triorthogonal directions from the same rock piece. Figures 13 and 14 give the stress-strain diagrams for such groups of prisms from two dam sites. It is noted that these diagrams are similar to those of the tests in galleries and that the differences in deformability of the three prisms are quite important.

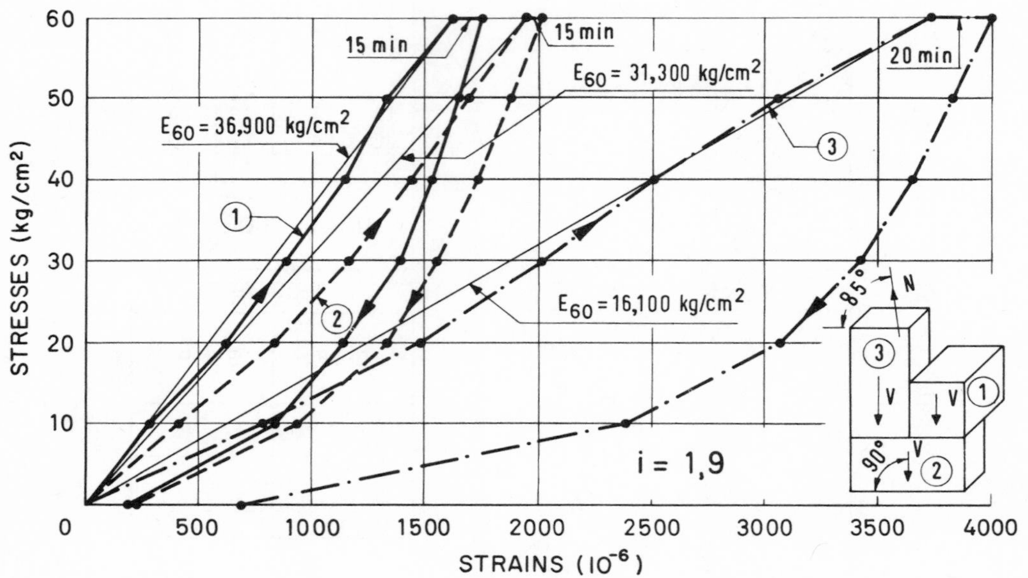

Fig. 13—Anisotropy of granites: Alto Rabagão Dam--stress-strain diagram.

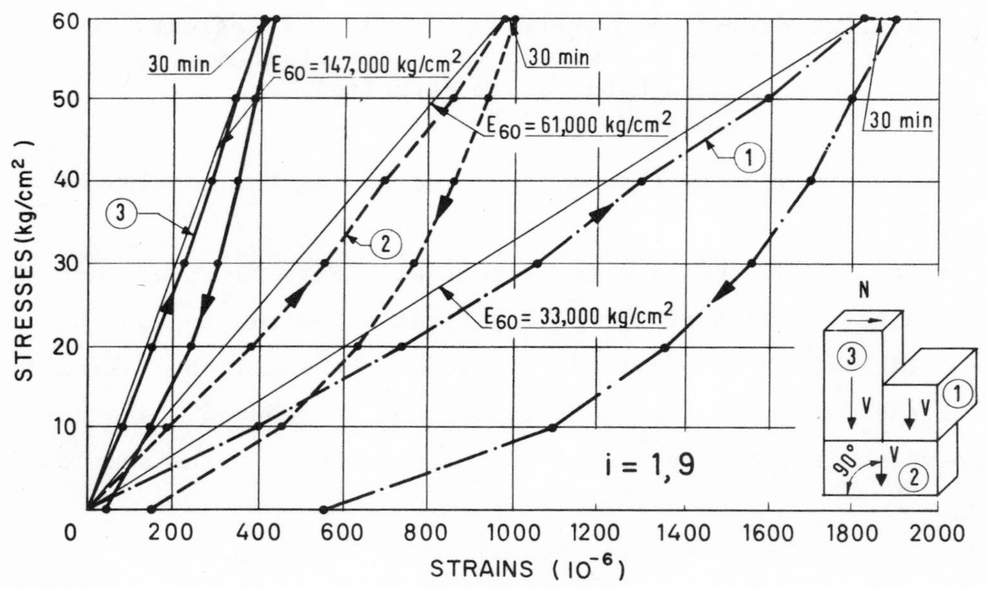

Fig. 14—Anisotropy of granites: Vilar Dam--stress-strain diagram.

Figure 15 indicates a correlation between the moduli of elasticity in the horizontal and the vertical directions on prisms from Alto Rabagão. These tests give an average anisotropy factor of about 2, which agrees with the results from tests in the galleries. One explanation is that horizontal microfissures due to tectonic forces or to the expansion of the rock near the surface can easily develop, which gives rise to a lower modulus in the vertical direction. There are indications that this factor of anisotropy decreases with depth.

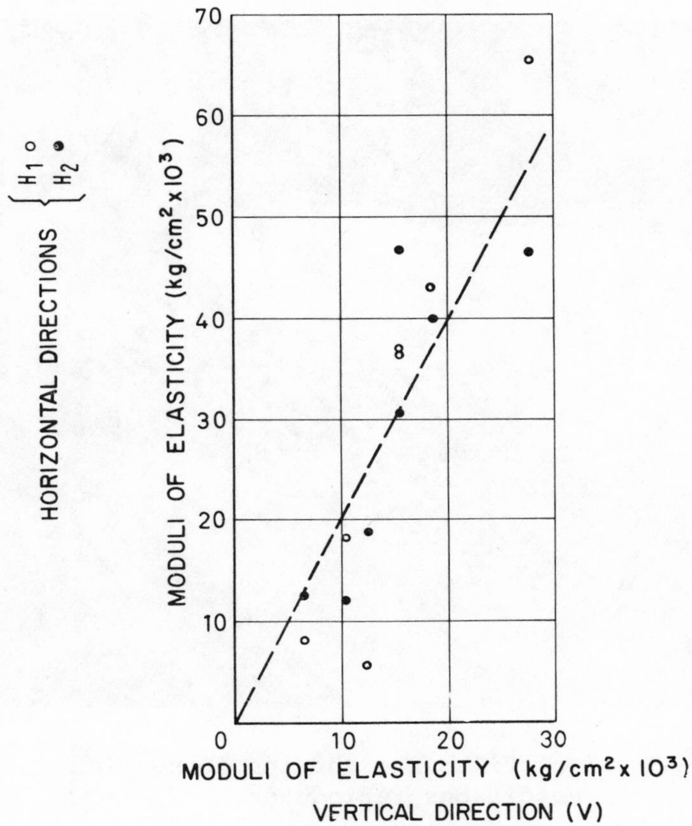

Fig. 15—Anisotropy of granites: Alto Rabagão Dam--moduli of elasticity.

In the case of Vilar Dam site (Fig. 14), it was observed that the modulus of elasticity in the horizontal direction along the valley systematically exceeds that along the horizontal direction normal to the river. This shows that deformability is lower in those directions along which the expansion of the rock mass was prevented.

Tests carried out in galleries and samples of schistous and stratified rock indicate much more complex conditions than in the case of granites.

The results above indicate the influence of the load direction (anisotropy) and of present and past internal stresses on the deformability of certain rocks.

Models of concrete dams are now being tested in which the heterogeneities and the anisotropy of the foundations are approximately reproduced (Fig. 16).

Fig. 16—Model of Vilar Dam showing foundation conditions reproduced.

Strength of Rocks

Shear Strength in Mass

The knowledge of the shear strength of rock masses is nowadays considered to be one of the most important problems in the design of concrete dams. The tests usually carried out in the laboratory, generally triaxial tests, can give no more than an approximate idea about the strength of some zones of the mass. *In situ* shear tests are usually carried out on rock blocks with areas up to about 1 sq m, giving a better picture of the resistance of the rock mass.

Tests carried out in Portugal[12] are usually performed on rock blocks, 70 x 70 cm, cut in the rock mass but kept attached to it in their base (Fig. 17). Tests are also carried out on concrete blocks of the same size cast against a plane surface of the gallery or trench or on the surface of a joint or fault after removing the rock on one

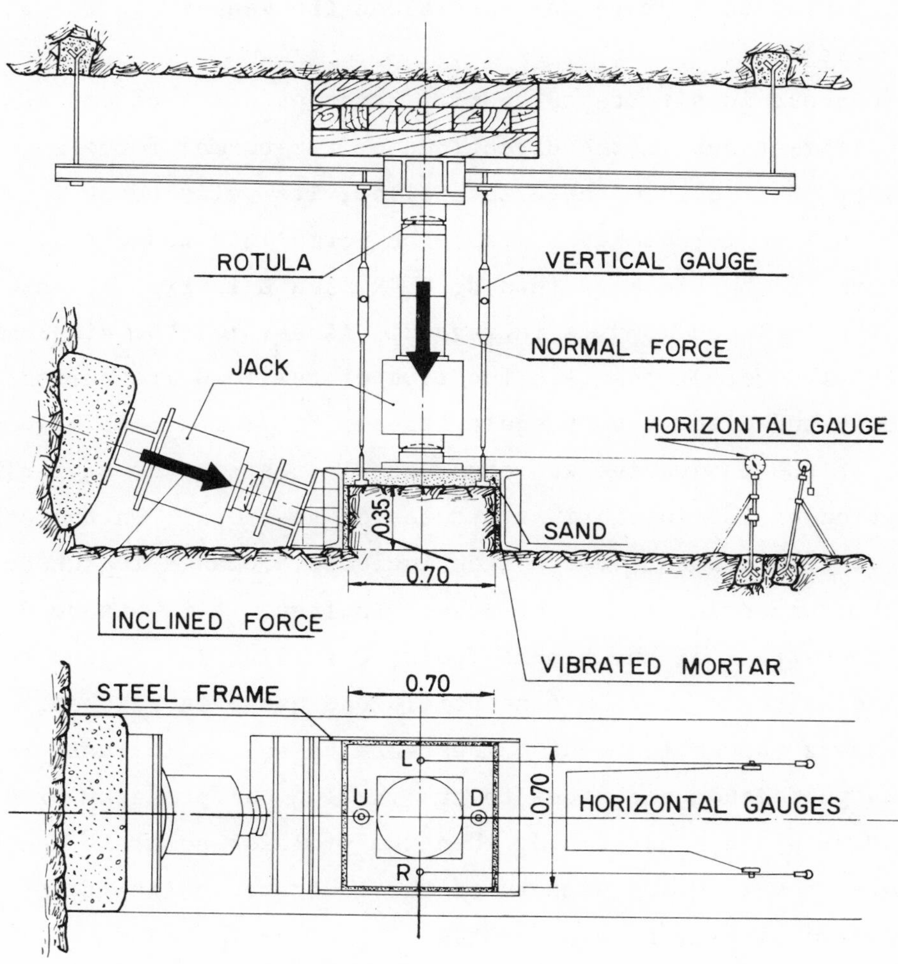

Fig. 17—Loading system for *in situ* shear test.

side. The tested zones are kept completely saturated during the tests.

By making tests in regions with different characteristics and states of alterations and along different directions, according to the bedding, schistosity planes, or the direction of joints, very useful information is obtained about the strength of the rock mass.

Yet, sometimes, these tests can not yield complete information about certain weak zones or can not indicate the over-all strength along planes of weakness of the rock. When the rock mass is very heterogeneous due to faults, joints, foldings, bedding, schistosity planes, microtectonics, alterations, etc., it may become difficult to obtain dependable results from tests on areas of 0.5 sq m. That is why shear tests on much larger areas, trying to rupture the rock along the weakest surfaces, are now being attempted.

In these *in situ* tests, normal stresses are kept constant and, after deformations in the directions of the normal force have reached stability, an inclined force is applied, its value being increased until final rupture takes place. The normal and tangential deformations of points of the block (points U, D, R, and L in Fig. 17) are recorded while the inclined load is increased. As a rule, the diagrams of tangential deformations as a function of the load are curved from the beginning and display no singularities.[12] Sometimes, however, a sharp change in the derivative was observed, indicating the beginning of disruption of the internal structures of the rock. On the other hand, vertical displacements very often indicate a change in the behavior when the tangential load increases. In fact, it is observed that the point downstream of the blocks (point D in Fig. 18) begins to move downward and, after a certain tangential load has been reached, upward. This indicates a change from volume decrease to volume increase in the sheared rock, a phenomenon analogous to the rheological property of dilatancy that occurs in a plastic body when the yielding point is reached. Typical diagrams of vertical movements of the indicated point of the sheared blocks have already been presented.[12]

The movement inversion being a singularity of the diagrams (points A in Fig. 18), it can be admitted that rupture began for the tangential load corresponding to that singularity.

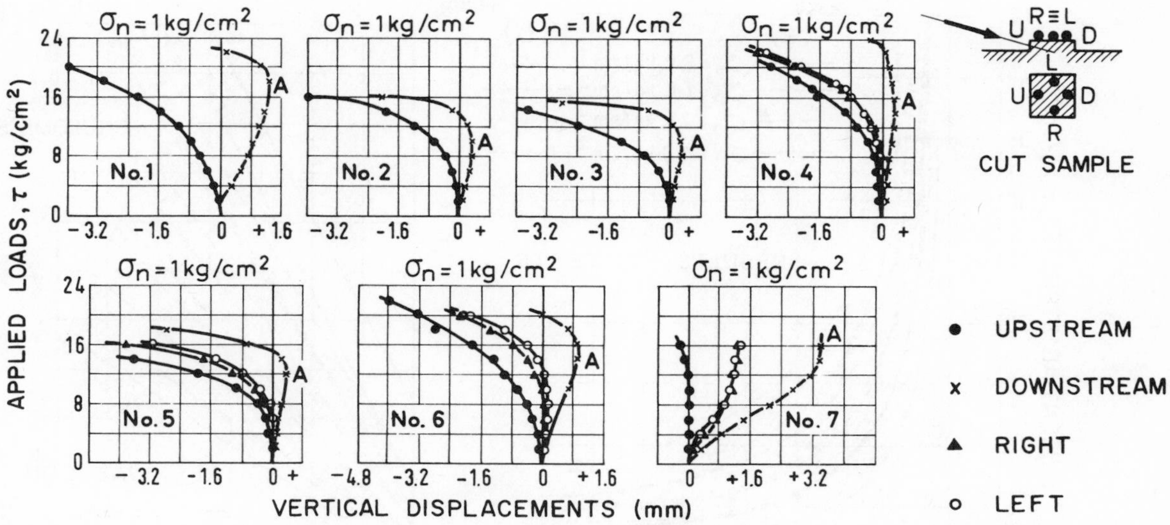

Fig. 18—Diagrams of vertical displacements of *in situ* shear tests: Alto Rabagão Dam (granite).

By adopting the criterion of maximum load and that of the change in movement of the points downstream, two Coulomb's lines can be plotted. In the case of granites with different degrees of alteration, the field results are previously plotted in terms of the tangential load for the various normal stresses, σ_n, as a function of the index of alteration. From these diagrams, Coulomb's lines for rocks having different degrees of alteration can be plotted assuming the two criteria of maximum load (Fig. 19). This procedure, which can only be followed when a large number of tests are carried out, considerably decreases the uncertainties and the dispersion of the results. Figure 19 concerns shearing tests carried out on granites inside galleries and trenches along horizontal planes. It is believed that if other tests were carried out in vertical planes by cutting blocks in the gallery walls, a higher strength would be obtained due to the anisotropy of these granites.

In the case of schists, tests are usually carried out by applying forces along three different directions: (1) normal force parallel to the schistosity planes, shear force also parallel to them ($/\!/$ $/\!/$); (2) normal force parallel to the schistosity planes, shear force perpendicular to them ($/\!/$ $\perp$); and (3) normal force perpendicular to the shear planes, tangential force parallel to them ($\perp$ $/\!/$). Tests in

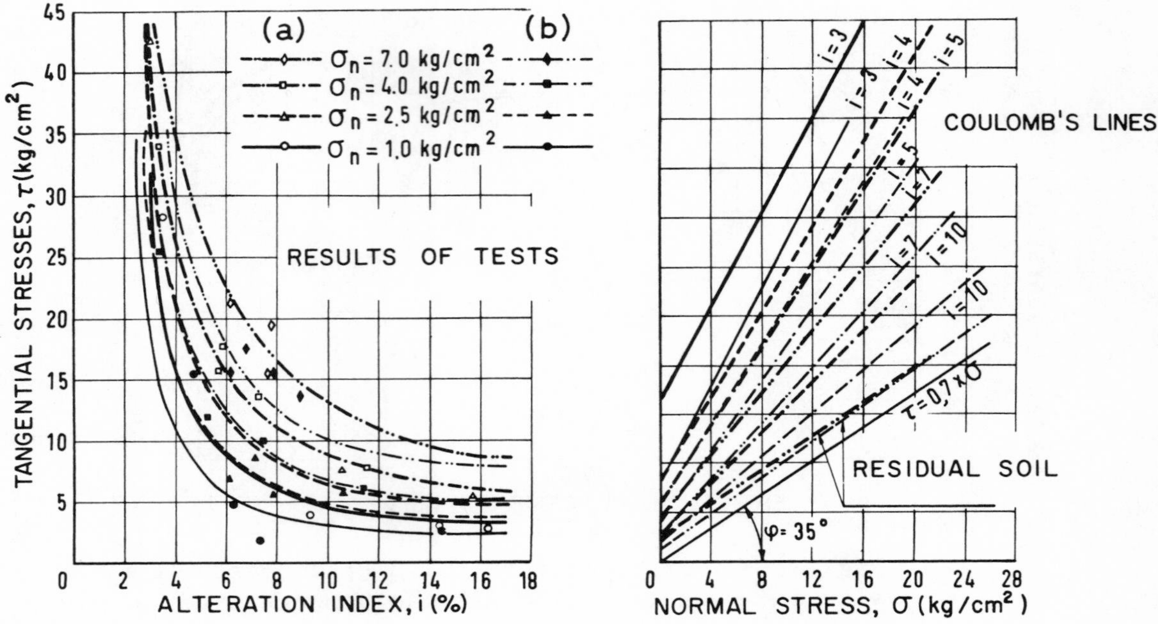

(a) Criterion of maximum load (b) Criterion of inversion of vertical displacements

Fig. 19—*In situ* shear test on granites: Alto Rabagão Dam.

various sites show that the first direction of the forces is the one for which the schist displays maximum strength, the third direction being the one of minimum strength. In some cases it has been found that the shear strength to forces along the second direction is about the same as in the second case.

Figure 20 presents the results of tests in two galleries excavated in schists having a certain difference in their degree of alteration, i. In this particular case it was impossible to perform a sufficient number of tests to obtain the variation of shear strength with the degree of alteration.

In a dam site, where faults are considered to affect the strength of the mass, tests are under way in which concrete prisms of 0.5 sq m are cast over the filling clay after the rock from one side of the fault has been removed. Tests carried out in two zones indicate negligible cohesion but angles of internal friction as high as 40 degrees. Such results seem to be in accordance with laboratory shear tests.

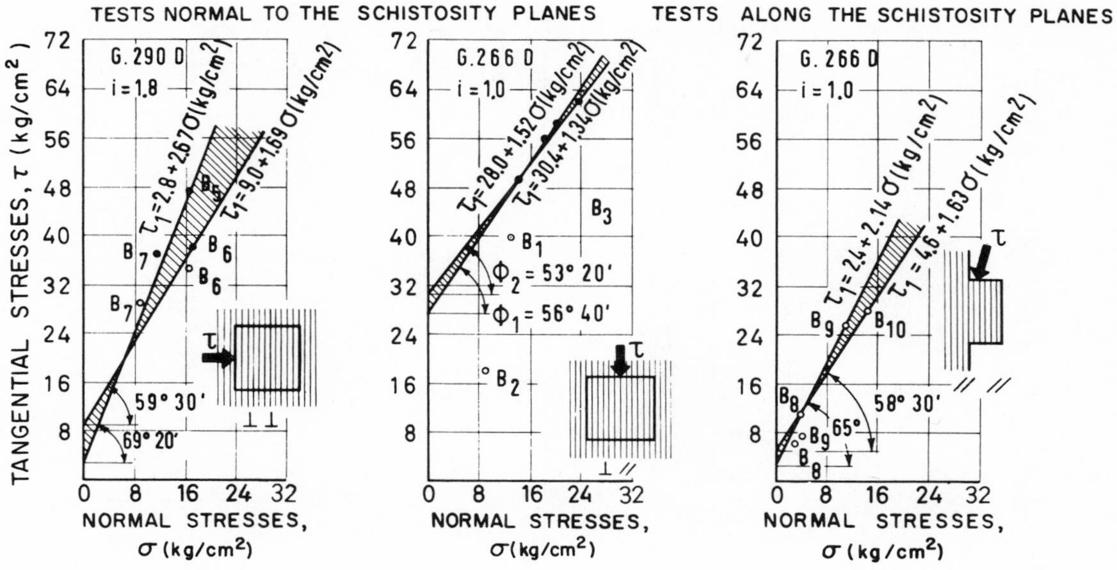

Fig. 20—*In situ* shear tests on schists: Valdecañas Dam--criterion of maximum load.

Triaxial and Compressive Strength

Results of triaxial, compressive, and tensile (diametral compression) tests on saturated granite cyclinders cut along the vertical direction from rocks of Alto Rabagão with various states of alteration have already been reported.[12] The strength considered was that based on the maximum load criterion. However, by means of measurements of the volume change of samples using strain gauges longitudinally and transversely bonded on the cylinder, it was also possible to interpret the results, assuming that when the volume of the sample starts to increase, internal disruption begins. The volume increase begins when the ratio between transverse and longitudinal strains (Poisson's ratio) exceeds 0.5. Diagrams of axial rupturing loads obtained in the tests previously reported,[12] and diagrams of the axial loads for which the volume began to increase, obtained in a few tests, are presented in Fig. 21.

Triaxial tests were also carried out, and others are under way on rocks with a very marked anisotropy, such as schists. In a study by F. Mello Mendes,[15] cylinders with their axes parallel to the schistosity

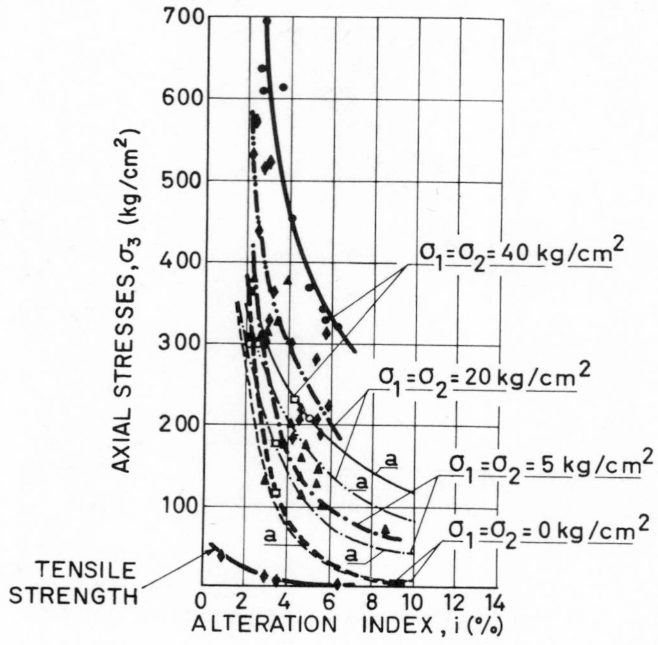

a = CRITERION OF VOLUME INCREASE

Fig. 21—Results of triaxial tests on granites.

planes were cut from very deep rocks. The results indicate a strength increase with the lateral pressures. An increase in length of the samples was systematically observed when they were loaded and unloaded before rupture. It is believed that high internal stresses in the petrographical structure are built up in deep rocks. By changing the state of stress, some bonds are eliminated, thus enabling the length of the samples to increase.(15)

Triaxial tests carried out on samples cut along directions having a certain angle with the bedding planes failed to show a reasonable strength increase, if any, with the lateral loads. In some cases, one Mohr's circle inside another is obtained; this means that in such cases the axial strength did not increase with the lateral pressure. Thus when planes of marked weakness are present in a rock (anisotropic strength), Mohr's theory of rupture can no longer be used.

By means of triaxial tests on cylinders with different angles between their axes and the bedding planes, it was possible to obtain very reasonable Coulomb's lines by computing the normal and tangential stresses in the planes of rupture when they coincided with the planes of schistosity. When these planes did not coincide, as happens when the axis of the cylinder makes a small angle with the normal to the schistosity planes, the points representing the stresses in the plane of rupture for the ultimate loads are much above the previous ones; but it is believed that, likewise, such points do not correspond to the maximum strength. This means that triaxial tests cannot yield complete information about the shear strength of anisotropic rocks along the various possible planes of rupture and that appropriate theories of rupture must be developed for such materials.

The anisotropy and the character of the rock strength are very important factors which must be taken into consideration when the stability of foundations of dams is being studied.

Compressive Strength and Bearing Strength

Compressive strength is usually determined when studying the rocks of a given dam site. Rocks of the same classification, apparently having the same structure, present sometimes a very marked difference in strength.

The diagram of Fig. 22 shows the compressive strength of granite cores from the Girabôlhos Dam site, which display much higher strengths for higher indices of alteration than those reported in Fig. 21 ($\sigma_1 = \sigma_2 = 0$). It also was observed that in comparison with the case of Alto Rabagão, these rocks presented a lower porosity for the nonweathered rocks. In fact, petrographical observations showed in that case a more marked microtectonics than in the case of Girabôlhos.

It is well known that the compressive strength of a rock is quite different from the strength observed by applying normal loads to the surface of a rock mass. Figure 23 shows results obtained in the case of Alto Rabagão by loading plane surfaces inside galleries with hard steel plates of various diameters. Besides a decrease of the bearing strength when the diameter of the loading plate increases, a strength (due to a load in a limited area) much higher than the compressive strength indicated in Fig. 21 ($\sigma_1 = \sigma_2 = 0$) is observed.

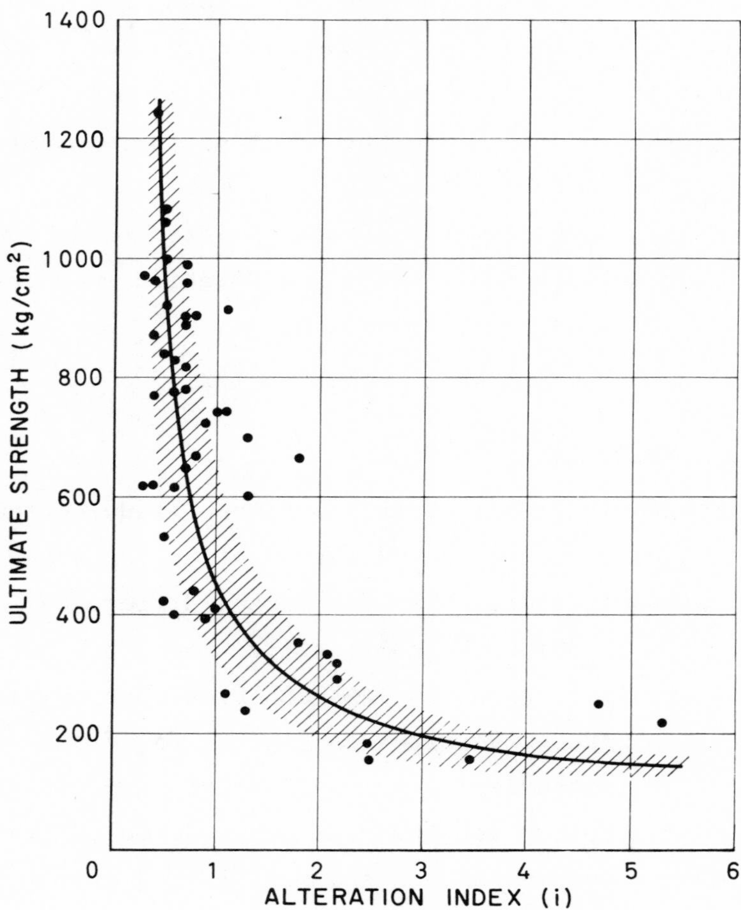

Fig. 22-Correlation between the compressive strength of core samples and the alteration index: Girabôlhos Dam.

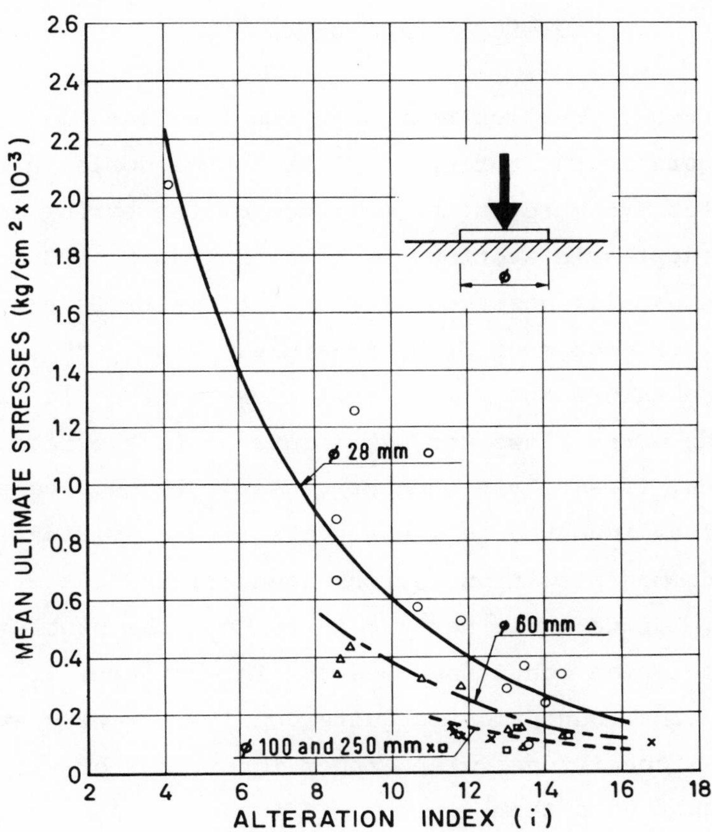

Fig. 23—Bearing capacity tests: Alto Rabagão Dam (granite).

Strength due to Inclined Loads

Since forces applied by a dam to its foundation are not normal to it, the idea arose of performing rupture tests by applying an inclined load to concrete blocks cast against the surface of the rock. Such tests[12] indicated that when the angle of the forces with the surface of the foundation is small, the strength curve is given by the Coulomb's lines; but when the forces become normal to the foundation, the strength increases considerably.

In order to better understand the conditions of rupture in such cases, laboratory loading tests are under way that rupture a homogeneous material representing a semi-infinite body with local loads of different inclinations.

Permeability of Rock Masses

It is the usual practice when studying the site of a dam to perform permeability tests in the borings. Water losses during a certain time for a given water pressure at the entrance of the boring are measured for different bore depths in sections not longer than 2 or 3 m. Three types of behavior are usually observed: the loss of water is proportional to, increases with, or decreases with, pressure. From such tests certain information is obtained about the water flow--whether it is turbulent or not, whether the water flows through joints or in the mass, and thus whether fissures, if any, are open or closed. It must be said, however, that the phenomena involved in these tests need clarification.

The most common situation in rock foundations is the flow of water through joints. Since their network is random, the problem of properly predicting percolation conditions and the interstitial pressures that will develop in the foundation is quite complex. Faults are very often zones of special conditions, with either impervious or very permeable zones being found near them.

It has often been observed that water losses during permeability tests take place along certain joints which are more permeable than the others. In order to obtain more information, it is considered advisable to make two or three borings near each other in order to properly identify the water percolation path and the joints along which the water flows. When joints are closed and filled and the rock is permeable, permeability tests yield diagrams similar to those of soils.

When designing the grouting curtains and the drainage systems of the foundation of a dam, such conditions of percolation must be known.

The grouting of foundations of dams is especially effective in reducing percolation along open joints, but grouting must often be carried out in two or three sets of borings with different directions in order for the grout to reach all the joints.

Even after grouting, water will flow along the joints and in the mass of the rock, thus developing interstitial pressures, which are a very important factor in the stability of the foundation of the dam. Permeability tests after grouting are thus advisable in cases of complicated foundation conditions.

The usual assumption that the rock masses can be considered as a homogeneous medium when determining the flow net must be checked by means of percolation tests in bores along different directions. Sometimes a vertical bore may present a very large loss of water, while a horizontal one may be more or less impermeable.

Another problem that must be kept in mind is the possibility of a permeability increase in the foundation with time, although this is not the ordinary case. Also, piping of the filling material of some joints can take place, giving rise to settlements of the rock mass and thus affecting the behavior of the dam. These are additional problems requiring investigation.

Internal Stresses in the Rock

According to what was previously said, internal stresses in a mass of rock have a very marked influence on its mechanical behavior.

It is believed that internal stresses can be due to the weight of the rock, to remaining tectonic forces, and to the rock expansion caused by alteration. Such alteration is probably one reason why higher internal stresses and more numerous rock bursts[16] are observed near faults, as the latter generally contain more altered zones. Reported measurements in the rock of the Picote underground power plant yielded much higher internal stresses in one side of a fault than in the opposite side.[17]

Important measurements of internal stresses by Nils Hast[18] at various distances from galleries showed that excavations can materially alter the internal state of stress, not always decreasing it. Other important observations by Nils Hast indicate the influence of joints and of the rock structure on the intensities and directions of internal stresses. Other measurements of internal stresses far from galleries were also recently reported.[19]

During excavations for foundations of dams in deep hard rock valleys, the appearance of important horizontal joints or of rock exfoliations at the valley bottom are often observed.

In the case of Picote Dam, an arched structure built in a narrow canyon of nonaltered granites at the international section of the Douro river, exfoliations and splitting rock were observed near the right slope. This was ascribed to a concentration of internal stresses, possibly due to the weight of the rock near the canyon wall and to remaining tectonic forces. In this case the rock presents many close-spaced horizontal joints at the valley bottom, but in the upper zones of the canyon, joints are much farther apart. These joints have a certain slant towards the river in both slopes, and some of them are filled with a thin layer of clay.

In a wall normal to the river bed it was possible to observe that the slanting joints of both sides join by curved joints at the bottom of the canyon (Fig. 24). By accepting the splitting theory of failure,[20] this path of the joints suggests that the mass of rock cracked along the isostatics of the internal state of stress in the crust, which was concentrated in the canyon. The isostatics were also curved due to the presence of the canyon.

The Aldeadavila Dam is located downstream in an area where it abruptly cuts the Iberian plateau and is about 600 m deep. During the exploration of the site, borings at the bottom indicated zones of complete loss of water through joints filled with river sand, as well as zones where the rock was considered impervious. During the excavations it was possible to observe, in a shaft, that important horizontal open joints were present. It is believed that they were produced by the state of stress in the rock. Because of these joints, the grouting of the river bed was an important task, which was carried out only after the blocks had reached a certain height. The absorption of grout was quite considerable to about 14 m in depth, indicating that the total opening of the various joints varied between 3 and 7 cm. The grouting was carried out through a large number

Fig. 24—Joints in a transverse vertical wall at the bottom of Picote Dam site.

of drill holes in the entire base of the dam. With this grouting treatment, joints were completely filled, and the rock mass became consolidated and almost impervious.

Conclusions

The mechanical behavior of large rock masses is a subject that needs much investigation. Such behavior has a marked influence on the state of stress in concrete dams. Since deformations of rock masses are not elastic and display consolidation and compaction due to load, it must be expected that the behavior of the dam changes, however slightly, year after year. Settlements and creep of the foundation due particularly to the dead load are to be expected during construction and for some years afterward. Such movements can open the joints of the dam unless they were already forced to open by a prestressing system. In poor rock, prestressing of dams and their foundations seems to be a very promising technique.

It must be borne in mind that, in many cases, excavation decompresses the rock and produces an internal fissuration that materially increases its deformability down to a certain depth.

Rock masses are usually anelastic, discontinuous, heterogeneous, and anisotropic. Thus, because of loads on the structure, the state of stress in dams and their foundations can be quite different from the one predicted by the usual theory of elasticity. Measurements on existing dams show deviations from purely elastic behavior, even though slight, in many cases. Theories are needed that will make it possible to determine states of stress in anisotropic foundations with elastic constants that vary with depth, time, and load. Such peculiarities can be approximately simulated in structural models of concrete dams.

Faults and joints are important factors in the deformability and shear strength of the rock masses. The degree of alteration of the rock, which can be quantitatively expressed as a function of, say, the density or the porosity of the rock, is also a determining factor of those properties. If, after excavation, the index of alteration of the rock is determined, a complete mapping of the mechanical properties of the surface of the foundations can be obtained.

Rocks, especially old ones, display stress-strain diagrams indicating internal microfissures. The unloading curves present a marked variation of deformability with stress. It seems that in the nonexecavated mass, most of the microfissures are closed, and the rock is compressed. This can be the reason why the seismic method indicates much higher moduli of elasticity than static tests in galleries during loading. When unloading begins, those tests indicate tangent moduli of elasticity close to the ones determined by the seismic method. Considerable increases of the modulus of elasticity with lateral pressure have been observed in triaxial tests.

Likewise, tests in galleries in altered rocks yield moduli of elasticity higher than those determined in samples, which seems to be due to the lateral restraint of the loaded mass in the galleries. This is contrary to what is often observed in good quality rocks, where the influence of joints is the main factor contributing to the *in situ* deformability.

Shear strength of rock masses being a basic property, it must be the object of careful investigation when designing a dam. *In situ* tests can yield the shear characteristics in the mass and along planes of joints and faults. They can also determine the shear characteristics of the filling material of the faults. Because of singularities in the rock, shearing tests must be carried out on very large surfaces of rock. The disruption of the structure of a rock mass when subjected to shear sometimes begins long before the maximum tangential load is reached. The volume increase of the sheared mass can be taken as a criterion for defining shear strength.

The shear strength of anisotropic rocks must be determined along various planes and directions in order to obtain complete information for the design.

Ordinary triaxial tests of rock samples can complement field shear tests. For stratified or schistous rocks, Mohr's theory of rupture must be abandoned. In such cases, triaxial tests can determine the strength along the weakest, but not along the strongest, planes.

Permeability tests of a rock mass are very important in the design of grouting curtains and drainage systems. It must be borne in mind that, in most cases, water percolation takes place along the joints, the mass of rock being generally quite impervious. The best effect of grouting to be expected is a complete filling of open joints and seams, thus reducing seepage in the foundations. Studies and observations of water percolation and interstitial pressures along fissured rocks are needed, and new methods must be developed for predicting the flow of water in rock foundations.

Internal stresses in rock masses affect their mechanical behavior. Such internal stresses appear to result from the weight of the underlying rock, from tectonic forces, and from the expansion of the rock due to alteration. In cases of deep valleys, exfoliation and open horizontal joints can frequently be observed.

Acknowledgments

The author is indebted to Mr. Manuel Rocha, Director of the Laboratório Nacional de Engenharia Civil, for his permission to publish this matter. His guidance and suggestions were very valuable at various stages of the rock foundation studies reported here.

Mr. J. Baptista Lopes, the engineer in charge of the group of rock foundation studies of the Dams Department, made important contributions to the work presented. Professor F. Mello Mendes and others also gave assistance in some of the tests.

The tests described were carried out for power companies in Portugal and Spain, owners of the dams. Their cooperation is gratefully acknowledged.

REFERENCES

1. Rocha, M. L., J. L. Serafim, and M. Cruz Azevedo, "Special Problems of Concrete Dams Studied by Models," RILEM Colloquium on Models of Structures, Bull. Rilem No. 12, Paris, 1961.
2. Lane, R. G. T., and J. L. Serafim, "The Structural Design of Tang-E Soleyman Dam," Proc. Inst. Civil Engrs., Vol. 22, London, 1962, pp. 257-290.
3. Oberti, G., "Experimentelle Untersuchungen über die Charakteristika der Verformbarkeit der Felsen," Geol. Bauw., Vol. 25, No. 2-3, Salzburg, 1960, pp. 95-113.
4. Müller, L., "Safety of Rock-abutment on Concrete Dams," VII Congress on Large Dams, Question 25, R. 90, Rome, 1961.
5. Gruner, E., "Dam Disasters," Proc. Inst. Civil Engrs., Vol. 24, January, 1963, pp. 49-60.
6. Rocha, M., J. L. Serafim, A. J. Fernandes, and J. Poole da Costa, "Experimental Studies of Buttress and Multiple Arch Dams," VII Congress on Large Dams, Question 26, Rome, 1961.
7. Zienkiewicz, O. C., and R. W. Gerstner, "Foundation Elasticity Effects in Gravity Dams," Proc. Inst. Civil Engrs., Vol. 19, June, 1961, pp. 209-216.

8. Serafim, J. L., Discussion of Question No. 26, "Techniques modernes relatives aux barrages en béton pour larges vallées et à leurs accessoires," VII Congress on Large Dams, Rome, 1961.

9. Rocha, M., J. L. Serafim, and A. Silveira, "Deformability of Foundation Rocks," V Congress on Large Dams, Rep. 75, Paris, 1955.

10. Bellier, J., J. F. Frey, and R. Marchand, "Compressibility of the Foundation Rock under Supports of Dams," _Proc. 3d Intern. Conf. Soil Mech. Found. Eng._, Vol. I, Switzerland, 1953, pp. 319-326.

11. Rocha, M., J. L. Serafim, and A. F. Silveira, Discussion of "Compressibility of the Foundation Rock under Supports of Dams," by J. Bellier, J. F. Frey, and R. Marchand, _Proc. 3d Intern. Conf. Soil Mech. Found. Eng._, Vol. III, Switzerland, 1953, pp. 167.

12. Serafim, J. L., and J. J. Baptista Lopes, "_In Situ_ Shear Tests and Triaxial Tests of Foundation Rocks of Concrete Dams," _Proc. 5th Intern. Conf. Soil Mech. Found. Eng._, Paper 2/18, Paris, 1961.

13. Serafim, J. L., Discussion on Question No. 25, "Les Travaux souterrains dans leur rapport avec l'étude et la construction des grands barrages," VII Congress on Large Dams, Rome, 1961.

14. Hamrol, A., "A Quantitative Classification of the Weathering and Weatherability of Rocks," _Proc. 5th Intern. Conf. Soil Mech. Found. Eng._, Vol. II, Paris, 1961, pp. 771-774.

15. Mendes, F. M., _Comportamento Mecanico de Rochas Xistosas_, Instituto Superior Tecnico, author's edition, Lisbon, 1960.

16. Grobbelaar, C., "A Statistical Study into the Influence of Dykes, Faults, and Rises on the Incidence of Rock Bursts," South African Council for Scientific and Industrial Research, R.N. 113, Pretoria, 1960.

17. Serafim, J. L., "Internal Stresses in Galleries," VII Congress on Large Dams, Question 25, R. 1, Rome, 1961.

18. Hast, N., "The Measurement of Rock Pressures in Mines," _Sveriges Geol. Undersökn._, Ser. C, No. 560, Stockholm, 1958.

19. May, A. N., "The Measurement of Rock Pressure Indice by Mineral Extraction," _Can. Mining Met. Bull._, October, 1960, pp. 747-753.

20. Terzaghi, K., "Stress Conditions for the Failure of Concrete and Rock," _Proc. A.S.T.M._, Vol. 25, 1945, p. 777.

DISCUSSION

J. FIDLER (USA):

Figure 18 of Dr. Serafim's paper giving results of in situ shear tests shows that the rock being sheared tends to expand until failure occurs. If this expansion was prevented by prestressing, it would appear that the rock would show a considerable increase in shear strength, probably far greater than would be computed from statics.

Dr. Serafim is asked whether tests have been conducted to determine the strengthening effect of prestressing in these circumstances, and whether he thinks it is feasible to use this method to improve the over-all strength of foundations, in particular the abutments of arch dams.

It may be of interest for the conference to know something of the work being done in Tasmania on the application of stressing to dam foundations.

The Hydro-Electric Commission presently has under construction a 150-ft-high massive buttress dam on a sedimentary foundation. The foundation consists of alternating layers of mudstone and siltstone. Between these layers are beds of brecciated material of very low shear strength.

In order to give the dam an adequate factor of safety in sliding, we proposed stressing both the dam and the foundations with cables some 100 ft long, inclined upstream and anchored approximately 80 ft below foundation level. The total stressing force applied is approximately 100 tons per lineal foot of dam.

J. L. SERAFIM in reply:

The question asked by Dr. Fidler of Tasmania is very interesting. In fact, I must say that recently I have been discussing this idea of prestressing rock masses; and, when the shear strength of a rock mass is not what is needed, we must think of prestressing the rock.

First of all, concerning the diagrams of Fig. 18 of my paper, i.e., the movement of point D when we shear the block of rock, I would say that the movement upward (negative values) indicates a volume increase in the sheared mass.

Everyone who has dealt with materials that have a yielding point knows that when they are subject to shear stresses beyond the yielding point, their volume increases. This is the very well-known property of dilatancy. So, we think that sheared rocks increase in volume after a certain value of tangential stress. The same thing has been observed in triaxial tests. This property can be used in such tests as a criterion for considering that rupture has started.

Now, of course, if, by means of prestressing, we increase the normal stress in a body under shear stress, we get a much better shear strength. Besides the usual vertical prestressing, we are suggesting the use of horizontal prestressing of foundations of arch dams having weak abutments and also of gravity dams founded in layered and weak formations. This horizontal prestressing is a good way of securing tridimensional strength of the dam and of the rock in the valley.

L. MÜLLER (Austria):

Dr. Serafim's remarks were very interesting indeed. We are right to assume that the volume of a rock mass increases near the flowing point or just before a failure. We can confirm it by practical experience. We have had only one opportunity, so far, to conduct large-scale rock tests with test blocks on the order of 30 cu m in volume. (These tests were in Japan.)

In these tests, with increasing load we observed the varying volume of the body, i.e., the Poisson's number. We found that the Poisson's number started with lower values than those usually assumed. As a rule this ratio is assumed to be 3 to 7 because that is the value of the rock; but, as I said, rock mass is different.

The Poisson's number during the tests started with values of 2.5 to 4, and reach 6 to 8. Then it suddenly dropped to values amounting to 2. The yield point as properly stated corresponded to this value.

Finally the number dropped below 2 and reached 1, a value strictly taboo according to the textbooks. Poisson's number (μ) has become much higher than 0.5. That means that an intact rock sample and a jointed body act differently.

Volume increase is a very significant criterion of the strength of rock masses. And, consequently, the confining pressure in rock masses is much more effective than in unjointed rock samples. This fact explains the efficacy of measures such as anchoring.

I. SHERMAN (USA):

Dr. Serafim, suppose you discover that the foundation material for a concrete dam is anisotropic, so that it has a higher modulus horizontally than vertically. How do you allow for this in the design of the dam?

J. L. SERAFIM in reply:

Well, I think I don't. At least we do not yet have any available theory for treating such a problem of the foundation of a concrete dam. Model tests can probably provide the first answer to the problem since we can reproduce anisotropic conditions in their foundations. When we speak of anisotropic materials, we must consider what we know from the theoretical point of view.

I think I shall not be too academic if I say that we have not yet considered applying existing theories to other important problems. We know, for instance, that in an orthotropic material, the concentrations of stresses around circular holes are much greater than in isotropic materials for certain directions of the loads. Such results can probably be used for interpreting results of tests in galleries and stresses near cracks in crystals.

J. A. TALOBRE (France):

Mr. Laginha Serafim's paper gives an example of the manner in which geotechnical surveys of dam abutments are conducted in some European countries.

I will not insist on the considerations developed by Mr. Serafim about the results of his hydraulic jack tests. We all can agree that the deformability of in-place rocks depends upon the joints, cracks, and voids existing in them, and that the modulus of elasticity of rocks is a function of the inverse of the void volume. It is clear too that the deflections measured with jacks are larger in the vertical direction (since joints are generally open wide at the roof of the adits) and that results of measurements of Young's modulus are larger in the direction of maximum internal rock stress. However, these discrepancies must not be confused with anisotropy.

On different occasions, I have insisted on the important drawbacks of shear tests for rocks. I will especially maintain today that most of the tests with shear boxes, as conducted by Mr. Serafim, are not really shear tests. The results of his tests therefore can not permit comparisons. Once more, on their merits alone, they do not authorize a firm conclusion about anisotropy of rocks.

Much care must be taken in order to prevent the disorganization of the rock during the test. When the tests can modify the strength of the rock at random, such irrational statements as those drawn from the triaxial tests of Mr. Mello Mendes, concerning the unfitness of the Mohr's criterion, are not surprising. Furthermore, no result of measurement of rock internal compression was given by Mr. Serafim. This is a gap.

Mr. Serafim uses an alteration index. Rocks, and specially granites under dam abutments, may decay under the action of percolating water. Professor Farran, in France, has for years studied the weathering of rocks.

It is possible today to figure scientifically the susceptibility of a rock to weathering and also to forecast where and when a rock will decay.

The alteration index of Mr. Serafim has no relation to the weathering index of Professor Farran. It should rather be called a "porosity coefficient."

But the main points on which the discussion has to bear are the usefulness of the alteration index, on one hand, and the reliability of model tests, on the other hand.

Any time one wants to obtain an average value of the properties of a rock mass, large-scale tests are necessary. But these tests are costly, and they destroy or disturb part of the rock of the site. In consequence, they can not be applied everywhere, and their reliability has to be checked by some convenient means. Mr. Müller proposes the use of diagrams of joints and of petrofabrics graphs. Mr. Serafim employs an alteration index, which is, in fact, a coefficient of porosity.

Which way is to be preferred? That is matter for discussion. In any case, the correlations established with the help of an alteration index seem very poor. At any rate, direct measurement of rock properties at any point of the site, with measuring devices suitable for this purpose, and with the use of statistical sampling methods, solves the problem far more satisfactorily.

Scale models also have important drawbacks. The behavior of rocks is far from elastic. Internal stresses, thermal expansion, variations of volume resulting from variations of water content or of pore pressure and timing are important factors that can hardly be taken into account with scale models.

Scale models are costly and can not be modified at will. In all respects, modern methods of analysis, and especially the joint analytical techniques method, are far more precise, efficient, and safe than scale models.

ABSTRACT

The object of underground protective construction is to provide resistance to explosively induced stress waves. It is reasonable that such constructions be made in very resistant media as characterized by hard rock. Such material is, however, the very kind most able to support and most likely to contain high residual stresses.

The phenomena and effects accompanying nuclear explosions on or in hard rock will be examined in relation to the possible response of underground openings. The possibilities for both resistant and isolating linings will be explored.

RÉSUMÉ

Le but des structures souterraines protectrices est d'assurer une résistance à des ondes de force produites par des explosions. Il est raisonnable de faire de telles constructions dans des milieux très résistants, tels que la roche dure. Ce máteriau est, cependant, celui même qui est le plus capable de supporter et le plus susceptible de contenir de grandes forces résiduelles.

Les phénomènes et les effets qui accompagnent les explosions nucléaires sur ou dans la roche dure seront examinés en fonction de la réponse possible des ouverture souterraines. Les possibilités pour les revêtements résistants ainsi qu'isolants seront étudiées.

AUSZUG

Der Zweck von unterirdischen Schutzbauten ist der Schutz gegen durch Explosionen erzeugte Spannungswellen. Es liegt nahe, solche Bauwerke in sehr widerstandsfähigem Material, wie es der feste Fels darstellt, anzulegen. Fels ist jedoch ein Material, welches imstande ist hohe Restspannungen aufrechtzuhalten und solche auch mit grosser Wahrscheinlichkeit aufweist.

Die Kernexplosionen begleitenden Erscheinungen und Auswirkungen auf oder im festen Fels werden im Bezug auf das mögliche Verhalten von unterirdischen Hohlräumen untersucht. Die Möglichkeiten von Auskleidungen, entweder als Schutz oder als Isolierung, werden erwogen.

ROCK MECHANICS CONSIDERATIONS IN THE DESIGN OF UNDERGROUND PROTECTIVE STRUCTURES

Harold L. Brode*

SINCE THE STATE of stress in the political world above the earth's crust is often much greater than the state of stress in the earth's crust, it is not illogical to look for protection below ground. The effects of explosives, in particular the effects of nuclear explosions, are vastly more damaging and far reaching above ground than they are below the surface. But however much safer it may be, the question of how much cover is adequate still demands an answer. In analyzing the adequacy of underground protective measures, it is helpful to have a quantitative picture of the explosion phenomena that accompanies a nuclear detonation.

In the nuclear reactions themselves, extremely high energy-densities and consequently unearthly temperatures are created. For a 1-MT explosion, some 10^{15} calories are released in less than a microsecond in a volume of at most a very few cubic meters. Temperatures measured in the tens of millions of degrees centigrade are created, and much of the energy of the explosion is transferred out of the weapon itself by radiation processes (even before the bomb can come apart).

If the explosion is contained underground, then the radiation energy is stopped in the immediately surrounding rock or soil, and within a microsecond, a very strong shock wave forms and begins compressing the earth and excavating a cavity--essentially blowing a bubble. While the shock is still very strong, the surrounding material must respond as a compressible fluid or dense gas, and its dynamics must be described in hydrodynamic terms. As the shock

---

*The RAND Corporation, Santa Monica, California.

weakens and the following bubble growth slows, the earth response must also begin to reflect some of its solid state properties. Depending on its nature, whether soil or rock, brittle or malleable, consolidated or granular, the subsequent phases may include plastic behavior, crushing or locking response, cracking, and slipping or shearing along planes or joints of weakness.

In the explosion history, many of these modes of response are in existence simultaneously. As the superheated interior material cools and expands in gaseous and liquid phases, some outer materials shocked to less exalted temperatures may flow or deform plastically, may simply crush or crack, while more distant material already engulfed by the weakening shock may be able to respond essentially elastically. Strict elastic response is an idealization which, on the large-scale dimensions of waves propagated from an explosive, can never be realized in natural materials. Although it may be adequate to view the response of small cavities in hard rock as elastic, it is both qualitatively and quantitatively invalid to expect the propagation through appreciable rock mass to conform to elasticity predictions. Even at very low stress levels, where laboratory rock samples behave elastically under static loading, such significant features as the peak stress versus distance or the peak particle velocity versus distance from the explosive source decreases much more rapidly than elasticity would allow.

The peak stress as a function of range for nuclear charges detonated in hard rock (granite), soft rock (tuff), and dry soil (desert alluvium) is illustrated in Fig. 1.[1] It is evident that in the megabar region, all materials show a decay with distance for peak stress of nearly an inverse cube of the radial distance, while at lower stresses in the 100-kb region (comparable to high explosive detonation pressures) the hard rock decay becomes more gradual ($\sim R^{-2}$). The alluvium stress continues to fall rapidly to 1/10 kb. The curves illustrated have been scaled to 1 MT. The soft rock case is based on indirect observations, surmise, and some early theory, but the hard rock and the soil cases are documented with considerable measurements. The stresses were in most cases derived from peak

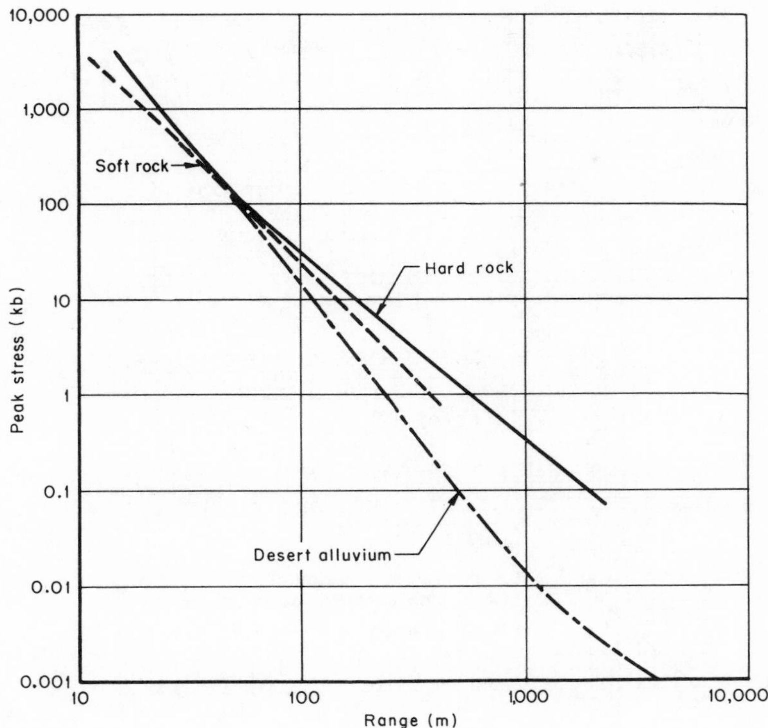

Fig. 1--1-MT nuclear explosions contained: peak stress versus range.

particle velocity measurements, but at the highest levels, direct stress readings with impedance techniques were available.

Theoretical treatment of the earliest phase requires hydrodynamic calculations and knowledge of equation of state or of thermodynamic properties of the very hot material. The current basis for providing equations of state for natural materials relies on both theoretical atomic structure properties and empirically derived or observed properties such as sound speed (i.e., seismic velocities), bulk modulus, heat of sublimation or of fusion, and dynamic compressibilities as derived from high explosive experiments. An example of the initial construction of one such equation of state is illustrated in Fig. 2 showing the pressure for cold compressions and for the shock compressions or for the Hugoniot.[2] Also shown is the zero temperature curve as derived from the Thomas-Fermi statistical atomic model. This latter model is appropriate for the high temperature, high compression region, but fails to include the necessary

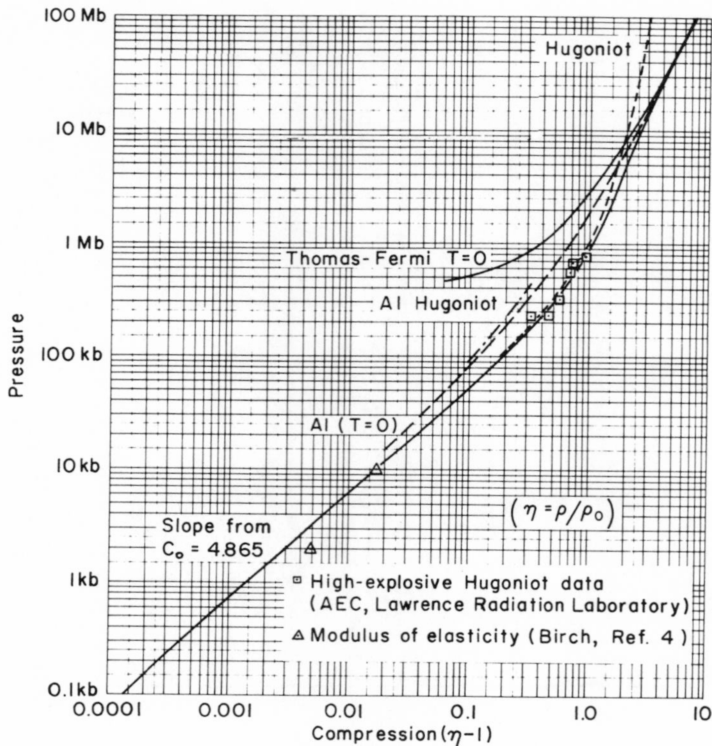

Fig. 2--Basalt compression characteristics.

nuclear and solid state effects of importance at lower temperatures or densities. The Thomas-Fermi model represents an account of only the electronic energies. Using all these inputs, and being guided by other well-founded cold compression and Hugoniot (shock) curves such as that for aluminum (illustrated), a consistent cold compression curve can be constructed and, when used as a correction to the temperature dependent Thomas-Fermi solution for an appropriate mixture of atomic elements, can lead to a prescription of the caloric equation of state. Such a case is illustrated in Fig. 3, where the ratio of the product of pressure and specific volume to specific internal energy of the rock is displayed for constant temperatures (in electron volts) as a function of the density relative to the standard density of the rock. The change of sign below normal density corresponds to a region of tension for lower temperatures and disappears at higher temperatures.

A simpler but similar equation of state has been used in hydrodynamic calculations modeling both the contained and a surface burst

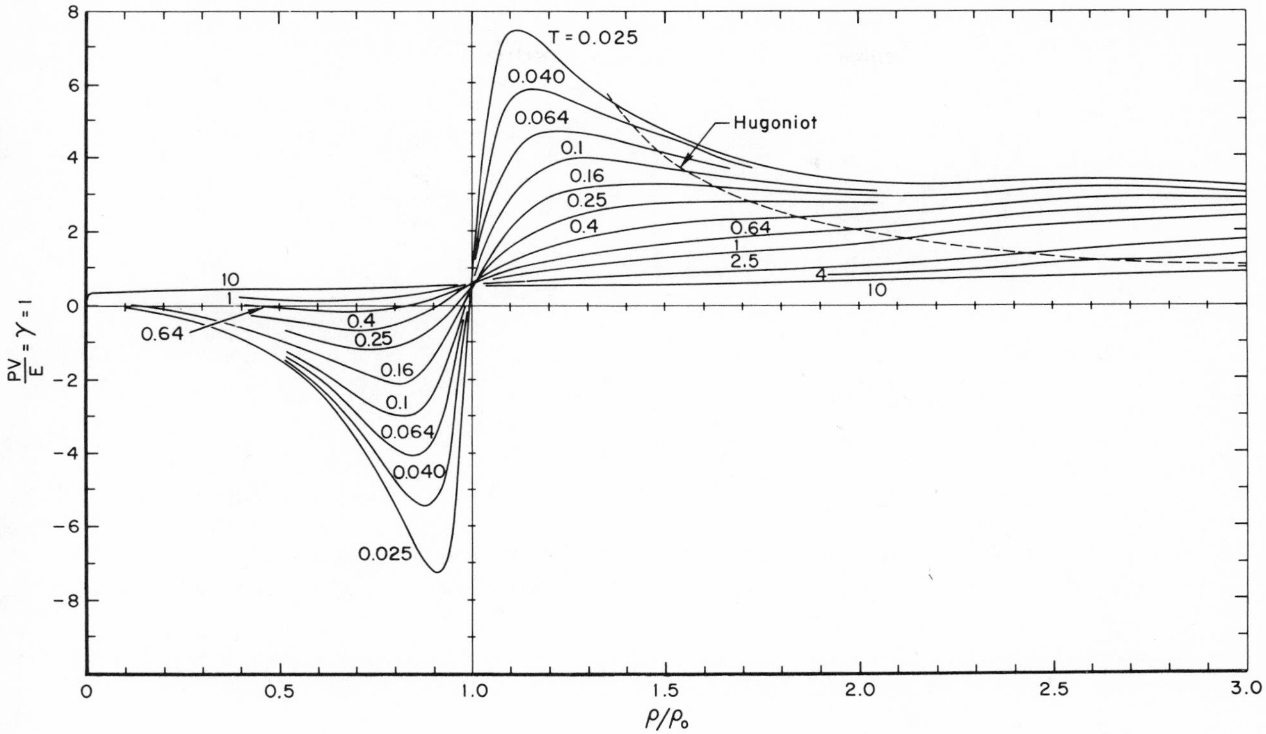

Fig. 3--Equation of state for basalt (isotherms).

nuclear explosion.[3] The surface burst calculations lead to shock patterns of the nature illustrated in Fig. 4, where isobars of pressure (in kilobars) are indicated at an instant in time early in the excursion of the ground shock but late in the crater formation. The final apparent crater is shown. At a similar time, the particle velocities accompanying the shock show the hemispherical expansion and show also the high speed ejection of the vaporized or pulverized material from the cratering region (Fig. 5). This latter phenomenon of material blowout attests to the important influence of the surface and to the consequent rarefactions generated behind the shock as the stress is relieved below the surface.

A proper calculation of the rock dynamics produced by a nuclear explosion on or near the earth's surface must include the complexities of such a hydrodynamic solution with its two-dimensional detail, but must also incorporate more realistic solid state properties at the lower stress level in order to extend the model to levels of interest in protective construction. The extension of predictions

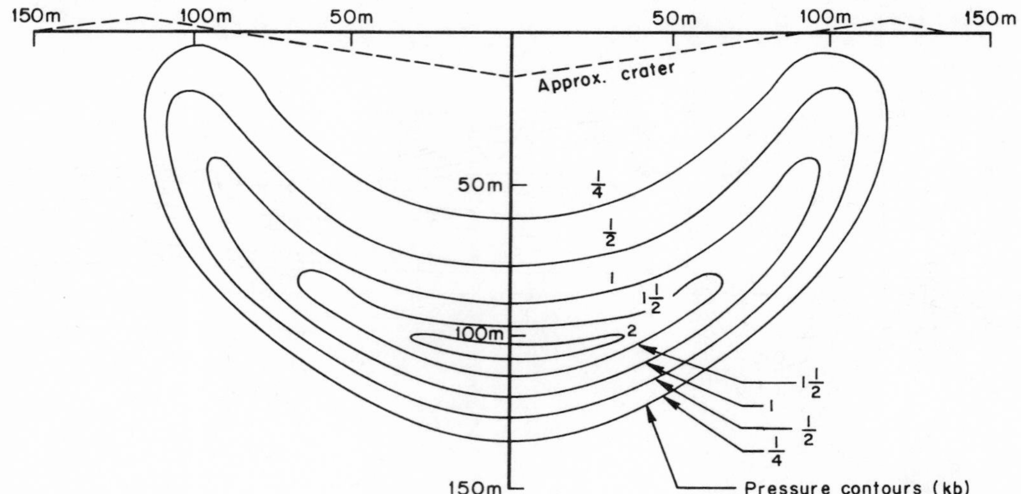

Fig. 4—Nature of ground shock from 1-MT surface burst (at 0.06 sec).

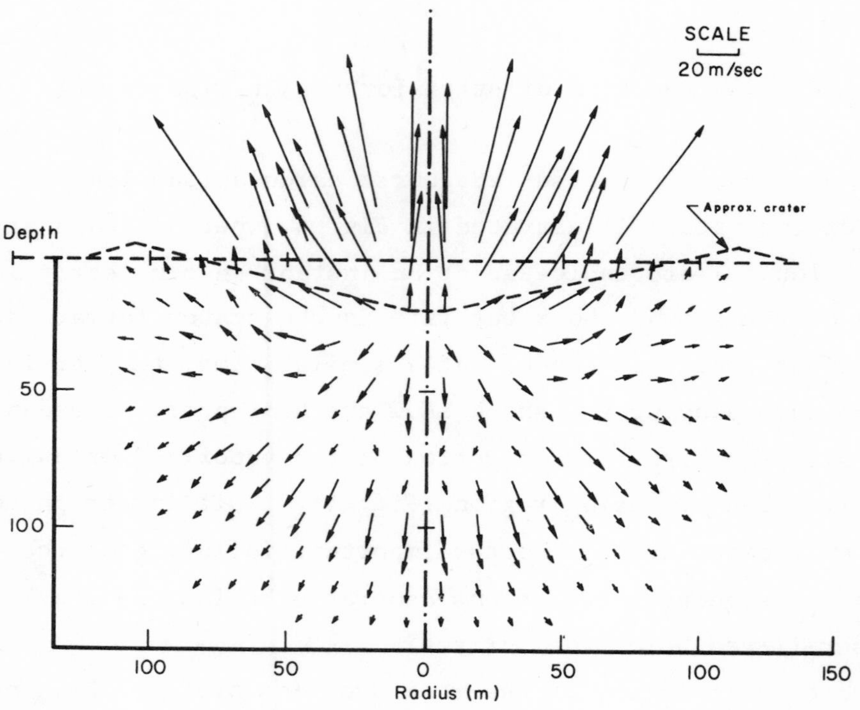

Fig. 5--Nature of particle velocities from 1-MT surface burst (at ~ 0.06 sec).

to lower stress levels than are believable from these fluid models may be approximately accomplished by comparisons of theory and observations for the contained explosions, together with the comparison of calculations for a surface burst with those for contained bursts. The latter comparison, between surface and contained bursts, leads to a measure of the escape of energy out of the ground into the air. The percentage of effective energy left in the ground from a surface burst is shown in Fig. 6 as a function of the shock depth as it expands into the ground.

Using this effective energy to scale the stress-distance curve for the empirical hard rock contained burst, one can derive an effective stress-distance relation for a surface burst, as is also illustrated in Fig. 6. Such a derivation is justified at least to the extent that hydrodynamics is appropriate. Although the hydrodynamic solutions are not likely to be valid in an absolute comparison below about 100 kb, the relative energy effectiveness predicted by the purely hydrodynamic solution may find some continued meaning at lower stress levels. The curve labeled "Surface burst" in Fig. 6 illustrates the vertical stress predictions reached in this manner. Figure 7 indicates the approximate stress distribution off the vertical, showing only slight contribution from the air shock on the surface.

The direct shock from the cratered region decreases rapidly with distance from the point of burst both because of the increasing volume of rock into which its energy must press, and also because of the rapid relief in rarefactions through the surface into the air. At the same time, the energy in the strong shock in the air is spread over a much wider area. At farther distances outside the crater and at shallow depths, the shock conditions induced by a near-surface explosion are largely determined by this air blast. The air blast pressures from a surface burst are less than 2 kb at the crater edge and less than 1 kb beyond the crater lip (Fig. 7). For a hard rock medium, such stress levels exhibit little permanent effect, and adequate shock insulation, together with designed resistance, can ensure survival outside this area immediately adjacent to the crater.

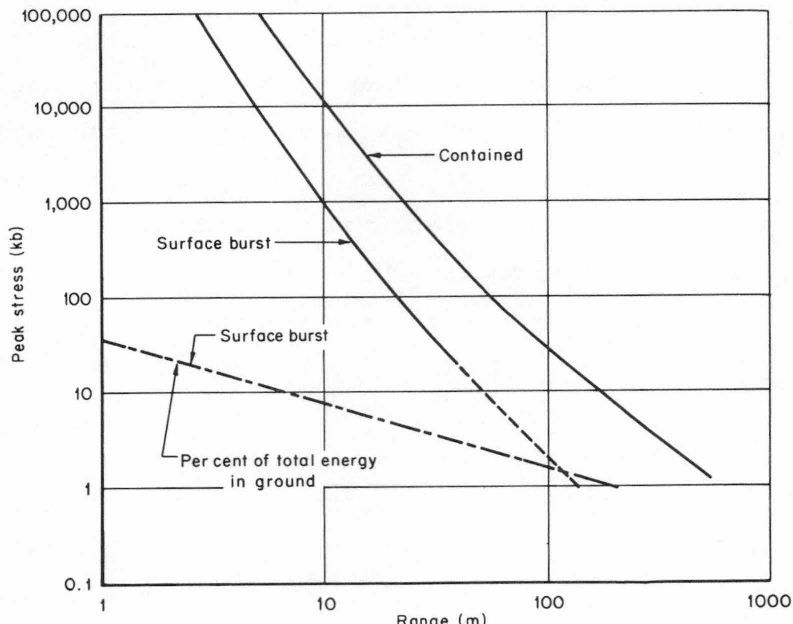

Fig. 6—Comparison of peak stress versus depth for 1-MT surface and contained explosions.

The limited nature of this region of air induced damage is due in part to the inability of the air to sustain for significant durations stresses that approach or exceed the elastic limits of the rock medium. Where the rock is capable of supporting elastically these air induced shock loads, an underground opening or structure requires little further protection in order to survive. (Most efforts to increase the survivability of an underground rock bound installation are constrained to this regime where the rock in the surroundings is not stressed beyond its elastic limits, but where the amplifications of stress at the boundaries of an opening in the rock can lead to failure.)

One feature of the strong air blast that further reduces its impact on buried structures is the very short duration of the high pressures. The very fast rise and decay of the strong air blast pulse puts much of the ground shock energy in high frequency components which experience considerable absorption and dispersion in the first few meters of earth cover. The normalized overpressure histories[5] of Fig. 8 dramatize the highly transient nature of the

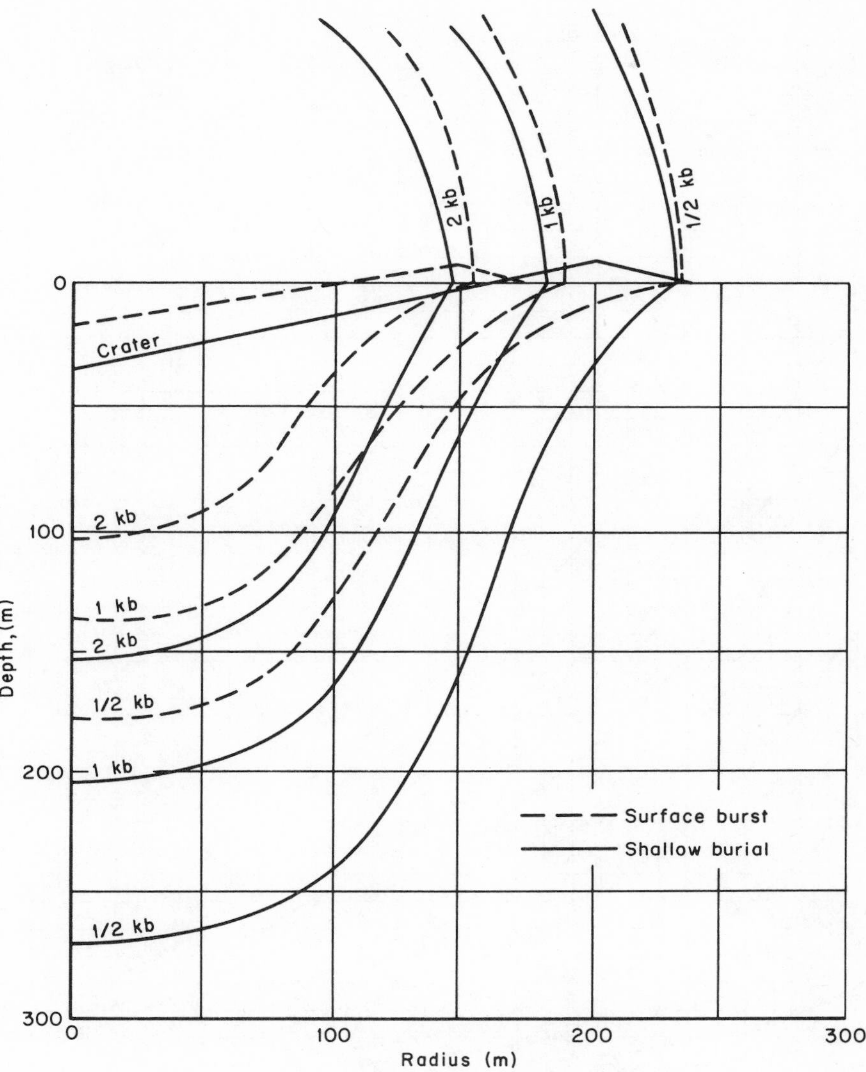

Fig. 7—Peak stress contours for 1 MT.

high overpressure loads. The duration of the positive overpressure for the high pressures from a 1-MT explosion runs for 1½ sec, but the air pressures nowhere remain more than 1/10 kb for longer than 30 ms. The pulse propagated into the ground must then begin with much of the energy in frequency components between thirty and a thousand cycles per second. (Finite rise times for the air shock preclude much higher frequency contributions.)

Under a nuclear shock, openings in rock may collapse, may spall, or may become disrupted at joints or faults. The modes of dynamic

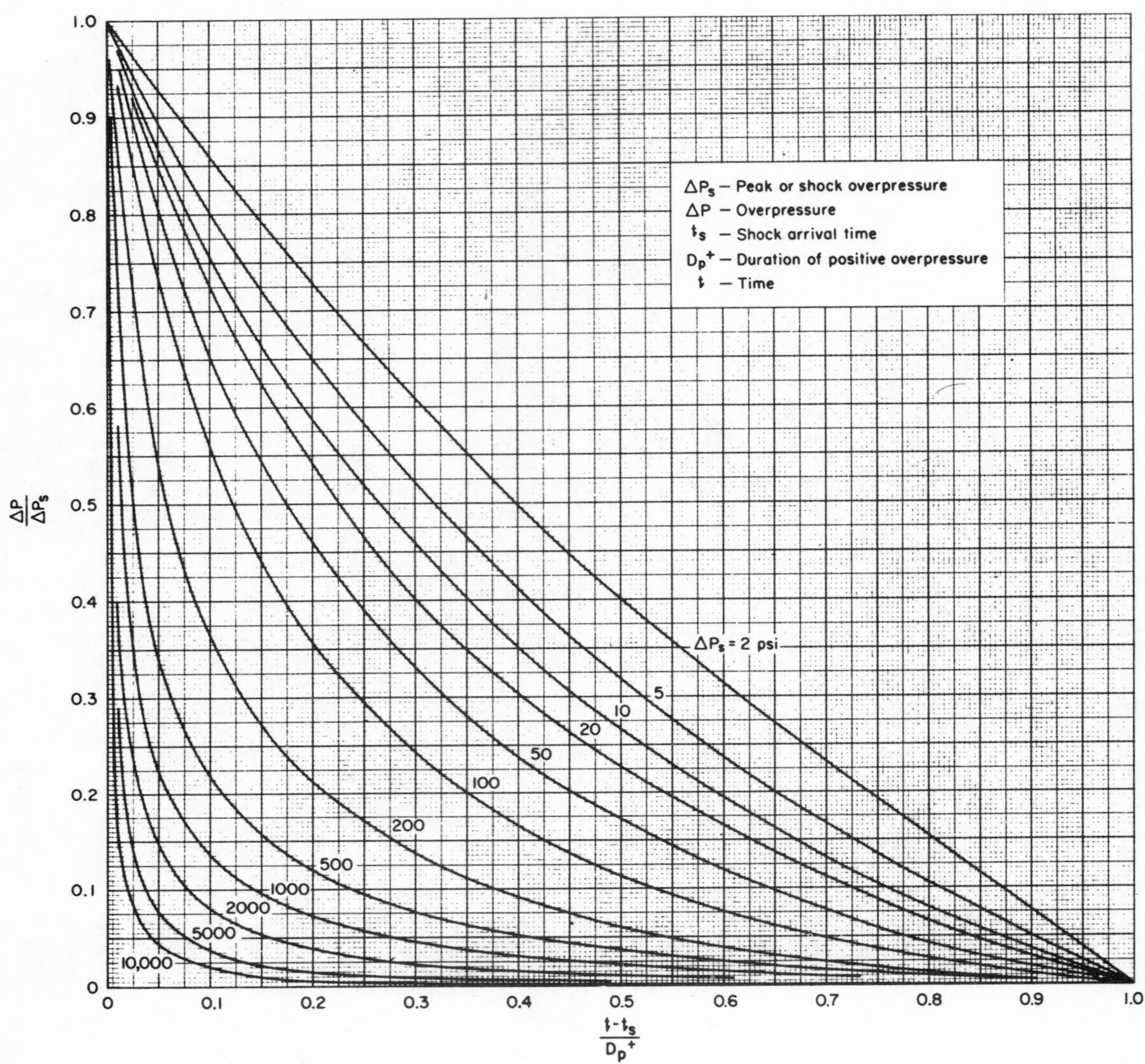

Fig. 8—Form factors for overpressure histories from nuclear explosions.

failure and the levels of transient stress at which collapse occurs have been the subject of considerable study. In most of these studies, the simplifying assumptions, made in order to facilitate analysis, have emphasized the linear elastic aspects and minimized the complications and changes in response character due to the less ideal properties, those leading to plastic deformations, cracking, and slabbing. Since nothing fails elastically, such calculations always leave something more to be done; they have, however, contributed much to the understanding of stress concentration and structure interaction. Further, the tendency is to simplify the applied stress patterns so as to be unrealistic in their relation to explosively derived loads. (Finite rise times and realistic decay rates are too readily overlooked.)

Typically, transient stress concentrations at boundaries of regular cavities having no sharp corners or angles are predicted not to exceed grossly the static amplifications. The classic example concerns a cylindrical opening in an ideal elastic medium loaded with a step wave of infinite duration. This opening experiences at one point a hoop stress about three times the applied stress.

Much can be done to alter the stress patterns in the vicinity of an opening, and some success can be anticipated in strengthening the cavity walls with special lining. The use of materials of low compressive strength between rock walls and resistant inner liners has the advantage of unifying the transient loads on the inner shell (as in Fig. 9) to reduce the likelihood of failure through buckling (i.e., of reducing the amplitudes of higher modes which could otherwise defeat the inherent strength of arched or curved liners); it also has the advantage of absorbing the high stress, high frequency aspects of the load, thus reducing spall and high shock input to the interior structure. Allowing for the care and expense of such elaborate insulation or isolation linings, cavities can be made to survive under loads which in the free field are close to the yield stress for the basic rock.

With such ultimate preparation, an underground structure in hard rock could be expected to survive at 1- or 2-kb free-field stress. Such stress levels occur (according to Fig. 7) beneath a 1-MT surface explosion at a depth of 100 or 200 m, and at comparable distances along the surface from the burst point. The sensitivity to some details of the initial nuclear explosive position is fairly obvious. Consequently, for other burst conditions and for other weapon geometries it is necessary to construct such stress envelopes with

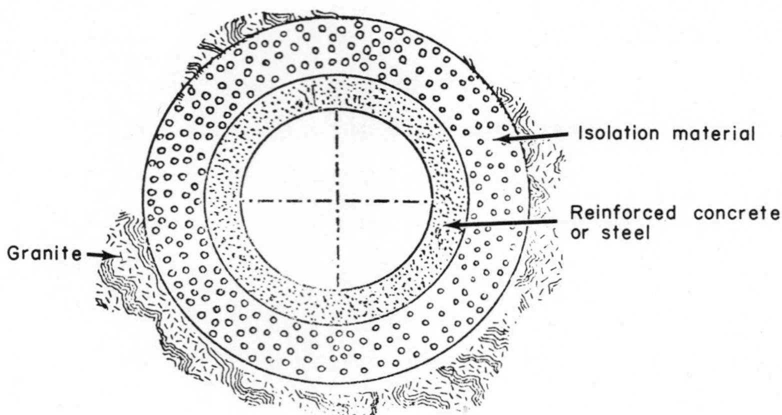

Fig. 9—Typical liner.

considerable care. In addition, some further caution in scaling to other yields must be recommended. In the absence of full scale test information, a simple volumetric or cube root of dimensions scaling with explosive energy has been commonly applied. However, the lack of scaling of gravity, of earth stratifications, of various material properties, as well as of explosion features, makes such procedures less than rigorous.

REFERENCES

1. Sandia Corporation, Close-in Phenomena of Buried Explosions, Second and Third Semiannual Reports, November 1, 1961, and May 1, 1962.
2. Brode, H. L., and A. C. Smith, Dynamic Properties of Matter under High Stress--Thermodynamic Descriptions, The RAND Corporation, RM-3361-PR, November, 1962.
3. Brode, H. L., and R. L. Bjork, Cratering from a Megaton Surface Burst, The RAND Corporation, RM-2600, June 30, 1960.
4. Birch, Francis, J. F. Schairer, and H. C. Spicer, "Handbook of Physical Constants," Geol. Soc. Am., Spec. Papers, No. 36, January, 1942.
5. Brode, H. L., Phenomena in the Immediate Vicinity of a Nuclear Explosion, The RAND Corporation, RM-3544-PR, February, 1963.

ABSTRACT

The effects of nuclear bursts on hard rock ground are considered with regard to protective tunnel design. The rock half-space is divided into five zones, for which various design requirements have to be considered. Results from TNT tests on granite models are reported, especially as concerns spalling effects. The large displacements in the rock seem to necessitate spring-mounted buildings in protected tunnels far from the explosion.

RÉSUMÉ

Les effets qui accompagnent les explosions nucléaires sur des terrains de roche dure sont examinés du point de vue du calcul du tunnel de protection. Le demi espace rocheux est divisé en cinq zones pour lesquelles il faut considérer des exigences de calcul variées. Les résultats de certains essais au Trinitrotoluol (TNT) sur modèles en granit sont donnés, en particulier dans le cas des effets de repoussement. Les déplacements considérables dans la roche nécessiteraient des structures montées sur ressorts à l'intérieur de tunnels protégés situés loin de l'explosion.

AUSZUG

Die Auswirkungen der Kernexplosionen im festen Fels werden im Hinblick auf unterirdische Schutzbauten untersucht. Der Halbraum des Felsen wird in fünf Zonen eingeteilt, die verschiedene Entwürfe verlangen. Ergebnisse von Modellversuchen mittels Trotylladungen in Granit werden gegeben, besonders im Bezug auf die Auswirkungen von Ausstössen. Es erscheint, dass die grossen Verschiebungen, die im Gebirge auftreten, im Schutztunnel federnd gelagerte Einbauten noch in grossen Abstand von dem Explosionsort verlangen.

NUCLEAR SURFACE BURSTS AND THE DESIGN OF PROTECTIVE CONSTRUCTION IN HARD ROCK

Sten G. A. Bergman*

NUCLEAR SURFACE BURSTS on hard rock ground will produce cratering and high pressure effects in the vicinity of the crater, as previously described by Brode.[1] The direct ground shock propagating outwards in the rock half-space also causes transient displacements in the rock. The displacement phenomena in granite have been studied experimentally and theoretically by Broberg,[2-4] among others. In the design of underground protective structures, the free-field ground shock pressure will decide the stability of the tunnel openings, and the transient displacements will decide what type of interior structure is needed to obtain the protection required.

Figure 1 was prepared to illustrate schematically the problems involved. The rock half-space in the vicinity of a nuclear surface burst is divided into zones that require various types of protective construction. These zones were calculated for hard granite rock for 1 MT and 50 KT, respectively. Formulas given for homogeneous granite by Broberg[2-4] have been used. It has been assumed that, in the vertical direction, all faults and joints are filled by water and that any extra attenuation of the shock by reflection in such rock faults may be considered negligible. On the other hand, such faults in connection with an irregular surface topography will probably attenuate the surface ground shock considerably. Since no data on this attenuation are available, it has, in Fig. 1, been quite arbitrarily assumed that the outer boundaries along the surface of zones 3 and 4 are about one-third the calculated values for homogeneous granite.

Thus, zone 1 is the rupture zone. It is not believed possible to construct tunnels that could withstand the stresses in this zone.

---

*Royal Swedish Fortification Administration, Stockholm, Sweden.

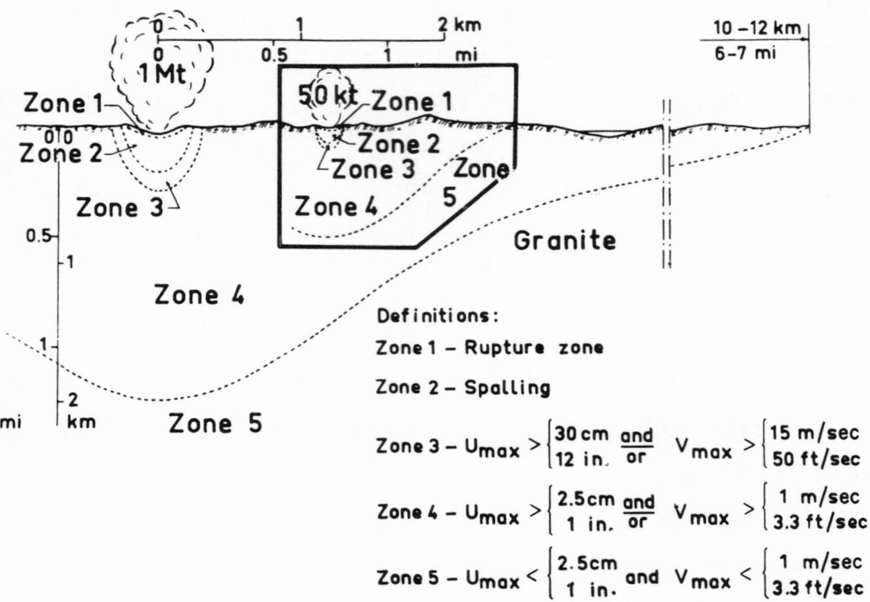

Fig. 1—Design zones in granite rock at nuclear surface bursts of 1 MT and 50 KT respectively.

Zone 2 is the spalling zone. In any unlined tunnel located within this zone, spalling of the roof and/or the walls will occur. At the upper boundary, the spalling will be very heavy, approaching a breakthrough of the tunnel roof. There has been some discussion about the possibilities of building for survival within the spalling zone. In the case of high-yield weapons, even big tunnels with, say, a 60-ft span will be small compared to the depth of zone 2. Since in this case the length of the ground shock compression wave will also be large compared to the tunnel span, it can be argued theoretically that the thickness of an eventual spall would be large enough to form a self-supporting confined arch over the tunnel span. If the idea of such an arch action is correct, it should be possible to demonstrate it in model tests.

Such model tests have been performed in Sweden with granite models of various scales,(5) the smallest being ½ in. in tunnel diameter. The biggest models included 5- x 7-ft tunnels subjected to surface bursts from charges of 2000 lb of TNT, which give a wave length of about seven times the tunnel span. All the tests showed that the spalling did actually occur as a multiple spalling with many thin layers about 2 in. thick, as shown in Fig. 2. We had predicted thicknesses of about 3 to 5 ft, thus, the model tests did not confirm the idea of confined arch action but, rather, they indicated that the shock front in hard rock is more intense than was anticipated. No relevant arguments have hitherto been presented which would show that the same phenomenon will not occur in the full scale nuclear case for hard rock with joints filled with water.

The model tests mentioned also included investigations on various methods to strengthen the rock against the spalling effect (Fig. 3). It was found that roof bolting, even with close spacing, could not prevent spalling and that it gave a rather poor strengthening effect, as shown in Fig. 4. A reinforced concrete layer (gunite) directly on the rock surface decreased the depth of the spalling as much as the concrete thickness, and the spalled concrete structure with a thickness of about one-twelfth of the tunnel span had enough strength to prevent the rock spall from falling down.

The conclusions seem to be that tunnels in hard, saturated rock which are intended to survive in the spalling zone should be strengthened by exceptionally heavy lining. The interior building must also be designed to take the very large displacements (U) and relative velocities (V) of the rock (see Fig. 1), and this will require very unusual spring-mounted systems.

Zone 3 is a zone with very large displacements, $U_{max} > 12$ in., and large relative velocities, $V_{max} > 50$ ft/sec. In this zone the ground shock may also cause large unstable blocks to loosen from tunnel roofs and walls. It is believed that in zone 3 the tunnel will be stable, although roof bolting and/or tunnel lining is recommended to prevent spalling of loose blocks. Very elaborate spring-mounted interior buildings are required. It may be noted that zone 3 tends to coincide with zone 2 for nuclear bursts in the low kiloton range (see Fig. 1).

Zone 4 is chosen so that the calculated maximum displacements U_{max} should be more than 1 in. and less than 12 in., or that the maximum relative velocity should be more than about 3 ft/sec and less than 50 ft/sec. Even in this zone there may be a definite risk that blocks loosen from tunnel roofs and walls.

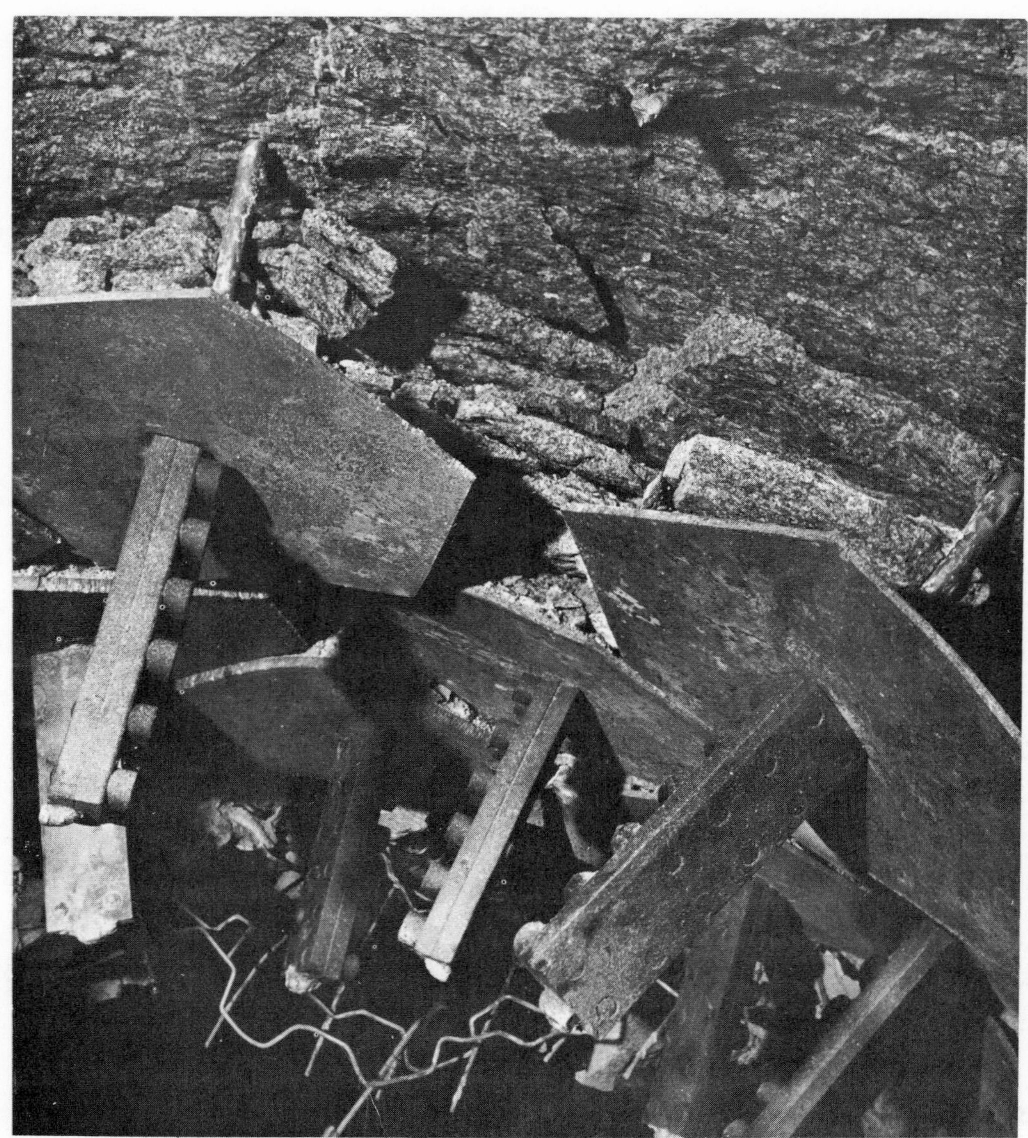

Fig. 2—Multiple spalling in thin layers in a 5- by 7-ft granite tunnel.

It is believed that the ground shock in zone 4 will not cause any great danger to the stability of tunnels at moderate depth, say at less than 1000 ft (300 m), if residual geologic stresses[6,7] or special fault zones do not complicate the problem. At greater depths, the ground shock stress should be added to the tectonic stress state, which alone may strain the stability of the tunnel considerably.[8] The possibilities of making good technical prognoses for very deep tunnels are so poor and the economical consequences

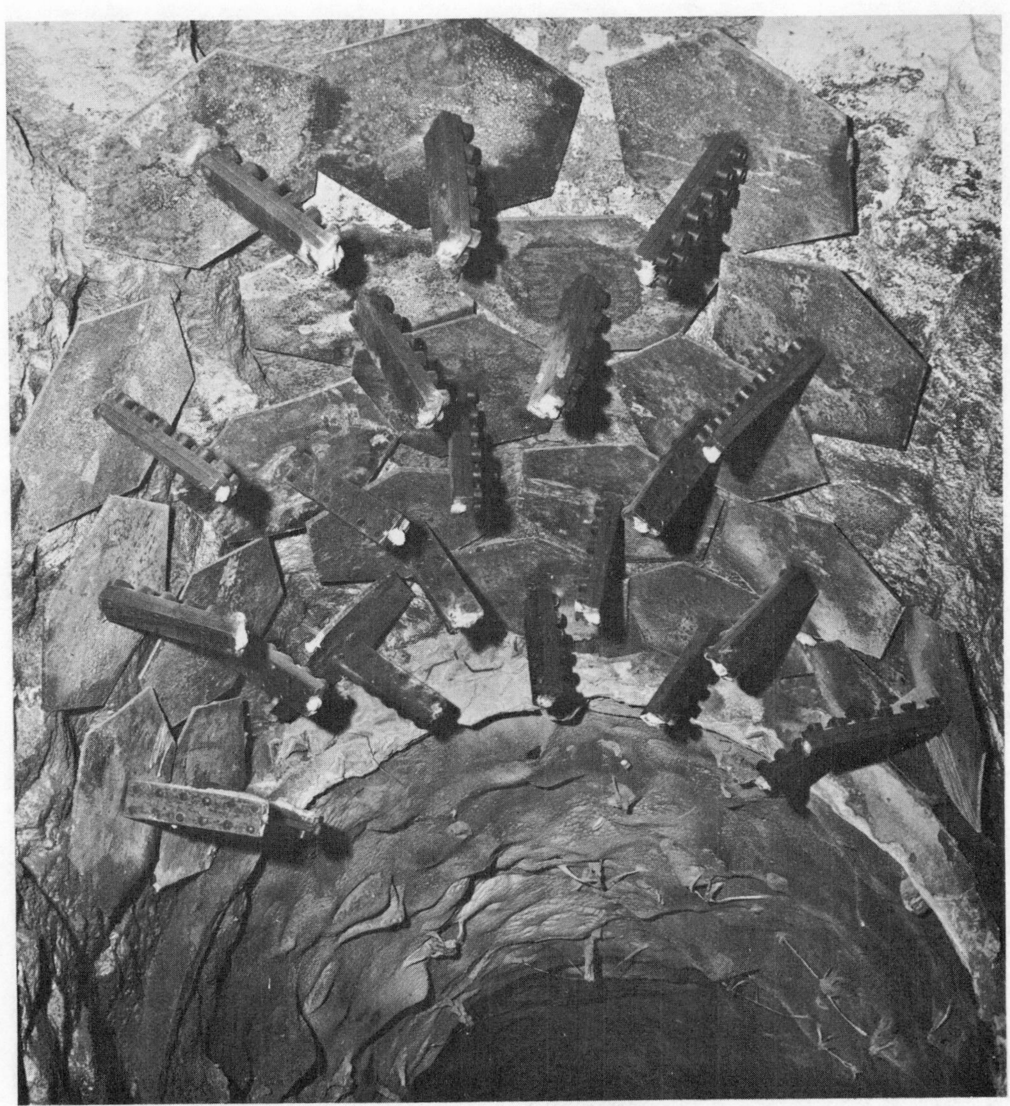

Fig. 3—Strengthening of tunnel roof with rock bolting (foreground) and reinforced gunite (background).

of bad estimates so serious that the very deep parts of zone 4 will probably not be used much--at least not for nuclear bursts in the megaton range.

The displacements and velocities caused by the ground shock in zone 4 are so large that most interior buildings in tunnels have to be spring-mounted. As indicated in Fig. 1, the displacements are attenuated rather weakly along the ground surface, especially for large nuclear bursts.

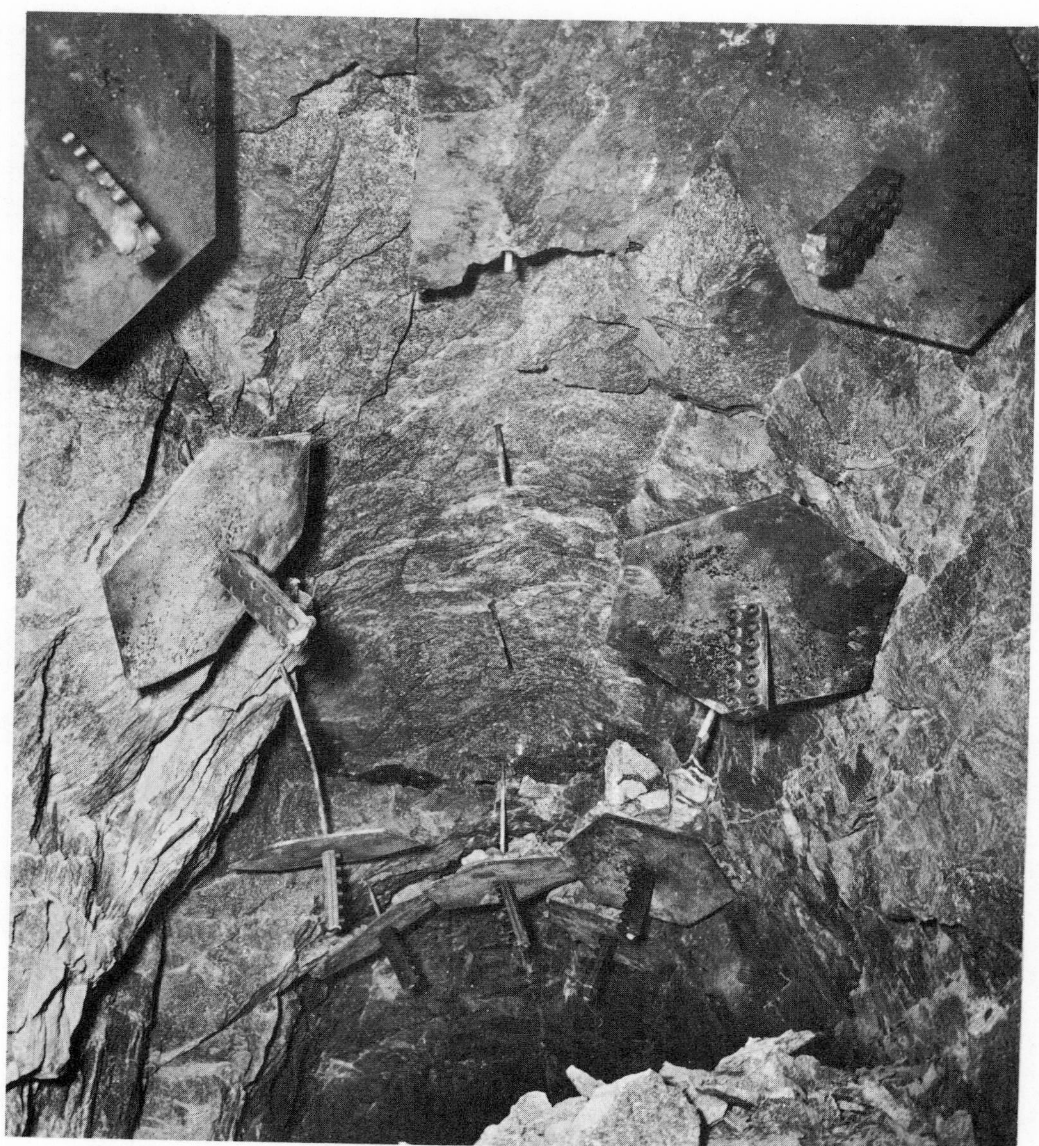

Fig. 4—Rock bolts after a spalling test.

In zone 5 the maximum displacement should not exceed 1 in., and the maximum relative velocity should not exceed about 3 ft/sec.

The boundary between zones 4 and 5 has been chosen so as to represent displacements and velocities which very carefully designed ductile structures would be able to withstand without any elaborate spring-mounting.[9] However, much of the equipment in such structures would probably have to be isolated from the ground motion.

The zones shown in Fig. 1 may be pessimistic in so far as they do not take into account any attenuation of high frequencies in vertical directions due to joints and faults always found in hard rock sites. However, it is believed that they show the right order of magnitude. It is obvious, then, that building underground establishments in hard rock for survival in case of nuclear surface bursts is a difficult and complicated task. The hard and competent rock, which is advantageous from a construction point of view, has definite disadvantages in other respects, since it does not attenuate the ground shock as well as is desired. A good site for underground openings should have a hard and competent rock covered by hundreds of feet of overburden with good attenuation characteristics, e.g., sand or gravel.[1,10]

I have used the expression "it is believed" quite a few times. I hope that this is not misunderstood as a firm belief in the numerical validity of the rather far-reaching extrapolations which have been made on the basis of model tests. The expression merely stems from the present working hypothesis of the research group that I represent. In fact, no one would be happier than I if someone presented a full-sized nuclear test in hard, saturated rock and showed that we have overestimated the effects by a factor of 10.*

REFERENCES

1. Brode, H. L., "Rock Mechanics Considerations in the Design of Underground Protective Structures," The RAND Corporation, RM-3583, May, 1963, pp. 15-114.
2. Broberg, K. B., "Sprängämnes verkan vid kontaktdetonation mot granitplattor (Effects of Contact Detonation on Granite)," Royal Swedish Fortification Administration, Rept. 109:15, Stockholm, 1960.

---

*This paragraph was obtained from the tape of the conference.

3. Broberg, K. B., "Utstötning i bergtunnlar (Spalling Phenomena in Rock Tunnels)," Royal Swedish Fortification Administration, Rept. 109:16, Stockholm, 1961.

4. Broberg, K. B., "Om skydd mot markskakningseffekter i berg (On Protection against Direct-transmitted Ground Shock in Rock)," Royal Swedish Fortification Administration, Rept. 109:18, Stockholm, 1963.

5. Broberg, K. B., "Om skydd mot utstötningsskador i bergtunnlar (On Protection against Spalling Effects in Rock Tunnels)," Royal Swedish Fortification Administration, Rept. 109:17, Stockholm, 1961.

6. Hast, N., "Bergtrycksmätningar i gruvor (Strain Measurements in Mines)," Jernkontorets Ann., Vol. 141, 1957, p. 601.

7. Coates, D. F., "Rock Mechanics Developed for Nuclear Defense Applied to Underground Openings," Can. Mining J., Vol. 83, No. 12, 1962, pp. 38-42.

8. Logcher, R. D., "A Method for the Study of Failure Mechanisms in Cylindrical Rock Cavities due to the Diffraction of a Pressure Wave," MIT Department of Civil Engineering, Technical Report T62-5, 1962.

9. Langefors, U., B. Kihlström, and H. Westerberg, "Ground Vibrations in Blasting," Water Power, F, 1958, pp. 335-338, 390-395, 421-424.

10. Genensky, S. M., and R. L. Loofbourow, Geological Covering Materials for Deep Underground Installations, The RAND Corporation, RM-2617, August, 1960.

DISCUSSION

H. BRODE (USA):

I would like to say I appreciate the work that the Swedish investigations represent, that we have followed them with interest before, and that I am also guilty of such--if there is guilt to be assigned--extrapolation of small explosions to large explosion effects, but there are the many cautions. I would just like to restate some of those cautions.

Almost all of the effects that one can imagine would work toward reducing the extent of large yield effects; that is, there are dissipative mechanisms. There are inefficiencies in the coupling, etc.

For instance, in the large-scale situation (and Dr. Bergman mentioned it), there is the problem that at greater depths the overburden stresses that must be present offer an added complication. In the response of openings there are many features which are avoided in the small-scale experiment but which cannot be omitted in the large-scale cases. For example, the transfer of the stress from, say, a surface burst through any large volumes or great distances of rock inevitably must encounter more inhomogeneous features than in any small-scale experiment. The variations in seismic velocities with overburden, the faulting, the jointing, the layering, and the normal stratification of rock must be present on a different scale and must offer greater interference with the stress transmission in the large-yield or very deep underground problems, whereas such imperfections are carefully avoided in the small experiments.

The dispersion which may come from the changes in seismic propagation velocities over large ranges, and particularly as the material comes under higher overburden loads, may also contribute to the decrease in stress in the vertical direction. The enhanced dispersion of the signal from horizontal bedding and from higher seismic velocities at depth work to decrease the vertical stresses and particle velocities. But probably the most important features have to do with the details of the source. Some features of the early explosion that I listed in my talk make nuclear explosion loads very distinct from high explosively derived loadings.

It can be very misleading to take (in a simple way) the energy from a high explosive experiment and scale its tonnage or its pounds of explosive in a linear fashion to a nuclear explosion yield. The very details of the coupling, particularly near a surface, are inevitably in the direction of very much less efficiency for a nuclear explosion. Similarity between the two can hardly occur until the nuclear burst is buried at least to a point where the stresses caused by the nuclear explosion are of the same order as those in the detonation process. For instance, in the explosion of a megaton nuclear charge, for similarity the charge should be buried at least 50 m in order for the shock wave that reaches the surface to be no stronger than the detonation wave from a megaton of high explosive at that point.

There are further differences in the nuclear explosion which have to do with the cooling of the air and the energy transfer out of the ground, but other large inefficiencies (independent of the surface interactions) already can be noted in the underground tests where one finds the shock from nuclear explosions, such as the Rainier shot of 1.7 KT, initially stronger but with a very rapid decay. Scaled to 1.7 KT, a high explosive shot would look very much different, i.e., you expect a much more efficient explosion from the high explosives because much less energy is left in the superheated rock.

The high explosive does not heat the rock nearly as much and does not make a large melt, or an initially large vaporized rock mass, so that at least this initial energy fraction is not left behind in the very earliest stages of a contained chemical explosion. This difference is equally valid for surface bursts.

S. G. A. BERGMAN in reply:

I think I may state that we agree with the arguments brought forward by Dr. Brode regarding the difficulties and the delicate nature of extrapolating experimental TNT results to nuclear applications. Thus, as can be found in the original test reports by Dr. Broberg, which I have used as a basis for my presentation, we have assumed that a 1-KT nuclear surface explosion will transfer only as much energy into the ground as a 0.2-KT TNT surface explosion. I believe that this conversion factor corresponds rather closely to recent estimates made by Dr. Brode.

The important question, as we see it, however, is whether the rapid decay of the ground shock effects found in dry Nevada rock, in which faults and joints probably are partially air filled, can be assumed to apply even to hard rock with all faults and joints completely filled with water. Since the acoustic impedance of dry, air filled faults may differ considerably from that of water filled faults, one may expect that the saturated rock will transmit the ground shock much better than the dry rock. Although all of Dr. Brode's arguments tend to show that our extrapolations probably overestimate the ground shock displacements--and I agree, they probably do--it would therefore be most desirable to obtain full-scale nuclear test results from saturated hard rock.

ABSTRACT

This presentation will include a description and some analysis of

(1) the upheaval of the floor of an open pit in competent rock at shallow depth,
(2) the horizontal deformation of a power tunnel at shallow depth, and
(3) a rock burst of unusual magnitude in an underground mine.

RÉSUMÉ

Cette présentation comprendra une description et une certaine analyse

(1) du soulèvement du fond d'un puits ouvert dans la roche capable à faible profondeur,
(2) de la déformation horizontale d'un tunnel de puissance à faible profondeur, et
(3) d'un éclatement de roche de grandeur peu commune dans une mine souterraine.

AUSZUG

Diese Arbeit bringt die Beschreibung und einige Berechnungen über

(1) Aufwölbung der Sohle einer offenen Baugrube geringer Tiefe in gesundem Fels,
(2) Horizontale Verformung eines Kraftwerksstollens in geringer Tiefe und
(3) Einen ungewöhnlich starken Bergschlag in einem Untertagebau.

SOME CASES OF RESIDUAL STRESS EFFECTS IN ENGINEERING WORK

D. F. Coates*

Introduction

AS STATED in its title, this conference is on the state of stress in the earth's crust. Information has been presented on fundamental concepts, methods of measurement, and principles of design. In addition, the chairman of this session requested that some actual cases be prepared that describe the effect on engineering projects from detrimental fracture and flow that are a result of residual stresses. The following cases occurred in Canada on both mining and civil engineering projects.

Floor Upheaval in an Open Pit

In the development of an open pit in Ontario it was necessary to strip up to 160 ft of overlying rock before the ore bearing formations were reached. Under about 10 to 15 ft of glacial till there was between 95 and 160 ft of Paleozoic limestone of the Ordovician Period in the Black River group of formations. Figure 1 is a photograph of the full depth of this limestone overlying the pre-Cambrian sediments and intrusives.

The basal deposits in the Black River formations consist of shales, some sandstones, dolomites, and limestone. Although the thickness of the basal formation varies, its order of magnitude is 100 ft. Above these basal deposits there is a middle phase in the group comprising pure, thick bedded, crystalline limestone about 150 ft thick. The average

*Professional research engineer, Ottawa, Canada.

Fig. 1—Top of open-pit wall in Paleozoic limestone.

compressive strength of the core samples obtained in the limestone at the open pit was found to be 16,900 psi, with a modulus of deformation of 6.1×10^6 psi.

The pre-Cambrian rock containing the ore consists of intensively intruded sediments. In general the rocks are hard and fairly massive, the actual types being granitic-gneiss, syenite, diorite, gabbro, limestone, and others.

During the stripping of the Paleozoic limestone the pit had been developed to the approximate dimensions of 1000 x 2000 ft, with a depth of 50 ft (see Fig. 2). Quite suddenly, in the month of January, the bottom of the pit cracked and heaved a maximum of about 8 ft within a few minutes. Figure 2 shows the location of the initial crack, which in time propagated itself.

Figure 3 shows the initial crack shortly after it occurred. It can be seen that about 50 ft of ground on either side of the crack had been deflected upwards. At this location the minimum thickness under

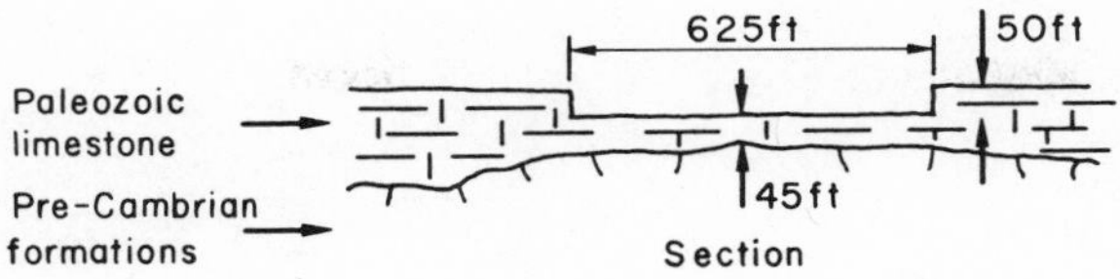

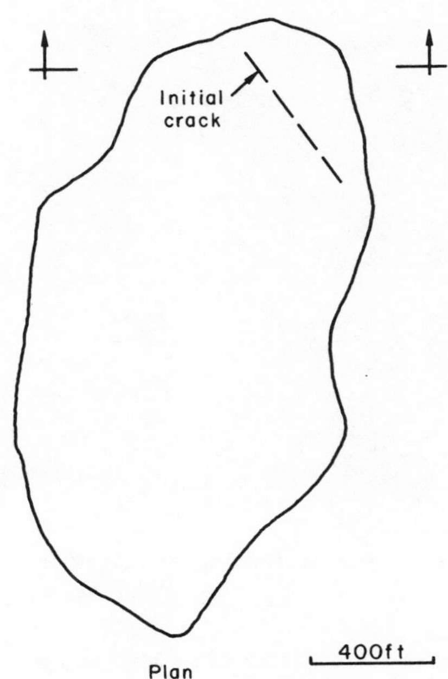

Fig. 2—Plan and section of open pit at time of floor upheaval.

the pit of the Paleozoic limestone over the pre-Cambrian rock was 45 ft, with an average thickness of about 55 ft.

If it is assumed that the upheaval was a case of elastic instability, the stress required to cause the cracking can be calculated. The geometry is rather awkward; however, the average width of excavation perpendicular to the crack can be used. This dimension would provide a member 1200 ft long.

The Euler formula for determining the critical stress for buckling has been used. The modulus of deformation of 6.1×10^6 psi is obtained from the tests on the core samples. The effective depth, visualizing a reduced moment of inertia due to natural joints in the rock, is assumed to be half the minimum thickness, or 22.5 ft. The end conditions of the formation as they pass under the walls can not be

Fig. 3—Upheaval of floor of open pit.

considered either fixed or pinned. An intermediate case that is commonly used for square-ended compression members is assumed, giving an effective length of 800 ft. Using these assumptions and ignoring side restraint, it would require a horizontal stress of approximately 4000 psi to cause buckling and upheaval.

By taking into account the concentration of horizontal stress below the excavation, the horizontal field stress is deduced to be of the order of 2000 psi in the Paleozoic limestone. By changing any of the above assumptions, of course, the resultant answer can be changed considerably.

Horizontal Deformation of a Power Tunnel

The second case is concerned with two 51-ft-diameter tunnels which were excavated in horizontally stratified rock with approximately 175 ft of rock over the roof, plus 130 ft of soil. The tunnels are 5½ mi long. Excavation of the tunnels was by the heading and bench method.[1]

The bedrock formations are shown in Fig. 4. The rock types are mainly dolomite, limestone, sandstone, and shale. The formations contain bedding and slip planes and tend to be heterogeneous in composition. At the elevation of the tunnels the rock formations, in spite of the test results on core samples, were considered to be of low strength, with the exception of the massive Irondequoit limestone that forms the roof.

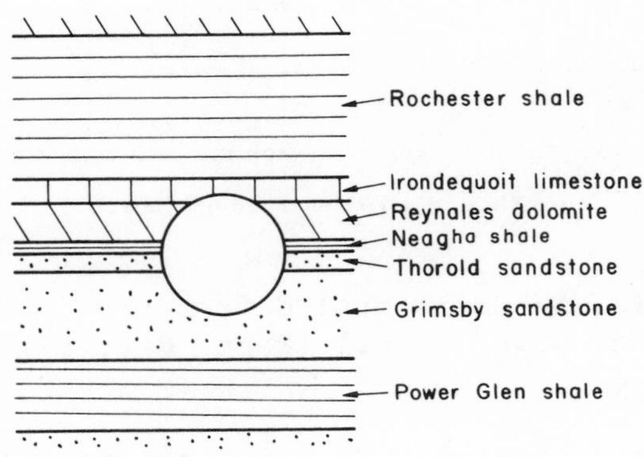

Fig. 4—Geological section at elevation of power tunnels (Hogg, Ref. 1).

The uniaxial compression strengths were as follows: Irondequoit limestone 15,000 psi, Reynales dolomite 12,000 psi, Thorold sandstone 15,600 psi, Grimsby sandstone 14,000 psi. The static moduli of deformation were as follows: Irondequoit limestone 7.28 x 10^6 psi, Reynales dolomite 5.95 x 10^6 psi, Thorold sandstone 2.62 x 10^6 psi, Grimsby sandstone 2.19 x 10^6 psi. It was not possible to obtain test values for the Rochester or Neagha shales as they disintegrated quickly on exposure to the atmosphere.

Measurements of the movement of the walls were begun during the excavation of the heading. Reference pins were anchored in the rock about 3 ft behind the working face. Horizontal diametral measurements were taken using an Invar micrometer tape.

The movement of the walls at one station in tunnel no. 1 are shown in Fig. 5. It can be seen that after an initial rapid movement the deformation rate decreased continuously until the excavation of the

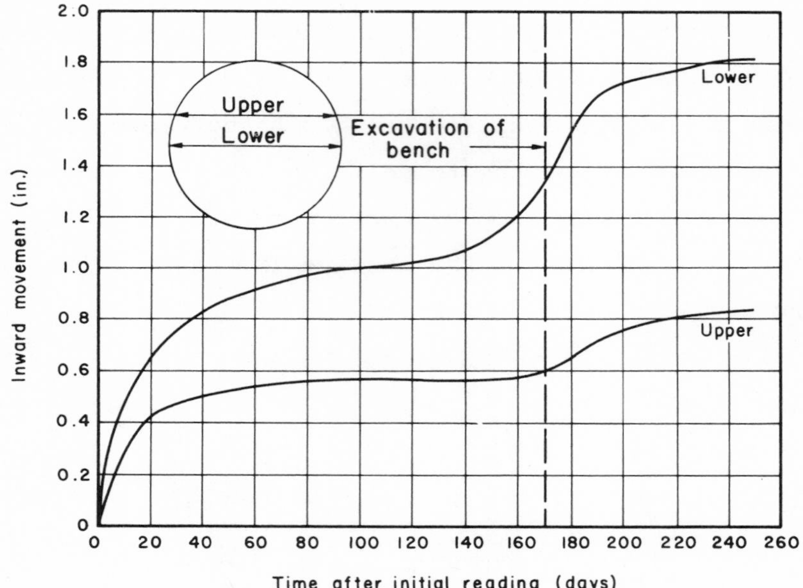

Fig. 5—Horizontal deformation of one power tunnel (Hogg, Ref. 1).

bench approached this station. When the bench was taken out, another large increment of deformation occurred followed by creep at a decreasing rate.

If the horizontal deformation resulting from the excavation of the heading or following the excavation of the bench is plotted against the logarithm of time, a straight line occurs.

Vertical movements in the tunnels were small. No measurements were made on the floor in tunnel no. 1; heaving, buckling, and cracking were observed, however. In tunnel no. 2, one measuring station indicated an upward movement of the floor of about 0.5 in. after the removal of the bench. Movements at the crown were of the order of 0.05 in.

The horizontal deformation of the walls of tunnel no. 2 varied from one-fourth to one-third of that measured in tunnel no. 1. The blasting and excavation of tunnel no. 2 caused the creep movements in tunnel no. 1 to increase for a period of time, after which they returned to the previous rate.

If it is assumed that the rock that has been deformed to produce the horizontal deformation of the walls has an average modulus of deformation of 5×10^6 psi, it is possible to calculate the order of magnitude of the residual stress in this area. By using an average vertical gravity stress of 340 psi and assuming that the horizontal deformation of 1.8 in. (see Fig. 5) is due to elastic strain, a horizontal stress normal to the axis of the tunnel of 4800 psi can be calculated. This, of course, is a very simple analysis.

Several questions arise from this trial analysis. Why, for example, did the second tunnel only have one-fourth to one-third of the deformation of the first tunnel? The determination of creep constants from the deformation-time measurements or a viscoelastic analysis, although complex, would also be of interest. In addition it would be interesting to determine the effect that the soft layers would have on this type of analysis.

A Large Rock Burst

The third case involves a very large rock burst in Northern Ontario which produced good seismic records at stations as far away as 920 km. It is possible that this rock burst may have released strain energy in the rock resulting from a high horizontal residual stress field.

The rock burst occurred in the mine shown in Fig. 6. Figure 6(a) is an elevation of the stoping area in the plane of the vein.[2] It can be seen that mining had proceeded to a depth greater than 4000 ft. Figure 6(b) is a cross section of the shaft showing the ore vein and two post-ore faults. Figure 6(c) is a plan of the vein and its relationship to the shaft at the 2450-ft level. Figure 6(d) is an idealized plan of the mined-out vein between the 1400- and 2800-ft levels.

The mineralization in this area is in pre-ore fault zones where the hangingwall has moved up 1300 to 1600 ft with respect to the footwall.[3,4] This faulting produced a considerable amount of gouge, brecciation, and alteration in the wall rocks. In addition, considerable post-ore cross and strike faulting has occurred.

Syenite porphyry is the most common host rock of the veins. Some augite syenite also occurs in the vein zone. Porphyry core samples were tested and gave the following mechanical properties: uniaxial

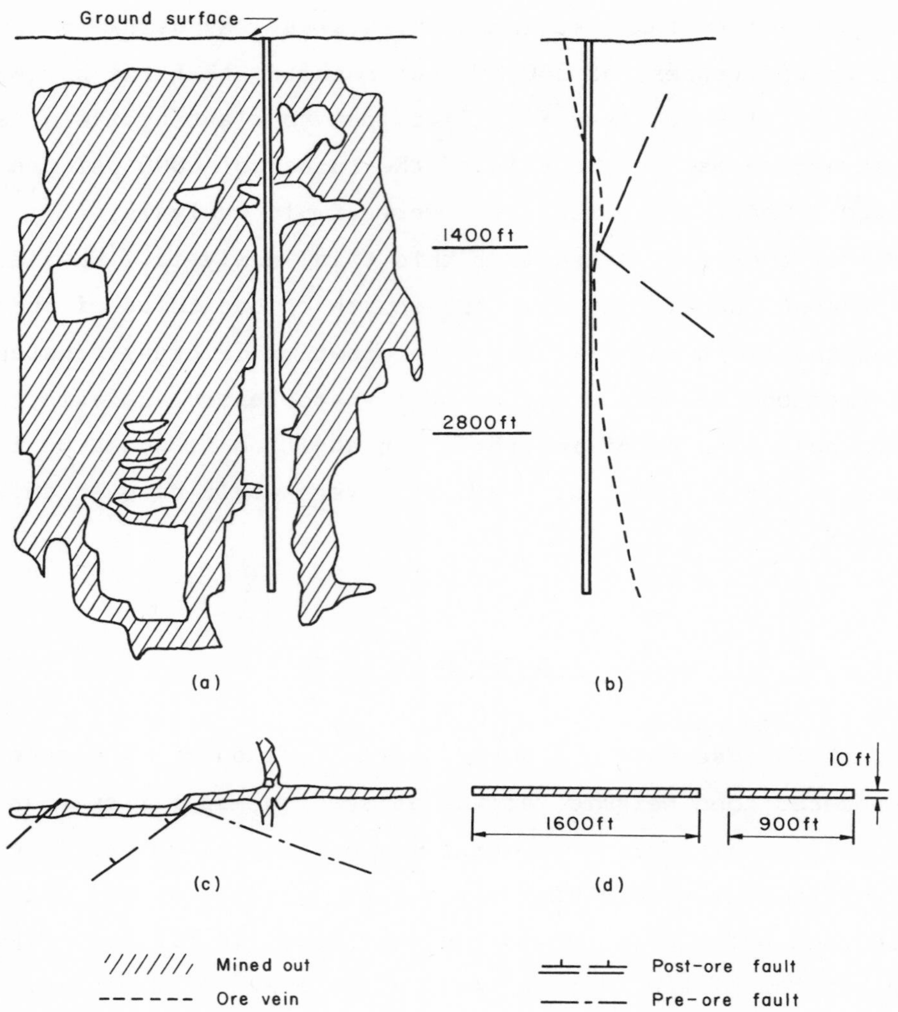

Fig. 6—Plans and sections of mine at time of large rock burst.

compression strength 36,300 psi, modulus of deformation 9.4×10^6 psi, Poisson's ratio 0.21, uniaxial tension strength 1900 psi, modulus of rigidity 3.8×10^6 psi, and specific gravity 2.7.(5,6)

The vein generally dips between 75 ft and 85 degrees. Its width varies from small stringers to a maximum of 70 ft. The vein rock consists of quartz and pyrite as well as the host rock.

During the driving of development drifts and crosscuts, many small bursts had occurred, producing falls of rock off the backs and walls. Neither the frequency nor the severity of these bursts increased with depth. They were particularly numerous, among other zones, on the 2825-ft horizon.(6)

When the large rock burst occurred, the shaft was closed completely from the 1400- to the 2800-ft level. By examining the seismic records recorded at Ottawa, Shawinigan Falls, Williams College, and Harvard University, the amount of energy transmitted to the ground surrounding the mine from the burst was estimated (using Jeffreys' method) at 5×10^{17} ergs.(7,8)

If the disturbance represents the strain energy released from the crushed pillar, the volume of ground included in the pillar would be of interest. With a height of 1400 ft, a width of 200 ft and an effective thickness of 50 ft, about 14×10^{6} cubic feet of rock might have been involved. This would be equivalent to a release of 2600 ft-lb of strain energy per cubic foot of rock.

On the other hand, if the disturbance represents the increased stress that is suddenly applied to the surrounding rock that must support the forces hitherto resisted by the rock in the pillar, the above calculation would not be valid.

The amount of strain energy that would be in the shaft pillar if only gravitational stresses were acting can be calculated. As the vein between the 1400- and the 2800-ft level is almost vertical, the horizontal field stress in this rock before mining, assuming no residual stress, would be about one-fourth of the vertical stress. The average horizontal stress in the pillar after mining would be increased by a factor of about 7.25, based on the width of the stopes.

The average strain energy in the pillar rock owing to this horizontal stress, as well as the vertical gravitational stress, would vary from 80 ft-lb/cu ft at the 1400-ft level to 300 ft-lb/cu ft at the 2800-ft level, with an average of about 160 ft-lb/cu ft. Consequently, at the depth of the burst, using the above assumptions regarding the disturbance, an addition to the gravity stresses, or a residual stress, of 2500 psi would be required to provide the seismic waves that were recorded. There are, of course, many questionable assumptions in these simple calculations.

Acknowledgments

It is a pleasure to have the opportunity to thank those who have made these notes possible: Dr. G. L. Hole of the Bethlehem Steel Company, Inc.; Mr. H. O. Olsen of the Marmoraton Mining Company, Limited; Mr. G. E. Larocque, Mines Branch, Department of Mines and Technical Surveys; the Hydroelectric Power Commission of Ontario; and Professor R. G. K. Morrison of McGill University.

REFERENCES

1. Hogg, A. D., "Some Engineering Studies of Rock Movement in the Niagara Area," Geol. Soc. Am.--Eng. Geol. Case Histories, No. 3, 1959.
2. Morrison, R. G. K., "Report on the Rockburst Situation in Ontario Mines," Trans. Can. Inst. Mining Met., Vol. 45, 1942.
3. Charlewood, G., and J. Thomson, "Geology of the Lake Shore Mine," Ontario Department of Mines, Annual Report, Vol. 57, Part 5, 1946.
4. Robson, W. T., "Lake Shore Geology," Trans. Can. Inst. Mining Met., Vol. 39, 1936.
5. Coates, D., J. Udd, and R. Morrison, "Some Physical Properties of Rocks and Their Relationship to Uniaxial Compressive Strength," Proc. McGill Rock Mechanics Symposium, 1962.
6. Robson, W. T., "Rockburst Incidence, Research and Control Measures at Lake Shore Mines Limited," Trans. Can. Inst. Mining Met. Vol 49, 1946.
7. Willmore, P., Letter to R. G. K. Morrison, McGill University, March 25, 1957.
8. Jeffreys, H., The Earth, The Macmillan Co., Toronto, 1952.

DISCUSSION

R. KVAPIL (Czechoslovakia):*

Today's ways of attaining the stability of an underground working are characterized by the fact that the structure of timbering is the prime consideration, whereas the surrounding rocks have only a secondary rôle. The structure of timbering is employed for taking up the stresses from the surrounding rocks. It means that, from the standpoint of the behavior of rocks, we can speak of the passive way of attaining the stability as well as of the passive way of timbering.

There is also another way of attaining the stability of an underground working--an active one. It is distinguished by the fact that the rocks are of prime importance, and the structure of timbering is only of secondary importance.

The active way of attaining the stability of an underground working is characterized by the fact that the processes taking place at the deformation of the rocks around the underground working are affected and modified by an artificial intervention so as to promote stability. In other words, while the passive way is only an action against consequences, the active way is a well-thought-out technical intervention into the causes of the processes taking place in the rocks; and the consequences of these processes are employed for conditioning the stability of the underground working. Thus, in the active way there is an artificial regulation of deformation of the rocks around the underground working.

At present, the principle of the active way of attaining stability of an underground working is more convenient in regard to its technical function and the safety of the working than the passive way.

We are only at the beginning of this technique; nevertheless, the development of the principle of the active way appears to be one of the main facilities enabling us to build safe underground structures, e.g., in connection with the exploitation of mineral raw materials from great depths, or in connection with utilizing underground sources of heat energy of the earth, etc.

---

*This discussion was originally submitted as an independent paper entitled "Active Stability of Underground Workings." In the absence of the author, it was summarized at the conference by C. Fairhurst.

Nowadays, the principle of active timbering can be applied with advantage, for instance, for ensuring the stability of cross ways and forkings in mine roadways. When we apply the common, i.e., the passive, way of timbering, the timbering of a cross way or of a forking is often upset; it has to be rebuilt, and a lot of expense is involved. Failures may be prevented by applying the active timbering of cross ways.

The principle of the active way of attaining the stability is governed by definite valid relations. In order to define these relations, let us consider the simplified conditions in the manner illustrated in Fig. 1, in which the rock surroundings are denoted by R, the element inserted into the surroundings by N, and the prevailing loading by arrows.

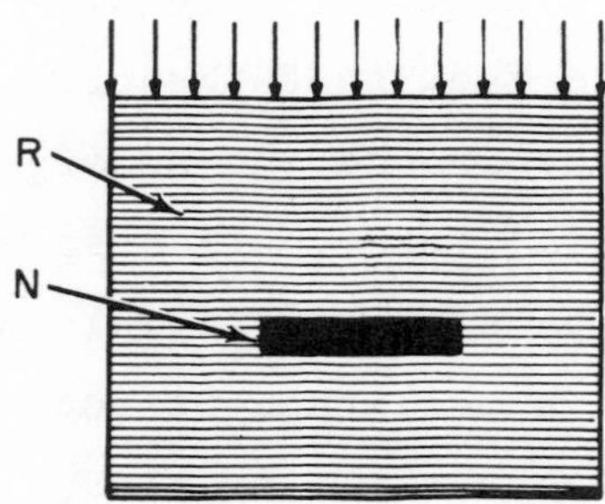

Fig. 1

The characteristic of the laying out of the stress depends upon mutual relations of the properties of the rock surroundings R and the inserted element N.

If those properties are expressed with the resistance against compression, wherein W_R is the resistance of the rock against the compression and W_N the resistance of the element against the compression, the basic valid relations may be defined by the ratios:

1. When resistance against the compression is equal in both the rock surroundings R and the inserted element N, so that $W_N = W_R$, then

$$W_N/W_R = 1 \ . \tag{1}$$

2. When the resistance against the compression is greater in the inserted element N than in the rock surroundings, i.e., $W_N > W_R$, the

resistance ratio is

$$W_N/W_R > 1 \ . \tag{2}$$

3. In the last case, the resistance against the compression in the inserted element N is smaller than that in the rock surroundings, i.e., $W_N < W_R$ and the ratio of the resistance is

$$W_N/W_R < 1 \ . \tag{3}$$

For each of the ratios (1), (2), and (3) there is a definite valid characteristic of the laying out of the stress. For the sake of a better visualization, Figs. 2(a), 2(b), and 2(c) depict these individual states.

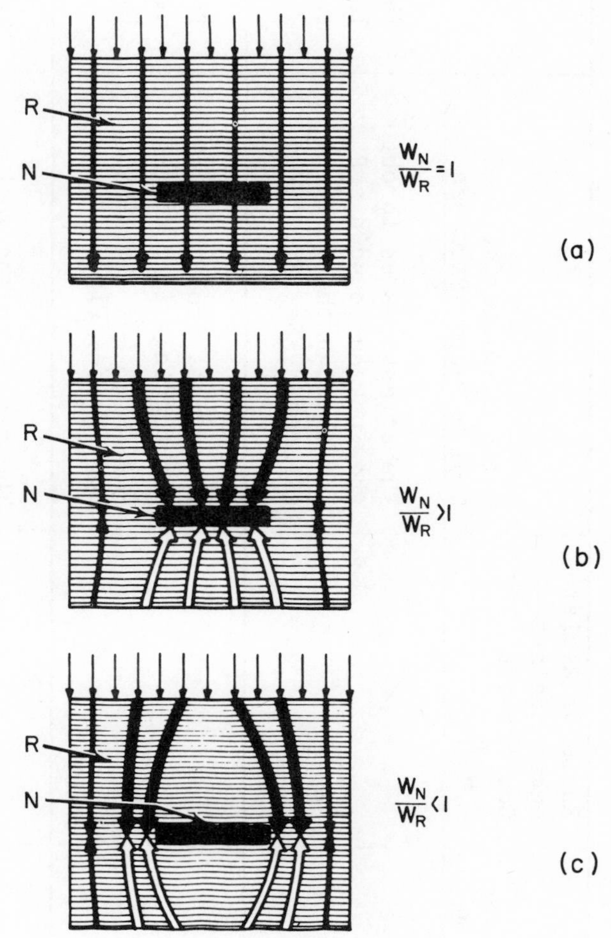

Fig. 2

The characteristic of the laying out of the stress is evident from the course of the thick arrows. The influence of the ratio W_N/W_R upon the characteristic of the laying out of the stresses is given in Table 1.

Table 1

Case No.	Ratio of Resistances against Compression, W_N/W_R	Valid Characteristic of Laying Out of Stresses	Characteristic of Loading
1.	$W_N/W_R = 1$	The stress passes through the rock surroundings R and through the inserted element N--See Fig. 2(a)	The inserted element loaded as much as the rock surroundings
2.	$W_N/W_R > 1$	The stress from the rock surroundings is concentrated on the inserted body N--See Fig. 2(b)	The inserted element is loaded more than the rock surroundings
3.	$W_N/W_R < 1$	The stress in the rock surroundings R passes round the inserted element N--See Fig. 2(c)	The inserted element is loaded less than the rock surroundings

For designing the stability of cross ways in mine roadways, case 3 in Table 1 is of particular importance. It is the state in which the stress passes round the inserted element.

The application of the principle of case 3 for attaining stability in cross ways is characterized by the fact that there is an excavation N hollowed in the rock R, as shown in Figs. 3 and 4.

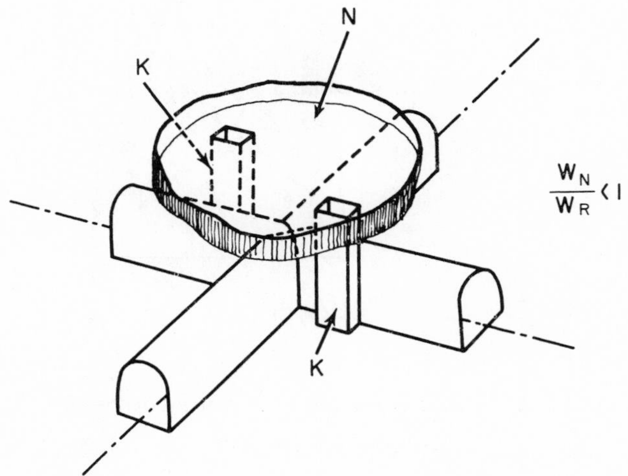

Fig. 3

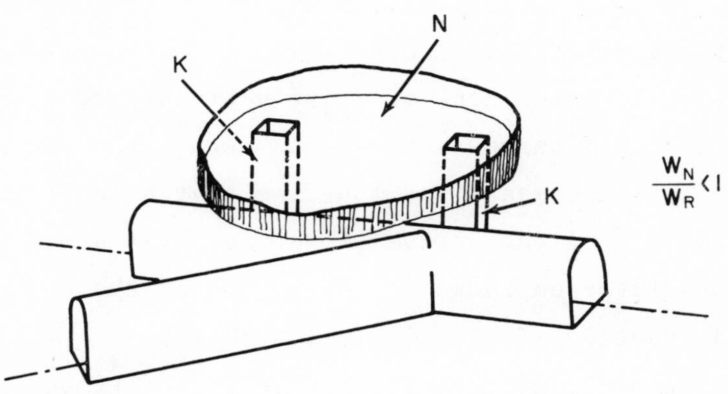

Fig. 4

The characteristic of the stress passing round the inserted element is apparent from Fig. 5, which is an upright cross section of the cross way of the roadway and of the excavation.

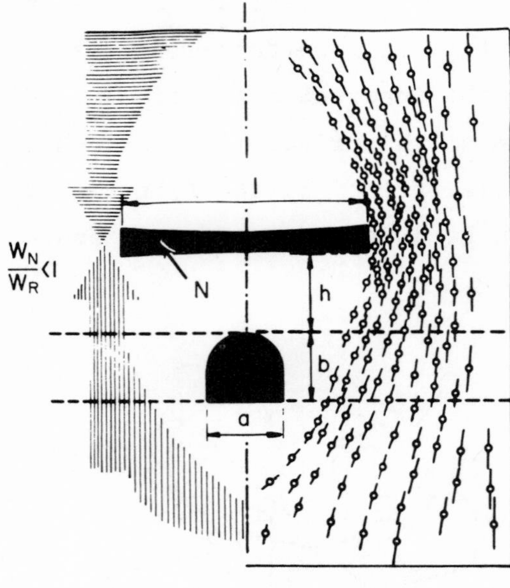

Fig. 5

The bottom of the excavation N is considerably relieved by that excavation and, consequently, the rock of the roof in the cross way is unloaded, too. It has a very favorable effect upon the active stability of the cross way. The width of the excavated space ℓ, as well as the thickness of the rock h in the roof of the cross way (see Fig. 5), depends both upon the acting natural conditions and those conditions created by mining. The cross way is under the excavated space N in the zone of greatly diminished stresses, and thus the rock itself serves actively to attain the stability of the cross way. It is necessary to emphasize that even if the space N is distorted, the aim of the active timbering is attained because the weight of the distorted rocks on the bottom of the space N brings forth only a fraction of the original stresses.

It is self-evident that the active way of timbering, based upon applying the effect ensuing from the ratio $W_N/W_R < 1$, may be employed in other cases too. For instance, the active stability may be attained in the mine roadway in a much stressed area under prevailing inclined loading, as shown in Fig. 6, in which N denotes the excavated space.

To attain the active stability of an underground working, it is not necessary to employ only the principle of the "passing-round" stress, which is caused by ratio (3), i.e., $W_N/W_R < 1$--case 3 in Table 1.

Under certain favorable conditions, the active stability of an underground working can also be attained by the application of the principle of concentration of stresses, which is caused by the effect ensuing from ratio (2), i.e., $W_N/W_R > 1$ (see Table 1, case 2).

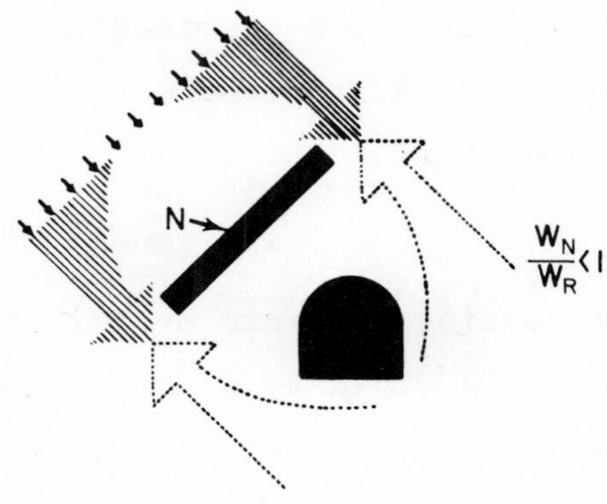

Fig. 6

The reasons for such a realization are presented in Fig. 7. There are holes driven at the sides of the underground working, and these holes are filled with expanding concrete. As a result of the increasing volume of the expanding concrete, the stress from the rock surroundings begins to concentrate on the body made of the expanding concrete N because the condition for case (2), $W_N/W_R > 1$, is fulfilled.

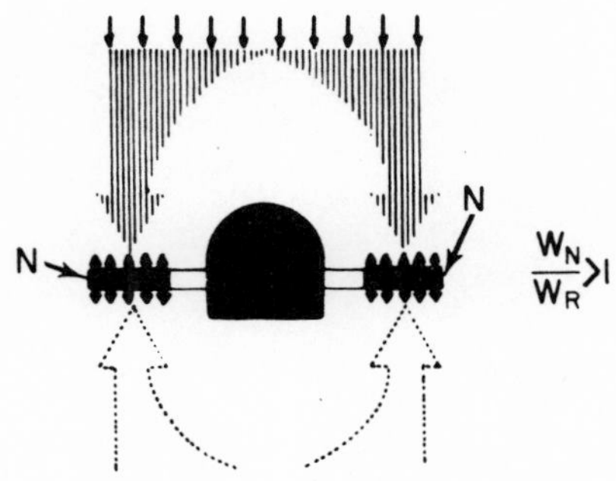

Fig. 7

When the properties of the rock and the original natural stress are favorable, active stability can be attained, since both the loading and the stress are appreciably diminished in the zone of the rock in the nearest surroundings of the working.

Various details of individual states of the active stability of underground workings, as well as various combinations of the basic principles, are being studied at present. For the sake of briefness only the basic principles have been examined here; nevertheless, the function of active stability and the chances of its future wide application are quite apparent.

C. FAIRHURST (USA):

Dr. Kvapil's contribution presents practical suggestions for reducing the stresses in the vicinity of underground workings. These suggestions may be particularly useful in deep mines or "in connection with stabilizing underground sources of heat energy of the earth, etc."

The term "active stability" is used to indicate that positive steps are taken to change (i.e., reduce) the stresses in the vicinity of the working compared with passive stability obtained by installation of supports to <u>sustain</u> existing stresses.

GENERAL DISCUSSION TO PART IV

S. CHAN (USA):

Based on data from the conventional methods of determination of stresses and the properties of the rock, would any gentleman like to suggest reasonable ranges of factors of safety for the design of underground openings of various depths in various environments? I appreciate that this is a very broad question, and therefore, I do not expect a single, complete answer.

D. F. COATES (Canada) in reply:

One consideration is the purpose of the underground opening. If it is a mining opening, no safety factor would be considered. If you are operating in unstable ground, you live with it and incur the expense of living with it.

Development openings connected with mining operations, being of a more permanent nature (e.g., shafts, conveyorways, etc.) should have incorporated in either the opening or the support system some safety factor greater than one. The actual appropriate number to use should depend on the type of ground and the support system being planned. However, the safety factor, if explicitly used, must at this stage be a matter for judgment.

Underground openings for construction projects (e.g., tunnels, underground power houses, and defense installations) are usually of a more permanent nature than mining development openings; thus the above comments also apply in these cases.

At the same time, it should be recognized that rock mechanics has not developed sufficiently so that hard rock failure can be predicted. Thus, it is difficult to use the normal civil engineering design process and include an explicit safety factor against this type of failure.

The design of a support system might follow the normal design procedure; however, rock pressures are also difficult to predict.

Another aspect of the topic is that if we knew the in situ strength of the rock that was to contain an underground opening, there is little that can normally be done to vary the stresses and thus incorporate some desired safety factor. The procedure is likely to be simply the determination of whether a margin of safety would exist, and if not, then a support system would be included. In the case of some rock structures such as pillars and stopes, this limitation would not necessarily apply.

A final observation on the subject of safety factors in rock is that when such a concept is used, the appropriate magnitude should be related to the amount of dispersion of strength or yield point about the mean. In other words, a large dispersion would logically indicate the need for a large safety factor to avoid failure or to reduce the probability of failure to some acceptable, small figure.

J. L. TALOBRE (France) in reply:

I think that a reliable factor of safety, i.e., the statistical risk of failure, may be determined for any underground work. I recently attempted such a determination for an underground nuclear power plant.

I previously showed how statistical methods could afford the possibility for computing the risk of failure when applied to rocks.

Failure represents, in each particular case, an amount of expense which can be roughly evaluated. In all cases, the risk can be lowered to a known degree by undertaking reinforcement works. These works result in a given amount of supplementary expense. Choice can be made of the solution which presents the major economic interest, that is to say, the minimum total expense, with the safety of the property being taken into account.

L. OBERT (USA) in reply:

We approached the problem of safety factors by making a study of 25 open stope mines in relatively competent rock. Areas of maximum extraction and/or room size were investigated. From the depth of the openings, the extraction, and geometrical considerations, the average stress in supporting pillars was computed. The strength of the pillar rock was determined by measuring the unconfined compressive strength of core specimens taken from the pillars. On the basis of rock strength to working stress, the safety factors varied from a minimum of 4 to a maximum of 8. Of course, this is not the true safety factor for the pillar, since the in situ strength of the rock is undoubtedly less than the corresponding strengths measured in the laboratory. On the same basis, it has been found that the safety factor for pillars in evaporite mineral deposits (salt, trona, borates) is very much lower. In a number of instances the calculated safety factor has been very close to unity, and this calculation is considered realistic in view of the fact that incipient and sometimes rather extensive failures were noted in areas in which the safety factor approached unity.

D. U. DEERE (USA):

I would like to have a comment regarding uplift in excavation in some of our shales in the Great Plains. Perhaps Mr. Nesbitt would give us a 2- or 3-minute résumé of some of their history.

R. NESBITT (USA) in reply:

I am familiar with the problem of rebound in shales to the extent that we recognize that the problem is not the same in different shales. Elastic rebound in some shales is interesting in that horizontal movements under the stimulus of forces or stresses which could be equal to, or greater than, those in the vertical direction do not express themselves until sometime after the vertical rebound has occurred.

In the section on Fundamentals, Dr. Kiersch asked if anyone at the conference had measurements on such movements, particularly vertical movements, which could be used in computing the rebound reaction of shales to excavation. I am in hopes that in the near future, through our office in Omaha, we can put into statistical records the experience we had at the Oake and Garrison dam projects, where, as some of us know, our shale foundations rebounded sharply during excavation. We had to make adjustments for such movements first, by letting most of the rebound dissipate before reloading and, finally, by anchoring structures deep enough so the movements were not great enough to seriously affect the structures founded in these shales.

Two weeks from now some of us are going to look at the Fort Peck Spillway, a structure which some 20 years ago was placed on the Bearpaw shale of Cretaceous age in Montana. The spillway slab, i.e., the floor or invert of that structure, has been observed over the years from both the engineering and scientific point of view because, far from experiencing a complete reduction of any movements shortly after the completion of the Fort Peck Spillway, steady movements have been in progress and have been recorded.

To what extent the concrete pavements on these shales have been affected by elastic rebound, or to what extent they might be formed by chemical processes going on within the shales attended by such rebound, will be an object of our studies in the future.

I have no other comments, other than to hope we eventually will be able to get some statistical information on the behavior of shales in different areas of the United States.

D. U. DEERE (USA):

Mr. Nesbitt, what kind of magnitude is involved in some of these vertical rebounds?

R. NESBITT (USA) in reply:

I think on the order, and I can stand corrected by anyone in the audience who has more accurate information, of from a few tenths of an inch to 3 or 4 ft during construction, this, of course, both dissipating with time and being of a lower order with increased depth.

D. U. DEERE (USA):

But these are time dependent and do continue at a low rate for months and years?

R. NESBITT (USA) in reply:

They are, and they vary with different areas and different shale chemistry, which we want to study further.

K. F. DALLMUS (Venezuela):

The so-called barreling effect has been visually observed in steep cuts where outward bulges develop in the face of the cut. If the base of the cut is in less competent material than the bottom of the excavation, the bulge will develop above the base of the cut. If the bottom of the excavation is also in incompetent materials, the bottom will also bulge upward as shown in Figs. 1 and 12 of Dr. Müller's paper. In road cuts where mechanical measurements are not warranted, periodic visual inspection will indicate when landslides are imminent.

Previous to making *in situ* measurements for large excavations, a regional geologist can generally predict the state of horizontal stress in any particular region based on the history of vertical movements which have affected the rocks in the area. This applies only to areas which are underlain by rocks whose bulk density approximates the grain density. Sedimentary rocks which have not been compacted to grain density and are uncemented can not store strain energy.

The problem of swelling shales is more a problem of chemistry than mechanics. Some clay minerals, especially montmorillinite, take up water in the crystal lattice with a large increase in volume. This was a serious problem in drilling deep wells for petroleum until heavy muds were developed which build up a thick cake with a low filtration rate on the wall of the hole.

O. J. OLSEN (USA):

My remarks will be rather general and brief. In the conference, fundamental theories have been discussed, consideration has been given to the degree to which these theories may be utilized to interpret measurements and test data so that we can expand our knowledge of rock behavior, and comments have been made on how to apply certain techniques to the design of structures that rest upon rock foundations as well as those excavated from within the rock masses. I feel we have made much progress but have only scratched the surface. Many tasks lie ahead in rock mechanics, which now has been officially recognized as a science. Lest we become complacent about what remains to be accomplished, let us consider a few of the areas in which we may concentrate our efforts.

The elastic theory, while useful as a starting point, has its definite limitations. But because of the need to know more about stresses that cannot be measured directly, we must establish a deformation modulus, or conversion factor if you prefer to call it that, for changing strains or deformations into stress. The various tests such as seismic velocity, jacking *in situ* tests, strain relief, and others depend to a certain extent directly or indirectly on elastic theory. So it becomes doubly important for us to investigate the factors affecting the relationships between stress and strain. Dr. Serafim tactfully referred to rock as having a "special kind of elasticity." It is true that rock having a strength of, say, 3000 psi in the ordinary sense of the word may, as the abutment for a dam, have imposed upon it a load of only one-third this amount. Confinement may further improve its adequacy and behavior, and we may think of it as being elastic under such a low stress. However, when we consider underground mine openings or power plants where there are many more unknown factors, or when there is a possibility that water may enter a fault zone or clay seam and weaken the rock surfaces under a dam, we must proceed with caution.

When a rock is loaded statically to one-fourth or one-third of its ultimate strength, it may be assumed to be elastic within a reasonable degree of accuracy. But when a structure is loaded to failure, we must also consider the plastic region of stress. Very little has been spoken of this or about any theories relating to plasticity. Perhaps we should concentrate more effort on the development and application of plastic theory. Mr. Merrill expressed an appropriate thought when he advocated, "Make the theory fit the rock, not rock fit the theory."

We also need more treatises on rock mechanics as well as more courses in our universities to attract students into this challenging field. Then there should be a plea for more conferences like this one, where information may be exchanged and where each organization represented can take back with them a problem which needs solving.

Some brief mention was made about the effect of wetting rocks. But can we say that we understand the mechanics of saturation of rock? Why is it that wet rocks are weaker or have a lower deformation modulus than dry ones? Why do seismic tests give higher values than static tests? Dr. Serafim finds that seismic velocity correlates best with tangent modulus of elasticity obtained from the unloading portion of the stress-strain curve. Mr. Wantland's comparison of seismic with other methods of measuring elastic properties in some instances has shown very good agreement, but in other cases there is a difference of as much as 4:1. Although moisture condition, stress level, and rate of loading account for some of the difference, a definite relationship has not been clearly established. Under perfect elastic behavior, both methods should theoretically give the same results, but the greater the inelasticity, the greater the difference. The Bureau of Reclamation has under way a long-range test program to compare and correlate these three methods of measuring elastic properties: seismic velocity, _in situ_ jacking, and laboratory core tests. We need a correlation factor between seismic and static properties for use in design so that advantage may be taken of the more economical methods in lieu of those which involve considerable expense.

The conditions of homogeneity and isotropy generally assumed in theoretical stress analysis are seldom realized in materials of the earth's crust. Bedding, cleavage, or schistosity can have pronounced effects upon both strength and deformation behavior of rocks. Professor Donath reported an experimental study currently in progress to determine the nature of these variations in slate rock. His results on cores tested at various orientations of bedding show that specimens stressed at an angle of about 30 degrees to cleavage have the least strength. Dr. Serafim observed that Mohr's theory of rupture is not valid for stratified or schistose rock. These are essentially the same as our conclusions based on a recent angle study of quartz mica schist from Reclamation's Morrow Point underground power plant. In such cases triaxial tests can give the strength along the weakest but not the strongest planes. Shear strength of anisotropic rocks, therefore, must be determined _in situ_ along various planes in order to obtain complete information for design. To evaluate shear characteristics in a rock mass and along joints and faults, tests must be made on large surfaces, such as proposed by Dr. Serafim. The Bureau of Reclamation is planning tests of this kind both in the laboratory and in the field.

Professor Brace, in his talk on brittle fracture of rocks, presented some interesting aspects of the general applicability of Mohr's and Griffith's theories of rock fracture. He reports that most investigators conclude that there is no existing failure law which holds for all rocks. But, it is generally found that the best approximation for "room temperature" tests is given by the Mohr criterion. Under certain types of loading, however, such as bending and compression, even Mohr's failure law does not correctly predict relative behavior. Griffith's original theory assumes that brittle materials contain sharp flaws or cracks which enlarge and spread through the material as stress is applied. The theory predicts fracture strength for a general state of stress in terms of tensile strength. Up to this point, I was not particularly impressed with anything new about the Griffith theory. A subsequent modification of it by McClintock and Walsh, which includes the closing of Griffith cracks and development of frictional forces across the crack surfaces, seems to represent a much more likely failure situation for material subjected to compressive stresses. This modification impresses me as being more logical in predicting fracture stresses than the original theory, which does not include the effect of frictional forces at crack surfaces. Upon closer study of the McClintock-Walsh-Griffith theory, I noticed a resemblance between it and the curvilinear analysis of what we in the Bureau of Reclamation refer to as the "PSR (principal stress relationship) curve." Further review of the modified theory and application of it to some of our triaxial test results appear to be justified. Theories must be put into practice to become useful.

These are a few of the challenges which I believe we have to meet in this new science of ours.

PANELISTS AND SPEAKERS

Bergman, Sten G. A.
Chief, Research Department
Royal Swedish Fortifications
Stockholm, Sweden

Birch, Francis
Department of Geological Sciences
Harvard University
24 Oxford Street
Cambridge 38, Massachusetts

Brace, William F.
Massachusetts Institute of Technology
Cambridge, Massachusetts

Brode, Harold L.
Physics Department
The RAND Corporation
1700 Main Street
Santa Monica, California

Clark, George B.
Director, Research Center
School of Mines and Metallurgy
University of Missouri
Rolla, Missouri

Coates, Don F.
Department of Mines and Technical Surveys
Ottawa, Ontario
Canada

Deere, Don U.
Departments of Geology and Civil Engineering
University of Illinois
Urbana, Illinois

Donath, Fred A.
Department of Geology
Columbia University
New York 27, New York

Duvall, Wilbur I.
Applied Physics Research Laboratory
U.S. Bureau of Mines
College Park, Maryland

Emery, Charles L.
Assistant Professor
Mining Engineering Department
Queen's University
Kingston, Ontario
Canada

Friedman, Melvin
Exploration and Production Research Division
Shell Development Company
P. O. Box 481
Houston 1, Texas

Handin, John W.
Exploration and Production Research Division
Shell Development Company
3737 Bellaire Boulevard
Houston 25, Texas

Hoek, Evert
National Mechanical Engineering Research Institute
P. O. Box 395
Pretoria, South Africa

Jaeger, J. C.
Department of Geophysics
Australian National University
Canberra, Australia

Judd, William R.
Aero-Astronautics Department
The RAND Corporation
1700 Main Street
Santa Monica, California

Merrill, Robert H.
Denver Mining Research Center
U.S. Bureau of Mines
Building 20, Federal Center
Denver, Colorado

Müller, Leopold
Ingenieurbüro für Geologie und Bauwesen
Franz-Josef-Str. 3
Salzburg, Austria

Olsen, Owen J.
Rock Mechanics Laboratory
U.S. Bureau of Reclamation
Federal Center
Denver 25, Colorado

Potts, Edward L. J.
Department of Mining Engineering
University of Newcastle upon Tyne 1
England

Robertson, Eugene C.
U. S. Geological Survey
Eastern Avenue and Newell Street
Silver Spring, Maryland

Serafim, J. Laginha
Consulting Engineer
Avenida Marquês de Tomar, 9.6°
Lisboa
Portugal

Talobre, Joseph A.
Consulting Engineer
50 Rue de Boulainvilliers
Paris 16
France

Wantland, Dart
Head, Geophysics Section
U.S. Bureau of Reclamation
Building 53, Federal Center
Denver, Colorado

Werth, Glenn C.
Plowshare Division
Lawrence Radiation Laboratory
P. O. Box 808
Livermore, California

DISCUSSANTS

Barron, Kenneth
Mining Research Laboratory
Department of Mines and Technical Surveys
Ottawa, Ontario
Canada

Biggs, Donald L.
Associate Professor
Department of Geology
Iowa State University
Ames, Iowa

Bollo, M. F.
Director
Société de Recherches Géophysiques S. A.
81 Rue Laugier
Paris 17, France

Burke, Harold W.
Tippetts, Abbett, McCarthy, Stratton Engineers
Box 3037
Stanford, California

Chan, Samuel S. M.
Geology Department
University of Idaho
1394 Walenta Drive
Moscow, Idaho

Coleman, Robert G.
U.S. Geological Survey
345 Middlefield Road
Menlo Park, California

Dallmus, Karl F.
Consulting Geologist
Dirección Postal
Apartado 10126, Sabana Grande
Caracas, Venezuela

Dickey, D. D.
U.S. Geological Survey
Federal Center
Denver 25, Colorado

Eaton, Gordon P.
Associate Professor of Geology
Department of Geology
University of California
Riverside, California

Everling, Georg (<u>in absentia</u>)
Steinkohlenbergbauverein
Essen, Friedrichstr. 2
Germany

Fairhurst, Charles
Associate Professor of Mineral Engineering
University of Minnesota
Minneapolis 14, Minnesota

Fidler, Jack
The Hydro-Electric Commission
Hobart, Tasmania
Australia

Goodman, Richard Edwin
Department of Mineral Technology
University of California
Berkeley, California

Höfer, Karl-Heinz (<u>in absentia</u>)
Internationales Büro für Gebirgsmechanik bei der Deutschen Akademie der Wissenschaften zu Berlin
Leipzig S 3, Friederikenstr. 60
Germany

Kiersch, George A.
Department of Geology
Cornell University
Ithaca, New York

Kvapil, Rudolf (<u>in absentia</u>)
Chief Expert for Mining
The State Institute for Designing Ore Mines
Rudny Projekt
Košice, Czechoslovakia

Laird, Raymond T.
Talson Engineering
646 Vernon Street
Oakland 10, California

Lang, Thomas A.
The Bechtel Corporation
220 Bush Street
San Francisco, California

Larocque, Glen E.
Mining Research Laboratory
Department of Mines and Technical Surveys
Ottawa, Ontario
Canada

Nesbitt, Robert H.
Chief Geologist
Office of the Chief of Engineers
Headquarters, Department of the Army
Washington 25, D. C.

Obert, Leonard
U.S. Bureau of Mines
College Park, Maryland

Oka, Yukitoshi (<u>in absentia</u>)
Assistant Professor
Kyoto University
Kyoto
Japan

Pincus, Howard J.
Professor
Department of Geology
Ohio State University
Columbus, Ohio

Serata, Shosei
Department of Civil Engineering
Michigan State University
East Lansing, Michigan

Sherman, Irving
Los Angeles County Flood Control District
13165 Welby Way
North Hollywood, California

Trump, Robert P.
Gulf Research and Development Company
P. O. Drawer 2038
Pittsburgh 30, Pennsylvania

Wiebenga, W. A. (<u>in absentia</u>)
Wentworth House
Bureau of Mineral Resources
203 Collins Street
Melbourne, Victoria
Australia

Wittke, Walter
Institut für Bodenmechanik und Grundbau
Technische Hochschule Karlsruhe
75 Karlsruhe
Kaiserstrasse 12
Germany

Woodruff, Seth D.
Civil Engineer-Engineer of Mines
2039 South Sixth Street
Alhambra, California

REGISTRANTS

AUSTRALIA

Fidler, Jack
Jaeger, J. C.
Wiebenga, W. A. (*in absentia*)

AUSTRIA

Müller, Leopold
Steinboeck, Wilhelm

CANADA

Ambrose, J. Willis
Bain, Ian
Barron, Kenneth
Brisbin, William C.
Coates, Don F.
Currie, John B.
Dolmage, Victor
Dymock, Thomas
Emery, Charles L.
Emery, Charles O.
Hardy, Henry Reginald, Jr.
Larocque, Glen E.
MacKenzie, Ian D.
McPherson, Robert Louis
Mylrea, Frank H.
Norris, Donald K.
Pasieka, Arnold Roy
Ross, John V.
Soles, James A.
Yamaguchi, Umetaro

CZECHOSLOVAKIA

Kvapil, Rudolf (*in absentia*)

ENGLAND

Potts, Edward L. J.

FRANCE

Baron, Guy A.
Bollo, M. F.
Lagarde, André Jean
Londe, Pierre
Maikovsky, Vinceslaw
Pakdaman, Khosrow
Talobre, Joseph A.
Vigier, Georges
Vincent, Alain

GERMANY

Everling, Georg (*in absentia*)
Höfer, Karl-Heinz (*in absentia*)
Leussink, Hans (*in absentia*)
Schardin, Hubert Reinhold
Wittke, Walter

ITALY

Berghinz, Carlo
Lotti, Carlo
Penta, Francesco A.
Tonini, Dino
Veder, Christian

JAPAN

Honsho, Shizumitsi
Nishihara, Masao

JAPAN (continued)

Oka, Yukitoshi (*in absentia*)
Takata, Takanobu

MEXICO

Covarrubias, Sergio W.
Flamand, Carlos L.
Isita, José Septién
Mooser, Frederic
Palacios, Miguel Angel
Ruiz, Mariano
Sánchez, Roberto
Santoyo, Enrique
Tamez, Enrique

NEW ZEALAND

Craddock, Campbell

PANAMA

Stewart, Robert H.

PORTUGAL

Serafim, J. Laginha

SOUTH AFRICA

Hoek, Evert
Sutherland, Robert Bruce

SWEDEN

Bergman, Sten G. A.

SWITZERLAND

Zimmerman, Fritz R.

UNITED STATES

Alabama

Bryan, Jack H.

Alaska

Ragan, Donal M.

Arizona

Hess, John D.
Komie, Earl E.
Motsinger, Richard N.
Nininger, H. H.

California

Adams, Herbert G.
Adams, Robert F.
Adams, William Mansfield
Ahrens, Thomas J.
Aldus, Arik
Allgood, Jay R.
Bean, Robert T.
Beck, Earl J., Jr.
Bjork, Robert L.
Blake, Wilson
Bliss, Percy H.
Boardman, Charles R.
Boothe, Roy
Boozer, George D.
Borg, Iris Y.
Brandt, Robert J.
Brode, Harold L.
Bronner, Finn Eyolf
Brooks, Nancy
Brown, Douglas R.
Burke, Harold W.
Bush, Robert Y.
Byerly, P. Edward
Campbell, Ian
Cherry, Jesse T.
Christie, John M.
Clark, Lorin D.
Coleman, Robert G.
Content, Charles S.
Converse, Frederick J.
Conwell, Fred R.
Cooper, Daniel C.
Crutchfield, William H., Jr.
Davis, Richard W.
Dobbs, R. O.
Eagen, Jack T.
Eaton, Gordon P.
Ehlig, Perry L.
Evans, Dennis, Sr.

Ewoldsen, Hans M.
Focke, Alfred B.
Frankian, Richard T.
Gardner, William I.
Gilman, Richard H.
Goodman, Richard Edwin
Hall, Charles E.
Hammer, J. Gordon
Hill, Jerry E.
Hoffman, Roy A.
Hollander, Margaret A.
Hollis, Edward P.
Hurley, Neal L.
Isenberg, Fred
Iwamura, Thomas I.
Jamison, Marshall V.
Jenkins, Roy W.
John, Klaus
Johnston, Elizabeth Jane
Jorgensen, Brandt D.
Judd, William R.
Kelly, Jack E.
Kern, John W.
Knopf, Adolph
Knopf, Eleanora B.
Kojan, Eugene
Kruse, George H.
Kues, Harry A.
Laird, Raymond T.
Lamar, Donald L.
Lang, Thomas A.
Laupa, Armas
Leeds, David J.
Levandowski, Donald W.
Low, Edward J.
Lung, Richard
Lynch, M. Elaine
MacDonald, Howard D.
Mackintosh, Charles
Macura, Daniel
Mann, John F., Jr.
Marliave, E. C.
Martin, Neill W.
Martin, Norman L.
Maurseth, Ray O.
McCauley, Marvin L.
McClure, Cole R., Jr.
McDonough, George F.
McGill, John T.
Misen, Robert T.
Mow, C. C.
Moye, Daniel G.
Murano, S. S.
Nalle, Peter B.
Nance, Olen A.
Nelson, Jerome S.
Nielsen, John P.
Nitta, Satoshi
Oberste-Lehn, Deane
Oertel, Gerhard
O'Neill, Alan L.
Paulson, John C.
Pentegoff, Vladimir P.
Peterson, Raymond A.
Petrie, Jerome B.
Plafker, George
Price, Walter H.
Randolph, Bernard G.
Raphael, Jerome M.
Reti, G. Andrew
Rieke, Herman Henry, III
Rittenhouse, Gary E.
Roberts, George D.
Rose, Don C.
Russell, Thomas L.
Sandoval, Charles A.
Sauer, Fred M.
Schilling, Adolf A.
Scholen, Douglas E.
Schoustra, Jack J.
Schroter, G. Austin
Scott, James B.
Scott, Ronald F.
Sherman, Irving
Sibley, William L.
Smith, Arthur C.
Smith, Raymond J.
Smith, Travis W.
Spellman, Howard A.
Spencer, Ralph W.
Starkey, John
Stolt, George
Stone, Robert
Takenouchi, Sukune
Thayer, Donald P.
Thomas, Robert G.
Thompson, Thomas F.
Tocher, Don
Tooley, Richard D.
Townsend, James Robert
Trantina, John A.
Trefzger, Robert E.
Tuttle, Jack K.
Voloshin, Vadim

California (continued)

Wahler, W. A.
Waisgerber, William
Weber, Ernest M.
Weide, David L.
Weiss, Lionel E.
Werth, Glenn C.
West, Lawrence J.
Wilder, Carl R.
Wing, Alan F.
Woodruff, Seth D.
Wool, J. M.
Workman, J. W.
Yelverton, Charles A.
Yoshihara, Takeshi
Zielbauer, Edward J.
Ziony, Joseph I.

Colorado

Braddock, William Alfred
Brandon, J. R.
Cannaday, Francis X.
Card, David C.
Cummings, David
Davis, Robert E.
Dickey, D. D.
Eckel, Edwin B.
Ege, John R.
Emerick, William L.
Fritts, Paul J.
Grosvenor, Niles E.
Harms, John C.
Hoskins, John R.
Houser, Fred N.
McKeown, Francis A.
Merideth, George T.
Merrill, Robert H.
Monk, Edward F.
Nichols, Thomas C., Jr.
Olsen, Owen J.
Osterwald, Frank W.
Robinson, Charles S.
Rouse, George C.
Varnes, David J.
Wantland, Dart
Williams, William P.
Wolf, William H.

Connecticut

Rodgers, John

District of Columbia

Barron, Reginald A.
Carswell, Bruce M.
Johnson, Wendell E.
Lewis, John G.
Nesbitt, Robert H.

Georgia

Conn, William V.

Idaho

Chan, Samuel S. M.

Illinois

Cording, Edward
Dally, James W.
Davisson, Melvin T.
Deere, Don U.
Miller, Raymond P.
Veltrop, Jan A.
Willis, Clifford L.
Wright, Fred D.

Indiana

Worth, Edward G.

Iowa

Biggs, Donald L.
Hoppin, Richard A.

Louisiana

Meyerhoff, Arthur A.

Maryland

Duvall, Wilbur I.
Morgan, Thomas A.

Obert, Leonard
Robertson, Eugene C.

Massachusetts

Becker, Herbert
Birch, Francis
Bombolakis, Emanuel G.
Brace, William F.
Riecker, Robert E.
Seifert, Karl E.
Walsh, Joseph B.

Michigan

Bacon, Lloyal O.
Bley, John
Ensign, Chester O.
Johnston, Bruce G.
Serata, Shosei

Minnesota

Atchison, Thomas O.
Buck, Robert O.
Fairhurst, Charles
Foerster, Hansgeorg J.
Fogelson, David E.
Gnirk, Paul F.
McWilliams, John R.

Mississippi

Hoff, George C.
Krinitzsky, E. L.
Sherman, Walter C.

Missouri

Caudle, Rodney D.
Clark, George B.
Moore, Bruce H.
Wagner, Warren R.

Nebraska

Bauman, Aaron H.
Distefano, Carl J.
Underwood, Lloyd B.

Nevada

Anthony, Micheal V.
Pack, Phillip D.
Price, Charles E.
Weskamp, Dale P.

New Jersey

Agron, Sam L.

New Mexico

Weart, Wendell D.
Zwoyer, Eugene M.

New York

Donath, Fred A.
Jaffé, Felice C.
Jonas, Ernest
Karpov, Alexander V.
Kiersch, George A.
Miller, Clifton W.
Oliver, Jack E.
Sherard, James L.
Watson, John F.

Ohio

Barnett, Robert E.
Pincus, Howard J.

Oklahoma

Byrne, Patrick J. S.
Masson, Peter H.

Oregon

Dodds, R. Kenneth
Snyder, Dell L.
Stuart, W. Harold
Thurber, Paul

Pennsylvania

Gray, Richard E.
Trump, Robert P.

South Dakota

Oshier, Edwin H.

Tennessee

Boegly, William J., Jr.
Bradshaw, R. Louis

Texas

Atkinson, David J.
Austin, Walter J.
Burchfiel, Burrell C.
Cheatham, John B.
DeFord, Ronald K.
Friedman, Melvin
Friedman, Robert H.
Handin, John W.
Heard, Hugh C.
Howard, John H.
Muehlberger, William R.
Odé, Helmer
Perkins, Thomas K.
Sowers, George M.
Stearns, David W.

Utah

Call, Richard D.
Morin, W. J.
Rausch, Donald O.
Rausher, Loren H.
Rollins, Ralph L.
Weiss, Alfred
Worley, Morris T.

Virginia

Black, Rudolph A.
Hanson, Roy E.
Nesbitt, Robert H.
Palmer, David A.
Smith, Carneal K.
Uhley, Robert P.

Washington

Ageton, Robert W.
Forrest, Merrill M.
Galster, Richard W.
Gresseth, Elbridge W.
Hjertberg, Svante E.
Jones, Glen D.
Monahan, Charles J.
Rodriguez, Ernest R.
Squier, L. Radley
Waddell, Galen G.

West Virginia

Foster, Julian M.

Wisconsin

Scott, James J.

VENEZUELA

Dallmus, Karl F.

SUBJECT INDEX

Reference abbreviations: *e = equation or formula*; *n = footnote*; *t = table*; *page number in boldface type = definition.*

ROCK AND SOIL INDEX

AUTHOR INDEX

Selected RAND Books

Bellman, Richard. *Dynamic Programming*. Princeton, N. J.: Princeton University Press, 1957.

Bellman, Richard E., Harriet H. Kagiwada, Robert E. Kalaba, and Marcia C. Prestrud. *Invariant Imbedding and Time-Dependent Transport Processes,* Modern Analytic and Computational Methods in Science and Mathematics, Vol. 2. New York: American Elsevier Publishing Company, Inc., 1964.

Bellman, Richard E., Robert E. Kalaba, and Marcia C. Prestrud. *Invariant Imbedding and Radiative Transfer in Slabs of Finite Thickness,* Modern Analytic and Computational Methods in Science and Mathematics, Vol. 1. New York: American Elsevier Publishing Company, Inc., 1963.

Buchheim, Robert W., and the Staff of The RAND Corporation. *Space Handbook: Astronautics and Its Applications.* New York: Random House, Inc., 1959.

Dole, Stephen H. *Habitable Planets for Man.* New York: Blaisdell Publishing Company, Inc., 1964.

Edelen, Dominic G. B. *The Structure of Field Space: An Axiomatic Formulation of Field Physics.* Berkeley and Los Angeles: University of California Press, 1962.

Gouré, Leon. *Civil Defense in the Soviet Union.* Berkeley and Los Angeles: University of California Press, 1962.

Harris, Theodore E. *The Theory of Branching Processes.* Berlin, Germany: Springer-Verlag, 1963; Englewood Cliffs, N. J.: Prentice-Hall, Inc., 1964.

Hirshleifer, Jack, James C. DeHaven, and Jerome W. Milliman. *Water Supply: Economics, Technology, and Policy.* Chicago: The University of Chicago Press, 1960.

Hitch, Charles J., and Roland McKean. *The Economics of Defense in the Nuclear Age.* Cambridge, Mass.: Harvard University Press, 1960.

Kramish, Arnold. *Atomic Energy in the Soviet Union.* Stanford, Calif.: Stanford University Press, 1959.

Lubell, Harold. *Middle East Oil Crises and Western Europe's Energy Supplies.* Baltimore, Maryland: The Johns Hopkins Press, 1963.

McKean, Roland N. *Efficiency in Government through Systems Analysis: With Emphasis on Water Resource Development.* New York: John Wiley & Sons, Inc., 1958.

O'Sullivan, J. J. (ed.). *Protective Construction in a Nuclear Age.* 2 vols. New York: The Macmillan Company, 1961.

Sokolovskii, V. D. *Soviet Military Strategy.* Translated and annotated by H. S. Dinerstein, L. Gouré, and T. W. Wolfe. Englewood Cliffs, N. J.: Prentice-Hall, Inc., 1963.

Williams, J. D. *The Compleat Strategyst: Being a Primer on the Theory of Games of Strategy.* New York: McGraw-Hill Book Company, Inc., 1954

Selected RAND Books